AF600839

Piezoelectric Electromechanical Transducers for Underwater Sound

Boris S. Aronov

This book is subject to a Creative Commons Attribution-NonCommercial 4.0 International Public License (CC BY-NC 4.0). To view a copy of this license, visit https://creativecommons.org/licenses/ by-nc/4.0/. Other than as provided by these licenses, no part of this book may be reproduced, transmitted, or displayed by any electronic or mechanical means without permission from the publisher or as permitted by law.

Copyright © Boris Aronov, author, 2022

ISBN 9781644698228 (hardback)

ISBN 9781644698266 (Open Access)

Published by Academic Studies Press

1577 Beacon Street

Brookline, MA 02446, USA

press@academicstudiespress.com

www.academicstudiespress.com

In memory of

my mentor, Lev Yakovlevich Gutin,

and my friends and colleagues, prominent Russian electroacousticians,

Lev Davidovich Lubavin and Vladimir Igorevich Pozern

PREFACE

This book is initiated by the engineering experience of the author. Throughout his career the author has encountered many problems known to others involved in the design of electroacoustic transducers. The fact of the matter is that the complexity of designing electroacoustic transducers is inherent in the multidisciplinary nature of the subject. Therefore, the developers and designers of the transducers must possess the knowledge of several different theoretical disciplines (such as the vibration of mechanical systems, electromechanical conversion by deformed piezoelectric bodies, and acoustic radiation) and be able to actively use this knowledge to derive equations that describe the performance of the transducers. Furthermore, creating practical transducer designs that meet certain requirements and can operate under realistic environmental conditions requires the knowledge of properties of materials used and a certain level of engineering intuition that cannot be developed without a clear understanding of the underlying physics. Hardly anyone may possess all these capabilities without having received a specially targeted education, which, to the best of the author's knowledge, is not commonly available in the academic world. Usually, the necessary skills may be acquired through self-education, which was the case for the author. The main difficulties that arise in this endeavor are not in the lack of available information. On the contrary, the theoretical disciplines listed above are very well developed and are well-represented in the literature. Nevertheless, all these disciplines employ different methods for solving their problems and the results obtained are usually presented in forms not suitable for direct use in concert for synthesizing equations that govern transducer performance. Thus, the results must be tailored accordingly.

Experiencing the above difficulties over several decades, the author gradually developed a special approach to treating transducers problems that allows one to overcome many of the obstacles. The essence of this approach is in the consistent application of the physics-based energy method for solving all the problems that arise in the course of treating electromechanical and electroacoustic transducers. The first attempt to describe this concept was undertaken in *Electromechanical transducers from piezoelectric ceramic* published in 1990 in Russia. This version has now been updated and expanded to the extent that it can be considered a completely different book. Only the underlying energy approach to solving the problems has remained unaltered. This book is written for students, applied scientists and engineers in a way that should prove fruitful both for those who have only begun to chart their careers in electroacoustics as well as for those at a more advanced level. The content of the book is split into four parts.

In Part I, titled "Introduction of energy method of treating the transducers," the main concepts of the method are considered (Chapter 1); applications of the method to calculating properties of transducers with single degree of freedom are illustrated (Chapter 2); and the study of problems for designing the transducers as a part of the transmit/receive channel is made (Chapter 3). The main concept is that of energy and following its transformation. Different types of energies involved in the electro-mechano-acoustic conversion during transducer operation are presented in the generalized coordinates. All the governing equations are derived from the energy principles, that is, from the Law of Conservation of Energy for transducers with a single mechanical degree of freedom, and from the Principle of Least Action for transducers with multiple degrees of freedom. Equations describing the electromechanical part of the problem are reinterpreted as Kirchhoff's equations for the corresponding equivalent electromechanical circuits. In Chapter 2, the general approach is applied towards calculating the properties of transducers of widely used types (spheres, cylinders, bars undergoing extensional vibration and for circular plates and rectangular beams vibrating in flexure) that may be considered as systems with single mechanical degree of freedom. In Chapter 3, the operating properties of transducers as a part of a transmit/receive channel are considered and some recommendations regarding a rational transducer designing are presented. Given that the single degree of freedom approximation covers many practical transducer designs, Part I can be regarded as a self-sufficient study of underwater electroacoustic transduction on a basic level and can be read independently from the rest of the book.

The general treatment of electroacoustic transduction requires an advanced knowledge of the vibration of mechanical systems, electromechanical conversion in deformed piezoceramic bodies and acoustic radiation. Information about these topics, which is necessary for the consideration of virtually all practical transducer types is presented in Chapters 4-6 of Part II under the title: "Subsystems of the Electroacoustic Transducers." All the constitutive equations are derived in these chapters from the Principle of Least Action as Euler's Equations in generalized coordinates. The obtained results are presented in the form of impedances, (including the radiation impedances), electromechanical transformation coefficients and acting forces (including those of acoustic origin) that can be directly substituted into the equivalent electromechanical circuits (multi contour in general) of the transducers. The diffraction coefficients and directional factors for differently configured transducer surfaces are also presented.

In Chapter 4, special attention is paid to the consideration of coupled vibrations in the generally two-dimensional mechanical systems. The results allow determining the range of aspect ratio, at which the system can be approximately considered as one-dimensional, where the problem can be simplified.

In Chapter 5, especial importance is ascribed to the theorem that sets the conditions at which the electromechanical conversion under the longitudinal and transverse piezoelectric effects can be treated qualitatively in the same way. This allows for the unifying calculation technique for the transducers that employ these types of ceramics polarization. Another important subject is the general analysis of optimizing the effective coupling coefficients in nonuniformly deformed piezoceramic bodies.

Chapter 6 touches upon several noteworthy issues. Besides solving the general radiation problems, it provides a detailed consideration of the effects of baffling parts of the surfaces of cylindrical and spherical transducers, which ensures their unidirectionality. The technique for the experimental investigation of the acoustic interaction between transducers (or between the mechanically isolated parts of the same transducer) is also analyzed. Since the baffles have an effect on the acoustic near field, the interactions can rarely be treated analytically for practical transducer configurations, hence more reliable characterization of the interaction can be obtained through an experimental investigation.

The results obtained in the Part II are used in Part III of the book titled “Calculating transducers of different types” for synthesizing equations that describe the detailed operation of transducers of various configurations: cylindrical (Chapter 7), spherical (Chapter 8), plates and beams vibrating in flexure (Chapter 9) and bar transducers (Chapter 10).

Chapter 7 presents a study of cylindrical transducers that employ multimode extensional and flexural vibration of complete and incomplete cylinders (slotted cylinder projectors are also considered) for various practical applications. Different modes of cylinder polarization are considered, including the tangential polarization (with striped electrodes). An extensive study is provided of the effects of coupled vibrations on the electromechanical and acoustic performance of transducers that employ cylindrical piezoelements having finite thickness to diameter aspect ratios. Chapter 8 covers transducers which employ general multimode extensional vibrations of complete and incomplete piezoceramic spherical shells, (hemispherical in particular). The baffling of parts of the surface is also considered, which allows for the use of multiple modes of vibration in unidirectional transducer operations

In Chapter 9, a general analysis is provided of transducers which feature flexural vibrations of circular and rectangular piezoceramic plates (beams), including non-uniform over thickness and radius (length) transducer designs. Optimizing the effective coupling coefficients of the transducers is considered making use of the nonuniformity of the distribution of deformations in the volume of the plates. Corrections for transducer parameters due to a finite thickness to radius (length) ratio of the plates are considered. It is then concluded that the accuracy with which the wave numbers can be predicted substantially depends on the aspect ratio (especially for the higher modes of vibration) and presenting their values without the notion of the aspect ratios is not appropriate.

In Chapter 10, the length expander bar transducers are considered Transitions of configurations of bars to thickness vibrating plates at different polarizations and related dependencies of their effective coupling coefficients on the aspect ratios are considered using the technique of coupled vibrations. Relatively small attention is paid to the widely used Tonpilz transducer designs because they have already been described in detail in the available literature.

Part IV (Chapters 11 through 15) is titled: "Some aspects of the transducers designing."

In Chapter 11, a review of the existing data and some new results is presented regarding effects of operating environmental conditions, such as the hydrostatic pressure, temperature, and drive level on the parameters of piezoceramics. It is emphasized that, under these conditions, the parameters of ceramics may deviate significantly from those that are given in specifications for normal conditions. Moreover, they may differ for samples of ceramics supplied by different (and even by the same) manufacturers. This must be kept in mind when calculating the operating parameters of transducers under real conditions and in estimating a reasonable accuracy of calculation of the parameters. The variations in the parameters of transducers intended for operating at great depths can be avoided by using designs which incorporate hydrostatic pressure compensation. Issues related to the practical implementation of the pressure compensation are examined in Chapter 12 (more general information), in Chapter 13 (regarding the liquid filled cylindrical projectors) and in Chapter 14 (regarding the hydrophones).

Chapter 13 presents some considerations regarding the practical challenges of the projectors design. Using the concept of the Reserves-of-Strength for improving parameters of the transducers of different types by optimizing their matching with the acoustic field is considered. The possibilities of increasing the dynamic and static mechanical strength of the projectors by prestressing and combining piezoceramic with passive materials in their mechanical systems are analyzed.

Chapter 14 is dedicated to the design of hydrophones and related issues. The hydrophones employing different transducer types are classified by the pressure and pressure-gradient hydrophones of the diffraction and motion types. Their properties as a source of energy of signal and internal noise for a receive channel are considered. Special attention is paid to the response of hydrophones and accelerometers to unwanted actions and to measures aimed at increasing their noise immunity.

Chapter 15 is crucial for the structure of the book because it introduces the practice of combining Finite Element Analysis (FEA) with analytical energy methods. This is illustrated with examples of flextensional and oval transducers. Combining powerful computer-based FEA techniques that are used to obtain results for vibration mode shapes with the energy method that yields great physical insight opens a new area of research collaboration for many transducer problems. FEA allows the determination of the vibration mode shapes for mechanical systems that cannot be approximated analytically due to the complexities of the mechanical system and its boundary conditions.

The book also contains appendices with information on the properties of the piezoelectric ceramics and passive materials that may be used in transducer designs, and on the properties of the special functions that are referred to throughout the book.

In summary, the book presents methods for calculating the properties of most common electroacoustic transducer problems with particular focus on underwater applications. Moreover, by combining the FEA technique to determine the prerequisite vibration mode shapes with the energy method, virtually any transducer type may be analyzed. Still, however, when it comes to choosing and designing a particular transducer for a particular application under demanding operating and environmental specifications – this remains somewhat of an art. Thus, recommendations of transduction choices for representative problems remain a guide and not a prescription for success.

It is inevitable that the book may contain typographical or content errors and thus the author would welcome the readers' comments and notifications of such.

Boris S. Aronov

Part II

Subsystems of the Electroacoustic Transducers

TABLE OF CONTENTS

CHAPTER 4

VIBRATION OF ELASTIC BODIES

4.1 Introduction

Equations of motion of elastic bodies will be derived throughout this treatment from the *Least Action Principle* regardless of the coordinate system used, as the corresponding Euler's equations. The forms of Euler's equations that depend on the type of function describing an energy status of the vibrating system were stated in Section 1.6.2.1. Thus, in the case that the kinetic energy, potential energy and the energies of external actions for a body (W_{kin}, W_{pot} and W_e) depend on the generalized velocities $\dot{\xi}_i$, and generalized coordinates (displacements) ξ_i, respectively, vibration of the body is described by the system of Euler-Lagrange equations (1.93)

$$\frac{d}{dt}\left(\frac{\partial W_{kin}}{\partial \dot{\xi}_i}\right)+\frac{\partial W_{pot}}{\partial \xi_i}=\frac{\partial W_e}{\partial \xi_i}=f_i, \quad (i=1,2,...). \tag{4.1}$$

If vibrations of a volume element of an elastic body in the geometry coordinates are concerned, in which case the energies of the element (kinetic w_{kin}, potential w_{pot} and energy w_e supplied to the volume element by external actions) depend on the velocity $\dot{\xi}$, displacement ξ and its derivative ξ_x' as functions of the geometry coordinates, the Euler's equations in the form (1.95) are applicable

$$\frac{d}{dt}\left(\frac{\partial w_{kin}}{\partial \dot{\xi}}\right)-\frac{\partial}{\partial x}\left(\frac{\partial w_{pot}}{\partial \xi_x'}\right)=\frac{\partial w_e}{\partial \xi}. \tag{4.2}$$

To further specify the Euler equations, an appropriate to a particular case coordinate system has to be chosen and explicit expressions for the energies involved in these equations must be obtained. At first, we will consider the problem in the geometrical coordinates.

4.2 Information on the Theory of Elasticity

4.2.1 Strain and Stress Tensors

External actions may result in certain changes in the form and volume of an elastic body. The changes are characterized by the elastic body deformations. A body volume element will be considered in orthogonal rectangular coordinate system with coordinates x_1, x_2, x_3 (Figure 4.1 (a)). Mathematically deformations are expressed through displacements of the points of the body $\boldsymbol{\xi}(x_1, x_2, x_3) = \mathbf{i}\xi_1 + \mathbf{j}\xi_2 + \mathbf{k}\xi_3$, where **i**, **j**, **k** are the unit vectors of the coordinate system. The expressions for the components of the tensor of deformations are of the form

$$S_{11} = \partial\xi_1/\partial x_1, \quad S_{22} = \partial\xi_2/\partial x_2, \quad S_{33} = \partial\xi_3/\partial x_3,$$
$$S_{23} = S_{32} = \frac{1}{2}\left(\frac{\partial\xi_2}{\partial x_3} + \frac{\partial\xi_3}{\partial x_2}\right), \quad S_{13} = S_{31} = \frac{1}{2}\left(\frac{\partial\xi_1}{\partial x_3} + \frac{\partial\xi_3}{\partial x_1}\right) \tag{4.3}$$
$$S_{12} = S_{21} = \frac{1}{2}\left(\frac{\partial\xi_1}{\partial x_2} + \frac{\partial\xi_2}{\partial x_1}\right).$$

Strains S_{11}, S_{22}, S_{33} characterize elongation in the directions of coordinate axes (Figure 4.1 (b)); S_{23}, S_{13}, S_{12} characterize a change in the angles between linear elements, which were parallel to the corresponding axes prior to deformations. As it is obvious from Figure 4.1 (c), $S_{13} = (\varphi_{13} + \varphi_{31})/2$. If a deformed element of the volume is turned by the angle φ_{31}, we obtain deformation due to a simple displacement, under which the cross section planes move in parallel to one of the coordinate planes. So, the displacement angle is $\varphi = (\varphi_{13} + \varphi_{31}) = 2S_{13}$.

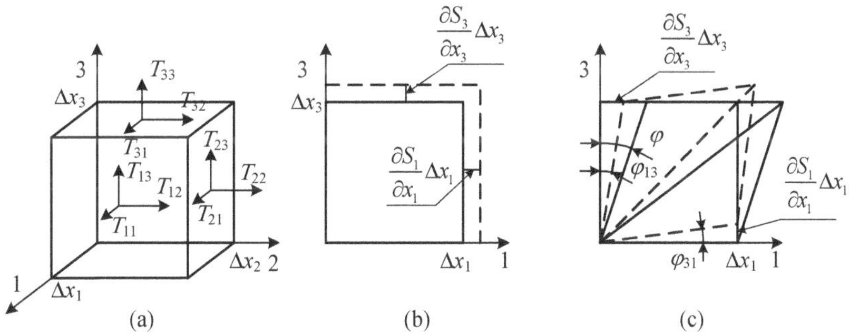

Figure 4.1: Components of tensors of the stress and strain in rectangular coordinates.

Further, along with the two number index designations of strain and stress, we will also use for brevity the single number designations, introduced as follows $S_{11} = S_1$, $S_{22} = S_2$, $S_{33} = S_3$, $2S_{23} = S_4$, $2S_{13} = S_5$, $2S_{12} = S_6$. In these designations S_4, S_5, S_6 are changes of angles between the respective sides of a volume element under deformation.

The potential energy of an elastic body made of passive materials is a function of its deformation, and the process of deformation under vibration is considered adiabatic. The change in the potential energy of a volume element under deformation is

$$\delta w_{pot} = \sum_{i=1}^{6} \frac{\partial w_{pot}}{\partial S_i} \delta S_i = T_i \delta S_i , \tag{4.4}$$

where $T_i = \partial w_{pot} / \partial S_i$ are the mechanical stresses. In relation (4.4) the rule is used, according to which the summation is supposed to be performed over the repeating indices in the products of the vector or tensor components, so that

$$T_i \delta S_i = \sum_{i=1}^{6} T_i \delta S_i . \tag{4.5}$$

The mechanical stresses are the forces that act on the sides of an elementary volume and that are reduced to their unit area. The stresses $T_1 = T_{11}$, $T_2 = T_{22}$, $T_3 = T_{33}$ act along the normal to the sides of the volume in the direction of the coordinate axis, which corresponds to the stress index. The stresses $T_4 = T_{23}$, $T_5 = T_{13}$, $T_6 = T_{12}$ act tangentially to the sides that are perpendicular to the axis corresponding to the first number and go in the direction of the axis corresponding to the second number.

It is established experimentally that under small deformation the mechanical stresses depend on deformations linearly, so that

$$T_1 = (\partial T_i / \partial S_j) S_j \quad (i, j = 1, 2, ..., 6) , \tag{4.6}$$

where $\partial T_i / \partial S_j = \partial^2 w_{pot} / \partial S_i \partial S_j = c_{ij}$ are the elastic moduli.

Equations (4.6) represent Hooke's law, the equations of state for an elastic body (in our case, under adiabatic conditions). The elastic moduli are determined experimentally. The number of independent moduli of elasticity depends on the symmetry of the material structure. In an isotropic body the independent terms are two constants λ and μ, which are called Lame constants and by means of which all the elasticity moduli can be expressed. In the general case, when it is impossible to make any additional assumptions about the relation between strains and stresses, Eqs. (4.6) for an isotropic body have the following form

$$\begin{gathered} T_{11} = T_1 = (\lambda + 2\mu) S_1 + \lambda S_2 + \lambda S_3, \\ T_{22} = T_2 = \lambda S_1 + (\lambda + 2\mu) S_2 + \lambda S_3, \\ T_{33} = T_3 = \lambda S_1 + \lambda S_2 + (\lambda + 2\mu) S_3, \\ T_{23} = T_4 = \mu S_4, \quad T_{13} = T_4 = \mu S_5, \quad T_{12} = T_6 = \mu S_6 . \end{gathered} \tag{4.7}$$

For particular conditions of deformation of elastic bodies having different geometry the relation between components of stress and strain tensors may be simplified. Examples of the body geometries that are of practical importance in the analysis of the common types of vibrational systems are shown in Figure 4.2. Consider deformation along the axis of a thin bar (along the

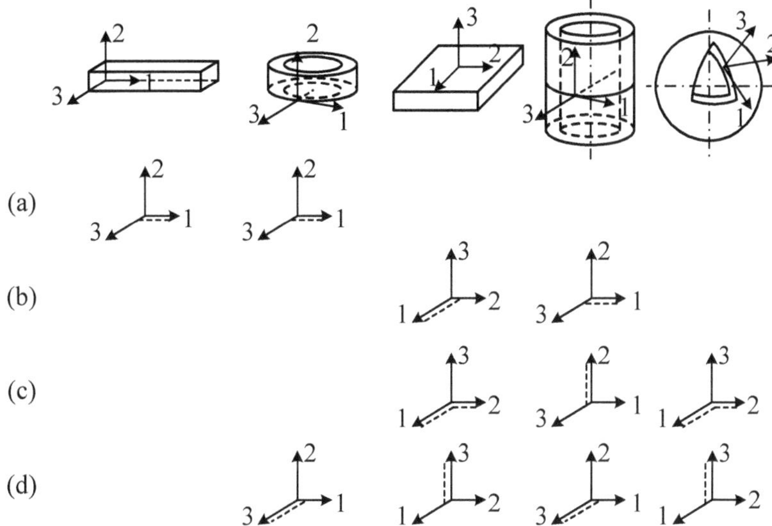

Figure 4.2: Images of typical mechanical systems of the transducers. The dashed lines in the coordinate systems under the images show directions of working deformations in the mechanical systems.

mean circumference of a thin ring with small height) (Figure 4.2 (a)). As the side surfaces of the bar are free of stress ($T_2 = T_3 = T_4 = T_5 = T_6 = 0$), these stresses are close to zero inside the bar due to small lateral dimensions. Under these conditions from the system (4.6) we obtain

$$\begin{aligned} S_2 &= S_3 = -\lambda S_1 / 2(\lambda + \mu) = -\sigma S_1, \\ T_1 &= [\mu(3\lambda + 2\mu)/(\lambda + \mu)]S_1 = YS_1. \end{aligned} \tag{4.8}$$

Here $\sigma = \lambda / 2(\lambda + \mu)$ is the Poisson's ratio, i.e., the ratio of the transverse relative compression (signified by the "minus" sign according to the rule of signs) to the longitudinal elongation; $Y = \mu(3\lambda + 2\mu)/(\lambda + \mu)$ is the Young's modulus, i.e., the ratio of the longitudinal stress to the longitudinal strain. It can be shown that $\lambda = Y\sigma / (1+\sigma)(1-2\sigma)$ and $\mu = Y / 2(1+\sigma)$.

Since the elastic constants of material have to be determined experimentally, and the

Young's modulus and Poisson's ratio can be obtained in a simple way, these constants are widely used to describe the state of elastic bodies along with the second Lame constant μ, which is the shear modulus (usually the term "modulus" is referred to the ratio of stress to strain).

It is noteworthy that values of the elastic constants can be obtained from the experimenting to 3 digits accuracy (see Table A1 of the elastic constants in Appendix A). Therefore, calculating parameters of transducers to a greater accuracy does not make sense.

Using constants Y and σ, Eqs. (4.7) can be rewritten as follows,

$$\begin{aligned} T_1 &= \frac{Y}{(1+\sigma)(1-2\sigma)}\left[(1-\sigma)S_1 + \sigma(S_2+S_3)\right], \\ T_2 &= \frac{Y}{(1+\sigma)(1-2\sigma)}\left[(1-\sigma)S_2 + \sigma(S_1+S_3)\right], \\ T_3 &= \frac{Y}{(1+\sigma)(1-2\sigma)}\left[(1-\sigma)S_3 + \sigma(S_1+S_2)\right], \\ T_i &= YS_i/2(1+\sigma) = \mu S_i \quad (i=4,5,6). \end{aligned} \tag{4.9}$$

Sometimes, and in the case under consideration in particular, when one or two of the stresses are zero, it becomes more convenient to use equations between stresses and strains, in which stresses are independent variables, namely,

$$\begin{aligned} S_1 &= [T_1 - \sigma(T_2+T_3)]/Y, \\ S_2 &= [T_2 - \sigma(T_1+T_3)]/Y, \\ S_3 &= [T_3 - \sigma(T_1+T_2)]/Y, \\ S_i &= T_i/\mu \quad (i=4,5,6). \end{aligned} \tag{4.10}$$

4.2.2 The Energy Densities in Particular Mechanical Systems

Example 1: Deformations of a Bar.

In the tensile deformation of a bar along axis 1 (Figure 4.2 (a)), the potential energy of a unit volume (potential energy density), as the energy accumulated under the change of strain from zero to the value of S_1, is equal to

$$w_{pot} = \int_0^{S_1} YS_1\,dS_1 = \frac{T_1S_1}{2} = \frac{YS_1^2}{2} = \frac{Y}{2}\left(\frac{\partial\xi_1}{\partial x_1}\right)^2. \tag{4.11}$$

The kinetic energy of the bar per unit length can be determined to the first approximation as follows

$$w_{kin} = \rho S_{cs}\dot{\xi}_1^2 / 2, \tag{4.12}$$

where $S_{c.s}$ is the area of a bar cross section.

The more precise value of w_{kin} can be obtained, if to consider transverse motion that accompanies the bar elongation. From Eqs. (4.8) and (4.3) for displacements in the plane of the bar cross section we obtain

$$\xi_2 = -\sigma x_2(\partial\xi_1 / \partial x_1), \quad \xi_3 = -\sigma x_3(\partial\xi_1 / \partial x_1). \tag{4.13}$$

Considering this motion, the correction to the kinetic energy per unit length of a bar is

$$\Delta w_{kin} = \frac{\rho}{2}\iint_{S_{cs}}\left(\dot{\xi}_2^2 + \dot{\xi}_3^2\right)dx_2dx_3 = \frac{\rho\sigma^2 J_p}{2}\left(\frac{\partial^2\xi_1}{\partial x_1\partial t}\right)^2, \quad J_p = \iint_{S_{cs}}\left(x_2^2 + x_3^2\right)dx_2dx_3, \tag{4.14}$$

where J_p is the polar moment of inertia for the bar cross section. The same conditions exist in the thin and short ring under circumferential deformation.

Example 2: Uniaxial Deformation in the Plane of a Thin Plate.

With uniaxial deformation in the plane of a thin plate, the dimension of which in the direction perpendicular to the tension is so large that, due to symmetry, deformation in this direction does not take place (Figure 4.2 (b)). In this case $T_3 = T_4 = T_5 = T_6 = 0$, $S_2 = 0$. The analogous conditions exist in the case of tension in the direction tangential to the mean circumference of a thin cylinder of a big height. From Eq. (4.10) it follows that $T_2 = \sigma T_1$, $S_3 - \sigma S_1 / (1-\sigma)$, $T_1 = YS_1 / (1-\sigma^2)$, and

$$w_{pot} = \frac{1}{2}\frac{Y}{1-\sigma^2}S_1^2 = \frac{1}{2}\frac{Y}{1-\sigma^2}\left(\frac{\partial\xi_1}{\partial x_1}\right)^2. \tag{4.15}$$

Example 3: Deformation in the Plane of a Thin Plate in Two Perpendicular Directions.

In the case of deformation in the plane of a thin plate in two mutually perpendicular directions (Figure 4.2 (c)), $T_3 = T_4 = T_5 = T_6 = 0$,

$$\begin{gathered} S_1 = -\sigma(T_1 + T_2)/Y = -\sigma(S_1 + S_2)/(1-\sigma), \\ T_1 = Y(S_1 + S_2)/(1-\sigma^2), \\ T_2 = Y(S_2 + S_3)/(1-\sigma^2), \\ w_{pot} = \frac{T_1S_1 + T_2S_2}{2} = \frac{1}{2}\frac{Y}{1-\sigma^2}(S_1^2 + 2\sigma S_1S_2 + S_2^2). \end{gathered} \tag{4.16}$$

Analogous conditions take place in the case of tension in a thin cylinder (ring) in the direction of its axis, taken as axis 1. If $S_1 = S_2 = S$, as for example, in the case of the spherical shell, then

$$w_{pot} = \frac{1}{2}\frac{2Y}{1-\sigma}S^2. \tag{4.17}$$

Example 4: Deformation Through the Thickness of a Plate with Large Transverse Dimensions.

In the case of deformation through the thickness of a plate with large transverse dimensions (Figure 4.2 (d)), due to symmetry it can be assumed that $\xi_1 = \xi_2 = 0$. Hence, $S_1 = S_2 = 0$, $S_3 = S_4 = S_5 = S_6 = 0$. From Eqs. (4.9) it follows that

$$T_3 = Y(1-\sigma)S_3 / (1+\sigma)(1-2\sigma), \quad T_1 = T_2 = -Y\sigma S_3 / (1+\sigma)(1-2\sigma), \tag{4.18}$$

where from

$$w_{pot} = \frac{1}{2}\frac{(1-\sigma)Y}{(1+\sigma)(1-2\sigma)}S_3^2. \tag{4.19}$$

Example 5: Deformation through the Thickness of a Tall Cylinder, Short Ring, and Spherical Shell.

With deformation through the thickness of a tall cylinder, short ring, and spherical shell (Figure 4.2 (d)), displacement ξ_1 in direction of the mean circumference and displacements $\xi_1 = \xi_2$ tangential to the middle surface of the spherical shell are equal to zero, due to symmetry. However, it follows from the expressions for strain in curvilinear coordinates (see Section 4.4.1) that strains S_1 and $S_1 = S_2$ have in these cases finite values. It can be shown that the following expressions are valid for the above bodies, respectively:

for a tall cylinder,

$$\begin{gathered} S_2 = 0, \quad T_1 = \frac{Y\left[(1-\sigma)S_1 + \sigma S_3\right]}{(1+\nu)(1-2\nu)}, \quad T_3 = \frac{Y\left[(1-\sigma)S_3 + \sigma S_1\right]}{(1+\sigma)(1-2\sigma)}, \\ w_{pot} = \frac{Y(1-\sigma)}{(1+\sigma)(1-2\sigma)}\left[S_1^2 + \frac{2\sigma}{1-\sigma}S_1S_3 + S_3^2\right]; \end{gathered} \tag{4.20}$$

for a short ring,

$$T_2 = 0, \quad T_1 = \frac{Y(S_1 + \sigma S_2)}{1-\sigma^2}, \quad T_3 = \frac{Y(S_2 + \sigma S_1)}{1-\sigma^2},$$
$$w_{pot} = \frac{1}{2}\frac{Y}{1-\sigma^2}(S_1^2 + 2\sigma S_1 S_3 + S_3^2); \tag{4.21}$$

for a spherical shell,

$$S_1 = S_2, \quad T_1 = T_2 = \frac{Y(S_1 + \sigma S_3)}{(1+\sigma)(1-2\sigma)}, \quad T_3 = \frac{Y\left[(1-\sigma)S_3 + 2\sigma S_1\right]}{(1+\sigma)(1-2\sigma)},$$
$$w_{pot} = \frac{2Y}{(1+\sigma)(1-2\sigma)}\left[2S_1^2 + 4\sigma S_1 S_3 + (1-\sigma)S_3^2\right]. \tag{4.22}$$

Example 6: Deformation of Torsion of a Bar.

Under this deformation each cross section turns by a certain angle $\partial\varphi = \tau\,\partial x_3$ relative to the axis of inertia of the bar (Figure 4.3). In this case, $\xi_1 = \tau x_2 x_3$, $\xi_2 = -\tau x_1 x_3$. Besides, at the deformation of torsion surface of a cross section experiences distortion, so that its points displace in the direction of axis 3, and $\xi_3 = \tau\psi(x_1, x_2)$. The function $\psi(x_1, x_2)$ depends on configuration of the cross section and must be determined for each case.

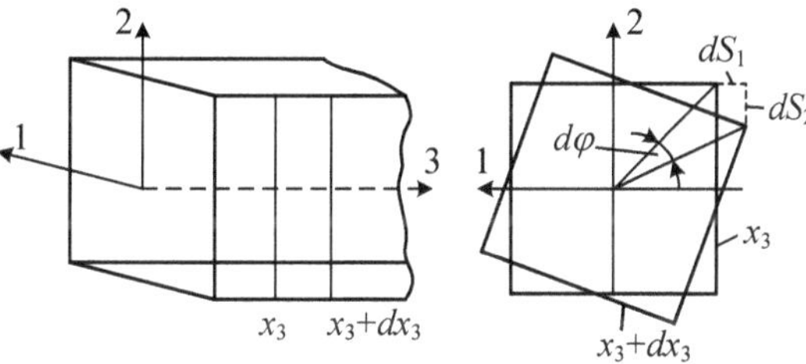

Figure 4.3: To the deformation of torsion.

It can be concluded using definitions (4.3) that in this case the strains are

$$S_1 = S_2 = S_3 = S_6 = 0,$$
$$S_4 = 2S_{23} = \frac{\ddot{a}\xi_2}{\ddot{a}x_3} + \frac{\ddot{a}\xi_3}{\ddot{a}x_2} = \tau\left(\frac{\partial\psi}{\partial x_2} - x_1\right),$$
$$S_5 = 2S_{13} = \frac{\ddot{a}\xi_1}{\ddot{a}x_3} + \frac{\ddot{a}\xi_3}{\ddot{a}x_1} = \tau\left(\frac{\partial\psi}{\partial x_1} + x_2\right). \tag{4.23}$$

The stress components that do not vanish are $T_4 = \mu S_4$ and $T_5 = \mu S_5$. It is known (Ref. 1) that for the circular and square cross sections $\psi \approx 0$. In these cases

$$S_4 = -x_1 \tau = -x_1 \frac{\partial \varphi}{\partial x_3}, \quad S_5 = x_2 \tau = x_2 \frac{\partial \varphi}{\partial x_3}. \tag{4.24}$$

The forces acting in the cross section of the bar are equivalent to a couple of moment $M_f = G\tau$ about the axis x_3, where G is the constant referred to as the torsional rigidity. For the circular and square cross sections $G = \mu J_p$. For a circle of radius $a, J_p = \pi a^4 / 2$, for a square with side w, $J_p = w^4 / 6$. For a rectangular cross section with aspect ratio not too large $G \approx \mu S_{cs}^4 / 4\pi^2 J_p$. The potential and kinetic energies for the unit length of a bar are, respectively,

$$w_{pot} = \frac{1}{2} M_f \frac{\partial \varphi}{\partial x_3} = \frac{1}{2} G\tau^2 = \frac{1}{2} G \left(\frac{\partial \varphi}{\partial x_3} \right)^2, \tag{4.25}$$

$$w_{kin} = \frac{1}{2} \rho J_p \dot{\varphi}^2. \tag{4.26}$$

Example 7: Bending of a Thin Beam

The bending of a thin ($t / l \ll 1$ and $w / l < 1$) beam (Figure 4.4) is already considered in Section 2.6, but here our goal is to get explicit expressions for the densities of potential and kinetic energy under the flexure that are necessary for deriving equations of motion.

Under the flexural deformation the neutral surface exists in a beam that does not experience either tension or compression, whereas the layers located on both sides of the neutral surface suffer deformation of opposite signs. The strain of the layer located at distance x_3 from the neutral surface is

$$S_1 = \frac{(R + x_3)\Delta\varphi - R\Delta\varphi}{R\Delta\varphi} = \frac{x_3}{R}, \tag{4.27}$$

where R is the radius of curvature of the neutral surface of the beam. As it follows from Figure 4.4

$$\Delta\varphi = \frac{\Delta x_1}{R} = \alpha_1 + \alpha_2 = -\left.\frac{\partial \xi_3}{\partial x_1}\right|_{x_1 + \Delta x_1} + \left.\frac{\partial \xi_3}{\partial x_1}\right|_{x_1} = -\frac{\partial^2 \xi_3}{\partial x_1^2} \Delta x_1, \tag{4.28}$$

where from $1 / R = -\partial^2 \xi_3 / \partial x_1^2$, where $\xi_3(x)$ is the normal displacement of the neutral surface.

Considering that flexure and displacement of the neutral surface are small (i.e., $t / 2R \ll 1$ and $\xi_3 \ll t$), we can assume that the cross sections of a beam remain flat and are displaced along the normal to its neutral plane [2]. From the fact that the side surfaces of the beam are free of stress follows that $T_2 \approx T_3 \approx 0$ and $T_4 \approx T_5 \approx T_6 \approx 0$ all over the volume, and

$$T_1 = YS_1 = -Yx_3 \frac{\partial^2 \xi_3}{\partial x_1^2}. \tag{4.29}$$

The potential energy per unit length of a beam is

$$w_{pot} = \frac{1}{2} \int_{-w/2}^{w/2} \int_{-t/2}^{t/2} T_1 S_1 dx_3 dx_2 = \frac{1}{2} Y \frac{wt^3}{12} \left(\frac{\partial^2 \xi_3}{\partial x_1^2} \right)^2, \tag{4.30}$$

where $(wt^3 / 12) = J_2$ is the moment of inertia of the section with respect to axis x_2. In general, if the cross- section has different configuration with moment of inertia J

$$w_{pot} = \frac{1}{2} YJ \left(\frac{\partial^2 \xi_3}{\partial x_1^2} \right)^2. \tag{4.31}$$

Thus, for example, for a circular of $r = a$ cross section $J = \pi a^4 / 4$.

The kinetic energy per unit length (ignoring the rotary inertia of the cross sections) is

$$w_{kin} = \frac{1}{2} \rho wt \dot{\xi}_3^2 = \frac{1}{2} \rho S_{cs} \dot{\xi}_3^2. \tag{4.32}$$

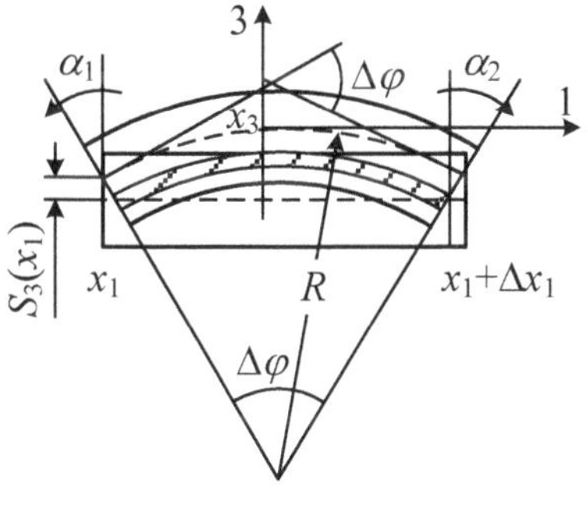

Figure 4.4: Flexure of a beam.

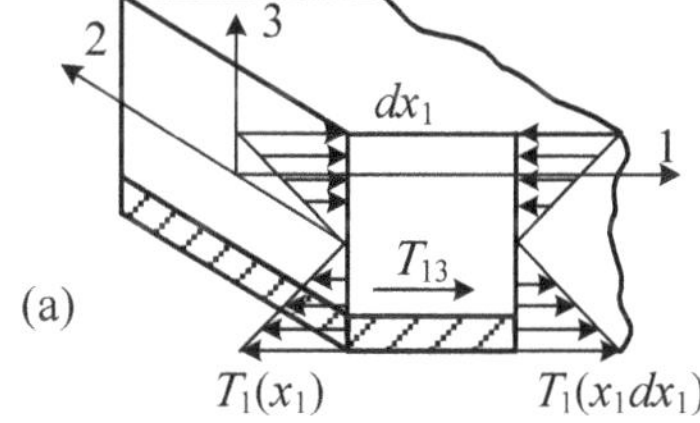

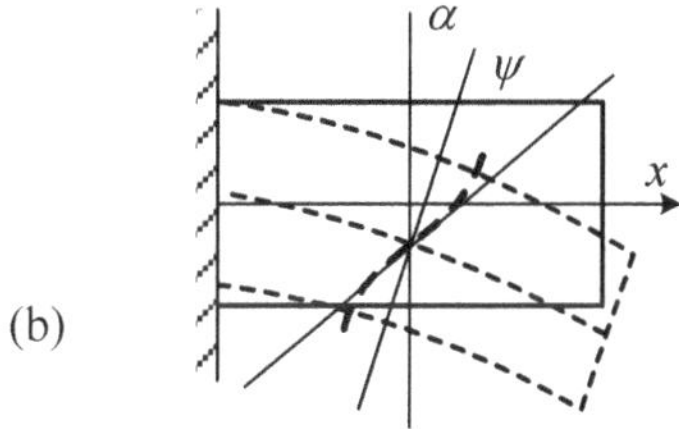

Figure 4.5: Depiction of shear deformation in a beam under flexure.

With increase of the relative thickness t/l of the beam or plate, the values of shear stresses, as well as the rotary inertia of the cross sections under flexure become significant, so that corrections must be made to the energies determined by formulae (4.30) and (4.32). The kinetic energy per unit length of the beam associated with turning of the sections by angle $\alpha = (\partial\xi_3 / \partial x_1)$ regarding axis 2 (Figure 4.4) is

$$\Delta w_{kin} = \frac{1}{2}\rho J_2 \alpha^2 = \frac{1}{2}\rho J_2 \left(\frac{\partial^2 \xi_3}{\partial x \partial t}\right)^2 . \tag{4.33}$$

Another correction is related to an additional deflection due to shear deformations, which takes place in a beam under its flexure, but which can be ignored at small t/l. If to consider the balance of forces acting on the shaded element of the volume (Figure 4.5 (a)) projected on axis 1, it will be found that the shear stresses $T_{13} = T_5$ are

$$S_5 = \frac{T_5}{\mu} = \frac{Yt^2}{8\mu}\frac{\partial^3 \xi_3}{\partial x_1^3} = \psi, \quad T_5 = \int_{t/2}^{x_3} \frac{\partial T_1}{\partial x_1} dx_3 = \frac{Y}{2}\left(\frac{t^2}{4} - x_3^2\right)\frac{\partial^3 \xi_3}{\partial x_1^3} . \tag{4.34}$$

The cross section of a beam becomes distorted due to shear. The shear strains, which are equal to angle ψ of the section element tilt relative to the original plane (Figure 4.5 (b)), have maximum value at the neutral axis, at $x_3 = 0$

$$S_5 = \frac{T_5}{\mu} = \frac{Yt^2}{8\mu}\frac{\partial^3 \xi_3}{\partial x_1^3} = \psi . \tag{4.35}$$

The additional deflection, ξ_{ad}, of the neutral axis can be considered as corresponding to this turning angle according to the relation $\psi = -(\partial\xi_{ad} / \partial x_1)$. Thus, the additional deflection is

$$\xi_{ad} = -\frac{Yt^2}{8\mu}\frac{\partial^2 \xi_3}{\partial x_1^2} \tag{4.36}$$

and the total deflection of the neutral plane becomes $\xi = \xi_3 + \xi_{ad}$. As in this approximation $T_5 \neq 0$, the correction should be applied to the value of w_{pot} expressed by formula (4.30), since this value was somewhat exaggerated (increase of deflection under invariable action shows reduction of the rigidity and results in decrease of the potential energy). Using expression (4.35) we obtain

$$\Delta w_{pot} = -\frac{w}{2\mu}\int_{-t/2}^{t/2} T_5^2 dx_3 = -\frac{wY^2 t^5}{240\mu}\left(\frac{\partial^3 \xi_3}{\partial x_1^3}\right)^2 . \tag{4.37}$$

4.3 Equations of Vibration in Rectangular Coordinates

Equations that describe movement of elastic bodies can be derived as the Euler's Eqs. (4.2) as result of substituting the corresponding expressions for the comprising energy densities. Thus, in examples 1, 2, 4, 6 of one-dimensional deformations the density of potential energy can be represented as

$$w_{pot} = K_\Delta (\xi'_x)^2 / 2, \tag{4.38}$$

where K_Δ is the specific rigidity of a volume element (in examples 1, 2, 4 it is equal to Y, $Y/(1-\sigma^2)$, and $Y(1-\sigma)/(1+\sigma)(1-2\sigma)$, respectively). And under torsion deformation (example 6) $\xi = \varphi$, $K_\Delta = G$ – the torsional rigidity. The density of kinetic energy can also be expressed in the generalized form as

$$w_{kin} = m_\Delta \dot{\xi}^2 / 2, \tag{4.39}$$

where m_Δ is the specific mass of a volume element, which is equal to ρ in examples 1, 2, 4 and to ρJ_p in example 6 (body under torsion).

If a force or a moment act on some cross section of a body, then the energy supplied to the body can be expressed as follows

$$w_e = f(x,t)\xi, \tag{4.40}$$

where under torsion $f(x, t)$ is the moment $M_f(x,t)$, and ξ is the turning angle φ. After substituting w_{pot}, w_{kin} and w_e from Eqs. (4.38), (4.39) and (4.40) into Eq. (4.2), for the cases under consideration will be obtained

$$c^2\xi'' - \ddot{\xi} = f(x,t)m_\Delta, \tag{4.41}$$

where $c = \sqrt{K_\Delta / m_\Delta}$ is the velocity of deformation propagation. In examples 1, 2, 4, $c = \sqrt{Y/\rho}$, $\sqrt{Y/\rho(1-\sigma^2)}$, $\sqrt{Y(1-\sigma)/\rho(1+\sigma)(1-2\sigma)}$, respectively, and in example 6 (under torsion) $c = \sqrt{G/\rho J_p}$ (for bars of circular and square cross sections $c = \sqrt{\mu/\rho}$). In order to obtain a definite solution for Eq. (4.41), the boundary conditions on the end surfaces of the bar at $x = 0$, $x = l$, and initial conditions at $t = 0$ have to be taken into account. Namely, values of $\xi(x,t)_{x=0,l}$ and/or $\xi'_x(x,t)_{x=0,l}$ for the longitudinal vibration, and $\varphi(x,t)_{x=0,l}$ and/or $\varphi'(x,t)_{x=0,l}$ for torsion, as well as $\xi(x,0)$, $\dot{\xi}(x,0)$ and $\varphi(x,0)$, $\dot{\varphi}(x,0)$, respectively.

The Euler's equation (4.2) cannot be applied to solving the flexural vibration problem (example 7), since w_{pot} depends on $s_x^{"}$, as it follows from expression (4.31). In this case the equation of vibration can be obtained directly from the variational principle in the form

$$\int_{t_1}^{t_2}\int_{x_1}^{x_2}(w_{kin}-w_{pot}+w_e)dtdx=0\,, \tag{4.42}$$

or, after substituting expressions (4.31), (4.32) and (4.40) for the energies involved

$$\delta\int_{t_1}^{t_2}\int_{x_1}^{x_2}\left[\dot{\xi}^2-\alpha^2(\xi_x^{"})^2+f_m\xi\right]dtdx=\delta J=0\,, \tag{4.43}$$

where $\alpha^2=Y\,t^2/12\rho$, $f_m=f/\rho S_{c.s.}$.

The equation of motion in this case can be derived in the following way. Let us assume that the function $\xi(x,t)$ satisfies Eq. (4.43). Let $\eta(x,t)$ be some function, which has the first and the second continuous derivatives, and which equals to zero together with its first derivative at values t_1, x_1 and t_2, x_2. Consider function $\xi(x,t)+\beta\eta(x,t)$. This function meets the same boundary conditions as $\xi(x,t)$ at the limits of integration. After substituting this function instead of $\xi(x,t)$ under the integral (4.43), the value of the integral at any small value of β should be greater than at β = 0, i.e., the integral has minimum at this point. Thus, it should be $J'_\beta(\xi+\beta\eta)=0$. From this condition the equation of motion can be found. The same result can be obtained, if to replace ξ with $\xi+\eta$ under the integral and, retaining the first-order terms in respect to η and its derivatives, to equate the integral to zero. In the latter case we arrive at

$$\int_{t_1}^{t_2}\int_{x_1}^{x_2}(\dot{\xi}\dot{\eta}-\alpha^2\xi_x''\eta_x''+f_m\eta)dxdt=0\,. \tag{4.44}$$

Performing the integration by parts, we obtain

$$\int_{t_1}^{t_2}\dot{\xi}\dot{\eta}dt=\dot{\xi}\eta\Big|_{t_1}^{t_2}-\int_{t_1}^{t_2}\ddot{\xi}\eta dt\,, \tag{4.45}$$

$$\int_{x_1}^{x_2}\xi_x''\eta_x''dx=\xi_x''\eta_x'\Big|_{x_1}^{x_2}-\xi_x'''\eta\Big|_{x_1}^{x_2}+\int_{x_1}^{x_2}\xi_x^{IV}\eta dx\,, \tag{4.46}$$

and taking into consideration, that $\eta\Big|_{t_1}^{t_2}=\eta\Big|_{x_1}^{x_2}=\eta'\Big|_{x_1}^{x_2}=0$, integral (4.44) becomes

$$\int_{t_1}^{t_2}\int_{x_1}^{x_2}(\ddot{\xi}+\alpha^2\xi_x^{IV}-f_m)\eta dxdt=0\,. \tag{4.47}$$

Since the function η is arbitrary within interval of integration, the expression in parenthesis has to vanish, and the following equation for function $\xi(x,t)$ should be satisfied

$$\alpha^2 \xi_x^{IV} + \ddot{\xi} = f_m = f / \rho S_{c.s.} \tag{4.48}$$

that presents the equation of flexural vibrations of a bar.

To make solution to the problem of vibration certain, the boundary conditions at the ends of the bar must be set (the boundary conditions on the other surfaces have been considered before, while deriving w_{pot}). To formulate the boundary conditions, the energy conservation law can be used in the form

$$\frac{d}{dt}(W_{kin} + W_{pot}) = \dot{W}_e, \tag{4.49}$$

where $\dot{W}_e$ is the energy flux through the ends of a bar. Thus,

$$\begin{aligned} &\frac{d}{dt}\int_{x_1}^{x_2} \frac{1}{2}\left[\dot{\xi}^2 + \alpha^2 (\xi_x'')^2\right] dx = \int_{x_1}^{x_2} (\dot{\xi}\ddot{\xi} + \alpha^2 \xi_x'' \dot{\xi}_x'') dx = \\ &= \alpha^2 \xi_x'' \dot{\xi}_x' \Big|_{x_1}^{x_2} - \alpha^2 \xi''' \dot{\xi} \Big|_{x_1}^{x_2} + \frac{1}{2}\int_{x_1}^{x_2} \dot{\xi}(\ddot{\xi} + \alpha^2 \xi_x^{IV}) dx = \dot{W}_e / \rho S_{cs}. \end{aligned} \tag{4.50}$$

(In course of obtaining this result the second term under the original integral was twice integrated by parts.) Since $\xi(x,t)$ satisfies Eq, (4.48) at $f_m = 0$,

$$\alpha^2 \xi_x'' \dot{\xi}_x' \Big|_{x_1}^{x_2} - \alpha^2 \xi_x''' \dot{\xi} \Big|_{x_1}^{x_2} = W_e / \rho S_{c.s.}. \tag{4.51}$$

In this expression $\dot{\xi}$ is the vibration velocity, $\alpha^2 \xi''' = Q$ is the shearing force acting at the cross section in the direction perpendicular to the middle surface of the bar, $\dot{\xi}_x'$ is the angular velocity, $\alpha^2 \xi_x'' = M_f$ is the bending moment.

If no energy flows through the ends of a bar $\dot{W}_e = 0$, we obtain ideal boundary conditions, under which

$$\xi_x'' \dot{\xi}_x' \Big|_{x_1}^{x_2} = 0, \quad \xi_x''' \dot{\xi} \Big|_{x_1}^{x_2} = 0. \tag{4.52}$$

For example, at the simply supported end the displacement and bending moment are equal to zero,

$$\xi = 0, \quad \xi_x'' = 0; \tag{4.53}$$

at the clamped end the displacement and slope are equal to zero,

$$\xi = 0, \quad \dot{\xi}'_x = 0; \tag{4.54}$$

at the free end the bending moment and the shearing force both vanish,

$$\xi''_x = 0, \quad \xi'''_x = 0. \tag{4.55}$$

In case that the energy flux through the ends is not equal to zero and is determined by the action of the external bending moments M_{fe} and/or forces Q_e, then

$$\alpha^2 \xi''_x = M_{fe}, \quad \alpha^2 \xi''' = Q_e. \tag{4.56}$$

It is noteworthy that not ideal boundary conditions for any mechanical vibrating system are due to a leakage or input of energy through the mechanical system boundaries.

Consider the most common method for solving the equations of elastic body vibration, the method based on separation of variables, with example of equations (4.41) and (4.48).

4.3.1 Separation of Variables and Normal Modes

Consider Eq. (4.41) at preset boundary and initial conditions, for example, at

$$\xi'_x(x,t)\big|_{x=0} = \xi''_x(x,t)\big|_{x=l} = 0, \tag{4.57}$$

$$\xi(x,0) = a(x), \ \dot{\xi}(x,0) = b(x), \tag{4.58}$$

where $a(x)$ and $b(x)$ are the arbitrary continuously differentiable functions. At first, consider solution to the auxiliary problem, namely, to the equation of free vibrations

$$c^2 \xi'' - \ddot{\xi} = 0 \tag{4.59}$$

that would satisfy the boundary conditions (4.57). We present the assumed solution as the product of two functions

$$\xi(x,t) = X(x)T(t), \tag{4.60}$$

where $X(x)$ is a function of x alone and $T(t)$ is a function of t alone. Substituting $\xi(x,t)$ in Eq. (4.59) and dividing the result by $X(x)T(t)$, we obtain

$$X''(x)/X(x) = \ddot{T}(t)/c^2 T(t) = -\lambda, \tag{4.61}$$

where λ is a constant. Thus, we have the following equations for determining $X(x)$ and $T(t)$

$$\ddot{T}(t)+\lambda c^2 T(t)=0\,, \tag{4.62}$$

$$X''(x)+\lambda X(x)=0\,. \tag{4.63}$$

The function $X(x)$ must satisfy the boundary conditions (4.57), therefore

$$X'(0)=X'(l)=0\,. \tag{4.64}$$

The values $\lambda=\lambda_i$ and the corresponding functions $X_i(x)$ that satisfy simultaneously Eq. (4.63) and conditions (4.64) are called the eigenvalues and eigenfunctions (or normal modes) of the problem, respectively. The non-zero solutions of Eq. (4.63) under conditions (4.64) exist only at positive values of λ_i. In this case according to Eq. (4.63)

$$X(x)=A_1\cos\sqrt{\lambda_i}\,x+A_2\sin\sqrt{\lambda_i}\,x\,. \tag{4.65}$$

To satisfy the boundary conditions (4.64) it should be $A_2=X'(0)=0$ and $A_1\sqrt{\lambda_i}\sin\sqrt{\lambda_i}\,l=X'(l)=0$. In order to get non-zero solutions, there must be $A_1\neq 0$ that results $\sin\sqrt{\lambda_i}\,l=0$ and $\sqrt{\lambda_i}=i\pi/l$. Thus, in this case the eigenvalues and corresponding normal modes are

$$\lambda_i=(i\pi)^2/l^2,\quad X_i(x)=A_{1i}\cos(i\pi x/l)\,. \tag{4.66}$$

The solutions of Eq. (4.62) that correspond to eigenvalues λ_i are

$$T_i(t)=B_{1i}\cos(i\pi ct/l)+B_{2i}\sin(i\pi ct/l)\,. \tag{4.67}$$

Here $(i\pi ct/l)=\omega_i$ is the natural frequency of vibration. Back to expression (4.60), we can see that the functions

$$\xi_i(x,t)=(B_{1i}\cos\omega_i t+B_{2i}\sin\omega_i t)\cos(i\pi x/l) \tag{4.68}$$

are the partial solutions of Eq. (4.59) under boundary conditions (4.64). Due to linearity of equation (4.59), the sum of partial solutions (4.68)

$$\xi(x,t)=\sum_{i=1}^{\infty}\xi_i(x,t)=\sum_{i=1}^{\infty}(B_{1i}\cos\omega_i t+B_{2i}\sin\omega_i t)\cos(i\pi x/l) \tag{4.69}$$

is also a solution, satisfying the same boundary conditions. This is the general solution for Eq. (4.59), because a proper selection of the constants B_{1i} and B_{2i} can result in satisfying the initial conditions (4.58) as well. By applying these conditions, we obtain

$$\xi(x,0) = a(x) = \sum_{i=1}^{\infty} B_{1i} \cos(i\pi x / l), \tag{4.70}$$

$$\dot{\xi}(x,0) = b(x) = \sum_{i=1}^{\infty} B_{2i}\omega_i \cos(i\pi x / l), \tag{4.71}$$

where from B_{1i} and B_{2i} can be derived as coefficients of expansion of functions $a(x)$ and $b(x)$ into Fourier series in terms of normal modes $X_i(x) = \cos(i\pi x / l)$, namely

$$B_{1i} = a_i(x) = \frac{2}{l}\int_0^l a(x)\cos\frac{i\pi x}{l}dx, B_{2i}\omega_i = b_i(x) = \frac{2}{l}\int_0^l b(x)\cos\frac{i\pi x}{l}dx. \tag{4.72}$$

Thus, the displacements in an elastic body that vibrates freely under specified boundary and initial conditions can be represented in the form of superposition

$$\xi(x,t) = \sum_{i=1}^{\infty} \xi_i(x,t) = \sum_{i=1}^{\infty} (a_i \cos\omega_i t + \frac{b_i}{\omega_i}\sin\omega_i t)\cos\frac{i\pi x}{l} = \sum_{i=1}^{\infty} \xi_{1i}(t)\theta_i(x) \tag{4.73}$$

of standing waves, which change harmonically with amplitude $\xi_i(t)$ and have time independent forms, $\theta_i(x)$, that are the eigenfunctions $\theta_i(x) = X_i(x)$. We will call $\theta_i(x)$ the normal modes of vibration and will assume that their maximum values are normalized to unity (in the particular case under consideration this is fulfilled automatically).

Expressions for displacements (4.73) at $\theta_i = \cos(i\pi x / l)$, $\omega_i = i\pi c / l$ determine vibrations under the end conditions (4.57), but in the general form they are also valid for the normal modes of vibration and natural frequencies corresponding to any other set of ideal conditions. Thus, if $\xi(0) = \xi(l) = 0$ (clamped ends), $\theta_i = \sin(i\pi x / l)$ and $\omega_i = i\pi c / l$. If $\xi(0) = 0$, $\xi'(l) = 0$ (one end is clamped, another is free), $\theta_i(x) = \sin[(2i-1)\pi x / l)]$ and $\omega_i = (2i-1)\pi c / l$.

4.3.2 Transient Vibration

Now we return to solving the inhomogeneous Eq. (4.41) under boundary conditions (4.57) assuming that at the moment of applying the force $f(x,t)$ conditions (4.58) are satisfied. As was previously noted, we assume that the force changes harmonically

$$f(x,t) = f(x)\cos\omega t. \tag{4.74}$$

We represent solution of Eq. (4.41) as the sum of the general solution of the homogeneous Eq. (4.73) (will be denoted $\xi_I(x,t)$), which satisfies the initial and boundary conditions, and of the partial solution $\xi_{II}(x,t)$ of the inhomogeneous equation under zero initial conditions (4.58), as follows

$$\xi(x,t) = \xi_I(x,t) + \xi_{II}(x,t), \tag{4.75}$$

$$\xi_{II}(x,t) = \xi_{II}(x)\cos\omega t. \tag{4.76}$$

After substituting the displacements from Eqs. (4.75) and (4.76) into Eq. (4.41), we obtain ordinary differential equation for the partial solution $\xi_{II}(x,t)$

$$c^2\xi''_{II}(x) + \omega^2\xi_{II}(x) = f(x)/m_\Delta. \tag{4.77}$$

We will be looking for a solution of Eq. (4.77) in the form of expansion into the Fourier series in terms of normal modes of vibrations $\theta_i(x) = \cos(i\pi x/l)$, namely, as

$$\xi_{II}(x) = \sum_{i=1}^{\infty}\xi_{II\,i}\theta_i(x) = \sum_{i=1}^{\infty}\xi_{II\,i}\cos(i\pi x/l). \tag{4.78}$$

Represent the active force as an expansion into the Fourier series in terms of the same normal modes

$$f(x) = \sum_{i=1}^{\infty} f_i\theta_i(x)_i = \sum_{i=1}^{\infty} f_i\cos(i\pi x/l), \tag{4.79}$$

where

$$f_i = \frac{2}{l}\int_0^l f(x)\cos\frac{i\pi x}{l}dx. \tag{4.80}$$

Upon substituting the assumed solution (4.78) and $f(x)$ from (4.79) into Eq. (4.77), we obtain

$$\xi_{II\,i}(\omega_i^2 - \omega^2) = f_i/m_\Delta = f_{i\,eqv}/M_i, \tag{4.81}$$

where

$$f_{i\,eqv} = \int_0^l f(x)\theta_i(x)dx, \quad M_i = m_\Delta\int_0^l \theta_i^2(x)dx. \tag{4.82}$$

Thus, the partial solution of Eq. (4.41) is

$$\xi_{II\,i}(x,t) = \frac{f_{i\,eqv}}{M_i(\omega_i^2 - \omega^2)}\sin\omega t \tag{4.83}$$

and the general solution (4.75) is

$$\xi(x,t) = \sum_{i=1}^{\infty}(\xi_{Ii} + \xi_{IIi})\cos(i\pi x / l)$$
$$= \sum_{i=1}^{\infty}\left\{\left[a_i \cos\omega_i t + \frac{b_i}{\omega_i}\sin\omega_i t\right] + \frac{f_{ieqv}\sin\omega t}{M_i(\omega_i^2 - \omega^2)}\right\}\cos\frac{i\pi x}{l}. \tag{4.84}$$

Here the first term in the brackets describes the free vibrations and the second term describes the forced vibrations of a body.

It is noteworthy that obtained solution is not in accord with practice. It follows from the solution that the normal vibration, once it starts, can last indefinitely long, and that with the frequency of an active force approaching one of the natural frequencies the amplitude of forced vibration increases indefinitely. The normal vibrations decay exponentially, whereas the amplitude of the forced vibration at $\omega = \omega_i$ is limited. The reason for such discrepancy lies in the fact that no internal losses of energy in course of deformation have been considered so far. Although these losses are small enough to treat the mechanical systems as absolutely elastic, they should be considered in the total energy balance, as it was done in previous chapters.

It is easy to make sure that the expansion coefficients in the formula (4.84) $\xi_i(x,t) = \xi_{Ii}(x,t) + \xi_{IIi}(x,t)$, which were obtained as a result of solution of Eqs. (4.59) and (4.41) for $\xi_{Ii}(x,t)$ and $\xi_{IIi}(x,t)$, respectively, can be derived from the following equations that describe these systems as systems with one degree of freedom

$$M_i\ddot{\xi}_i + K_i\xi_i = f_{ieqv} \quad (i = 1,2,...), \tag{4.85}$$

where $K_i = \omega_i^2 M_i$, $\xi_i(0) = a_i$, $\dot{\xi}_i(0) = b$ and a_i, b_i can be determined by formulae (4.72). After introducing in Eq. (4.85) an additional term $r_i\dot{\xi}$, which will account for energy losses, it becomes

$$M_i\ddot{\xi}_i + r_i\dot{\xi}_i + K_i\xi_i = f_{ieqv}. \tag{4.86}$$

In Eq. (4.86) the first term represents inertial force, the second term is frictional force and the third term stands for elastic force. Solution of Eq. (4.86) for free vibration (at $f_{ieqv} = 0$) under initial conditions (4.58) can be assumed in the form $\xi_i = Ae^{pt}$. Substituting this expression in the equation, we obtain

$$p^2 + 2a_i p + \omega_i^2 = 0, \tag{4.87}$$

where $2a_i = r_i / M_i$ and $\omega_i^2 = K_i / M_i$. From Eq. (4.87) we obtain $p = -\alpha_i \pm j\sqrt{\omega_i^2 - \alpha_i^2}$. With small losses $\omega_i^2 \gg \alpha_i^2 = r_i^2 / 4M_i^2$, the value $\omega_{1i}^2 = \omega_i^2 - \alpha_i^2$ is positive, and the general solution for the homogeneous Eq. (4.86) satisfying the initial conditions can be obtained in the following form (see formula (4.73))

$$\xi_i = e^{-\alpha_i t}\left[a_i \cos\omega_{1i}t + \left(\frac{b_i + \alpha_i a_i}{\omega_{1i}}\right)\sin\omega_{1i}t\right] = Ce^{-\alpha_i t}\cos(\omega_{1i}t + \varphi), \tag{4.88}$$

where C and φ are the initial amplitude and phase of vibration. Thus, taking losses into account we obtain damped free vibration that is precisely what is observed experimentally. Besides, the frequency of free vibrations has slightly changed compared to Eq. (4.84), namely, $\omega_{1i} = \omega_i\sqrt{1 - \alpha_i^2 / \omega_i^2}$. Since usually $\alpha_i^2 \ll \omega_i^2$, this change can typically be ignored.

For examining the mode of forced vibrations, we will switch to the complex form $f_i(x,t) \to F_i(x,t) = F_i(x)e^{j\omega t}$, $\dot{\xi}_i(x,t) \to U_i(x,t) = U_i(x)e^{j\omega t}$. After substituting these expressions into Eq. (4.86) we obtain

$$U_i\left[r_i + j(\omega M_i - K_i / \omega)\right] = F_i . \tag{4.89}$$

At frequencies far from the natural frequency ω_i the value of r_i can be ignored, so that $U_i = F_i / M_i(\omega_i^2 - \omega^2)$, as it was obtained by formula (4.83) without involving losses. At $\omega = \omega_i$ the amplitude of vibration is limited by the value $U_i = F_i / r_i$.

The stage of motion, at which both damped normal vibrations and forced vibrations exist simultaneously, is referred to as the transient process. At a certain point, when it becomes possible to ignore the normal vibrations, a steady state of forced vibrations begins. Further we will consider the steady state vibrations only.

Thus, displacements in a vibrating body can be represented as expansion (4.84) in terms of normal modes both in the transient and in the steady state vibration. Since under fixed distribution of displacements a mechanical system has one degree of freedom (such systems were considered in Chapter 2), this means that any complicated vibrations of an elastic body can be represented as superposition of simple vibrations of the body, each having one degree of freedom and can be considered independently.

4.3.3 Equivalent T-Network of a Longitudinally Vibrating Bar

Under the steady state of vibration the solution of Eq. (4.59) can be represented in the complex form. Let the complex quantity of displacement be denoted $\xi(x,t) \to \bar{\xi}(x,t) = \bar{\xi}(x)e^{j\omega t}$ substituting this function into Eq. (4.59) we arrive at the equation

$$\bar{\xi}''(x) + k^2\bar{\xi}(x) = 0, \tag{4.90}$$

where $k^2 = \omega^2 / c^2$. This is the one-dimensional wave equation. It matches in form with Eq. (4.63) at $\lambda = k^2$, and its solution in the form of expansion in terms of the normal modes of vibration was previously considered. It is another form of solution of Eq. (4.90). Under this solution the arbitrary constants in expression for $\bar{\xi}(x)$,

$$\bar{\xi}(x) = A_1 \sin kx + A_2 \cos kx, \tag{4.91}$$

can be determined directly from the conditions at the ends of a bar. In setting conditions at the ends, we must follow the adopted rule of signs, the effect of which was illustrated in Figure 1.11 (a). Since we consider tensile strains and displacements that generate such strains as conventionally positive, the displacements at the right end, $x = l$, should be considered with their signs, while those at the left end, $x = 0$, with the opposite signs (the positive directions of displacement and of the axis x are opposite in this case). The forces F_0 and F_l shown in Figure 4.6 (c) are positive because they coincide in direction with the displacements. As the compress ive forces should result in negative values of stress, the conditions at the ends are as follows

$$YS_{cs}\bar{\xi}'\big|_{x=0} = -F_0, \quad YS_{cs}\bar{\xi}'\big|_{x=l} = -F_l. \tag{4.92}$$

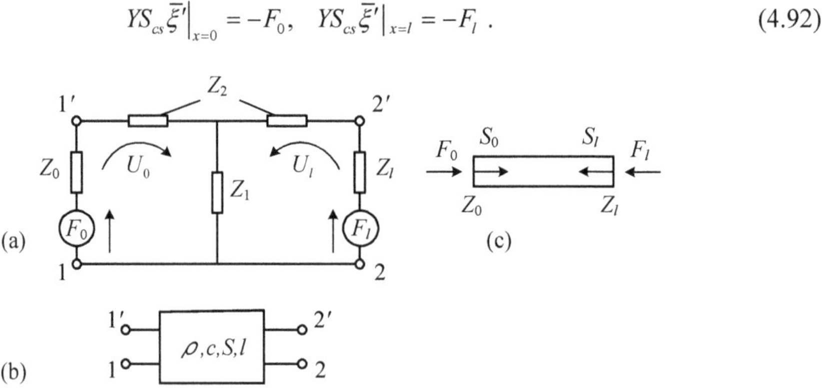

Figure 4.6: Equivalent circuit of a longitudinally vibrating bar loaded on the ends: (a) T-network representation, (b) equivalent two-port block representation, (c) illustration for the rule of signs.

If the ends of a bar are loaded with mechanical impedances Z_0 and Z_l, then the forces of reaction applied to the ends are $F = -Z\dot{\bar{\xi}} = -ZU$. At directions of displacements looking inside a bar, as shown in Figure 4.6 (c), $\bar{\xi}_0$ and $\bar{\xi}_l$ are negative, so the responsive forces are tensile and must lead to positive stress at the ends, so that the end conditions should be written as follows

$$YS_{cs}\bar{\xi}'\big|_{x=o} = -Z_0 U_0, \quad YS_{cs}\bar{\xi}'\big|_{x=l} = -Z_l U_l \,. \tag{4.93}$$

If we consider the displacements $\bar{\xi}_0$ and $\bar{\xi}_l$ as given, then putting in formula (4.91) $\bar{\xi}(0) = -\bar{\xi}_0$ and $\bar{\xi}(l) = -\bar{\xi}_l$ we obtain

$$A_2 = -\xi_0, \quad A_1 = \left(\bar{\xi}_l + \bar{\xi}_0 \cos kl\right) / \sin kl \tag{4.94}$$

and

$$\bar{\xi}(x) = (\bar{\xi}_l + \bar{\xi}_0 \cos kl)\sin kx / \sin kl - \bar{\xi}_0 \cos kl \,. \tag{4.95}$$

Thus, the displacements in a bar prove to be completely determined by two values $\bar{\xi}_0$ and $\bar{\xi}_l$. Therefore, a longitudinal vibrating bar can be considered as a system with two degrees of freedom with the generalized coordinates $\bar{\xi}_0$ and $\bar{\xi}_l$. In this system the losses can be accounted for, if to assume that the sound speed and the wave number are complex quantities. Namely, $\bar{c} = c\left(1 + j/2Q_m\right)$ and $\bar{k} = \omega/\bar{c} = k\left(1 - j/2Q_m\right)$, where Q_m is the quality factor of material.

If forces F_0, F_l and impedances Z_0 and Z_l are applied to the ends of a bar simultaneously, then taking into consideration expressions (4.92) and (4.93) the following conditions should be satisfied

$$YS_{c.s}\bar{\xi}'\big|_{x=0} = -Z_0 U_0 - F_0 \,, \tag{4.96}$$

$$YS_{c.s}\bar{\xi}'\big|_{x=l} = -Z_l U_l - F_l \,. \tag{4.97}$$

After substituting expression for $\bar{\xi}(x)$ from (4.95) and $Y = \rho c^2$ in these equations we arrive at

$$\frac{\rho c S_{c.s}(U_0 + U_l)}{j \sin kl} + j\rho c S_{c.s} U_0 \tan\frac{kl}{2} + Z_0 U_0 + F_0 = 0 \,, \tag{4.98}$$

$$\frac{\rho c S_{c.s}(U_0 + U_l)}{j \sin kl} + j\rho c S_{c.s} U_l \tan\frac{kl}{2} + Z_l U_l + F_l = 0 \,. \tag{4.99}$$

It can be verified that Eqs. (4.98) and (4.99) are the Kirchhoff's equations for "currents" U_0 and U_l in the circuit in Figure 4.6 (a), where the bar is represented by the two-port T-network between points 1, 1' and 2, 2' with impedances

$$Z_1 = -j\rho c S_{c.s} \sin kl, \quad Z_2 = Z_3 = j\rho c S_{c.s} \tan(kl/2) . \tag{4.100}$$

For brevity, this two-port network may be presented, as the equivalent block shown in Figure 4.6 (b).

The values of velocities of the ends can be determined using the circuit of Figure 4.6 (a), Consequently the distribution of vibrations in a bar at various combinations of loads and acting forces on the ends can be found by formula (4.95). If it is necessary to determine velocity of vibration of any point inside of a bar, the bar can be represented by a cascade connection of two T-networks that correspond to the parts of the bar on the left and right sides of this point.

The two-port circuit in Figure 4.6 (a) can be used for determining value of the mechanical input impedance Z_{in} of a bar. With load impedance Z_L at the opposite end it is

$$Z_{in} = \frac{j\rho c S_{c.s} \tan kl + Z_L}{1 + j(Z_L / \rho c S_{c.s}) \tan kl} . \tag{4.101}$$

In the cases that $Z_L = 0$ (the opposite end is free, i.e., mechanical short circuited)

$$Z_{in} = j\rho c S_{c.s.} \tan kl , \tag{4.102}$$

and at $Z_L \to \infty$ (the opposite end is fixed, i.e., mechanical open circuited)

$$Z_{in} = j\rho c S_{c.s.} / \tan kl . \tag{4.103}$$

Since Eq. (4.90) originated from Eq. (4.59), which pertains to the general case of one-dimensional longitudinal and torsional vibrations, its solution in the form of Eq. (4.95), where $\bar{\xi}_0$ and $\bar{\xi}_l$ are determined by expressions (4.98) and (4.99), and the equivalent circuit of Figure 4.6 (a) are valid for all the above considered particular cases. It is sufficient only in the case of longitudinal vibrations to use in the final result $c = \sqrt{Y / \rho(1-\sigma^2)}$ for vibrations through the width of a long stripe (see Figure 4.2 (b)) and $c = \sqrt{Y(1-\sigma) / \rho(1+\sigma)(1-2\sigma)}$ for vibrations through the thickness of a plate with large transverse dimensions (see Figure 4.6 (d)). In the case of the torsional vibrations (see Figure 4.5) ρS_{cs} must be replaced with ρJ_p and the sound speed must be calculated by formula $c = \sqrt{G / \rho J_p}$, which yields the value $c = \sqrt{\mu / \rho}$ for the bars of a circular and square cross sections. In the case of torsion, the generalized active forces are the moments, and the generalized displacements are the turning angles.

Presenting solution to the problems of vibration of a transducer mechanical system in the form of Eq. (4.95) is especially advantageous if it is considerably loaded at the ends, and/or when a relatively broad frequency range of operation is concerned. It is much more convenient to employ the solution in form of expansion in terms of normal modes of vibration for analyzing the transducers operating at small loads and in relatively narrow frequency ranges around the resonances.

4.3.4 Normal Modes of the Transverse Vibrating Beams

The method of expansion in terms of normal modes may be applied to solving the equation of transverse vibrations of a beam (4.48) under the boundary conditions (4.53)–(4.55) in the mode of the steady state harmonic vibrations,

$$\alpha^2 \xi_x^{IV} + \ddot{\xi} = f(x,t) / \rho S_{cs} . \tag{4.48}$$

Here $\alpha^2 = EYJ / \rho S_{c.s}$, $f(x,t) = f(x)\cos\omega t$.

The equation of free vibrations corresponding to Eq. (4.48), in which the notations $\xi(x,t) = X(x)e^{j\omega t}$ and $\omega^2 / \alpha^2 = k^4$ are used, can be presented in the form

$$X^{IV} - k^4 X = 0 . \tag{4.104}$$

Substituting the assumed solution in the form of $X(x) = e^{\lambda x}$ in this equation we will find four values of λ : $\pm jk, \pm k$. Thus, the general solution of Eq. (4.48) can be represented as

$$X(x) = A_1 e^{jkx} + A_2 e^{-jkx} + A_3 e^{kx} + A_4 e^{-kx} \tag{4.105}$$

or

$$X(x) = C_1 \sin kx + C_2 \cos kx + C_3 \sinh kx + C_4 \cosh kx . \tag{4.106}$$

Relations between the arbitrary constants as well as the equations for determining natural frequencies can be found by applying boundary conditions. In this way we also obtain expressions for normal modes. Consider the following variants of the ideal boundary conditions. In the case of a beam with simply supported ends the boundary conditions are of the form of (4.53)

$$X(0) = X(l) = 0, \quad X'\big|_{x=0,\, x=l} = 0 . \tag{4.53}$$

The equation to determine the natural frequencies is $\sin k_i l = 0$, where from $k_i l = \pi i$,

$\omega_i = \alpha k_i^2 = \alpha(i\pi / l)^2$. The normal modes are

$$X_i(x) = \sin(i\pi x / l) . \tag{4.107}$$

In the case that both ends are free it follows from (4.55) that

$$X'' |_{x=0,l} = 0, \quad X''' |_{x=0,l} = 0 . \tag{4.55}$$

From Eq. (4.104) will be obtained that

$$\cos k_i l \cdot \cosh k_i l = 1 . \tag{4.108}$$

The successive values of $k_i l$ are 0, 4.73, 7.85, and

$$\omega_0 = 0, \quad \omega_1 = \alpha(4.73/1)^2, \quad \omega_2 = \alpha(7.85/1)^2 . \tag{4.109}$$

The presence of the natural frequency $\omega_0 = 0$ signifies that the beam can be displaced as a solid body.

For a beam with one end clamped (at $x = 0$) and another end free (at $x = l$) the boundary conditions are

$$X(0) = 0, \quad X'(x)_{x=0} = 0, \quad X''(x)_{x=l} = 0, \quad X'''(x)_{x=l} = 0 . \tag{4.110}$$

The frequency equation is

$$\cos k_i l \cdot \cosh k_i l = -1 , \tag{4.111}$$

and the successive roots of this equation are

$$k_i l = 1.87, \ 4.69 . \tag{4.112}$$

At $i > 2$ they are the same as for the beam with free ends. The first natural frequency is

$$\omega_1 = \alpha(1.87 / l)^2 . \tag{4.113}$$

For a beam clamped at both ends the boundary conditions are

$$X(x)\big|_{x=0,l} = 0, \quad X'(x)\big|_{x=0,l} = 0 \tag{4.114}$$

and the same frequency equation (4.108), as for the beam with free ends is valid. This results in the same successive natural frequencies (4.109) except for absence of the natural frequency $\omega_0 = 0$, because no motion as a solid body is possible in this case.

The beams having boundary conditions that can be approximated by the considered ideal conditions are used as parts of mechanical systems of different transducer designs, mainly of those operating in the frequency range below their first resonance frequency. In this frequency range the first mode of vibration dominates. With very good approximation the modes of static deflection of the beams under uniformly distributed loads can be used instead of the normal modes for solving vibration problems in this frequency range by Rayleigh's method (definitely, except for vibration of beams with free ends). Following Ref. 2 they are for the beams:

with simply supported ends

$$\theta(x) = (16/5l)(x - 2x^3/l^2 + x^4/l^3); \tag{4.115}$$

with clamped ends

$$\theta(x) = (16x^2/l^2)(1 - 2x/l + x^2/l^2); \tag{4.116}$$

with one clamped and another free ends

$$\theta(x) = 2(x/l)^2(1 - 2x/3l + x^2/6l^2). \tag{4.117}$$

The modes of static deflection for the beams are presented in Figure 4.7. In the same Figure are presented the normal mode for the simply supported beam for comparison, and the normal mode for the beam with free ends. Coordinates of important points on the modes of vibration are: the inflection points, at which signs of curvature change of the mode for the beam with clamped ends, at $x_1 = 0.24l$ and $x_2 = 0.76l$; the nodal points, at which displacements of beam with free ends are zeros, at $x_1 = 0.22l$ and $x_2 = 0.78l$. If the beam with free ends is simply supported at the nodal points, then it can move at low frequencies predominantly as a solid body (in the piston like mode), i.e., vibration in the first mode is suppressed.

Before proceeding with representing solution for a vibration problem as expansion in terms of the normal modes, the following general properties of the normal modes must be noted. They can be derived using equations of separation (4.63) and (4.104). Consider for example Eqs. (4.104) after substituting the normal modes X_n and X_m with corresponding eigenvalues k_n and k_m

$$X_n^{IV} - k_n^4 X_n = 0, \tag{4.118}$$

$$X_m^{IV} - k_m^4 X_m = 0. \tag{4.119}$$

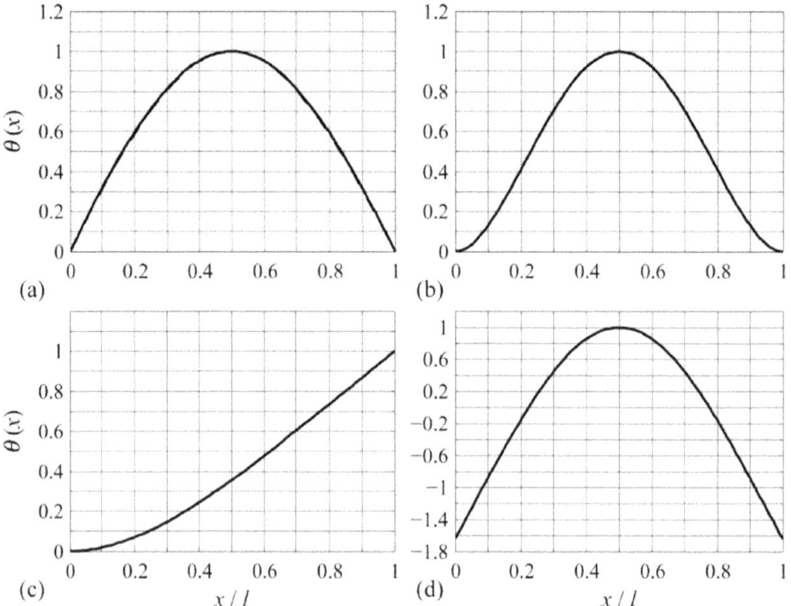

Figure 4.7: The modes of static deflections for the beams: (a) with simply supported ends, (b) with both ends clamped, (c) with one end clamped and another free, (d) normal mode of the beam with free ends. The normal mode for the simply supported beam, $\theta(x) = \sin(\pi x / l)$, coincides with the static deflection shown in (a) within the thickness of the line.

Multiplying the first equation by X_m and second by X_n, subtracting equations and integrating over interval [0, l] we obtain

$$(k_n^4 - k_m^4)\int_0^l X_n(x)X_m(x)dx = \int_0^l \left(X_m X_n^{IV} - X_n X_m^{IV}\right)dx\,. \tag{4.120}$$

After integrating the right-side integral by parts and applying different ideal boundary conditions we arrive at conclusion that in all the cases the normal modes are orthogonal, i.e.,

$$\int_0^l X_n(x)X_m(x)dx = 0 \quad (n \neq m)\,. \tag{4.121}$$

An arbitrary continuously differentiable function $g(x)$ that satisfies boundary conditions of the problem can be expanded in terms of normal modes $X_i(x)$ as follows

$$g(x) = \sum_{i=1}^{\infty} g_i X_i(x)\,, \text{ where } g_i = \frac{1}{N_i}\int_o^l g(x)X_i(x)dx \text{ and } N_i = \int_0^l X_i^2(x)dx\,. \tag{4.122}$$

Now we return to Eq. (4.48). Remember that under steady state vibration the solution can be represented in the complex form, in which case it is denoted $\xi(x,t) \to \overline{\xi}(x,t) = \overline{\xi}(x)e^{j\omega t}$, and $f(x,t) \to F(x)e^{j\omega t}$. After representing the assumed solution and the acting force as expansions into series in terms of the normal modes that correspond to a particular type of boundary conditions,

$$\overline{\xi}(x) = \sum_{i=1}^{\infty} \overline{\xi}_i X_i(x) \tag{4.123}$$

and

$$F(x) = \sum_{i=1}^{\infty} F_i X_i(x) \tag{4.124}$$

After substituting these series into Eq. (4.48), we obtain the solution in the form of expression

$$\overline{\xi}_i = \frac{F_i}{\rho S_{c.s.}(\omega_i^2 - \omega^2)} = \frac{F_{eqvi}}{\rho S_{c.s.} N_i(\omega_i^2 - \omega^2)}. \tag{4.125}$$

In the course of manipulations, it was taken into account that due to Eq. (4.106)

$$X^{IV} = k^4 X = (\omega_i^2 / \alpha^2) X . \tag{4.126}$$

Note that following relations (4.122)

$$F_{eqvi} = \int_0^l F(x) X_i(x) dx . \tag{4.127}$$

Expression (4.125) represents magnitude of the normal mode X_i in expansion of transverse steady state vibration of a beam. It is analogous to expression (4.83) for longitudinal vibrating bar, and the same considerations regarding accounting for the energy losses in real situation are applicable.

Correlation between magnitudes of modes of vibration, and number of terms in the series (4.123) that have to be taken into account depend essentially on distribution of the active force $F(x)$. If $F(x) = F(x_o)$ is a concentrated force applied at point x_o on the side surface of a simply supported beam, then by formula (4.127) $F_{eqvi} = F(x_o) \cdot X_i(x_o)$, and the modes of vibration, for which the force is applied to a node of mode of vibration, are not excited. If the force distribution coincides with any of the normal modes, $f(x) \sim X_m(x)$, then in series (4.123) only the term corresponding to this mode of vibration will remain, because the force distribution in this case is orthogonal to all the other vibration modes, and $f_{i\,eqv} = 0$ at $i \neq m$. In reality, distributions of active forces require for their representation only several terms of expansion, therefore the series (4.123) proves to be limited as well.

4.3.5 Corrections Due to Finite Thickness of Bars and Beams

In the case that dimensions of cross sections of bars and beams are not too small compared to their length, corrections (4.14), (4.33) and (4.37) to the energies of their vibration have to be taken into account, when deriving equations of motion.

At first consider the kinetic and potential energies of vibrating bars and beams having small relative thickness, assuming that distribution of displacements is represented by series (4.123),

$$W_{kin} = \frac{1}{2}\rho S_{c.s}\int_0^l \left(\sum_{i=1}^{\infty}\dot{\xi}_i X_i\right)^2 dx = \frac{1}{2}\sum_{i=1}^{\infty}\dot{\xi}_i^2 M_{eqvi}\,, \tag{4.128}$$

where

$$M_{eqvi} = \rho S_{c.s}\int_0^l X_i^2 dx\,. \tag{4.129}$$

The potential energy of deformation of a beam is

$$W_{pot} = \frac{1}{2}YJ\int_0^l \left(\sum_{i=1}^{\infty}\xi_i X_i''\right)^2 dx = \frac{1}{2}\sum_{i=1}^{\infty}\xi_i^2 YJ\left(\frac{\omega_i}{\alpha}\right)^2\int_0^l X_i^2 dx = \frac{1}{2}\sum_{i=1}^{\infty}K_{eqvi}\xi_i^2\,, \tag{4.130}$$

where

$$K_{eqvi} = YJ\left(\frac{\omega_i}{\alpha}\right)^2\int_0^l X_i^2 dx\,. \tag{4.131}$$

In the expressions for the energies M_{eqvi} and K_{eqvi} are the equivalent mass and equivalent rigidity that correspond to normal mode X_i (the modal mass and rigidity). In the process of manipulations the orthogonal property of the normal modes was used, and the expression

$$\int_0^l (X_i'')^2 dx = \left(\frac{\omega_i}{\alpha}\right)^2\int_0^l X_i^2 dx\,, \tag{4.132}$$

which can be obtained by multiplying both parts of Eq. (4.122) by X_i, integrating by parts over the beam length, and applying the ideal boundary conditions.

Expressions (4.128) and (4.130) for the kinetic and potential energies are general for the beams under ideal boundary conditions and the values of M_{eqvi}, K_{eqvi}, and ω_i depend on the particular boundary conditions.

For the longitudinally vibrating bar with free ends ($X_i = \cos(i\pi x/l)$) from expressions for the kinetic and potential energies will be obtained that

$$M_{eqvi} = \rho S_{cs} l/2 = M/2, \quad K_{eqvi} = (i\pi)^2 YS_{cs}/2l, \quad \omega_i^2 = (i\pi)^2 Y/l^2\rho. \tag{4.133}$$

For the transversely vibrating beam with simply supported ends ($X_i = \sin(i\pi x/l)$)

$$M_{eqvi} = M/2, \quad K_{eqvi} = \frac{(i\pi)^4 wt^3 Y}{12l^3}, \quad \omega_i^2 = (i\pi)^4 YJ/(\rho S_{cs} l^4). \tag{4.134}$$

Consider now effect of correction to the kinetic energy by formula (4.14) that accounts for the energy of the transverse motion under the longitudinal vibration of a bar with free ends (Rayleigh's correction). Upon substituting displacement in the form of expansion in terms of normal modes into expression (4.14) and integrating over the length of the bar we obtain

$$\Delta W_{kin} = \frac{\rho\sigma^2 J_p}{2}\int_0^l \left(\sum_{i=1}^{\infty} \dot{\xi}_i \frac{i\pi}{l}\sin\frac{i\pi x}{l}\right)^2 dx = \frac{1}{2}\sum_{i=1}^{\infty} \dot{\xi}_i^2 \rho\sigma^2 J_p \frac{(i\pi)^2}{2l}. \tag{4.135}$$

Adding ΔW_{kin} to the kinetic energy (4.128) for the longitudinally vibrating bar does not change the form of the series that represents $W'_{kin} = W_{kin} + \Delta W_{kin}$. The modal equivalent masses change only, and this results in changing the natural frequencies. New expression for the modal masses and resonance frequencies will be found to be

$$M'_{eqvi} = M_{eqvi}\left[1 + (i\pi\sigma)^2 J_p / l^2 S_{cs}\right], \tag{4.136}$$

$$\omega_i' = \omega_i / \sqrt{1 + (i\pi\sigma)^2 J_p / l^2 S_{cs}} \approx \omega_i \left[1 - (i\pi\sigma)^2 J_p / 2l^2 S_{cs}\right]. \tag{4.137}$$

For the bars of the circular and rectangular cross sections $J_p = \pi a^4/2$ and $J_p = wt^3/6$, respectively. The correction increases with increase of the ratio of transverse dimension to the half-wave of deformation ($\lambda \sim 2l/i$). For a bar with the square cross section this ratio is equal to $(i\pi\sigma w)^2/12l^2$, and at $w/l < 1/3$ the correction term in expression (4.137) is less than 0.01 for the first mode of vibration.

For the transversely vibrating beams the effects of shear deformation and rotary inertia that result in changing densities of the kinetic and potential energies must be taken into account. In the expansion (4.123) will be denoted $X_i = \theta_i(x)$ in order to distinguish from the case of longitudinal vibration. Upon substituting the displacement $\xi(x,t) = \xi_1(t)\theta_1(x)$ into expressions (4.33) for Δw_{kin} and (4.37) for Δw_{pot} and integrating over length of a beam we arrive at

$$\Delta W_{kin} = \frac{1}{2}\dot{\xi}_1^2 \rho J_2 \int_0^l \left(\frac{d\theta_1}{dx}\right)^2 dx\,, \tag{4.138}$$

$$\Delta W_{pot} = -\frac{1}{2}\xi_i^2 \frac{wY^2t^5}{120\mu}\int_0^l \left(\frac{\partial^3\theta_1}{\partial x^3}\right)^2 dx\,. \tag{4.139}$$

For a beam with simply supported ends the following corrections will be obtained to parameters M_{eqv} and K_{eqv} originally defined by the formulae (4.134) for the first mode of vibration, which is the most usable for practical applications,

$$M'_{eqv} = M_{eqv}\left(1+\frac{\pi^2}{12}\frac{t^2}{l^2}\right), \quad K'_{eqv} = K_{eqv}\left[1-\frac{(\pi t)^2 Y}{20\mu l^2}\right]. \tag{4.140}$$

So far the equations of vibration of bars and beams were considered. Other important piezoelement geometries are plates. As to the rectangular plates with commensurable lateral dimensions, they are rarely used in transducer designs. A rectangular surface can be assembled of the beams or strips, if necessary. The width of these elements may be small enough in comparison with their length. By contrast, vibration systems in the form of the circular plates are widely used. Solving the problems of vibration of the circular plates and of any bodies with curvilinear surfaces (cylindrical and spherical) require introducing the curvilinear coordinates to be able to match coordinate system with surfaces, on which boundary conditions are defined.

4.4 Equations of Vibration in Curvilinear Coordinates

4.4.1 Curvilinear Coordinates

Consider a volume element in a curvilinear orthogonal system of coordinates $\boldsymbol{q}_1$, $\boldsymbol{q}_2$, $\boldsymbol{q}_3$ with coordinate surfaces $q_1 = f_1(x_1,x_2,x_3)$, $q_2 = f_2(x_1,x_2,x_3)$, $q_3 = f_3(x_1,x_2,x_3)$ the volume element being formed by intersection of pairs of surfaces corresponding to the values of q_i and $q_i + dq_i$ (Figure 4.8 (a)). Unit vectors $\boldsymbol{q}_1$, $\boldsymbol{q}_2$, $\boldsymbol{q}_3$ are tangential to coordinate lines (lines of intersection of coordinate surfaces), which converge in one point. The position of the coordinate vectors relative to the former rectangular coordinates with unit vectors $\boldsymbol{x}_1$, $\boldsymbol{x}_2$, $\boldsymbol{x}_3$ is determined by cosines of angles between directions of the vectors q_1 and x_1, $\cos(q_1, x_1)$. Suppose that the rectangular coordinates x_1, x_2, x_3 are expressed through the curvilinear coordinates q_1, q_2, q_3 as follows: $x_1 = \varphi_1(q_1, q_2, q_3)$, $x_2 = \varphi_2(q_1, q_2, q_3)$, $x_3 = \varphi_3(q_1, q_2, q_3)$. Then the elementary length along the rectangular coordinate lines will be

$$dx_i = \frac{\partial \varphi_i}{\partial q_1} dq_1 + \frac{\partial \varphi_i}{\partial q_2} dq_2 + \frac{\partial \varphi_i}{\partial q_3} dq_3, \tag{4.141}$$

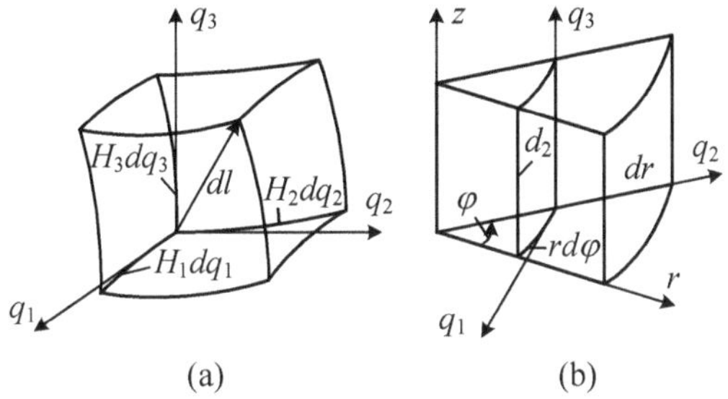

Figure 4.8: Curvilinear systems of coordinates: (a) general type, (b) cylindrical.

and the direction cosines of unit vectors, $\cos(q_l, x_i)$, will be proportional to $\ddot{a}\varphi_i / \ddot{a}q_l$. Conditions of orthogonality of the curvilinear coordinate system are

$$\frac{\partial \varphi_1}{\partial q_l}\frac{\partial \varphi_1}{\partial q_k} + \frac{\partial \varphi_2}{\partial q_l}\frac{\partial \varphi_2}{\partial q_k} + \frac{\partial \varphi_3}{\partial q_l}\frac{\partial \varphi_3}{\partial q_k} = 0, \quad l \neq k. \tag{4.142}$$

Calculating the elementary length in curvilinear coordinates (see Figure 4.8) with taking into consideration Eqs. (4.141) and (4.142) yields

$$\begin{aligned} dl^2 &= dx_1^2 + dx_2^2 + dx_3^2 \\ &= \sum_{i=1}^{3}\left[\left(\frac{\partial \varphi_i}{\partial q_1}\right)^2 dq_1^2 + \left(\frac{\partial \varphi_i}{\partial q_2}\right)^2 dq_2^2 + \left(\frac{\partial \varphi_i}{\partial q_3}\right)^2 dq_3^2\right] \\ &= H_1^2 dq_1^2 + H_2^2 dq_2^2 + H_3^2 dq_3^2. \end{aligned} \tag{4.143}$$

Here H_i are the Lame coefficients of the curvilinear system of coordinates,

$$H_i^2 = \left(\frac{\partial \varphi_1}{\partial q_i}\right)^2 + \left(\frac{\partial \varphi_2}{\partial q_i}\right)^2 + \left(\frac{\partial \varphi_3}{\partial q_i}\right)^2 = \left(\frac{\partial x_1}{\partial q_i}\right)^2 + \left(\frac{\partial x_2}{\partial q_i}\right)^2 + \left(\frac{\partial x_3}{\partial q_i}\right)^2. \tag{4.144}$$

It follows from formula (4.143) that elementary lengths along the axes of curvilinear system of coordinates are $dl_1 = H_1 dq_1$, $dl_2 = H_2 dq_2$, $dl_3 = H_3 dq_3$.

When curvilinear coordinates are used, a convention has to be established regarding designations of the coordinate axes, components of the displacements and components of the tensors of strain and stress. The elementary lengths along all the axes of rectangular coordinates are equal ($H_1 = H_2 = H_3 = 1$), therefore it is possible to introduce numerical designations for all the above components. The peculiarity of mechanical systems made from piezoelectric ceramics that axis 3 is commonly directed along the poling vector in this case can be considered in advance. During this treatment the cylindrical coordinates will be predominantly used. In case of the cylindrical coordinates $r,\ \varphi,\ z$ (Figure 4.8 (b)) $x_1 = r\cos\varphi$, $x_2 = r\sin\varphi$, $x_3 = z$. The Lame coefficients for the cylindrical coordinates are $H_r = 1,\ H_\varphi = r,\ H_z = 1$ following expressions (4.144). The expressions for the components of the strain tensor are

$$S_{rr} = \frac{\partial \xi_r}{\partial r},\quad S_{\varphi\varphi} = \frac{1}{r}\frac{\partial \xi_\varphi}{\partial \varphi} + \frac{\xi_r}{r},\quad S_{zz} = \frac{\partial \xi_z}{\partial z}, \tag{4.145}$$

$$S_{\varphi z} = \frac{\partial \xi_\varphi}{\partial z} + \frac{1}{r}\frac{\partial \xi_z}{\partial \varphi},\quad S_{rz} = \frac{\partial \xi_z}{\partial r} + \frac{\partial \xi_r}{\partial z},\quad S_{r\varphi} = \frac{1}{r}\frac{\partial \xi_r}{\partial \varphi} + r\frac{\partial}{\partial r}\left(\frac{\xi_\varphi}{r}\right). \tag{4.146}$$

In the axial symmetric case $\xi_\varphi = 0,\ \partial\xi_r / \partial\varphi = 0,\ \partial\xi_z / \partial\varphi = 0,$ and

$$S_{rr} = \frac{\partial \xi_r}{\partial r},\quad S_{\varphi\varphi} = \frac{\xi_r}{r},\quad S_{zz} = \frac{\partial \xi_z}{\partial z}, \tag{4.147}$$

$$S_{\varphi z} = S_{r\varphi} = 0,\quad S_{rz} = \frac{\partial \xi_z}{\partial r} + \frac{\partial \xi_r}{\partial z}. \tag{4.148}$$

In case of the spherical coordinates $r,\ \varphi,\ \theta$ $x_1 = r\sin\varphi\cos\theta$, $x_2 = r\sin\varphi\sin\theta$, $x_3 = r\cos\varphi$. The Lame coefficients for the spherical coordinates are $H_r = 1$, $H_\varphi = r$, $H_\theta = r\sin\varphi$ following expressions (4.144). The expressions for the components of the tensor of strain in the axially symmetrical case are

$$S_{rr} = \frac{\partial \xi_r}{\partial r},\quad S_{\theta\theta} = \frac{\xi_r}{r},\quad S_{\varphi\varphi} = \frac{1}{r\sin\theta}\frac{\partial \xi_\varphi}{\partial \varphi} + \frac{\xi_\theta}{r}\cot\theta + \frac{\xi_r}{r}, \tag{4.149}$$

$$2S_{\theta\varphi} = \frac{1}{r\sin\theta}\frac{\partial \xi_\theta}{\partial \varphi} - \frac{\xi_\varphi}{r}\cot\theta,\quad 2S_{r\theta} = \frac{\partial \xi_\varphi}{\partial r} - \frac{\xi_\theta}{r}, \tag{4.150}$$

$$2S_{r\varphi} = \frac{1}{r\sin\theta}\frac{\partial\xi_r}{\partial\varphi} + \frac{\partial\xi_\varphi}{\partial r} - \frac{\xi_\varphi}{r}. \tag{4.151}$$

All the relations between stresses and strains expressed in rectangular coordinates will be valid, if coordinates r, φ, z (r, φ, θ for spherical coordinate system) correspond to designations 1, 2, 3, respectively, so far as the isotropic bodies are considered. In case that the body is made of piezoceramics, its volume element is related to the orthogonal crystallographic coordinate system, in which the direction of axis 3 (unit vector $\boldsymbol{q}_3$) coincides with the direction of poling vector ***P***. Directions of the unit vectors $\boldsymbol{q}_1$ and $\boldsymbol{q}_2$ may be arbitrary, but a common convention is that all the unit vectors form the right-hand system (i.e., rotation of $\boldsymbol{q}_1$ to the coincidence with $\boldsymbol{q}_2$ must be seen as counter-clockwise from the end of vector $\boldsymbol{q}_3$).

4.4.2 Vibrations in the Plane of Circular Disks of Small Height

Note that within this Section it is assumed that the height is dimension of a disk in direction perpendicular to its plane (axis 3) and the thickness of an annual disk is (*a-b*), where *a* is the outer and *b* is the inner radius of the disk.

4.4.2.1 Deriving Equation of Motion

Consider the axial symmetric vibration in the plane of the circular disks (Figure 4.9) that are the widespread piezoelement configurations in transducer designing. Relations between the stresses and strains in this case, as well as formula (4.16) for w_{pot}, can be presented in the cylindrical coordinates by substituting expressions for strain from (4.147)

$$S_1 = S_{rr} = \ddot{a}\xi_r \,/\, \ddot{a}r, \quad S_2 = S_{\varphi\varphi} = \xi_r \,/\, r. \tag{4.152}$$

As a result, we obtain

$$T_1 = T_{rr} = \frac{Y}{1-\sigma^2}\left(\frac{\partial\xi_r}{\partial r} + \sigma\frac{\xi_r}{r}\right), \quad T_2 = T_{\varphi\varphi} = \frac{Y}{1-\sigma^2}\left(\frac{\xi_r}{r} + \sigma\frac{\partial\xi_r}{\partial r}\right), \tag{4.153}$$

$$w_{pot} = \frac{1}{2}\frac{Y}{1-\sigma^2}\left[\left(\frac{\partial\xi_r}{\partial r}\right)^2 + 2\sigma\frac{\xi_r}{r}\frac{\partial\xi_r}{\partial r} + \left(\frac{\xi_r}{r}\right)^2\right]. \tag{4.154}$$

The density of kinetic energy is

$$w_{kin} = \rho\dot{\xi}_r^2 / 2. \tag{4.155}$$

Note that w_{kin} and w_{pot} are related to volume element $d\tilde{V}$. In this case $d\tilde{V} = r\,dr\,d\varphi\,dz$. To obtain the equations of vibration we will use the same procedure, as was used for treating the transverse vibrations of a beam. Following the variational principle, we obtain

$$\delta\int_{t_1}^{t_2}\int_{\tilde{V}}(w_{kin} - w_{pot})d\tilde{V}dt =$$
$$= \delta\int_{t_1}^{t_2}\int_{0}^{a}\left\{\rho\dot{\xi}_r^2 - \frac{Y}{1-\sigma^2}\left[\left(\frac{\partial\xi_r}{\partial r}\right)^2 + 2\sigma\frac{\xi_r}{r}\frac{\partial\xi_r}{\partial r} + \left(\frac{\xi_r}{r}\right)^2\right]\right\}r\,dr\,dt = 0. \tag{4.156}$$

Upon substituting under the integral function $\xi_r + \eta(r,t)$ for ξ_r, where $\eta\,(r, t)$ is an arbitrary continuously differentiable function, which takes zero values at the boundaries of integration interval. Displacement ξ_r inside of the disk must satisfy the condition

$$\int_{t_1}^{t_2}\int_{0}^{a}\left\{\dot{\xi}_r\dot{\eta} - c^2\left[\frac{\partial\xi_r}{\partial r}\frac{\partial\eta}{\partial r} + \sigma\left(\frac{\eta}{r}\frac{\partial\xi_r}{\partial r} + \frac{\xi_r}{r}\frac{\partial\eta}{\partial r}\right) + \frac{\xi_r}{r^2}\eta\right]\right\}rdrdt = 0\,, \tag{4.157}$$

where

$$c^2 = Y/\rho(1-\sigma^2)\,. \tag{4.158}$$

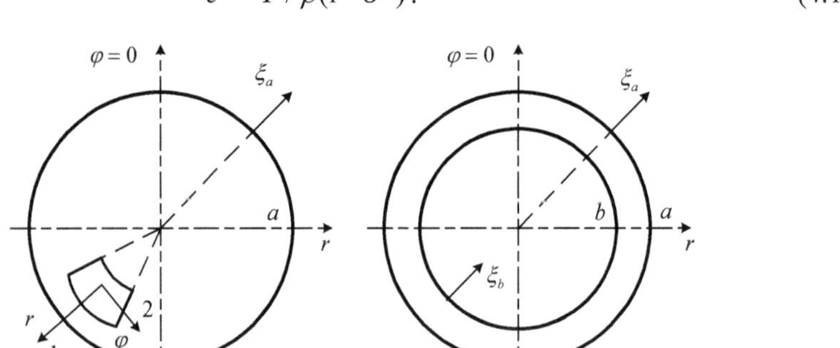

Figure 4.9: Circular disks in the extensional vibration in the plane: (a) solid disk, (b) annular disk (ring) of a finite thickness, (a-b).

After integrating by parts, the terms containing $\dot{\eta}$ and η_r', the expression will be obtained

$$\int_{t_1}^{t_2}\int_{0}^{a}\left\{c^2\left[\frac{\partial}{\partial r}\left(r\frac{\partial\xi_r}{\partial r}\right) - \frac{\xi_r}{r}\right] - \ddot{\xi}_r r\right\}\eta drdt = 0\,. \tag{4.159}$$

As the function η is arbitrary, this condition can be held only, when ξ_r satisfies the equation

$$\frac{1}{r}\frac{\partial}{\partial r}\left(r\frac{\partial \xi_r}{\partial r}\right)-\frac{\xi_r}{r^2}-\frac{1}{c^2}\ddot{\xi}_r=0 . \tag{4.160}$$

Besides, the boundary conditions must be fulfilled on the edge of the disk. Under the ideal boundary conditions, i.e., with no energy flux passing through the edge, $T_1 \cdot \xi_r\big|_{r=a} = 0$. Thus, either $\xi_r(a) = 0$ at the clamped edge, or

$$T_1\big|_{r=a} = \frac{Y}{1-\sigma^2}\left(\frac{\partial \xi_r}{\partial r}+\sigma\frac{\xi_r}{r}\right)_{r=a} = 0 \tag{4.161}$$

at the free edge.

Under the steady state harmonic vibrations Eq. (4.160) turns into the Bessel equation of the first order

$$\frac{d^2\xi}{dr^2}+\frac{1}{r}\frac{d\xi}{dr}+\left(k^2-\frac{1}{r^2}\right)\xi = 0 , \tag{4.162}$$

where $k^2 = \omega^2 / c^2$ and $c^2 = Y / \rho(1-\sigma^2)$ (the subscript r in ξ_r is omitted for brevity). Its general solution is

$$\xi = AJ_1(kr) + BN_1(kr) , \tag{4.163}$$

where $J_1(kr)$ and $N_1(kr)$ are the Bessel and Neumann functions of the first order. Description of the Bessel functions can be found, for example, in Ref. 3. Some of the properties of these functions are given in Appendix C.1. Coefficients A and B are the arbitrary constants that must be determined using the boundary conditions.

4.4.2.2 The Case of the Solid Disk

From condition of zero radial displacement in the center of the disk it should be $B = 0$. Thus, the normal modes of the problem are $X_i(r) = J_1(k_i r)$. The eigenvalues, $\lambda_i = k_i a$, and corresponding natural frequencies

$$\omega_i = (\lambda_i / a)\sqrt{Y / \rho(1-\sigma^2)} . \tag{4.164}$$

Using the boundary condition (4.162) for the free edge we obtain equation

$$\frac{dJ_1(k_i r)}{dr}\Big|_{r=a} + \sigma\frac{J_1(k_i a)}{a} = 0 . \tag{4.165}$$

The following relations for the Bessel functions are known (see Appendix C. 1)

$$J_1'(x) = J_0(x) - J_1(x)/x, \quad N_1'(x) = N_0(x) - N_1(x)/x. \tag{4.166}$$

Considering the first of the relations (4.166), the equation (4.165) becomes

$$J_0(k_i a) = (1-\sigma)\frac{J_1(k_i a)}{k_i a}. \tag{4.167}$$

The solutions of this equation at $\sigma = 0.3$ are

$$k_i a = \lambda_i = 2.05,\ 5.38,\ 8.57, \text{ and } \lambda_i \approx i\pi - 0.9 \text{ at } i > 3. \tag{4.168}$$

The resonance frequencies that correspond to the modes of vibration $X_i(r) = J_1(k_i r)$ will be obtained by formula (4.164).

It follows from Eq. (4.167) that the eigenvalues and normal modes depend on Poisson's ratio σ to a certain extent. In case that the discs are made from piezoceramics, Y and σ must be replaced by $Y_1^E = 1/s_{11}^E$ and $\sigma_1^E = -s_{12}^E/s_{11}^E$. Although for different piezoceramic materials $\sigma_1^E \neq 0.3$, within limits of values of σ_1^E for the most usable PZT compositions (σ_1^E is approximately between 0.27 and 0.35) the values of λ_i change less than by 1% from their values at $\sigma = 0.3$, as it can be verified using the equation (4.167). Thus, formula (4.168) for the resonance frequencies remains valid for the piezoceramic discs within this accuracy.

Equation (4.162) describes also radial vibrations of an infinitely long cylinder. The only difference is that in this case the sound speed $c = \sqrt{Y(1-\sigma)/\rho(1+\sigma)(1-2\sigma)}$ must be used. Therefore, the formula for calculating the natural frequencies will be

$$f_i = \frac{\lambda_i}{2\pi a}\sqrt{\frac{Y(1-\sigma)}{\rho(1+\sigma)(1-2\sigma)}} \tag{4.169}$$

with the same values of λ_i, as those in the case of a disk.

4.4.2.3 Annular Disk or Isotropic Ring of a Finite Thickness

The assumption of isotropic properties of the ring, strictly speaking, is exactly applicable to the piezoelements that are axial poled. Peculiarities arising at different directions of polarization will be considered in Chapter 7.

For the case of free vibration, we assume that the outer and inner side surfaces of the ring are free of stress, i.e.,

$$T_1 = \frac{Y}{1-\sigma^2}\left(\frac{d\xi}{dr}+\sigma\frac{\xi}{r}\right) = 0 \text{ at } r = a \text{ and at } r = b. \tag{4.170}$$

After applying these boundary conditions to the general solution (4.163) the frequency equation and mode shape of vibration, $\theta(r) = \xi(r)/\xi(a)$, will be obtained as follows.

The set of equations that correspond to the boundary conditions will be

$$A\left[ka\,J_0(ka)-(1-\sigma)J_1(ka)\right]+B\left[ka\,N_0(ka)-(1-\sigma)N_1(ka)\right]=0, \tag{4.171}$$

$$A\left[kb\,J_0(kb)-(1-\sigma)J_1(kb)\right]+B\left[kb\,N_0(kb)-(1-\sigma)N_1(kb)\right]=0. \tag{4.172}$$

The frequency equation for this system of equations is

$$\frac{ka\,J_0(ka)-(1-\sigma)J_1(ka)}{ka\,N_0(ka)-(1-\sigma)N_1(ka)} = \frac{kb\,J_0(kb)-(1-\sigma)J_1(kb)}{kb\,N_0(kb)-(1-\sigma)N_1(kb)}. \tag{4.173}$$

Spectrum of the wave numbers k_i can be found from this equation for various relations

$$\frac{b}{a} = 1-\frac{w}{a}, \tag{4.174}$$

where $w = a-b$ is the thickness of a ring, by a straightforward calculation. Less formal procedure of determining the lower resonance frequencies and the mode shapes of vibration of a ring vs. its relative thickness was suggested in Ref. 4, as follows.

Let us denote

$$\psi(kr) = \frac{kr\,J_0(kr)-(1-\sigma)J_1(kr)}{kr\,N_0(kr)-(1-\sigma)N_1(kr)}. \tag{4.175}$$

Plot of this function for $\sigma = 0.3$ up to its first null at value $kr = 2.05$ is depicted in Figure 4.10.

It can be shown that function $\psi(kr)$ has maximum at $kr = \sqrt{1-\sigma^2}$. In vicinity of this point $r \approx b \approx a$ and $r = (a+b)/2$ that corresponds to the case of a thin ring. Given that for a disk

$$c = \sqrt{\frac{Y}{\rho(1-\sigma^2)}}, \tag{4.176}$$

for the resonance frequency in the region of $(b/a) \approx 1$ we obtain expression

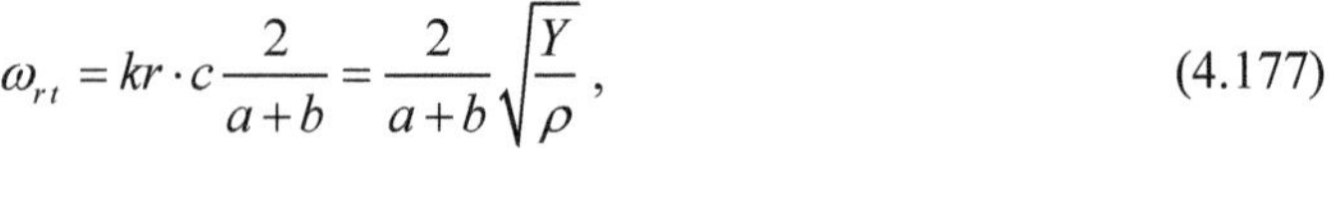

$$\omega_{rt} = kr \cdot c \frac{2}{a+b} = \frac{2}{a+b}\sqrt{\frac{Y}{\rho}}, \tag{4.177}$$

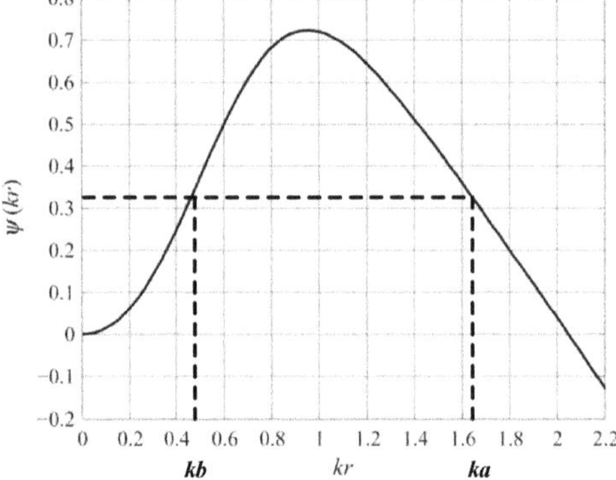

Figure 4.10: Plot of the function $\psi(kr)$.

which is the formula for resonance frequency of the thin ring. The ratio of values *ka* and *kb* in the points on *kr* axis that correspond to abscissa $\psi(ka) = \psi(kb)$ determines ratio of the inner to outer radius b/a, and their difference is $k(a-b) = kw$, or

$$ka\left(1 - \frac{b}{a}\right) = k\frac{w}{a}. \tag{4.178}$$

With ratio *b*/*a* known from the plot in Figure 4.10 the corresponding resonance frequency can be found from this relation for a ring having outer radius *a*. Let us represent expression for the resonance frequencies of the thick rings as

$$\omega_r(b/a) = \frac{2}{a+b}\sqrt{\frac{Y}{\rho}}F(b/a) = \omega_{rt}(r_{av})F(b/a). \tag{4.179}$$

Here $\omega_{rt}(r_{av})$ is the resonance frequency of the thin ring having radius equal to average radius of the thick ring under consideration, and $F(b/a)$ is a correction factor. Obviously, $F(1) = 1$ and $F(0) = 1.025/\sqrt{1-\sigma^2} = 1.07$, as the resonance frequency at $b = 0$ must correspond to those for radially vibrating circular disk by formula (4.164) at $\lambda = 2.05$. Plot of the correction factor vs. ratio of inner to outer radius is presented in Figure 4.11. It is seen from Figure 4.11 that the resonance frequency of a thick ring up to ratio $b/a \approx 0.5$ can be calculated by formula for the thin ring having the same average radius with accuracy greater than 5%. After the resonance frequency and thus the wave number $k_{b/a} = \omega_r(b/a)/c$ is determined, the ratio *B*/*A* of arbitrary constants can be obtained from either of Eqs. (4.172) or (4.173) as

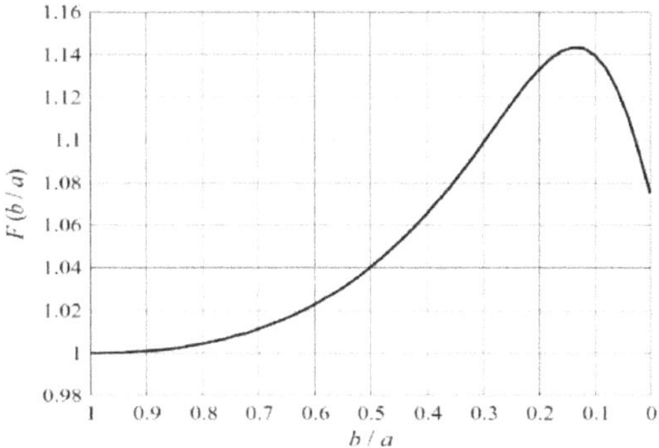

Figure 4.11: Correction factor that relates the resonance frequency of a thick ring to the frequency of a thin ring having the same average radius.

$$\frac{B}{A}=\frac{k_{b/a}a\,J_0(k_{b/a}a)-(1-\sigma)J_1(k_{b/a}a)}{k_{b/a}a\,N_0(k_{b/a}a)-(1-\sigma)N_1(k_{b/a}a)}=-\frac{k_{b/a}b\,J_0(k_{b/a}b)-(1-\sigma)J_1(k_{b/a}b)}{k_{b/a}b\,N_0(k_{b/a}b)-(1-\sigma)N_1(k_{b/a}b)}. \quad (4.180)$$

And the mode shape of vibration will be found using the general expression (4.163) for the radial displacement as

$$\theta(r)=\frac{\xi(r)}{\xi(a)}=\frac{J_1(k_{b/a}r)+(B/A)N_1(k_{b/a}r)}{J_1(k_{b/a}a)+(B/A)N_1(k_{b/a}a)}. \quad (4.181)$$

The mode shapes of vibration for the rings having different thickness are plotted in Figure 4.12

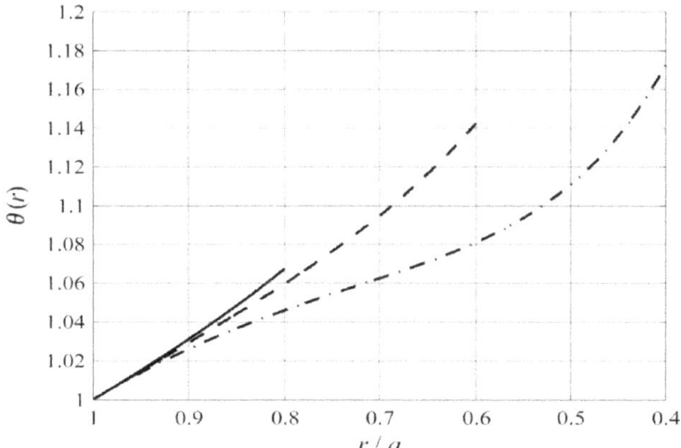

Figure 4.12: The mode shapes of vibration for the rings having different thickness: solid line – $b/a = 0.8$, dashed line – $b/a = 0.6$, and dot-dash line – $b/a = 0.4$.

It is noteworthy that the mode shapes determined for the ring isotropic in its plane (which is the case for axial poled piezoelement) remain the same to great accuracy for the piezoelements poled in radial and circumferential directions and can be used for calculating equivalent parameters of transducers that employ vibration of the rings regardless of the mode of their polarization. Results of calculating the parameters for different modes of polarization will be presented in Section 7.2.2.1.

4.4.3 Axial Symmetric Flexural Vibrations of a Thin Circular Plate

Under the same assumptions, as those made in the case of the flexural vibration of a rectangular beam in Section 4.3.4, we obtain

$$S_{rr} = -z\partial^2\xi_z / \partial r^2 . \tag{4.182}$$

Since at the same time $S_{rr} = \partial\xi_r / \partial r$, it follows that $\xi_r = -z\partial\xi_z / \partial r$ and

$$S_{\varphi\varphi} = \frac{\xi_r}{r} = -\frac{z}{r}\frac{\partial\xi_z}{\partial r} . \tag{4.183}$$

Because of the condition that in a thin plate $T_z = 0$, the relations (4.16) for stress, strain and density of the potential energy can be used, if to change the coordinate axes 1 and 2 to *rr* and $\varphi\varphi$, respectively. Thus,

$$T_{rr} = \frac{Y}{1-\sigma^2}(S_{rr} + \sigma S_{\varphi\varphi}) = -\frac{zY}{1-\sigma^2}\left(\frac{\partial^2\xi_z}{\partial r^2} + \frac{\sigma}{r}\frac{\partial\xi_z}{\partial r}\right), \tag{4.184}$$

$$T_{\varphi\varphi} = \frac{E}{1-\nu^2}(S_{\varphi\varphi} + \sigma S_{rr}) = -\frac{zY}{1-\sigma^2}\left(\frac{1}{r}\frac{\partial\xi_z}{\partial r} + \sigma\frac{\partial^2\xi_z}{\partial r^2}\right), \tag{4.185}$$

$$w_{pot} = \frac{D}{2}\left[\left(\frac{\partial^2\xi_z}{\partial r^2} + \frac{1}{r}\frac{\partial\xi_z}{\partial r}\right)^2 - 2(1-\sigma)\frac{1}{r}\frac{\partial\xi_z}{\partial r}\frac{\partial^2\xi_z}{\partial r^2}\right] . \tag{4.186}$$

Here $D = Yt^3 / 12(1-\sigma^2)$ is the flexural rigidity of a plate. Using expressions for w_{pot} and

$$w_{kin} = \rho\dot{\xi}_z^2 / 2 , \tag{4.187}$$

the following equation of the flexural vibration of the circular plate will be obtained from variational principle in the way completely analogous to those used in the case of the flexural vibration of a beam,

$$\nabla^4 \xi_Z + (\rho t a^4 / D)\ddot{\xi}_Z = 0, \tag{4.188}$$

where $\nabla^4 = \left[\partial^2 / \partial r^2 + (1/r)\partial / \partial r\right]^2$ is the differential operator that should be applied to ξ_z.

The ideal boundary conditions per unit length of the plate edge are analogous to those expressed by relations (4.53) through (4.55). They are formulated as follows:

for the simply supported edge

$$\xi_Z(a) = 0, \quad M_r = -D\left(\frac{\partial^2 \xi_Z}{\partial r^2} + \frac{\sigma}{r}\frac{\partial \xi_Z}{\partial r}\right)_{r=a} = 0; \tag{4.189}$$

for the clamped edge

$$\xi_Z(a) = 0, \quad \left.\frac{\partial \xi_Z}{\partial r}\right|_{r=a} = 0; \tag{4.190}$$

for the free edge

$$\left.M_r\right|_{r=a} = 0, \quad Q = \frac{\partial M_r}{\partial r} = -D\frac{\partial}{\partial r}\left(\frac{\partial^2 \xi_Z}{\partial r^2} + \frac{\sigma}{r}\frac{\partial \xi_Z}{\partial r}\right)_{r=a} = 0. \tag{4.191}$$

In case of harmonic vibration, the equation of separation is

$$(\nabla^4 - \lambda_i^4)\xi_Z = 0, \tag{4.192}$$

where

$$\lambda_i^2 = \omega_i a^2 / \sqrt{D/\rho t}. \tag{4.193}$$

Equations for determining the normal modes of the problem will be obtained from Eq. (4.192) after substituting λ_i^2 in the following forms

$$(\nabla^2 + \lambda_i^2)X_i(r) = \frac{d^2 X_i}{dr^2} + \frac{1}{r}\frac{dX_i}{dr} + \lambda_i^2 X_i = 0, \tag{4.194}$$

$$(\nabla^2 - \lambda_i^2)X_i(r) = \frac{d^2 X_i}{dr^2} + \frac{1}{r}\frac{dX_i}{dr} - \lambda_i^2 X_i = 0. \tag{4.195}$$

Solutions of these equations are the cylindrical functions $J_0(\lambda_i r/a)$ and $I_0(\lambda_i r/a)$ of the real and imaginary variables, respectively. Their linear combinations are the normal modes for the boundary problem under consideration, so that

$$X_i(r) = AJ_0(\lambda_i r/a) + BI_0(\lambda_i r/a). \tag{4.196}$$

The eigenvalues λ_i^2 and relations between coefficients A and B have to be found by applying the boundary conditions (4.173) through (4.177). Information on these values is available in Ref. 5. We reproduce some of them at $\sigma = 0.3$.

In the case of the simply supported edge

$$\theta_i(r) = \frac{J_0(k_i r) I_0(k_i a) - J_0(k_i a) I_0(k_i r)}{I_0(k_i a) - J_0(k_i a)}, \tag{4.197}$$

$$\lambda_i = k_i a = 2.23,\ 5.45,\ 8.61, \tag{4.198}$$

$$\omega_1 = (2.23 / a)^2 \sqrt{D / \rho t}\,. \tag{4.199}$$

There are i nodal circles on the surface of a plate, at which $\theta_i(r) = 0$. At $i = 1$ the nodal circle is on the edge ($r_1 = a$), at $i = 2$ $r_1 = a$ and $r_2 = 0.44a$.

In the case of the free edge

$$\theta_i(r) = \frac{I_1(k_i a) J_0(k_i r) - J_1(k_i a) I_0(k_i r)}{I_1(k_i a) - J_1(k_i a)}. \tag{4.200}$$

There exists the normal mode of vibration $\theta_0(r) = 1$ at $\lambda_0 = 0$, which means that movement of a plate is possible without deformation. Otherwise $\lambda_i = 3.01, 6.2, 9.3, \ldots$. Radii of nodal circles are: $r_1 = 0.68a$ at $i = 1$; $r_1 = 0.84a$ and $r_2 = 0.39a$ at $i = 2$.

In case that the edge is clamped expression for the normal mode is the same as (4.197), but the eigenvalues are

$$\lambda_i = k_i a = 3.2,\ 6.3,\ 9.4, \ldots, \tag{4.201}$$

$$\omega_1 = (10.2 / a^2)\sqrt{D / \rho t}, \quad \omega_2 = (39.6 / a^2)\sqrt{D / \rho t}\,. \tag{4.202}$$

Radii of the nodal circles are $r_1 = 0.38a$ and $r_2 = a$ at $i = 2$.

4.4.3.1 Corrections to the Energy Densities for the Plates due to Finite Thickness

With increase of the relative thickness t/a of the circular plates the values of shear stresses, as well as the rotary inertia of the cross sections under flexure, become significant, so that corrections must be made to the energies determined by formulae (4.30) and (4.32) analogous to those made for the beams in Section 4.3.5. With this goal the expressions for the related energy densities formulated in Section 4.3 for the element of volume in the rectangular coordinates must be rewritten for the element of volume in the axial symmetric polar coordinates.

The additional kinetic energy per unit volume along the radius of plate associated with turning of the cross sections by angle $\alpha = (\partial\xi_z / \partial r)$ in regard to axis 2 (Figure 4.4) is according to (4.33)

$$\Delta w_{kin} = \frac{1}{2}\rho J_2 \dot{\alpha}^2 = \frac{1}{2}\rho J_2 \left(\frac{\partial^2 \xi_z}{\partial r \partial t}\right)^2, \tag{4.203}$$

where

$$J_2 = r d\varphi t^3 / 12 . \tag{4.204}$$

Another correction is related to an additional deflection due to shear deformations that takes place in a plate under flexure but can be ignored at small t/a. If to consider the balance of forces acting on the shaded element of the volume (Figure 4.5 (a)) projected on axis 1, according to (4.34) it will be found that the shear stresses $T_{13} = T_5$ are

$$T_5 = \int_{t/2}^{z} \frac{\partial T_1}{\partial r} dz = \frac{Y}{2(1-\sigma^2)}\left(\frac{t^2}{4} - z^2\right)\frac{\partial^3 \xi_z}{\partial r^3} . \tag{4.205}$$

The cross section of a plate perpendicular to the radius becomes distorted due to shear. The shear strains, which are equal to angle

$$\psi_s = S_5 = T_5 / \mu \tag{4.206}$$

of tilt of the section element relative to the original plane (Figure 4.5 (b)) has maximum value at the neutral axis at $z = 0$,

$$\psi_s(0,r) = S_5 = T_5 / \mu = \frac{Yt^2}{8\mu(1-\sigma^2)}\frac{\partial^3 \xi_z}{\partial r^3} . \tag{4.207}$$

An additional deflection of the neutral axis, ξ_{ad}, corresponds to this turning angle. It can be determined according to the relation $\psi_s = -(\partial\xi_{ad} / \partial r)$. Thus, the total deflection of the neutral plane becomes $\xi = \xi_z + \xi_{ad}$. As in this approximation $T_5 \neq 0$, the correction must be applied to the value of w_{pot} expressed by formula (4.186). Using expression (4.205) we obtain

$$\Delta w_{pot} = -\frac{r d\varphi}{2\mu}\int_{-t/2}^{t/2} T_5^2 dz = -\frac{r d\varphi Y^2 t^5}{240\mu(1-\sigma^2)^2}\left(\frac{\partial^3 \xi_z}{\partial r^3}\right)^2 . \tag{4.208}$$

The sign minus is since the value of potential energy was somewhat exaggerated. Increase of deflection under invariable action shows reduction of the rigidity and results in decrease of the potential energy.

Due to shear deformation distortion of the cross section takes place that is associated with additional displacements in the radial direction,

$$\xi_{rs}(z,r) = \psi_s(z,r)z \ . \tag{4.209}$$

The related radial strain is

$$S_{rs}(z,r) = d\xi_{rs}(z,r)/dr \ . \tag{4.210}$$

Following expressions (4.206) and (4.205)

$$\xi_{rs}(z,r) = \psi_s(z,r)z = \xi_o \frac{Y}{2(1-\sigma^2)\mu}\left(\frac{t^2}{4} - z^2\right) z \frac{d^3\theta}{dr^3} \tag{4.211}$$

and, consequently, the additional radial strain due to the shear deformation is

$$S_{rs}(z,r) = \frac{d\xi_{rs}}{dr} = \xi_o \frac{Y}{2(1-\sigma^2)\mu}\left(\frac{t^2}{4} - z^2\right) z \frac{d^4\theta}{dr^4} \ . \tag{4.212}$$

As follows from this expression, $S_{rs}(z = \pm t/2) = S_{rs}(0) = 0$, and it has the maximum value at $z \approx \pm 0.3t$.

The corrections of energy densities do not influence the mode shapes of vibrations. They must be taken into consideration, when calculating the resonance frequencies and equivalent parameters of the mechanical systems of a transducer employing the circular plate. The additional radial strain by formula (4.212) results in changing the electromechanical conversion. All these effects will be considered in application to the corresponding transducer types in Part III.

4.4.4 Vibration of a Circular Ring in its Plane

The axisymmetric radial vibration of a thin circular ring as one degree of freedom system was described on Section 2.3. Here vibration of a general type in the plane of a ring will be considered. The geometry of a thin circular ring is shown in Figure 4.13. The thickness (t) and height (h) of the ring will be assumed to be small compared to the radius (a) of the middle surface of the ring. As the surfaces of a ring are free of stress, it can be assumed that all the stress in the volume of the ring vanish except for the extensional stress in the circumferential direction that is determined by product of Y and the corresponding deformation in the circumferential direction. In the general case the cross sections of the vibrating ring may move in the radial and circumferential directions. This may cause both the extensional deformations in the ring and bending of its middle surface. The displacement of a cross section in the radial direction is denoted as $\xi_r(\varphi)$. Note that under assumption of a thin ring the elementary theory of bending is applicable, which states that the cross sections remain undistorted and perpendicular to the neutral surface while turning at some angle. Displacements in the direction

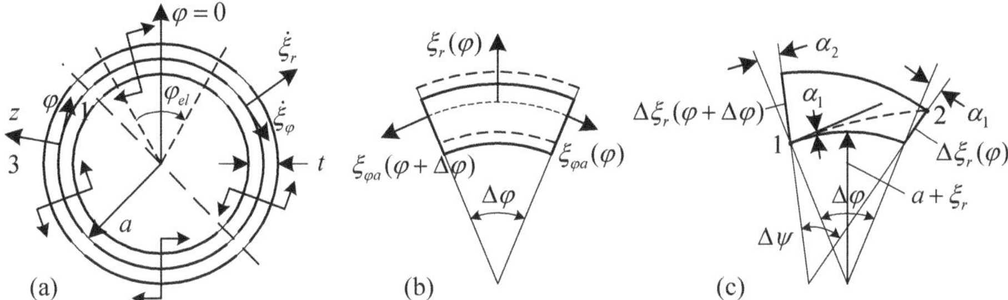

Figure 4.13: (a) Geometry of a ring under deformation, (b) radial displacement of a segment of the ring that causes its elongation, and (c) turns of the cross sections of the segment that cause change of the curvature.

of the normal to the cross section will be denoted as $\xi_\varphi(\varphi)$. They may be presented in the form

$$\xi_\varphi(r,\varphi) = \xi_{\varphi a}(\varphi) + \xi_{\varphi b}(r,\varphi), \tag{4.213}$$

where $\xi_{\varphi a}$ is the averaged over cross section displacement, and $\xi_{\varphi b}$ is the displacement due to turning of cross section under bending of the middle (neutral) surface. This displacement change through the thickness, and $\xi_{\varphi b}(a,\varphi) = 0$. Displacements that take place in a small element of a ring are shown in Figure 4.13 (b). Expression for the strain in the circumferential direction in polar coordinates is

$$S_{\varphi\varphi} = \frac{\partial \xi_\varphi}{r\partial\varphi} + \frac{\xi_r}{a} = \frac{\partial \xi_{\varphi a}}{a\partial\varphi} + \frac{\xi_r}{a} + \frac{\partial \xi_{\varphi b}}{a\partial\varphi}, \tag{4.214}$$

where it is considered that $t \ll a$, and therefore $1/r \approx 1/a$. The sum of the first two terms in expression (4.214) represents the relative elongation of the elements of the ring, and it will be denoted S_φ. The term ξ_r / a is due to change of the mean radius under displacement in the radial direction (the displacement ξ_r is positive, if it leads to the tension, i.e., when is directed from the ring's center). The term $\partial\xi_{\varphi a} / a\partial\varphi$ presents longitudinal deformation of the element of a ring as a bar. The last term, $\partial\xi_{\varphi b} / a\partial\varphi$, is due to change of curvature of an arc element of the ring, as illustrated in Figure 4.13 (c). To identify the meaning of this term, consider transition of an arc element of the middle surface into a curved state as a result of two successive displacements of the arc ends with coordinates φ and $\varphi+\Delta\varphi$ in the radial direction. Namely, $\Delta\xi_r(\varphi)$ and $\Delta\xi_r(\varphi+\Delta\varphi)$ (displacement in the circumferential direction does not lead to changes in the curvature). At displacement $\Delta\xi_r(\varphi)$ the arc turns about the location $\varphi+\Delta\varphi$ by the angle $\alpha_1 = \Delta\xi_r(\varphi)/(a+\xi_r)\Delta\varphi \approx \Delta\xi_r(\varphi)/a\Delta\varphi$ in the positive (anti-clockwise) direction. This is the angle, at which the normal to the arc turns at location φ. At displacement $\Delta\xi_r(\varphi+\Delta\varphi)$ the arc turns about φ at the angle $\alpha_2 = \Delta\xi_r(\varphi+\Delta\varphi)/a\Delta\varphi$ in the negative (clockwise) direction, and the normal to the arc in the point $\varphi+\Delta\varphi$ turns at this angle. Thus, the change of the angle between the normals to the curved arc element at extreme points or, in other words, change of the angle $\Delta\psi$ between the new curvature radii R drawn to these points in comparison with angle $\Delta\varphi$ is

$$\Delta\psi - \Delta\varphi = \alpha_1 - \alpha_2 = -\left[\frac{\Delta\xi_r(\varphi+\Delta\varphi)}{a\Delta\varphi} - \frac{\Delta\xi_r(\varphi)}{a\Delta\varphi}\right] = -\frac{1}{a}\frac{\partial^2\xi_r}{\partial\varphi^2}\Delta\varphi . \tag{4.215}$$

As the length of the arc element does not change, $R\Delta\psi = (a+s_r)\Delta\varphi$, and the curvature of the arc element after bending is

$$\frac{1}{R} = \frac{\Delta\psi}{(a+\xi_r)\Delta\varphi} = \frac{1}{a+\xi_r}\left(1 - \frac{1}{a}\frac{\partial^2\xi_r}{\partial\varphi^2}\right) \approx \frac{1}{a} - \frac{1}{a^2}\left(\frac{\partial^2\xi_r}{\partial\varphi^2} + \xi_r\right). \tag{4.216}$$

Manipulations made in expression (4.216) involve the replacement $1/(a+\xi_r) \approx (a-\xi_r)/a^2$ and the dropping of the second order term $(\xi_r/a^2)(\partial^2\xi_r/\partial\varphi^2)$. From expression (4.216) we obtain change of curvature of the arc element under bending as

$$\frac{1}{R} - \frac{1}{a} = \chi_\varphi = -\frac{1}{a^2}\left(\frac{\partial^2\xi_r}{\partial\varphi^2} + \xi_r\right). \tag{4.217}$$

The deformation of a layer having radius r due to the change of curvature can be determined by comparing its lengths, l_2 and l_1, after and before the bending, respectively. This is illustrated by Figure 4.14. We denote separation of the layer under consideration from the middle line

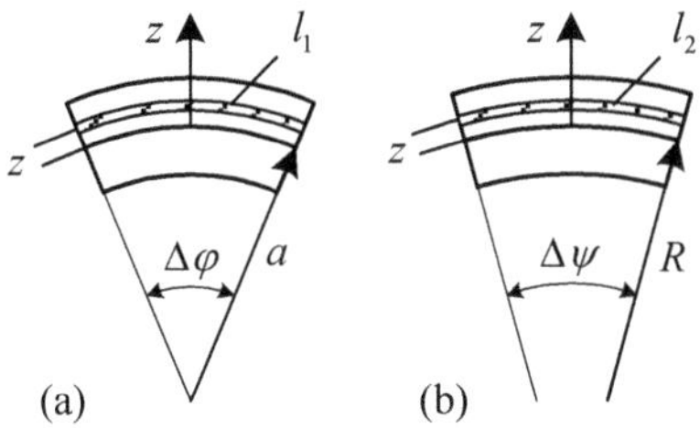

Figure 4.14: Segment of a ring before (a) and after (b) bending.

as $r-a=z$. This separation does not change under bending. The length of the middle line does not change as well, where from follows that

$$R\cdot\Delta\psi = a\cdot\Delta\varphi\,. \tag{4.218}$$

Thus, the deformation of the layer due to bending is

$$S_{\varphi b}(r) = \frac{l_2 - l_1}{l_1} = \frac{(R+z)\Delta\psi - (a+z)\Delta\varphi}{(a+z)\Delta\varphi}\,. \tag{4.219}$$

After some manipulations with expression (4.219) that take into consideration relation (4.218) for the thin rings (under the assumption that $t \ll a$, and therefore $a+z \approx a$ in denominator) will be obtained

$$S_{\varphi b} = -\frac{r-a}{a^2}\left(\frac{\partial^2 \xi_r}{\partial \varphi^2} + \xi_r\right). \tag{4.220}$$

Finally, the expression (4.214) for the overall deformation may be presented as

$$S_{\varphi\varphi} = S_\varphi + S_{\varphi b} = \frac{1}{a}\left(\frac{\partial \xi_\varphi}{\partial \varphi} + \xi_r\right) - \frac{r-a}{a^2}\left(\frac{\partial^2 \xi_r}{\partial \varphi^2} + \xi_r\right). \tag{4.221}$$

After expression for deformations in vibrating ring is determined, the potential energy per element of volume of the ring can be represented as

$$\begin{aligned} &= \frac{1}{2}\int_{\tilde V} Y(S_\varphi^2 + S_{\varphi b}^2)\,d\tilde V = \frac{1}{2}Yah\int_{-t/2}^{t/2}(S_\varphi^2 + S_{\varphi b}^2)dz d\varphi = \\ &= \frac{Yth}{2a^2}\left[\left(\frac{\partial \xi_\varphi}{\partial \varphi} + \xi_r\right)^2 + \frac{t^2}{12a^2}\left(\frac{\partial^2 \xi_r}{\partial \varphi^2} + \xi_r\right)^2\right]. \end{aligned} \tag{4.222}$$

Here $\tilde V$ denotes the element of volume of the ring ($\tilde V = ah\cdot dz d\varphi$) and the coordinate system is introduced that is shown in Figure 4.11, in which $r-a=z$. The first term under the last integral corresponds to energy of the extensional vibration. They take place under the condition that

$$\frac{\partial^2 \xi_r}{\partial \varphi^2} + \xi_r = 0. \tag{4.223}$$

The second term corresponds to energy of the flexural vibration of a ring that takes place under the condition that

$$\frac{\partial \xi_\varphi}{\partial \varphi} + \xi_r = 0. \tag{4.224}$$

As will be shown further, the natural frequencies of vibrations related to these two kinds of deformation differ greatly. Besides, when employed in transducer designs, these two kinds of vibration involve different conditions of excitation. Hence, as an approximation, it is possible to consider these vibrations separately by either ignoring the energy of flexural deformation, or by considering the flexure to occur without tension of the middle surface of a ring.

The differential equations of the extensional vibration of a ring along the circumference can be derived as the Euler equations (4.2) under the condition (4.200), using expressions for the potential and kinetic energy per element of volume of the ring

$$w_{pot} = \frac{Yth}{2a^2}\left(\frac{\partial \xi_\varphi}{\partial \varphi} + \xi_r\right)^2, \tag{4.225}$$

$$w_{kin} = \frac{1}{2}\rho(\dot{\xi}_r^2 + \dot{\xi}_\varphi^2)d\tilde{V} = \frac{1}{2}\rho S_{cs}\left[\dot{\xi}_\varphi^2 + \left(\partial\dot{\xi}_\varphi / \partial\varphi\right)^2\right]. \tag{4.226}$$

Thus, we arrive at equations

$$\frac{d}{dt}\left(\frac{\partial w_{kin}}{\partial \dot{\xi}_r}\right) + \frac{\partial w_{pot}}{\partial \xi_r} = \rho S_{cs}\ddot{\xi}_r + \frac{YS_{cs}}{a^2}\left(\frac{\partial \xi_\varphi}{\partial \varphi} + \xi_r\right) = 0, \tag{4.227}$$

$$\frac{d}{dt}\left(\frac{\partial w_{kin}}{\partial \dot{\xi}_\varphi}\right) - \frac{\partial}{\partial \varphi}\left(\frac{\partial w_{pot}}{\partial \xi'_\varphi}\right) = \rho S_{c.s}\ddot{\xi}_\varphi + \frac{YS_{c.s}}{a^2}\frac{\partial}{\partial \varphi}\left(\frac{\partial \xi_\varphi}{\partial \varphi} + \xi_r\right) = 0. \tag{4.228}$$

After differentiating the first equation with respect to φ and adding it to the second, we will find that in this case

$$\frac{\partial \xi_r}{\partial \varphi} + \xi_\varphi = 0. \tag{4.229}$$

Under this condition the equation for ξ_r will be

$$\ddot{\xi}_r - \frac{Y}{\rho a^2}\left(\frac{\partial^2 \xi_r}{\partial \varphi^2} - \xi_r\right) = 0. \tag{4.230}$$

The boundary condition for equation (4.230) is 2π periodicity of function $\xi_r(\varphi)$.

Deriving equation of the flexural vibration of a ring and analyzing solutions for both extensional and flexural vibrations of the rings is easier to perform in the generalized coordinates. This will be done in Sections 4.5.2 and 4.7.

4.5 Equations of Vibration in the Generalized Coordinates

4.5.1 The General Outline of Solving Vibration Problems in the Generalized Coordinates

An alternative to deriving differential equations of motion is solving the vibration problems in the generalized coordinates. This approach proves to be more appropriate in many cases in respect to mechanical systems of transducers. A brief description of the method is given in Section 1.6. The equations of motion in this case are of the form of Euler-Lagrange equations (4.1). The general outline of solving the vibration problems using generalized coordinates can be laid out as follows.

Let us assume that $\theta_i(\boldsymbol{r})$ is a certain complete system of functions, which are defined within a vibrating body and satisfy the boundary conditions. It will be referred to as the supporting system of functions. Displacements in the vibrating body can be represented as expansion into series

$$\xi(\boldsymbol{r},t) = \sum_{i=1}^{\infty} \xi_i(t)\theta_i(\boldsymbol{r}) \tag{4.231}$$

with unknown coefficients $\xi_i(t)$. Since distribution of displacements and therefore a state of the mechanical system becomes determinate provided the values of coefficients ξ_i are found, they can be considered as the generalized coordinates. For determining thus introduced generalized coordinates the variational principle in the form of Eq. (1.91) can be applied, assuming that function L that characterizes state of the body is expressed by relation (1.94). In the general case expressions for the kinetic and potential energies of a vibrating body can be represented as follows

$$W_{kin} = \frac{1}{2}\sum_{i=1}^{\infty}\sum_{l=1}^{\infty} M_{il}\dot{\xi}_i\dot{\xi}_l\,, \tag{4.232}$$

$$W_{pot} = \frac{1}{2}\sum_{i=1}^{\infty}\sum_{l=1}^{\infty} K_{il}\xi_i\xi_l\,, \tag{4.233}$$

where

$$M_{il} = \frac{1}{\dot{\xi}_i}\frac{\partial W_{kin}}{\partial \dot{\xi}_l} \text{ and } K_{il} = \frac{1}{\xi_i}\frac{\partial W_{pot}}{\partial \xi_l} \tag{4.234}$$

have dimensions of masses and rigidities. For example, following relations (4.38) and (4.39) for one-dimensional longitudinal and torsional vibrations expressions for the energies are

$$W_{kin} = \frac{1}{2}\int_{\tilde{V}} m_\Delta \dot{\xi}^2 d\tilde{V}, \quad W_{pot} = \frac{1}{2}\int_{\tilde{V}} K_\Delta (\xi')^2 d\tilde{V}\,. \tag{4.235}$$

Upon substituting $\xi(x,t)$ from Eq. (4.231) we arrive at the expressions for the masses and rigidities

$$M_{il} = \int_{\tilde{V}} m_\Delta \theta_i(x)\theta_l(x)d\tilde{V}, \quad K_{il} = \int_{\tilde{V}} K_\Delta \theta'_i(x)\theta'_l(x)d\tilde{V}\,. \tag{4.236}$$

The term W_e in expression (4.1) represents the total energy of external actions, which comprises the energy due to action of external mechanical source, designated as W_m, and due to reaction of a mechanical load, W_L, into which a part of energy flows. Since energy W_m flows into the body, while W_L flows out of it (see Section 1.6), $W_e = W_m - W_L$. If we consider the external actions in the most general case as the forces $f_m(\boldsymbol{r}_\Sigma)$ and $f_L(\boldsymbol{r}_\Sigma)$ distributed over surface of the body, then $W_e = \xi[f_m(\boldsymbol{r}_\Sigma) - f_L(\boldsymbol{r}_\Sigma)]$ and using expansion (4.231) we obtain

$$W_e = \sum_{i=1}^{\infty}\xi_i\int_{\Sigma}[f_m(\boldsymbol{r}_\Sigma) - f_L(\boldsymbol{r}_\Sigma)]\theta_i(\boldsymbol{r}_\Sigma)d\Sigma = \sum_{i=1}^{\infty}\xi_i(f_{mi} - f_{Li}), \tag{4.237}$$

where f_{mi} and f_{Li} have the meaning of equivalent forces (see Eq. (4.82)) that correspond to the modes of vibration θ_i. Upon switching to the complex form and introducing distributed impedance of a load, $z_L(\mathbf{r}_\Sigma)$, we arrive at the expression for the equivalent reaction of the load

$$F_L(\boldsymbol{r}_\Sigma) = z_L(\boldsymbol{r}_\Sigma)\bar{\dot{\boldsymbol{\xi}}} = z_L(\boldsymbol{r}_\Sigma)\sum_{i=1}^{\infty}\bar{\dot{\boldsymbol{\xi}}}_i\theta_i(\boldsymbol{r}_\Sigma), \tag{4.238}$$

where $\bar{\dot{\xi}}$ is complex quantity that corresponds to instantaneous value of $\dot{\xi}$ from formula (4.231). From Eq. (4.237) we obtain

$$\bar{\dot{W}}_m = \sum_{i=1}^{\infty} F_{mi}\bar{\dot{\xi}}_i^*, \quad F_{mi} = \int_\Sigma F_m(\boldsymbol{r}_\Sigma)\theta_i(\boldsymbol{r}_\Sigma)d\Sigma. \tag{4.239}$$

$$\bar{\dot{W}}_L = \sum_{i=1}^{\infty} F_{Li}\bar{\dot{\xi}}_i^*, \quad F_{Li} = \int_\Sigma F_L(\boldsymbol{r}_\Sigma)\theta_i(\boldsymbol{r}_\Sigma)d\Sigma. \tag{4.240}$$

Substituting expression (4.238) for $F_L(\boldsymbol{r}_\Sigma)$ under the integral (4.240) results in

$$F_{Li} = \sum_{n=1}^{\infty}\bar{\dot{\xi}}_n\int_\Sigma z_L(\boldsymbol{r}_\Sigma)\theta_n(\boldsymbol{r}_\Sigma)\theta_i(\boldsymbol{r}_\Sigma)d\Sigma = Z_{Li}\bar{\dot{\xi}}_i, \tag{4.241}$$

where

$$Z_{Li} = Z_{Lii} + \sum_{n\neq i}^{\infty} z_{Lni}(\bar{\dot{\xi}}_n / \bar{\dot{\xi}}_i) \tag{4.242}$$

is the equivalent impedance of a load; Z_{Lii} is the self-impedance of a load for vibration mode θ_i, z_{Lni} is the mutual impedance that correspond to interaction between vibration modes θ_i and θ_n. The energy flux (4.240) that flows into the load can now be represented as

$$\bar{\dot{W}}_L = \sum_{i=1}^{\infty} Z_{Li}\left|\bar{\dot{\xi}}_i\right|^2. \tag{4.243}$$

Consider several examples of the external actions.

Action of the lumped force, $F_m(\boldsymbol{r}_x) = T(\boldsymbol{r}_x)d\Sigma_x$, and load, $Z_L(\boldsymbol{r}_x) = z_L(\boldsymbol{r}_x)d\Sigma_x$, where $T(\boldsymbol{r}_x)$ and $z_L(\boldsymbol{r}_x)$ are the mechanical stress and density of a load that vanish outside of the unit area $d\Sigma_x$. In this case

$$Z_{Lni} = \int_\Sigma z_L(\boldsymbol{r}_x)\theta_i(\boldsymbol{r}_x)\theta_n(\boldsymbol{r}_x)d\Sigma = z_L(\boldsymbol{r}_x)\theta_i(\boldsymbol{r}_x)\theta_n(\boldsymbol{r}_x)d\Sigma_x = Z_L(\boldsymbol{r}_x)\theta_i(\boldsymbol{r}_x)\theta_n(\boldsymbol{r}_x), \tag{4.244}$$

$$F_{mi} = \int_\Sigma T(\boldsymbol{r}_x)\theta_i(\boldsymbol{r}_x)d\Sigma = T(\boldsymbol{r}_x)\theta_i(\boldsymbol{r}_x)d\Sigma_x = F_m(\boldsymbol{r}_x)\theta_i(\boldsymbol{r}_x). \tag{4.245}$$

If $\theta_i(\boldsymbol{r}_x) = 0$, then $F_{mi} = 0$ and $Z_{li} = 0$, i.e., actions applied to points on the nodal line don't affect the mode of vibrations $\theta_i(\mathrm{r})$. Consequently, the elements of fastening a mechanical system preferably must be placed close to the nodal lines.

Action of a force distributed uniformly with density ΔF_m. The equivalent acting force is

$$F_{mi} = \int_\Sigma \Delta F_m \theta_i(\boldsymbol{r}_\Sigma) d\Sigma = \Delta F_m \int_\Sigma \theta_i(\boldsymbol{r}_\Sigma) d\Sigma = \Delta F_m S_{avi}, \tag{4.246}$$

where

$$S_{avi} = \int_\Sigma \theta_i(\boldsymbol{r}_\Sigma) d\Sigma \tag{4.247}$$

is the average area of the surface vibrating in the mode $\theta_i(\boldsymbol{r}_\Sigma)$. In particular, the acoustic field produces action of this kind, if dimensions of the transducer body are considerably smaller than the wavelength of sound. In this case the sound pressure $P(t) \approx P_o$, where P_o is the acoustic pressure in the free field, and $F_{mi} \approx P_o S_{avi}$. The volume velocity, $U_{\tilde{V}}$, produced by a surface vibrating with distribution of displacement $\bar{\xi}(\boldsymbol{r}) = \bar{\xi}_0 \theta_i(\boldsymbol{r})$ is $U_{\tilde{V}} = \dot{\bar{\xi}}_0 S_{avi}$. Obviously, $S_{avi} \le S_r$ (S_r is the area of the vibrating surface), where equality is achieved at $\theta_i(\boldsymbol{r})=1$. If some parts of the surface vibrate in anti-phase, the values of S_{av} may be considerably reduced and even may drop to zero. Thus, under the transverse vibrations of a beam with simply supported ends $\theta_i(x) = \sin(i\pi x / l)$, $S_{avi} = 2S_r / i\pi$ at uneven *i*, and $S_{avi} = 0$ at even *i*. If $S_{avi} = 0$, the corresponding mode of vibration cannot be excited under an uniform action over the surface (in this case $F_{mi} = 0$), and the surface vibrating in such a mode does not produce a volume velocity ($U_{\tilde{V}} = 0$). Because of this it is not expedient to use mechanical systems with small values of the average area for mechanoacoustic conversion, especially if they have a small wave size. The values of an average area can be significantly increased by baffling the parts of the surface vibrating in anti-phase.

Action of force intended for exciting a single mode of vibration. Distribution of the force must be orthogonal to all the other modes of vibration, as it follows from formula (4.239). In fact, if distribution of the acting force replicates distribution of the displacements in the desirable mode of vibration $\theta_i(\boldsymbol{r}_\Sigma)$, i.e., $F_m(\boldsymbol{r}_\Sigma) = \Delta F_m \theta_i(\boldsymbol{r}_\Sigma)$, where ΔF_m does not change over the surface, then

$$F_{mi} = \int_\Sigma \Delta F_m \theta_i^2(\boldsymbol{r}_\Sigma) d\Sigma = \Delta F_m S_{eff\,i} \tag{4.248}$$

and $F_{mn} = 0$ at $n \ne i$ due to orthogonality of vibration modes θ_i and θ_n. The quantity

$$S_{eff\,i} = \int_{\Sigma} \theta_i^2(\boldsymbol{r})d\Sigma \tag{4.249}$$

will be called the effective surface area of vibration mode θ_i.

Thus, the values of the average and effective surface areas S_{avi} and $S_{eff\,i}$ defined by formulae (4.247) and (4.249), and location of the nodal lines of the modes of vibration are the important properties along with other equivalent parameters of the mechanical system of a transducer.

Upon substituting expressions (4.233) and (4.237) for the energies (on this stage without applying of a mechanical load, i.e., at $f_L = 0$) the system of Euler equations (4.1) becomes

$$M_{ii}\ddot{\xi}_i + K_{ii}\xi_i + \sum_{n\neq i}^{\infty}(M_{ni}\ddot{\xi}_n + K_{ni}\xi_n) = f_{mi}, \quad (i = 1, 2, ...). \tag{4.250}$$

If the mechanical system is uniform (in the expressions (4.236) m_Δ and K_Δ are constant) and the normal modes X_i are used as the supporting functions, then from orthogonality of the normal modes follows that $M_{ni} = 0$ and $K_{ni} = 0$, and Eqs. (4.250) become independent. In this case the generalized coordinates ξ_i are called normal coordinates and Eqs. (4.250) coincide with Eqs. (4.85), which were obtained as a result of solving the vibration problem for the mechanical system by method of expansion in terms of normal modes. Considering losses of energy, we have brought Eqs. (4.85) to the form of Eqs. (4.86). In the analogous way we will account for energy losses in Eqs. (4.250), simultaneously changing to the general case, in which W_e is determined by relation (4.238). Then equations (4.250) become

$$M_{ii}\ddot{\xi}_i + r_i\dot{\xi}_i + K_{ii}\xi_i + \sum_{n\neq i}^{\infty}(M_{ni}\ddot{\xi}_n + K_{ni}\xi_n) = f_{mi} - f_{Li} \quad (i = 1, 2...). \tag{4.251}$$

Under harmonic vibrations Eqs. (4.251) present an infinite system of algebraic equations with constant coefficients relative to the generalized coordinates ξ_i. Changing to the complex form and taking into consideration expressions (4.241) and (4.242) for F_{Li} and Z_{Li}, we arrive at

$$\begin{aligned}(j\omega M_{ii} + K_{ii}/j\omega + r_i)U_i + \sum_{n\neq i}^{\infty}(j\omega M_{ni} + K_{ni}/j\omega)U_n + \\ + \left[Z_{Lii} + \sum_{n\neq i}^{\infty} z_{Lni}(U_n/U_i)\right]U_i = F_{mi}, \quad (i = 1, 2, 3, ...).\end{aligned} \tag{4.252}$$

It is convenient to denote for brevity

$$j\omega M_{ii} + K_{ii} / j\omega + r_i = Z_{m\,ii}, \quad j\omega M_{ni} + K_{ni} / j\omega = z_{m\,ni},$$
$$Z_{m\,ii} + \sum_{n \neq i}^{\infty} z_{mni} (U_n / U_i) = Z_{m\,i}, \tag{4.253}$$

where Z_{mii} is the mechanical self-impedance of vibration mode θ_i, $z_{m\,ni}$ is the mutual mechanical impedance, which characterizes interaction between vibration modes θ_i and θ_n. Using these designations, Eqs. (4.252) can be finally represented as

$$(Z_{m\,ii} + Z_{Lii})U_i + \sum_{n \neq i}^{\infty} (z_{mni} + z_{Lni})U_n = F_{m\,i}\,. \tag{4.254}$$

In the case that the equivalent forces F_{mi} are of electromechanical origin (for example, forces acting on the body surface from the side of electro-dynamic, electromagnetic, or other electromechanical generators), Eqs. (4.254) describe vibration of the mechanical system of the corresponding transducers in the generalized coordinates (velocities). Procedure for solving the systems of equations such as (4.254) is well known (e.g., see Ref. 5). Thus, the problem to be solved in each particular case is in determining the impedances involved in the equations.

Solution of the infinite systems of equations can be obtained by the method of reduction, as it is demonstrated in Ref. 5. But the same problem can be solved differently, if to reduce the number of degrees of freedom for the mechanical system under consideration in advance, and to present the displacements as the finite series

$$\xi_N(x) = \sum_{i=1}^{N} \xi_i \theta_i(x)\,. \tag{4.255}$$

Once an approximate solution, $\xi_N(x)$, is obtained, it can be made more precise by gradual increasing the number of equations N. Such technique of solving the problem of elastic body vibration is known as the Ritz's method.

With exception for the particular case of normal coordinates, the equations of the system (4.224) are coupled. The degree of coupling the equations, calculating difficulties in solving vibration problem and clarity of physical interpretation of results to a great extent depend on selection of the supporting functions. Solution of the system of equations is fairly simple, if the normal modes of the vibration problem for a real body are selected as the supporting functions.

Sometimes it proves possible to guess the form of normal modes based on symmetry of mechanical system. But in general, the problem of determining normal modes is equivalent in terms of complexity to that of determining vibration of the mechanical system.

Another practical way of analyzing a real system is to use the normal modes of an idealized vibrational system, for which they are known, provided that the latter does not significantly differ from the real system. In order to illustrate the above-mentioned approaches, consider seeveral examples. The first group of examples illustrates employing the normal modes of vibration.

4.5.2 Extensional Vibration of the Complete Rings

The transducer with mechanical system in the shape of a thin piezoelectric ceramic ring was considered in Section 2.3 under assumption of uniform excitation and uniform acoustic loading, as a typical example of transducer with one mechanical degree of freedom. If to reject these assumptions, then the symmetry considerations are not valid and the ring may vibrate in the extensional modes with displacements taking place in the radial and circumferential directions, as shown in Figure 4.15. The general expressions for displacements in the extensional vibration of a ring are presented in Section 4.4.4.

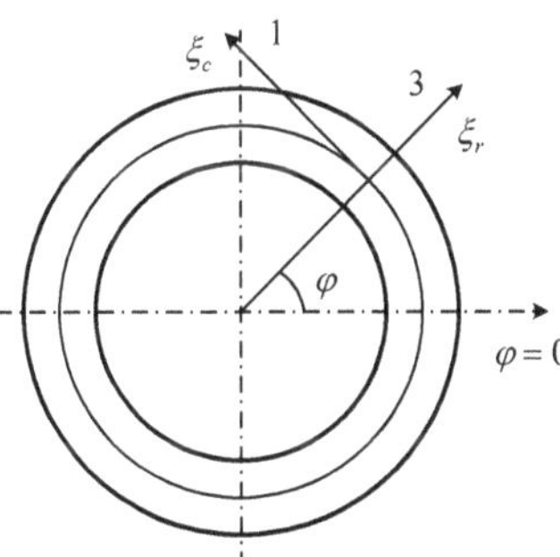

Figure 4.15: The extensional vibrations of a general type in the plane of a ring: ξ_r and ξ_c are the displacements in the radial and circumferential directions.

The potential and kinetic energies of the ring following expressions (4.225) and (4.226) are

$$W_{pot} = \frac{Yth}{2a^2}\int_0^{2\pi}\left(\frac{\partial \xi_\varphi}{\partial \varphi} + \xi_r\right)^2 d\varphi, \tag{4.256}$$

$$W_{kin} = \frac{1}{2}\rho tha \int_0^{2\pi} \left[\dot{\xi}_\varphi^2 + \left(\partial \dot{\xi}_\varphi / \partial \varphi \right)^2 \right] d\varphi . \tag{4.257}$$

Taking into consideration 2π periodicity of function $\xi_r(\varphi)$, an arbitrary distribution of the

the radial displacement on the ring surface can be represented as

$$\xi_r(\varphi) = \sum_{i=0}^{\infty} \xi_i \cos i\varphi . \tag{4.258}$$

According to Eq. (4.213)

$$\xi_\varphi(\varphi) = \sum_{i=1}^{\infty} i\xi_i \sin i\varphi . \tag{4.259}$$

After substituting the functions $\xi_r(\varphi)$ and $\xi_\varphi(\varphi)$ into expressions for the energies (4.256) and (4.257) we obtain:

$$\begin{aligned} W_{kin} &= \frac{\rho tha}{2} \int_0^{2\pi} \left[\left(\sum_{i=0}^{\infty} \dot{\xi}_i \cos i\varphi \right)^2 + \left(\sum_{i=1}^{\infty} i\dot{\xi}_i \sin i\varphi \right)^2 \right] d\varphi = \\ &= \frac{1}{2} M \left[\dot{\xi}_0^2 + \frac{1}{2} \sum_{i=1}^{\infty} \dot{\xi}_i^2 \left(1 + i^2 \right) \right] = \frac{1}{2} M_{eqv\,o} + \sum_{i=1}^{\infty} \frac{1}{2} M_{eqvi} \dot{\xi}_i^2 , \end{aligned} \tag{4.260}$$

where $M = 2\pi ath\rho$ is the mass of the ring and M_{eqvi} is the equivalent mass that corresponds to the mode of vibration θ_i.

$$\begin{aligned} W_{pot} &= \frac{Yth}{2a} \int_0^{2\pi} \left[\sum_{i=0}^{\infty} \xi_i \left(1 + i^2 \right) \cos i\varphi \right]^2 d\varphi = \\ &= \frac{1}{2} \frac{2\pi thY}{a} \left[\xi_0^2 + \frac{1}{2} \sum_{i=1}^{\infty} \xi_i^2 \left(1 + i^2 \right) \right] = \frac{1}{2} K_{eqv0} + \sum_{i=1}^{\infty} \frac{1}{2} K_{eqvi} \xi_i^2 . \end{aligned} \quad \text{Her} \tag{4.261}$$

Here K_{eqvi} is the equivalent mass that corresponds to the mode of vibration θ_i. When calculating the integrals, orthogonality of functions $\cos i\varphi$ and $\sin i\varphi$ on the interval 0 to 2π was used. It follows from Eqs. (4.260) and (4.261) that

$$M_{eqv0} = M, \quad M_{eqvi} = M\left(1 + i^2\right)/2 \tag{4.262}$$

and

$$K_{eqv0} = \frac{2\pi thY}{a}, \quad K_{eqvi} = \frac{\pi thY}{a}\left(1 + i^2\right)^2 . \tag{4.263}$$

After substituting the parameters by formulas (4.262) and (4.263) into the general expression for the natural resonance frequencies,

$$\omega_i = \sqrt{K_{eqvi} / M_{eqvi}}\,, \tag{4.264}$$

we obtain

$$\omega_i = \omega_0 \sqrt{1+i^2} \tag{4.265}$$

as the resonance frequencies of the extensional vibrations of different order, where φ $\omega_0 = (1/a)\sqrt{Y/\rho}$ is the natural frequency of zero order (at $i = 0$), or of the pulsating mode of vibration. Under extensional vibrations of this kind the displacements of cross sections of a ring are

$$\boldsymbol{\xi} = \sqrt{\xi_{ri}^2 + \xi_{\varphi i}^2} \cdot e^{j\gamma} = \xi_i \sqrt{\cos^2 i\varphi + i^2 \sin^2 i\varphi} \cdot e^{j\gamma}\,, \tag{4.266}$$

where $\gamma =$ arctan(i tan$i\varphi$). As the angle φ changes, vector $\boldsymbol{\xi}$ turns by angle γ relative to axis $\varphi = 0$. Positions of the vector at $\pi/4$ increments of angle φ for $i = 1$ are shown in Figure 4.15. The cross sections of ring that don't move in radial direction have maximum displacement tangential to the circumference. As an order of the vibration mode increases, natural frequencies become multiples of i, while the motion becomes more concentrated in the tangential direction ($|\xi_{\varphi i}| = i|\xi_{ri}|$). Using expression (4.221) the value of mechanical stress in the ring can be obtained as

$$T_\varphi = YS_\varphi = \frac{Y}{a}\sum_{i=0}^{\infty} \xi_i (1+i^2)\cos i\varphi\,. \tag{4.267}$$

4.5.3 Extensional Vibration of Incomplete Rings

Consider extensional vibration of an incomplete ring with free ends. Geometry of the ring is shown in Figure 4.16. We assume that the ring is thin and short, that is $(t/a) \ll 1$ and $(h/a) \ll 1$. Therefore, the only nonzero stress in the ring is $T_\varphi = YS_\varphi$. Using expression for strain S_φ from Eq. (4.70) and condition (4.76) of existing the extensional vibration only, we obtain

$$S_\varphi = \frac{1}{a}\left(\frac{\partial \xi_\varphi}{\partial \varphi} + \xi_r\right) = \frac{1}{a}\left(\xi_r - \frac{\partial^2 \xi_r}{\partial \varphi^2}\right). \tag{4.268}$$

The conditions at free ends are

$$T_{\varphi} = YS_{\varphi}\big|_{\varphi=\pm\beta} = 0 \text{ i.e., } \left(\xi_r - \frac{\partial^2 \xi_r}{\partial \varphi^2}\right)_{\varphi=\pm\beta} = 0\,. \tag{4.269}$$

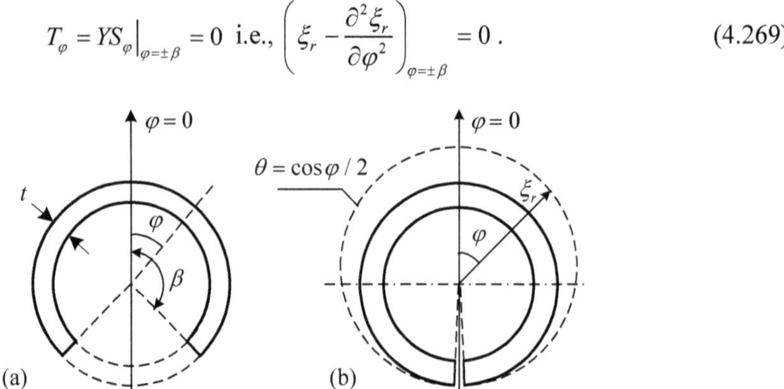

Figure 4.16: Geometry of an incomplete circular ring with opening angle β. Dashed line shows the lowest mode of vibration of the ring with a slot - $\theta_0(\varphi,\pi) = \cos(\varphi/2)$.

For transducer application purposes the symmetric in respect to axis $\varphi = 0$ vibrations are of interest. This condition together with conditions (4.269) are satisfied by the functions

$$\theta_i(\varphi,\beta) = \cos(2i+1)\frac{\pi}{2\beta}\varphi \quad (i = 0,1,2,...)\,. \tag{4.270}$$

They form a complete orthogonal set of functions in the interval $-\beta \le \varphi \le \beta$. Indeed,

$$\int_{-\beta}^{\beta}\left[\cos(2i+1)\frac{\pi}{2\beta}\varphi\right]\left[\cos(2l+1)\frac{\pi}{2\beta}\varphi\right]d\varphi = \begin{cases} 0, & i \ne l \\ \beta, & i = l \end{cases} \tag{4.271}$$

The functions (4.270) can be used as the set of supporting functions, and the general solution to the problem of extensional vibration of an incomplete ring that satisfies the boundary conditions (4.269) can be represented as series

$$\xi_r(\varphi) = \sum_{i=0}^{\infty}\xi_i \cos\left[(2i+1)\frac{\pi}{2\beta}\varphi\right], \tag{4.272}$$

$$\xi_{\varphi}(\varphi) = \sum_{i=0}^{\infty}\xi_i\left[(2i+1)\frac{\pi}{2\beta}\right]\sin\left[(2i+1)\frac{\pi}{2\beta}\varphi\right]. \tag{4.273}$$

After substituting these expressions for displacements into formulas for the potential and kinetic energies per element of a ring (4.225) and (4.226), will be obtained:

$$W_{pot} = \frac{thY}{a} \sum_{i=0}^{\infty} \xi_i^2 \left\{ 1 + \left[\frac{\pi}{2\beta}(2i+1) \right]^2 \right\}^2 = \frac{1}{2} \sum_{i=0}^{\infty} \xi_i^2 K_{eqvi} , \tag{4.274}$$

where

$$K_{eqvi} = \frac{thY}{a} \beta \left\{ 1 + \left[\frac{\pi}{2\beta}(2i+1) \right]^2 \right\}^2 \tag{4.275}$$

$$W_{kin} = \frac{1}{2} tha\rho\beta \sum_{i=0}^{\infty} \dot{\xi}_i^2 \left\{ 1 + \left[\frac{\pi}{2\beta}(2i+1) \right]^2 \right\} = \frac{1}{2} \sum_{i=0}^{\infty} \dot{\xi}_i^2 M_{eqvi} , \tag{4.276}$$

where

$$M_{eqvi} = tha\rho\beta \left\{ 1 + \left[\frac{\pi}{2\beta}(2i+1) \right]^2 \right\} \tag{4.277}$$

is the equivalent mass that corresponds to the mode of vibration θ_i.

The resonance frequencies of free vibrations are $f_i = (1/2\pi)\sqrt{K_i / M_i}$, and after substituting expressions (4.248) and (4.249) will be obtained that

$$f_i = \frac{1}{2\pi a} \sqrt{\frac{Y}{\rho}} \sqrt{1 + \left[\frac{\pi}{2\beta}(2i+1) \right]^2} = f_{r0} \sqrt{1 + \left[\frac{\pi}{2\beta}(2i+1) \right]^2} , \tag{4.278}$$

where $f_{r0} = (1/2\pi a)\sqrt{Y/\rho}$ is the lowest resonance frequency of a complete ring of radius a. Thus, for example, the ring having a thin slot or a crack (in this case $\beta \approx \pi$) has the resonance frequencies $f_0 = f_{r0}\sqrt{1.25}$, $f_1 = 1.8 f_{r0}$, $f_2 = 2.7 f_{r0}$ as compared with the resonance frequencies of the complete ring f_{r0}, $1.4 f_{r0}$ and $2.2 f_{r0}$, respectively. The mode shape that corresponds to the lowest resonance frequency of a ring with a slot is $\theta_0(\varphi, \pi) = \cos(\varphi/2)$. It is shown in Figure 4.16. The modal distribution of stress in the ring is

$$T_\varphi = \xi_{ri} \left[1 + (2i+1)^2 \left(\frac{\pi}{2\beta} \right)^2 \right] \left[\cos(2i+1) \frac{\pi}{2\beta} \varphi \right] . \tag{4.279}$$

4.5.4 Flexural Vibration of the Complete Rings

Equation of the flexural vibration of a ring can be derived under the condition (4.260)

$$\frac{\partial \xi_\varphi}{\partial \varphi} + \xi_r = 0 \tag{4.280}$$

that the middle surface of the ring does not elongate. Under this condition expressions (4.225) and (4.226) for the potential and kinetic energies per element of the ring can be modified to the following form

$$w_{kin} = \frac{1}{2}\rho S_{cs}(\dot{s}_r^2 + \dot{s}_\varphi^2) = \frac{1}{2}\rho S_{cs}\left[\dot{\xi}_\varphi^2 + \left(\frac{\dot{\xi}_\varphi}{\partial \varphi}\right)^2\right], \tag{4.281}$$

$$w_{pot} = \frac{1}{2}\frac{Yht^3}{12a^4}\left(\frac{\partial^3 \xi_\varphi}{\partial \varphi^3} + \frac{\partial \xi_\varphi}{\partial \varphi}_r\right)^2. \tag{4.282}$$

The same system of supporting functions $\theta_i(\varphi) = \cos i\varphi$ can be accepted, as in the case of the extensional vibrations, considering 2π periodicity of solution for the flexural vibration of a complete ring. Thus, the displacements in the radial and circumferential directions that meet the condition (4.231) will be presented as follows

$$\xi_r(\varphi) = \sum_{i=0}^{\infty} \xi_i \cos i\varphi, \tag{4.283}$$

$$\xi_\varphi(\varphi) = -\sum_{i=1}^{\infty} (\xi_i / i)\sin i\varphi. \tag{4.284}$$

After substituting expressions for the displacements into formulas (4.281) and (4.282), we obtain the following representations for the potential and kinetic energies of the ring:

$$W_{pot} = \frac{1}{2}\frac{\pi Yt^3 h}{12a^3}\sum_{i=2}^{\infty}(1-i^2)^2\xi_i^2 = \frac{1}{2}\sum_{2}^{\infty} K_{eqvi}\xi_i^2 \tag{4.285}$$

(note that at $i = 1$ the potential energy cannot exist, therefore the series (4.283) and (4.284) have to be started from $i = 2$); and

$$W_{kin} = \frac{1}{2}\frac{M}{2}\sum_{i=2}^{\infty}\left(1+\frac{1}{i^2}\right)+\dot{\xi}_i^2 = \frac{1}{2}\sum_{i=2}^{\infty} M_{eqvi}\dot{\xi}_i^2. \tag{4.286}$$

Expressions for the equivalent masses, M_{eqvi}, and rigidities, K_{eqvi}, (compliances $C_{eqvi} = 1/K_{eqvi}$) that follow from the relations (4.285) and (4.286) are

$$M_{eqvi} = \frac{M(1+i^2)}{2i^2}, \quad K_{eqvi} = \frac{1}{C_{eqvi}} = \frac{\pi Yt^3 h(i^2-1)^2}{12a^3} \quad (i = 2,3,...). \tag{4.287}$$

Thus, the resonance frequencies of the flexural vibration of a ring are

$$f_i = \frac{1}{2\pi\sqrt{M_{eqvi}C_{eqvi}}} = \frac{t}{2\pi\sqrt{12}a^2}\sqrt{Y/\rho}\frac{i(i^2-1)}{\sqrt{i^2+1}}. \tag{4.288}$$

The lowest resonance frequency occurs at $i = 2$ and is

$$f_2 = 0.12\frac{t}{a^2}\sqrt{Y/\rho}. \tag{4.289}$$

The next resonance frequency is $f_3 \approx 8f_2$, therefore the mode of vibration $\xi_r = \xi_{ro}\cos 2\varphi$ can be considered as dominant in the frequency range below and around frequency f_2. Ratio of the lowest resonance frequency of extensional vibration of a ring, $f_0 = \left(\sqrt{Y/\rho}\right)/2\pi a$, to frequency f_2 is $(f_0/f_2) = 1.3\,a/t$. Thus, for the relatively thin rings free flexural and extensional vibrations can be treated as independent. It is noteworthy that the case at i=1 corresponds formally to translation of the ring as a rigid body in the direction $\varphi = 0$ with mass $M_1 = M$ and "resonance frequency" $\omega_1 = 0$.

Distribution of the mechanical stress in a ring can be found using Eq. (4.221), in which the second term represents the flexural deformations, namely

$$S_\varphi = -\frac{r-a}{a^2}\left(\frac{\partial^2\xi_r}{\partial\varphi^2} + \xi_r\right), \tag{4.290}$$

$$T_\varphi = YS_\varphi = \frac{Y(r-a)}{a^2}\sum_{i=2}^{\infty}\xi_i(i^2-1)\cos i\varphi. \tag{4.291}$$

4.5.5 Vibration of the Spherical Shells

4.5.5.1 Introduction

Piezoceramic spherical shells are common in underwater acoustics predominantly as omnidirectional (zero mode) projectors and hydrophones. Example of a transducer that can be used in this capacity is considered in Section 2.2. Spherical shell transducers can also be used as directional by combining different modes of vibration, or by baffling of parts of their surfaces. Such applications will be considered in Chapter 8. They require more general analysis of vibration of the shells. Vibration of the passive spherical shells has been examined by many authors. An extensive bibliography on this issue is provided in Ref. 6.

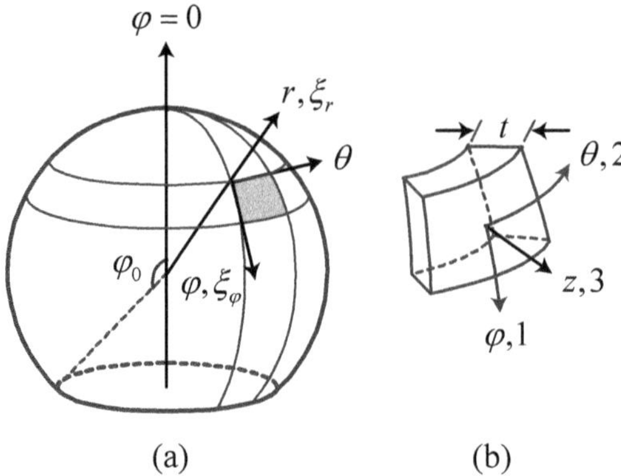

Figure 4.17: (a) Geometry of the spherical shell and (b) its differential element and variables used.

The geometry of the spherical shell in the general case of an open shell is shown in Figure 4.17. We will consider the axisymmetric vibrations of thin elastic spherical shells only. Therefore, the components of motion are the radial and tangential displacements of the middle surface, ξ_r and ξ_φ, which are independent of the azimuthal angle θ. The opening angle for the incomplete sphere is denoted φ_0. The thickness of the shell, t, is assumed to be small enough to neglect the radial stress through the thickness ($T_z \approx 0$).

In course of analysis of the spherical mechanical system a set of supporting functions that define the distribution of radial displacements for the normal modes of vibration must be determined and the generalized coordinates introduced. After this is done, the modal equivalent parameters of the spherical mechanical system and corresponding resonance frequencies must be calculated. All these goals will be achieved with reference to work [8], in which the general case of axisymmetric vibration of the finite thickness spherical shells open at one pole and effects of the shell bending are considered. It was s found that the frequency spectrum of vibration of the complete shells corresponds with two coupled sets of modes, namely, the "membrane" modes, which are associated with extension of the shell, and the bending modes. Their related resonance frequencies form the upper and lower branches, respectively. Two modes constitute the exceptions: the zero-order ("breathing") mode, where the displacements are purely radial, and the first-order mode, where the distribution of radial displacements is

$\xi_r = \xi_0 \cos\varphi$. In both cases only the extensional modes of deformation and corresponding resonance frequencies exist. In the case of the first-order vibration, formally, the resonance frequency $f = 0$ also exists, which can be ascribed to translation of the shell as a rigid body in the direction $\varphi = 0$ without deformation. We denote the corresponding displacement as ξ_{1t}. This rigid body movement must be considered by adding the term $M_{sph}\dot{\xi}_{1t}^2/2$ (M_{sph} is the mass of the sphere) to the total kinetic energy of the shell. The resonance frequency corresponding to the zero mode of vibration is considered as the fundamental mode, although it is not the lowest resonance frequency (the lowest are the resonance frequencies of some of the bending modes).

The membrane modes are independent of the shell thickness so far as the classical theory of thin shells is applicable. The general convention is that this is valid for the range of the relative thicknesses (thickness to radius ratio) $t/a \leq 0.05$. At this condition the terms on the order of z/a can be neglected in any expression for strain, S, by replacing $S/(1+z/a) \approx S$. It is of note that the spherical shells for transducer applications usually have relative thickness in the range of $t/a = 0.05$ to 0.2. The factor $1/(1+z/a) \approx 1 - z/a$ could be retained to increase the accuracy of calculations, but this would unjustifiably complicate the general analysis. It will be shown later by comparing the results of calculations made with and without this factor that difference in estimating transducer parameters is insignificant up to $t/a = 0.2$.

The bending modes vary with thickness. According to the membrane theory (Ref. 2) the bending mode related resonances are confined within a finite interval at low frequencies, whereas the bending theory predicts that they may extend to the high frequencies for all values of the thickness. Thus, the resonances of the bending modes can appear in the operating frequency range of the transducer and may cause a corruption of its frequency response and directional factor. Therefore, it is appropriate to use this more general approach, when considering broadband operation of a multimode spherical transducer.

For vibration of incomplete spherical shells, especially those with great opening angles, the most comprehensive analysis available is also given in Ref. 8. Thus, all the calculations of the spherical shell parameters will be produced based on the mode shapes of vibration determined by following the results presented in this work without repeating their derivation. The brief outline of the procedures performed in this work follows.

As it is shown in Ref. 8, the expressions for strains of the middle surface of a shell are

$$S_\varphi = \frac{1}{a}[\xi'_\varphi(\varphi) + \xi_r(\varphi)], \quad S_\theta = \frac{1}{a}[\xi_\varphi(\varphi)\cot\varphi + \xi_r(\varphi)], \tag{4.292}$$

where the prime sign (') denotes the derivative with respect to φ. Changes of curvature of the middle surface in directions of the meridian and azimuth are

$$\chi_\varphi = \frac{1}{a^2}[\xi'_\varphi(\varphi) - \xi''_r(\varphi)], \quad \chi_\theta = \frac{1}{a^2}\cot\varphi[\xi_\varphi(\varphi) - \xi'_r(\varphi)]. \tag{4.293}$$

The total strain across the thickness is

$$S_{\varphi\varphi} = S_\varphi + z\chi_\varphi, \quad S_{\theta\theta} = S_\theta + z\chi_\theta. \tag{4.294}$$

The expressions for stress are

$$T_{\varphi\varphi} = \frac{Y}{1-\sigma^2}(S_{\varphi\varphi} + \sigma S_{\theta\theta}), \quad T_{\theta\theta} = \frac{Y}{1-\sigma^2}(S_{\theta\theta} + \sigma S_{\varphi\varphi}). \tag{4.295}$$

The expression for the kinetic energy of deformation for a sphere with the opening angle φ_0 is

$$W_{kin} = \pi a^2 t\rho \int_0^{\varphi_0} \left(\dot{\xi}_r^2 + \dot{\xi}_\varphi^2\right)\sin\varphi d\varphi. \tag{4.296}$$

The expression for the potential energy is

$$\begin{aligned} W_{pot} &= \frac{1}{2}\int_{\tilde{V}}(T_{\varphi\varphi}S_{\varphi\varphi} + T_{\theta\theta}S_{\theta\theta})d\tilde{V} = \\ &= \frac{Yt\pi a^2}{1-\sigma^2}\int_0^{\varphi_0}[(S_\varphi^2 + 2\sigma S_\varphi S_\theta + S_\theta^2)\sin\varphi d\varphi + \frac{t^2}{12}(\chi_\varphi^2 + 2\sigma\chi_\varphi\chi_\theta + \chi_\theta^2)\sin\varphi d\varphi]. \end{aligned} \tag{4.297}$$

Analysis for the complete and incomplete spherical shells will be fulfilled separately.

4.5.5.2 Complete Spherical Shells

The displacements of complete spherical shells, being periodic functions with period 2π, can be presented by Legendre polynomials, $P_i\ (\cos\varphi)$, which are solutions of the differential equation (Appendix C.3)

$$(x^2 - 1)\frac{d^2P_i(x)}{dx^2} + 2x\frac{dP_i(x)}{dx} - i(i+1)P_i(x) = 0 \tag{4.298}$$

at $x = \cos\varphi$. The Legendre polynomials present an appropriate orthogonal set of supporting functions in the interval [1, -1] for solving the transducer problems, and the radial and tangential displacements can be presented in the form

$$\xi_r(\varphi) = \sum_{i=0}^{\infty} \xi_{ri} P_i(\cos\varphi), \quad \xi_\varphi(\varphi) = \sum_{i=1}^{\infty} \xi_{\varphi i} P_i'(\cos\varphi). \tag{4.299}$$

The modal displacements ξ_{ri} and $\xi_{\varphi i}$ are related in the following way[8],

$$\xi_{ri}(\varphi) = \xi_{ri} P_i(\cos\varphi), \quad \xi_{\varphi i}(\varphi) = -\xi_{ri}(1+\sigma)C_i P_i'(\cos\varphi) \quad (i = 1,2,3,...). \tag{4.300}$$

Here is denoted $\xi_{ri} = \xi_{ri}(\varphi)_{\varphi=0}$ and

$$C_i = \frac{(1+\sigma)(1+\delta)+\delta(\lambda_i - 2)}{(1-\sigma-\lambda_i)(1+\sigma)(1+\delta)+\Omega^2(1-\sigma^2)(1+\sigma)}, \tag{4.301}$$

$$\lambda_i = i(i+1), \quad \delta = \frac{t^2}{12a^2}, \quad \Omega^2 = \frac{\rho\omega^2 a^2}{Y}. \tag{4.302}$$

After the modal resonance frequencies ω_i are determined from frequency equations, the nondimensional frequency parameters Ω^2 and coefficients C_i can be calculated. At $i \geq 2$ ω_i, Ω_i and C_i have two values: higher and lower, which correspond to the membrane and bending modes, respectively. They will be denoted as ω_{im}, Ω_{im}, C_{im} and ω_{ib}, Ω_{ib}, C_{ib}. Substituting the modal displacements given by Eq. (4.300) into expressions (4.292) and (4.293) results in determining the strains and curvature change and, subsequently, in calculating the kinetic and potential energies associated with deformation of a shell. Note that according to relations (4.300) all the quantities appear to be expressed through the modal displacements of the pole, ξ_{ri}, which can be taken for the generalized coordinates. Given that at $i \geq 2$ two modes of vibration exist for each number *i*, the corresponding generalized coordinates ξ_{rim} and ξ_{rib} must be introduced. It is also of note that the radial component of translation of the shell as a rigid body, which we denote ξ_{r1t}, has the same dependence on φ, as the radial displacement in the first mode. Therefore, the total radial displacement associated with the mode shape at $i = 1$ should be represented as $\xi_{r1\Sigma}(\varphi) = (\xi_{r1} + \xi_{1t})\cos\varphi$, when considering the radiation related problems.

The subscript r will be further omitted for brevity and the respective set of generalized coordinates ξ_0; ξ_1, ξ_{1t}; ξ_{im}, ξ_{ib} will be used, where $i = 2, 3,...$. The displacements and all the displacement dependent functions with numbers $i \geq 2$ will be represented as

$$\begin{gathered} \xi_i = \xi_{im} + \xi_{ib}, \quad S_{\varphi i} = S_{\varphi im} + S_{\varphi ib}, \quad S_{\theta i} = S_{\theta im} + S_{\theta ib}; \\ \chi_{\varphi i} = \chi_{\varphi im} + \chi_{\varphi ib}, \quad \chi_{\theta i} = \chi_{\theta im} + \chi_{\theta ib}. \end{gathered} \tag{4.303}$$

The displacement at $i = 1$ is $\xi_{1\Sigma} = \xi_1 + \xi_{1t}$, but it must be remembered that no stress is associated with the generalized coordinate ξ_{1t}.

Taking into consideration orthogonality of Legendre polynomials on the interval $0 \le \varphi \le \pi$, the kinetic and potential energies can be represented as superposition of the modal energies. Thus,

$$W_{kin} = \sum_i W_{kin\,i} = \frac{M_{sph}}{2}\dot{\xi}_{1t}^2 + \pi a^2 t\rho \sum_i \int_0^{\pi} [\dot{\xi}_{ri}^2(\varphi) + \dot{\xi}_{\varphi i}^2(\varphi)]\sin\varphi d\varphi = \\ = \frac{1}{2}[M_{eqv0}\dot{\xi}_0^2 + M_{sph}\dot{\xi}_{1t}^2 + M_{eqv1}\dot{\xi}_1^2 + \sum_{i\ge 2}(M_{eqvi\,m}\dot{\xi}_{i\,m}^2 + M_{eqvi\,b}\dot{\xi}_{i\,b}^2)]. \tag{4.304}$$

$$W_{pot} = \sum_i W_{pot\,i} = \frac{Yt\pi a^2}{1-\sigma^2}\sum_i \int_0^{\pi} [(S_{\varphi i}^2 + 2\sigma S_{\varphi i}S_{\theta i} + S_{\theta i}^2) + \\ + \frac{t^2}{12}(\chi_{\varphi i}{}^2 + 2\sigma\chi_{\varphi i}\chi_{\theta i} + \chi_{\theta i}{}^2)]\sin\varphi d\varphi = \\ = \frac{1}{2}[K_{eqv0}\xi_0^2 + K_{eqv1}\xi_1^2 + \sum_{i\ge 2}(K_{eqvi\,m}\xi_{i\,m}^2 + 2K_{i\,mb}\xi_{i\,m}\xi_{i\,b} + K_{eqvi\,b}\xi_{i\,b}^2)], \tag{4.305}$$

Table 4.1: Equivalent parameters of the complete spherical shells.

Parameter	i	0	1	2[2)]	3[2)]	5[2)]
Ω_i [1)]	m	$\sqrt{2/(1-\sigma)}$	$\sqrt{3/(1-\sigma)}$	2.9	3.9	6.2
	b	-	-	0.75	0.92	1.24
M_{eqvi}, $\times 4\pi a^2\rho t$	m	1	0.5	0.60	0.62	0.26
	b	-	-	0.29	0.17	0.10
K_{eqvi}, $\times\frac{2\pi Yt}{1-\sigma^2}$	m	4(1+σ)	3(1+σ)	8.9	16.3	14.3
	b	-	-	0.28	0.23	0.16
K_{imb}		-	-	-0.01	0.0016	0.027
C_i	m	-	1/2(1+σ)	0.45	0.41	0.19
	b	-	-	-0.21	-0.1	-0.036

where M_{eqvi} and K_{eqvi} ($K_{eqvi\,m}$ and $K_{eqvi\,b}$ at $i \ge 2$) are the modal equivalent masses and rigidities. Parameters $K_{i\,mb}$ are the mutual rigidities, which are accounted for the elastic coupling between the membrane and bending modes at $i \geq 2$. Expressions for the nondimensional resonance frequencies and for the equivalent parameters are presented in Table 4.1.

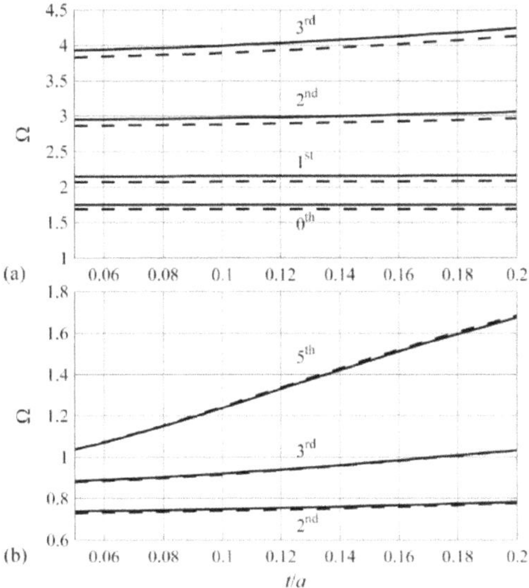

Figure 4.18: Dependences of the normalized resonance frequencies Ω_i of a spherical shell as functions of t/a and σ: (a) for the membrane modes, and (b) for the bending modes. Solid lines – $\sigma = 0.33$, dotted lines – $\sigma = 0.30$.

The calculations of parameter values for modes at $i \geq 2$ are made for Poisson's ratio σ = 0.3 and t/a = 0.1. The dependence of the resonance frequencies on ratio t/a at different values of σ is shown in Figure 4.18. As can be seen from Figure 4.198 (a), the resonance frequencies of the membrane modes are nearly independent of t/a for $t/a < 0.2$. Also, the resonance frequencies of the bending modes are practically independent of σ. In order to simplify the general analysis, we will intentionally sacrifice some accuracy by performing the calculations at σ = 0.3 and t/a = 0.1, unless particular transducers are considered. The accuracy of calculations can be increased, if needed, by using real properties of a particular spherical shell.

The modal radial displacement distributions in the membrane modes for the complete shell at i = 0, 1, 2, 3 are presented in Figure 4.19 in comparison with the mode shapes of incomplete (hemispherical) shell in the next section.

4.5.5.3 Incomplete Spherical Shells

The solution to the equation of free vibrations of a spherical shell open at one pole results in Legendre functions of complex order, i.e., conical functions $P_{m_i}(\cos\varphi)$ of the first kind, where $m_i = 0.5 + \sqrt{0.25 + \lambda_i}$, and λ_i can be a complex quantity. The modal displacements in this case can be presented as[8]

$$\xi_{rm}(\varphi) = \sum_{i=1,2,3} A_i P_{m_i} \cos\varphi, \quad \xi_{\varphi m}(\varphi) = -(1+\sigma)\sum_{i=1,2,3} C_i A_i P'_{m_i}(\cos\varphi). \tag{4.306}$$

The frequency equations for determining the normalized resonance frequencies Ω in this case have to be formulated according to boundary conditions that exist on the open edge of a shell at $\varphi = \varphi_0$. Only free boundary conditions will be considered here because they can be realized exactly, and they are the most realistic and practical boundary conditions for the intended transducer designs. For the free boundary to exist, the stress, moment, and shearing force must vanish.

Once the resonance frequencies are determined, the values of λ_i can be calculated from the corresponding cubic equation[8] that produces three values of λ (one real and two complex conjugate) for each resonance frequency. The coefficients C_i may then be found from Eq. (4.301) and the mode shape coefficients A_i can be found following the procedure described in Ref. 8. Given that $P_{m_i}(1) = 1$, the pole's radial displacement can be represented as

$$\xi_r(0) = \xi_r = A_1 + A_2 + A_3 = A_1(1 + A_2/A_1 + A_3/A_1), \tag{4.307}$$

where it can be assumed $A_1 = 1$. The resonance frequencies Ω of the first membrane mode and bending modes of vibration for a hemisphere ($\varphi_0 = \pi/2$) and for a spherical segment with opening angle $\varphi_0 = \pi/3$, together with values of the quantities for determining the corresponding mode shapes (m_i, C_i, A_i), are presented in Table 4.2. Data presented in Table 4.2 show that the resonance frequencies of the membrane modes, which are usable for transducer applications, are preceded by those of the bending modes. A relatively noticeable contribution to mechano-acoustic conversion can be expected from the lowest bending modes, judging by the comparison of the mode shapes of the radial displacements shown in Figure 4.19. Subsequently, we will only consider parameters of the membrane modes and of the lowest bending modes.

The equivalent modal masses and rigidities for the spherical shells were introduced and determined in Ref. 9 from the general expressions for the kinetic and potential energies (4.296) and (4.297). After substituting expressions (4.263) for the displacements $\xi_r(\varphi)$ and $\xi_\varphi(\varphi)$, performing all the calculations and considering that in accordance to relation (4.307) $\xi_r(0) = 1 + A_2 + A_3$, the modal potential and kinetic energies of deformation may be represented in the form

$$W_{kin} = \frac{1}{2} M_{eqv} \dot{\xi}_r^2(0), \quad W_{pot} = \frac{1}{2} K_{eqv} \xi_r^2(0) . \tag{4.308}$$

For the case of hemispherical shell ($\varphi_0 = \pi / 2$) it follows from Table 4.2 that the terms in Eq. (4.306) corresponding to A_2 are dominant for the membrane mode. Thus, we may assume that

$$\xi_{rm}(\varphi) \cong \xi_r P_1(\cos\varphi), \quad \xi_{\varphi m}(\varphi) \cong -0.38(1+\sigma)\xi_r P_1'(\cos\varphi) . \tag{4.309}$$

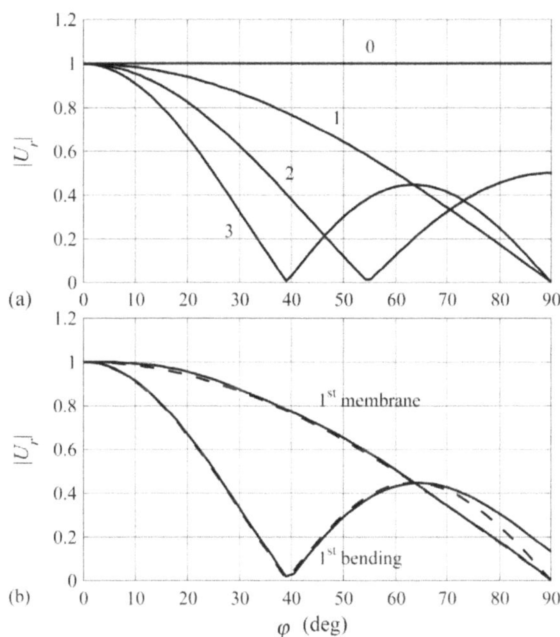

Figure 4.19: The modal radial velocity distributions: (a) for the complete spherical shells, $P_n(\cos\varphi)$, $n = 0, 1, 2, 3$; (b) for the hemispheres' first membrane and first bending modes. Solid lines – approximate solution, dashed lines – exact solution.

For the first bending mode the terms related to A_1 are dominant, and therefore

$$\xi_{rb}(\varphi) \cong \xi_r P_3(\cos\varphi), \quad \xi_{\varphi b}(\varphi) \cong 0.1(1+\sigma)\xi_r P_3'(\cos\varphi) . \tag{4.310}$$

The approximate distributions of displacements in accordance with expressions (4.309) and (4.310), and the exact results that take into account all the terms from Table 4.2, are presented in Figure 4.19 (b) and show sufficiently good agreement.

Table 4.2: Coefficients related to calculation of parameters of incomplete spherical shells.

		Hemisphere ($\varphi_0 = \pi / 2$)				Semisphere ($\varphi_0 = \pi / 3$)		
	Mode	$1m$	$2m$	$1b$	$2b$	$1m$	$1b$	$2b$
	Ω	2.07	2.82	0.9	1.03	2.55	0.95	1.4
A_i [1)]	1	1	1	1	1	1	1	1
	2	11.7	-15.5	0.006 -0.003i	-0.01 -0.004i	-10.8	0.04 -0.05i	0.01
	3	10^{-5}	$-4\cdot10^{-8}$	0.006 +0.003i	-0.01 +0.004i	-14	0.04 +0.05i	0.01
m_i [1)]	1	7.5	10.7	2.9	3.8	8.3	3.2	5.5
	2	1	3	2.1 -3.2i	1.3 -3.1i	1.7	1.8 +3.0i	-0.5 -1.3i
	3	-0.5 -7.5i	-0.5 -10.9i	2.1 +3.2i	1.3 +3.1i	-0.5 +8.5i	1.8 -3.0i	-0.5 +1.3i
C_i [1)]	1	-0.02	-0.01	-0.1	-0.06	-0.015	-0.08	-0.03
	2	0.39	0.52	0.015 -0.016i	0.04 -0.06i	0.45	0.02 +0.06i	0.22
	3	0.02	0.01	0.015 +0.016i	0.04 +0.06i	0.012	0.02 -0.06i	0.03

The expressions (4.309) and (4.310) coincide with expressions (4.300) for displacements in one half of a complete sphere vibrating in the first membrane mode and in the first bending mode (at $i = 3b$), respectively. Therefore, the resonance frequencies of a hemisphere are the same as for a complete sphere vibrating in the first mode, and all the equivalent parameters are twice smaller than the analogous parameters of the complete sphere of the same geometry vibrating in the first mode.

It is noteworthy that these results for hemisphere could be predicted without calculations due to the fact that the boundary conditions for a hemisphere with free edge (stress T_θ, moment M_θ, and shearing force Q are zero) virtually coincide with the analogous conditions in the cross section at $\varphi = \pi / 2$ of the complete sphere vibrating in the first mode. The only difference is that a finite shearing force exists in the last case. However, for the thin shells, the shear-related energy is negligible compared with the "membrane" energy and significantly smaller than the bending energy. The comparison of vibration distributions for the hemisphere presented in Figure 4.19 (b) shows that neglecting the bending energy does not change the membrane modes and only slightly changes the bending modes at the angles close to $\varphi = \pi/2$.

Table 4.3: Equivalent parameters of incomplete spherical shells.

Geometry	Parameter	Bending		Membrane	
		1	2	1	2
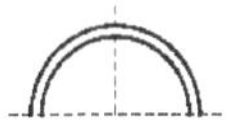	Ω	0.90	1.03	2.07	3.82
	M_{eqvi}, $\times 4\pi a^2 \rho t$	0.12	0.11	0.21	0.51
	K_{eqvi}, $\times \frac{2\pi Yt}{1-\sigma^2}$	0.18	0.22	1.60	13.6
	Ω	0.95	1.40	2.55	-
	M_{eqvi}, $\times 4\pi a^2 \rho t$	0.09	0.03	0.24	-
	K_{eqvi}, $\times \frac{2\pi Yt}{1-\sigma^2}$	0.15	0.10	2.75	-

In the case of the opening angle $\varphi_0 = \pi/3$, the following simplified expressions for the displacements may be obtained:

for the membrane mode

$$\xi_{rm}(\varphi) = \xi_r P_{1.6}(\cos\varphi), \quad \xi_{\varphi m}(\varphi) = -0.45(1+\sigma)\xi_r P'_{1.6}(\cos\varphi)\,; \tag{4.311}$$

and for the first bending mode

$$\xi_{rb}(\varphi) = \xi_r P_{3.6}(\cos\varphi), \quad \xi_{\varphi b}(\varphi) = -0.07(1+\sigma)\xi_r P'_{3.6}(\cos\varphi)\,. \tag{4.312}$$

Equivalent parameters of the shells at opening angles $\varphi_0 = \pi/2$ and $\varphi_0 = \pi/3$ that are calculated from expressions (4.296), (4.297) for the kinetic and potential energies using the above expressions for the generalized displacements are presented in Table 4.3.

4.5.6 Flexural Vibration of Nonuniform Beams

Vibration of the piezoceramic uniform bimorph beams and circular plates were considered in Section 2.6. As it will be shown in Chapter 9, the parts of active material of the piezoceramic bender transducers employing rectangular and circular plates that contribute the least to the electromechanical conversion can be replaced by passive material (metal predominantly) in order to optimize the operating properties of the transducer. This requires considering peculiarities of flexural vibration of these mechanical systems with properties nonuniform through their thickness and over length (radius). The procedure of calculating vibration of various nonuniform mechanical systems remains virtually the same, but their numerical illustration looks in the simplest way in application to the rectangular beams. Considering this example is moreover reasonable that the results obtained regarding the effect of nonuniformity through the thickness are valid for all the mechanical systems vibrating in flexure.

Practical modifications of the nonuniform beam designs are shown in Figure 4.20 and Figure 4.21. The symmetric trilaminar beam is composed of two piezoelectric ceramic layers having equal thickness that are cemented to the central laminate made of passive material.

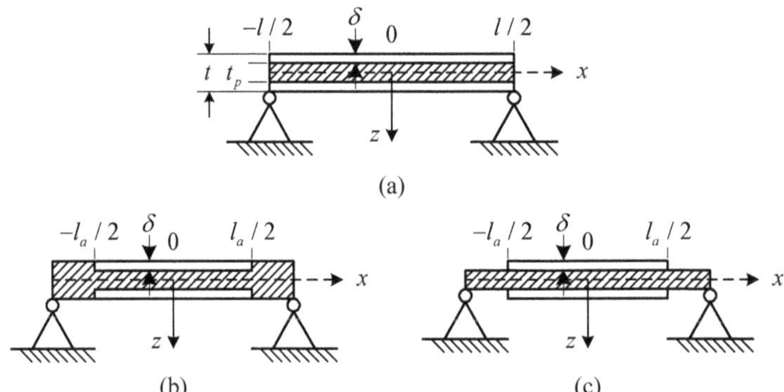

Figure 4.20: Modifications of the symmetric trilaminar beams: (a) uniform over the length, (b) with parts of active layers replaced by a passive material, (c) with parts of the active layers removed. Passive material is shown as dashed.

Bilaminar beam is composed of active and passive parts having different mechanical properties.

We will start the treatment from the symmetric trilaminar beams. The position of the neutral surface under bending in the trilaminar beams remains the same, as for the uniform beam due to symmetry. The position of the neutral surface in the bilaminar beams depends on the relation between the thicknesses and elastic properties of the active and passive layers, and this complicates the problem. Considering vibration of nonuniform beams that are composed of materials with different properties require using expressions for the potential and kinetic energies in the following general forms.

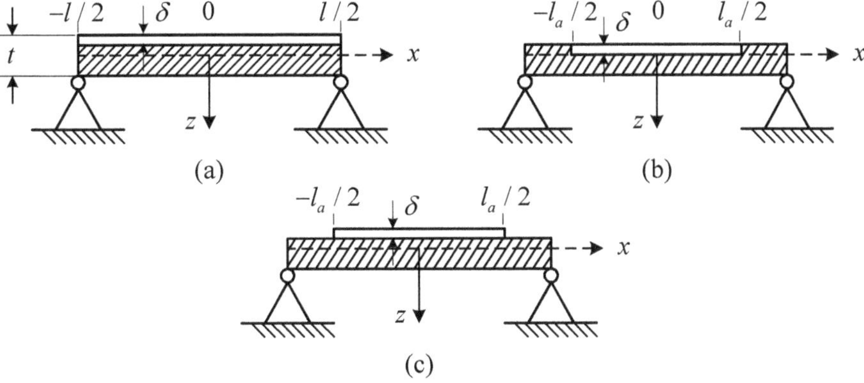

Figure 4.21: Modifications of the bilaminar beams: (a) uniform over the length, (b) with parts of active layer replaced by a passive material, (c) with parts of the active layers removed. Passive material is shown as dashed.

$$W_{pot} = \frac{1}{2} w \int_{-l/2}^{l/2} \int_{-t/2}^{t/2} z^2 Y(x,z) \xi_z^2 dxdz \,, \tag{4.313}$$

$$W_{kin} = \frac{1}{2} w \int_{-l/2}^{l/2} \int_{-t/2}^{t/2} \rho(x,z) \dot{\xi}_z^2 dxdz \,, \tag{4.314}$$

where the density and Young's modulus may change over the volume.

4.5.6.1 Trilaminar Beam Uniform over the Length

Variant of the beam in Figure 4.20 (a) can be considered as uniform over the length with some equivalent parameters of material that can be determined as follows. The stresses across the cross section are determined as $T = YS_1$, where S_1 is described by Eq. (2.113), namely, given that $\xi(x) = \xi_o \theta(x)$,

$$T_p = zY_p \xi \frac{d^2\theta}{dx^2} \text{ at } |z| < \frac{t_p}{2} \,, \tag{4.315}$$

$$T_a = zY_a^E \xi \frac{d^2\theta}{dx^2} \text{ at } |z| \geq \frac{t_p}{2} \,, \tag{4.316}$$

where the Young's moduli of the passive and active laminates are denoted by Y_p and Y_a^E.

The densities of the active and passive laminates will be denoted ρ_a and ρ_p. The following notations will be used for brevity to characterize properties of the trilaminar beams

$$\frac{Y_p}{Y_a^E} = \gamma_Y, \quad \frac{\rho_p}{\rho_a} = \gamma_\rho, \quad \frac{\delta}{t} = y. \tag{4.317}$$

Here $y = 0.5$ corresponds to the fully active bimorph design. Next, the energy status of the vibrating beam has to be considered. The potential energy of the beam at $E = 0$ will be determined after substituting the Young's moduli for the active and passive parts of the beam into expression (4.313). As result of integrating through the thickness we obtain

$$W_{pot\,t}^E = \frac{\xi_o^2}{2} \frac{wt^3}{12} Y_{eqvt} \int_{-1/2}^{1/2} \left(\frac{d^2\theta}{dx^2} \right)^2 dx, \tag{4.318}$$

where Y_{eqvt} denotes the equivalent Young's modulus of the trilaminar beam, and

$$Y_{eqvt} = Y_a^E + \left(Y_p - Y_a^E\right)\left(1 - \frac{2\delta}{t}\right)^3 = Y_a^E[1 + (\gamma_Y - 1)(1 - 2y)^3]. \tag{4.319}$$

The kinetic energy of the beam will be determined after substituting the densities for the active and passive parts of the beam into expression (4.314). As result of integrating over the thickness we obtain

$$W_{kin\,t} = \frac{\dot{\xi}_o^2}{2} \rho_{eqvt} tw \int_{-1/2}^{1/2} \theta^2(x) dx, \tag{4.320}$$

where ρ_{eqvt} denotes the equivalent density of the beam

$$\rho_{eqvt} = \frac{\rho_a 2\delta + \rho_p t_p}{t} = \rho_a[\gamma_\rho - 2(\gamma_\rho - 1)y]. \tag{4.321}$$

Thus, a trilaminar beam can be considered as uniform over the length beam having the equivalent Young's modulus, Y_{eqvt}, and density, ρ_{eqvt}, that are presented by formulas (4.319) and (4.321). Further, only the beams with simply supported ends will be considered. The normal modes of vibration in this case are the same, as for any uniform beam with simply supported ends, i.e., $\theta_i(x) = \cos(i\pi x / l)$. Therefore, the expressions for the equivalent mass and rigidity of the trilaminar beam coincide with the analogous expressions for the bimorph beam (see Section 2.6.1), if ρ_a and $1/s_{11}^E$ are replaced by the equivalent density ρ_{eqvt} and equivalent Young's modulus Y_{eqvt}, namely, for the first mode

$$K_{eqvt}^{E}=\frac{1}{C_{eqvt}^{E}}=\frac{\pi^{4}wt^{3}Y_{eqvt}}{24l^{3}},\quad M_{eqvt}=\frac{1}{2}wtl\rho_{eqvt}\,. \tag{4.322}$$

(In regard to the nonuniform over length beams these notations will be shortened to K_{1t} and M_{1t}.)

The resonance frequency of the trilaminar beam is

$$f_{t}=0.45\frac{t}{l^{2}}\sqrt{\frac{Y_{eqvt}}{\rho_{eqvt}}}\,. \tag{4.323}$$

4.5.6.2 Trilaminar Beam Nonuniform over the Length

The beams of variants shown in Figure 4.20 (b) and (c) must be treated as nonuniform over the length. We will distinguish them by subscripts A (the case that parts of ceramics are replaced by a passive material) and B (the case that the parts of ceramics are removed). The length and thickness of layers of active material are denoted l_{a} and δ.

The unknown distribution of displacement of the neutral surface in these cases can be represented by the series in terms of the normal modes of vibration of uniform bar. Assuming that the beam is simply supported

$$\xi(x)=\sum_{i=1}^{2m+1}\xi_{i}\cos(i\pi x/l)=\sum_{i=1}^{2m+1}\xi_{i}\theta_{i}(x)\quad(m=1,2,...) \tag{4.324}$$

For transducers designing applications the lowest mode of vibration is of primary interest. It is logical to suggest that to the first approximation this mode should be close to the first mode of the uniform beam. This is moreover that the ratio l_{a}/l cannot be too small from the practical considerations. In order to find out whether this assumption leads to sufficiently accurate results, or a contribution of the higher modes is significant, we will at first take into calculation the first two terms of the series (4.324). Thus, we assume that

$$\xi(x)=\xi_{1}\cos(\pi x/l)+\xi_{3}\cos(3\pi x/l)\,. \tag{4.325}$$

After substituting this expression for $\xi(x)$ into general formulas (4.313) and (4.314) for the potential and kinetic energies the following results will be obtained.

The kinetic energy of the beam in the case A can be represented as

$$W_{kin\,tA} = \frac{1}{2} wt\rho_{eqvt} \int_{-l/2}^{l/2} [\dot{\xi}_1 \cos(\pi x/l) + \dot{\xi}_3 \cos(3\pi x/l)]^2 dx +$$
$$+4w\delta(\rho_p - \rho_a) \int_{-l_a/2}^{l/2} [\dot{\xi}_1 \cos(\pi x/l) + \dot{\xi}_3 \cos(3\pi x/l)]^2 dx. \quad (4.326)$$

After performing calculations of the integrals, the expression for the kinetic energy will be

$$W_{kin\,tA} = \frac{1}{2}\left[\frac{wtl}{2}\rho_{eqvt}(\dot{\xi}_1^2 + \dot{\xi}_3^2) + wl\delta(\rho_p - \rho_a)(\dot{\xi}_1^2 F_1 + \dot{\xi}_3^2 F_3 - \frac{2}{\pi}\dot{\xi}_1\dot{\xi}_3 F_{13})\right], \quad (4.327)$$

where

$$F_1(l_a/l) = 1 - (l_a/l) - (1/\pi)\sin(\pi l_a/l), \quad (4.328)$$

$$F_3 = 1 - (l_a/l) - (1/3\pi)\sin(3\pi l_a/l), \quad (4.329)$$

$$F_{13} = \sin(\pi l_a/l) + (1/2)\sin(2\pi l_a/l). \quad (4.330)$$

The kinetic energy by expression (4.326) can be presented in the form

$$W_{kin\,tA} = \frac{1}{2} M_{1tA}\dot{\xi}_1^2 + M_{13tA}\dot{\xi}_1\dot{\xi}_3 + \frac{1}{2} M_{3tA}\dot{\xi}_3^2. \quad (4.331)$$

Here and further the following notations are introduced for the equivalent parameters of the trilaminar beams: K_{1t}, M_{1t} and K_{3t}, M_{3t} for the first and third modes of vibration of uniform over length beam; K_{1tA}, M_{1tA} and K_{3tA}, M_{3tA} - for the beam of modification *A*; K_{1tB}, M_{1tB} and K_{3tB}, M_{3tB} - for the beam of modification *B*. Notations with subscripts 13*t A* and 13*t B* are used for the mutual rigidities and masses between the first and third modes

Thus, in the expression (4.331)

$$M_{1tA} = M_{1t}\left[1 + \frac{2(\gamma_\rho - 1)y}{\gamma_\rho - 2(\gamma_\rho - 1)y} F_1\right], \quad (4.332)$$

$$M_{3tA} = M_{3t}\left[1 + \frac{2(\gamma_\rho - 1)y}{\gamma_\rho - 2(\gamma_\rho - 1)y} F_3\right], \quad (4.333)$$

note that for a beam with simply supported ends $M_{3t} = M_{1t}$.

$$M_{13tA} = -M_{1t}\frac{2y(\gamma_\rho - 1)}{\pi[\gamma_\rho - 2(\gamma_\rho - 1)y]} F_{13}. \quad (4.334)$$

In expressions for masses in variant *B* subscript *A* should be replaced by *B*, and γ_ρ in the numerators must be set to zero in Eqs. (4.331)-(4.334).

Expression for the potential energy of the beam of variant A can be represented as

$$W_{pot\,tA} = \frac{wt^3}{24} Y_{eqvt} \int_{-l/2}^{l/2} \left[\xi_1 (\pi / l)^2 \cos(\pi x / l) + \xi_3 (3\pi / l)^2 \cos(3\pi x / l \right]^2 dx +$$

$$+4(Y_p - Y_a^E) \frac{wt^3}{24} [1-(1-2y)^3] \times \tag{4.335}$$

$$\times \int_{l_a/2}^{l/2} \left[\xi_1 (\pi / l)^2 \cos(\pi x / l) + \xi_3 (3\pi / l)^2 \cos(3\pi x / l) \right]^2 dx$$

This expression can be presented after manipulations analogous to those performed for the kinetic energy in the form

$$W_{pot\,tA} = \frac{1}{2} K_{1tA} \xi_1^2 + K_{13tA} \xi_1 \xi_3 + \frac{1}{2} K_{3tA} \xi_3^2, \tag{4.336}$$

where

$$K_{1tA} = K_{1t} \left[1 + \frac{(\gamma_Y - 1)[1-(1-2y)^3]}{1+(\gamma_Y - 1)(1-2y)^3} F_1 \right], \tag{4.337}$$

$$K_{3tA} = 81 K_{1t} \left[1 + \frac{(\gamma_Y - 1)[1-(1-2y)^3]}{1+(\gamma_Y - 1)(1-2y)^3} F_3 \right], \tag{4.338}$$

$$K_{13tA} = -\frac{9}{\pi} K_{1t} \frac{(\gamma_Y - 1)[1-(1-2y)^3]}{1+(\gamma_Y - 1)(1-2y)^3} F_{13}. \tag{4.339}$$

In the expressions for rigidities for variant B subscripts A should be replaced by B, and γ_Y in the numerators of formulas (4.337)-(4.339) must be set to zero.

Given that the kinetic and potential energies of the nonuniform beams are known, the Lagrange equations of free vibration of the beams can be derived, and the lower resonance frequency and mode of vibration to the second approximation can be determined. Namely, from

$$\frac{d}{dt}\left(\frac{\partial W_{kin}}{\partial \dot{\xi}_i} \right) - \frac{\partial W_{pot}}{\partial \xi_i} = 0 \quad (i = 1, 2, \ldots), \tag{4.340}$$

after substituting the expressions (4.331) and (4.336) for the kinetic and potential energies we arrive into equations for variant A

$$(K_{1tA} - \omega^2 M_{1tA}) \xi_1 + (K_{13tA} - \omega^2 M_{13tA}) \xi_3 = 0, \tag{4.341}$$

$$(K_{13tA} - \omega^2 M_{13tA}) \xi_1 + (K_{3tA} - \omega^2 M_{3tA}) \xi_3 = 0, \tag{4.342}$$

and analogous set of equations for variant B with replacement of subscript A by B.

The frequency equation, from which the lower resonance frequency to the second approximation may be found, is

$$(\omega_{1tA}^2-\omega^2)(\omega_{3tA}^2-\omega^2)-\frac{1}{M_{1tA}M_{3tA}}\left(K_{13tA}-\omega^2 M_{13tA}\right)^2=0\,. \tag{4.343}$$

Here,

$$\omega_{1tA}^2=K_{1tA}/M_{1tA}\text{ and }\omega_{3tA}^2=K_{3tA}/M_{3tA}\,. \tag{4.344}$$

After the lower resonance frequency ω_I is obtained, the ratio of displacements (or the mode shape coefficient, ms) can be determined from either of Eqs. (4.341) or (4.342), as

$$ms_I=\left.\frac{\xi_3}{\xi_1}\right|_{\omega=\omega_I}=-\frac{K_{1tA}-\omega_I^2 M_{1tA}}{K_{13tA}-\omega_I^2 M_{13tA}}\,. \tag{4.345}$$

The distribution of vibration at the resonance frequency ω_I now can be represented as

$$\xi(x)=\xi_1[\cos(\pi x/l)+ms_I\cos(3\pi x/l)]\,. \tag{4.346}$$

Thus, the mode shape of vibration to the second approximation is

$$\theta(x)=\frac{\xi(x)}{\xi(0)}=\frac{1}{1+ms_I}[\cos(\pi x/l)+ms_I\cos(3\pi x/l)]\,. \tag{4.347}$$

Analogous calculations may be performed for variant B.

Results of calculating parameters of nonuniform trilaminar beams made according to the above analysis are presented in Figure 4.22 and Figure 4.23 for the first and second approximations.

All the calculations are performed for the relative thicknesses of the active material $y=\delta/t$ that correspond to the maximum of the coupling coefficients, as this will be illustrated in Chapter 9 for different passive materials used. The reason behind this is that the reduction of the volume of active material presupposes that these transducers are not intended for radiation of the maximum possible power in case it is limited by the electric field. At this condition optimizing the coupling coefficient is the reasonable choice. Further reduction of piezoelement thickness may also be appropriate depending on requirements for transducer operation, but in this case the effect of nonuniformity on the results of calculating the parameters obviously will be lesser.

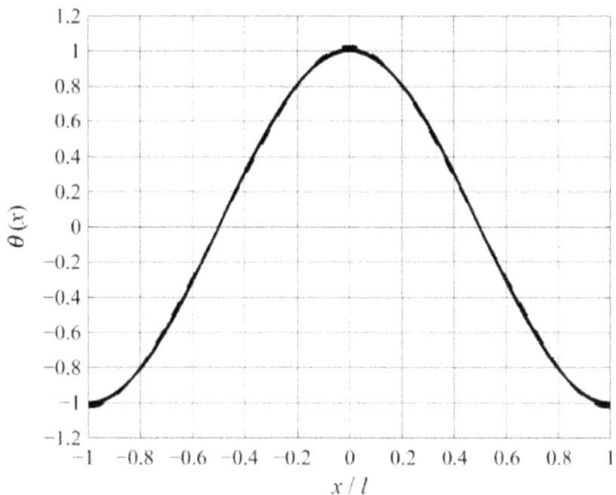

Figure 4.22: The distribution of displacements (mode shapes $\theta(x)$) determined to the second approximation at $l_a/l = 0.6$ in comparison with $\theta(x) = \cos(\pi x/l)$ for the first approximation. Calculations are made for case *B* and combinations of PZT-4 with aluminum and steel. Deviations of the results are within the shaded areas.

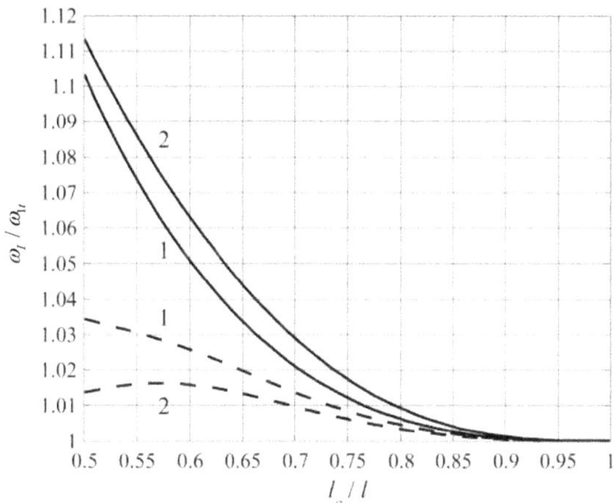

Figure 4.23: Ratio of the resonance frequencies of nonuniform over the length trilaminar beams calculated to the second (ω_I) and to the first (ω_{1t}) approximations. Solid lines – $(\omega_I/\omega_{1t})_A$, dashed lines – $(\omega_I/\omega_{1t})_B$. Combinations of aluminum – PZT-4 are labeled as number 1, and combinations of steel – PZT-4 are labeled as number 2.

The following conclusions can be drawn from the results presented in Figure 4.22 and Figure 4.23. The main result is that the mode shape of the trilaminar nonuniform beam to the second approximation remains practically the same as for the uniform beam at least up to values

$l_a/l = 0.6$. This is illustrated in Figure 4.22 for the thickness of an active layer that is optimal in terms of the effective coupling coefficient of a corresponding piezoelement, even for the most critical case that the parts of active material are removed from the ends of the beam (variant B). The mode shape coefficients by formula (4.347) that are presented in Figure 4.22 support this conclusion, as their plots show that the contribution of the third mode of vibration is very small. As the result the resonance frequency of the beams calculated to the second approximation don't deviate from those determined to the first approximation by more than 5% up to values $l_a/l \approx 0.6$. It is of note that at $l_a/l < 0.6$ the effective coupling coefficients starts to drop, as it will be shown in Chapter 9, and further reducing of the length of piezoelement does not make sense. Thus, the results of calculating the equivalent parameters M_{1tA}, K_{1tA} and M_{1tB}, K_{1tB} of the nonuniform beams obtained to the first approximation from formulas (4.332) and (4.337), and the resonance frequencies determined with their application are valid in all the practically reasonable transducer designing range of changing the length of the active element.

4.5.6.3 Bilaminar Beam Uniform over the Length

The peculiarity of this case is that the neutral plane under bending does not coincide with the middle plane, as it was in the case of the symmetrical trilaminar beam. Thus, as the first step the location of the neutral plane (coordinate z_0) must be determined. By definition the neutral plane should be free of stress, and therefore its coordinate z_0 may be found from the condition that

$$\int_0^{z_0} T_x dz + \int_{z_0}^{t} T_x dz = 0 , \tag{4.348}$$

where

$$T_x = -(z - z_0)\xi_0 Y(z)\frac{d^2\theta}{dx^2} . \tag{4.349}$$

If the thickness of the active part is δ and the modules of the active and passive parts are Y_a^E and Y_p, then condition (4.348) is equivalent to

$$\int_0^{\delta} Y_a^E (z - z_0)dz + \int_{\delta}^{t} Y_p (z - z_0)dz = 0 , \tag{4.350}$$

where from

$$\frac{z_0}{t} = \frac{1}{2}\frac{\gamma_Y + (1-\gamma_Y)y^2}{\gamma_Y + (1-\gamma_Y)y}. \tag{4.351}$$

Here $\gamma_Y = Y_p / Y_a^E$ and $y = \delta / t$, as denoted by expressions (4.317).

For a rational transducer design z_0 should be greater than δ (otherwise the electromechanical effects in the piezoelectric element above and below the neutral plane would be in opposite phase). We denote the value of z_0 that is equal to δ as z_{0m}. Obviously, $z_{0m} = \delta_{\max}$ is the maximum reasonable thickness of piezoceramic layer for a given combination of active and passive materials. From equation (4.351) we obtain

$$\frac{z_{0m}}{t} = \frac{\sqrt{\gamma_Y}}{1+\sqrt{\gamma_Y}}. \tag{4.352}$$

The ratio z_{0m}/t for different combinations of active and passive materials is given in Table 4.4. After the position of the neutral plane is determined, the equivalent parameters of a transducer with the piezoelectric element of different relative thickness can be calculated.

The potential energy of a bilaminar beam is

$$\begin{aligned} W_{potb}^E &= \frac{1}{2}\xi_0^2 w\left[\int_0^\delta (z-z_0)^2 Y_a^E dz + \int_\delta^t (z-z_0)^2 Y_p dz\right]\int_{-l/2}^{l/2}\left(\frac{d^2\theta}{dx^2}\right)^2 dx = \\ &= \frac{\xi_0^2}{2}\frac{w}{3}\left\{\left[(\delta-z_0)^3 + z_0^3\right]Y_a^E + \left[(t-z_0)^3 + (z_0-\delta)^3\right]Y_p\right\}\int_{-l/2}^{l/2}\left(\frac{d^2\theta}{dx^2}\right)^2 dx. \end{aligned} \tag{4.353}$$

Table 4.4: Ratio $(z_{0m}/t) = y_{\max}$ for different combination of materials.

			PZT-4		
Passive Material	Aluminum	Steel	Glass	G-10	Alumina
Y_a^E, 10^9 N/m^2			For PZT-4 is 81		
Y_p, 10^9 N/m^2	70	210	62	24	300
$(z_{0m}/t) = y_{\max}$	0.48	0.62	0.47	0.35	0.66

We denote the equivalent Young's modulus as

$$Y_{eqvb}(z_0) = 4\left\{\left[\left(\frac{\delta}{t}-\frac{z_0}{t}\right)^3 + \left(\frac{z_0}{t}\right)^3\right]Y_a^E + \left[\left(1-\frac{z_0}{t}\right)^3 + \left(\frac{z_0}{t}-\frac{\delta}{t}\right)^3\right]Y_p\right\}. \tag{4.354}$$

In the particular case that $z_0 = z_{0m} = \delta_{\max}$

$$Y_{eqvb}(z_{0m}) = 4\left[Y_a^E\left(\frac{z_{0m}}{t}\right)^3 + Y_p\left(1-\frac{z_{0m}}{t}\right)^3\right] = 4Y_a^E[y^3 + (1-y)^3\gamma_Y]. \tag{4.355}$$

For the combinations of the active and passive materials listed in Table 4.4, the plots for $Y_{eqvb}(z_0/t)$ vs. δ/t are represented in Figure 4.24. Expression (4.322) for the equivalent rigidity of the uniform over the length trilaminar beam is valid for a bilaminar beam under the assumption that Y_{eqvt} is replaced by Y_{eqvb} that is given by formula (4.354) (or (4.355) in the case that $z_0 = z_{0m}$).

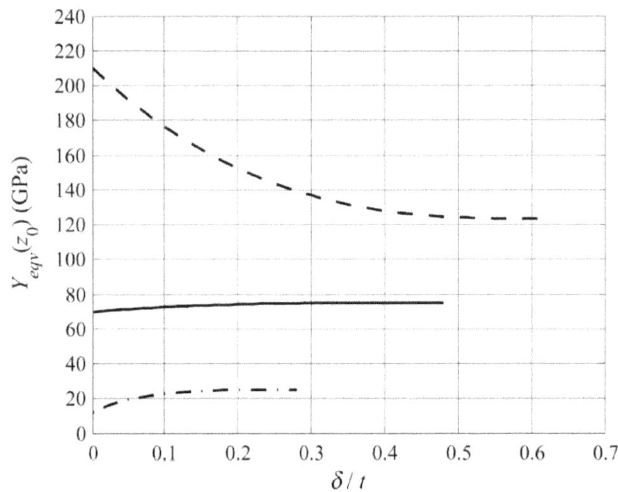

Figure 4.24: The equivalent Young's modulus as a function of δ/t (up to $\delta = z_{0m}$) for different combinations of PZT-4 ceramics with: aluminum (solid line), steel (dashed line) and G-10 (dot-dash line).

The kinetic energy of a bilaminar beam is

$$W_{kinb} = \frac{\dot{\xi}_0^2}{2} w \int_{-l/2}^{l/2} \left[\rho_a\delta + \rho_p(t-\delta)\right]\theta^2(x)dx = \frac{\dot{\xi}_0^2}{2}\rho_{eqvb} t S_{eff} = \frac{\dot{\xi}_0^2}{2} M_{eqv}, \tag{4.356}$$

where

$$\rho_{eqvb} = \frac{\rho_a\delta + \rho_p(t-\delta)}{t} = \rho_a[y + (1-y)\gamma_\rho] = \rho_a[\gamma_\rho - (\gamma_\rho - 1)y] \tag{4.357}$$

and S_{eff} is defined by formula

$$S_{eff} = w \int_{-l/2}^{l/2} \theta^2(x)dx \,. \tag{4.358}$$

For the beams with simply supported ends at the first mode of vibration

$$[\theta(x) = \theta_1(x) = \cos(\pi x / l)] \quad S_{eff} = S_{eff1} = 0.5wl \,. \tag{4.359}$$

Using expressions (4.354) and (4.357) for the equivalent rigidity and mass we arrive at the general formula for the resonance frequency of the bilaminar beam

$$f_b(z_0) = f_{ra}\sqrt{\frac{\rho_a}{Y_a^E}} \cdot \sqrt{\frac{Y_{eqvb}(z_0)}{\rho_{eqvb}(z_0)}} \,, \tag{4.360}$$

where f_{ra} is the resonance frequency of the fully active beam. In the case that $z_0 = z_{0m}$

$$f_{rb}(z_{0m}) = f_{ra}\frac{2\sqrt{\gamma_Y}}{\sqrt{\left(1+\sqrt{\gamma_Y}\right)\left[\sqrt{\gamma_Y}+\gamma_\rho\right]}} \,. \tag{4.361}$$

4.5.6.4 Bilaminar Beams Nonuniform over the Length

The same considerations on optimizing effective coupling coefficient can be followed regarding the nonuniform bilaminar beam design, as in the case of the nonuniform trilaminar symmetric beam. The main peculiarity of this case is that exact position of the neutral surface throughout the length of the beam can't be predicted. We will assume that the neutral surfaces are positioned individually within the passive ends and bilaminar part of the beam. Namely, in the bilaminar part, as they were determined for the uniform bilaminar beam, and in the end parts, as they were determined for uniform passive beam. This assumption ignores the irregularities that take place on the boarders between the parts, which may be a source of additional errors. The assumption is based on belief that influence of these irregularities may be noticeable in a close proximity to the boarders only. But, anyway, the results obtained under this assumption require experimental or Finite Element Analysis verification.

It is noteworthy that determining position of the neutral surface in a particular cross section of a beam, as well as determining an optimal correlation between thicknesses of the active and passive layers in terms of maximizing the effective coupling coefficient depend from distribution of properties of materials over the thickness only, so far as elementary theory of bending is applicable. Therefore, all the results obtained in this regard for uniform bilaminar beam in the preceding section remain valid.

After the assumption regarding location of the neutral surfaces in the parts of a beam is accepted, procedure of successive approximations to calculating the electromechanical parameters of the partially active bilaminar beam can be applied in the same way as it was done in the case of trilaminar beam. Two variants of nonuniform over the length design that are shown in Figure 4.21 (b) and (c) have practical sense. They will be labeled, as variants *A* and *B*.

At first, consider the first approximation assuming that distribution of displacement is $\xi(x) = \xi_0 \cos(\pi x / l)$. Determine the energy state of the vibrating beam at this mode of vibration. The kinetic energy is:

for variant *A* (parts of ceramic are replaced by a passive material)

$$W_{kin\,bA} = \frac{\dot{\xi}_0^2}{2} w \left[t\rho_{eqvb} \int_{-l/2}^{l/2} \cos^2(\pi x / l) dx + 2\delta(\rho_p - \rho_a) \int_{l_a/2}^{l/2} \cos^2(\pi x / l) dx \right], \tag{4.362}$$

and for variant *B* (parts of ceramic are removed)

$$W_{kin\,bB} = \frac{\dot{\xi}_0^2}{2} w \left[t\rho_{eqvb} \int_{-l/2}^{l/2} \cos^2(\pi x / l) dx - 2\delta\rho_a \int_{l_a/2}^{l/2} \cos^2(\pi x / l) dx \right]. \tag{4.363}$$

After performing integrations and some manipulations we obtain the following expressions for the equivalent masses:

$$M_{b1A} = M_{b1} \left[1 + 2y \frac{\gamma_\rho - 1}{\gamma_\rho - (\gamma_\rho - 1)y} F_1 \right], \tag{4.364}$$

$$M_{b1B} = M_{b1} \left[1 - \frac{2y}{\gamma_\rho - (\gamma_\rho - 1)y} F_1 \right], \tag{4.365}$$

where

$$M_{b1} = M_{eqv} = [\gamma_\rho - (\gamma_\rho - 1)y] t S_{eff1}. \tag{4.366}$$

The potential energy for the variant *A* is

$$W_{pot\,bA} = \frac{\xi_0^2}{2} w \left[\frac{\pi^4 t^3}{24 l^3} Y_{eqvb} + 2(Y_p - Y_a^E) \int_0^{\delta} z^2 dz \int_{l_a/2}^{l/2} \frac{\pi^4}{l^4} \cos^2(\pi x / l) dx \right]. \tag{4.367}$$

For the variant *B* it must be set $Y_p = 0$ in expression (4.303), where Y_{eqvb} is in general determined by formula (4.354). In the case that $z_0 = z_{0m} = \delta$, the expression for Y_{eqvb} simplifiers to formula (4.355). It is noteworthy that this latter case, in which the most of amount of active material is replaced by a passive, is the most representative in terms of margins of accuracy of the first approximation in comparison with the second approximation to a real mode of vibration. Determining these margins is one of the goals of our treatment. For this case after performing integrations and some manipulations we obtain the following expressions for the equivalent rigidities: for the variant A

$$K_{b1A} = K_{b1}\left[1+\frac{(\gamma_Y-1)y^3}{y^3+(1-y)^3\gamma_Y}F_1\right], \tag{4.368}$$

and for the variant B

$$K_{b1B} = K_{b1}\left[1-\frac{y^3}{y^3+(1-y)^3\gamma_Y}F_1\right], \tag{4.369}$$

where

$$K_{b1} = \frac{\pi^4 wt^3 Y_{eqvb}}{24l^3}. \tag{4.370}$$

Consider now the second approximation assuming that distribution of displacement on the surface of a vibrating beam is

$$\xi(x) = \xi_1\cos(\pi x/l)+\xi_3\cos(3\pi x/l). \tag{4.371}$$

As the most significant changes of the distribution of displacements and hence changes in values of the equivalent parameters can be expected for the design of Figure 4.21 (c), in which case the nonuniformity is the strongest, we will consider this modification of nonuniform beam at first. In the same manner, as it was done for the nonuniform trilaminar beam, after substituting the two-term expression for the displacement into relations for the corresponding energies and performing manipulations we will arrive at the following expressions for the equivalent parameters of a bilaminar nonuniform beam to the second approximation.

For the equivalent masses:

$$M_{b1B} = M_{b1}\left[1-\frac{2y}{\gamma_\rho-(\gamma_\rho-1)y}F_1\right], \tag{4.372}$$

$$M_{b3B} = M_{b3}\left[1 - \frac{2y}{\gamma_\rho - (\gamma_\rho - 1)y} F_3\right] \tag{4.373}$$

(note that for the simply supported beams $M_{b3} = M_{b1}$), and

$$M_{b13B} = M_{b1} \frac{2y}{\pi[\gamma_\rho - (\gamma_\rho - 1)y]} F_{13}\,. \tag{4.374}$$

For the equivalent rigidities:

$$K_{b1B} = K_{b1}\left[1 - \frac{y^3}{y^3 + (1-y)^3 \gamma_Y} F_1\right], \tag{4.375}$$

$$K_{b3B} = 81K_{b1}\left[1 - \frac{y^3}{y^3 + (1-y)^3 \gamma_Y} F_3\right], \tag{4.376}$$

$$K_{b13B} = \frac{9}{2\pi} K_{b1} \frac{y^3}{y^3 + (1-y)^3 \gamma_Y} F_{13}\,. \tag{4.377}$$

As far as all of the equivalent parameters of a nonuniform bilaminar beam are determined to the second approximation, the frequency equation (4.343), formulas for calculating mode shape coefficients and resulting mode of vibration (4.344) and (4.346) can be used for calculating all the characteristics of the beam and for comparing with values of analogous characteristics determined to the first approximation. Results of the calculations are presented in Figure 4.25 and Figure 4.26. (Note that Figure 4.25 is practically the same as Figure 4.22 for the trilaminar beam).

The same conclusion as for the nonuniform trilaminar beams can be made following the results of calculations presented in the figures that the mode shapes of vibration of bilaminar nonuniform beams remain the same as for a uniform beam. This conclusion is valid for all the practically reasonable values of relative thickness of active material. $y < z_{0m}/t$, even to greater extend of ratios l_a/l than for the nonuniform trilaminar beams. And all the parameters of transducers that employ the nonuniform bilaminar beams including the resonance frequencies and effective coupling coefficients can be calculated with sufficient accuracy using the equivalent parameters determined to the first approximation.

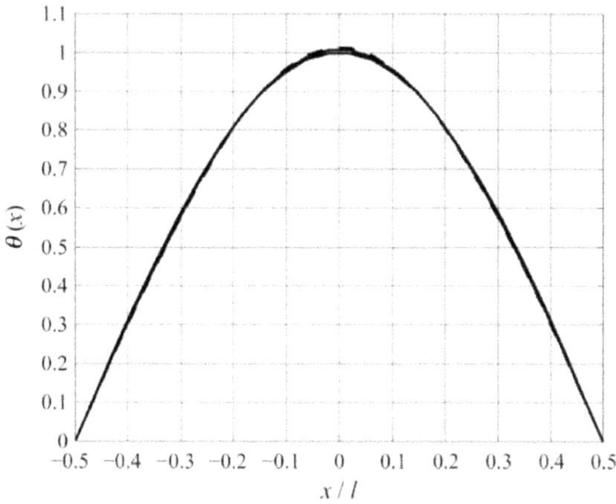

Figure 4.25: The distribution of displacements (mode shapes $\theta(x)$) determined to the second approximation at $l_a / l = 0.6$ in comparison with $\theta(x) = \cos x$ used to the first approximation.

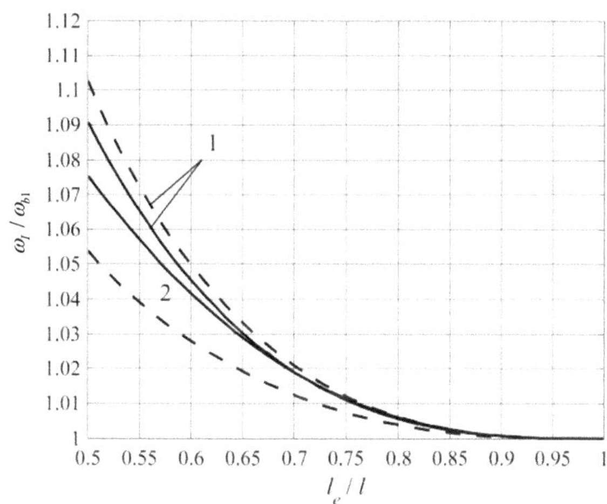

Figure 4.26: Ratio ω_l / ω_{b1} of the resonance frequencies of nonuniform bilaminar beams calculated to the second (ω_l) and to the first (ω_{b1}) approximations. Solid lines – $(\omega_l / \omega_{b1})_A$, dashed lines – $(\omega_l / \omega_{b1})_B$. Combinations of aluminum – PZT-4 are labeled as number 1, and combinations of steel – PZT-4 are labeled as number 2.

4.5.7 Flexural Vibration of Nonuniform Circular Plates

The general expressions for the kinetic and potential energies of a vibrating plate having nonuniform properties over the volume are as follows.

The kinetic energy is

$$W_{kin} = \frac{1}{2}\int_{\tilde{V}} \rho\dot{\xi}(r)^2 d\tilde{V} = \frac{1}{2}\dot{\xi}_0^2 2\pi \int_0^a \left[\int_0^t \rho(z,r)dz\right]\theta^2(r/a)rdr = \frac{1}{2}\dot{\xi}_0^2 M_{eqv}, \tag{4.378}$$

where $\rho(z,r)$ is the density of material, and $\rho(z,r)=\rho_a$ for the active parts and $\rho(z,r)=\rho_p$ for the passive parts of the plate.

The potential energy is

$$W_{pot}^E = \frac{1}{2}\xi_0^2 2\pi \int_0^a\int_0^t (z-z_0)^2 Y_\sigma(z,r)\times \\ \times\left[\left(\frac{d^2\theta}{dr^2}\right)^2 + 2\sigma(z,r)\frac{1}{r}\frac{d\theta}{dr}\frac{d^2\theta}{dr^2} + \left(\frac{1}{r}\frac{d\theta}{dr}\right)^2\right]rdrdz, \tag{4.379}$$

where for brevity the modified elastic modulus is introduced,

$$Y_\sigma = \frac{Y}{1-\sigma^2}, \tag{4.380}$$

as the combination of the Young's modulus and Poison's ratio for the material. On the passive and active parts of a plate the modified elastic moduli are, respectively,

$$Y_{\sigma p} = \frac{Y_p}{1-\sigma_p^2} \text{ and } Y_{\sigma a} = \frac{Y_1^E}{1-\sigma_1^{E2}}. \tag{4.381}$$

The integrals over volume $\tilde{V}$ in the expressions (4.378) and (4.379) are the general representations for the energies. Further they are specified for the circular plates vibrating in the first mode, i.e., at $\xi(r)=\xi_0\theta(r/a)$.

4.5.7.1 Vibration of the Radially Uniform Plates

In this section we will consider radially uniform though may be nonuniform by the thickness circular plates. That is $Y_\sigma(z,r)=Y_\sigma(z)$, $\sigma(z,r)=\sigma(z)$, and $\rho(z,r)=\rho(z)$. Peculiarity of expression (4.379) for the potential energy is that the function within brackets strictly speaking may be different depending on what values of Poisson's ratios have the passive and active materials used in the mechanical system. But it can be shown that these expressions can be unified to a great accuracy for the modern compositions of PZT piezoceramics and for the passive materials used in the transducer designs intended for underwater applications. Thus, it can be adopted value $\sigma = 0.3$ for Poison's ratio for all the ideal boundary conditions. With this goal consider integral

$$L(\sigma) = 2\pi \int_0^a \left[\left(\frac{d^2\theta}{dr^2} \right)^2 + 2\sigma \frac{1}{r} \frac{d\theta}{dr} \frac{d^2\theta}{dr^2} + \left(\frac{1}{r} \frac{d\theta}{dr} \right)^2 \right] rdr \, . \tag{4.382}$$

Let us denote for brevity

$$L_I = \int_0^a \left[\left(\frac{d^2\theta}{dr^2} \right)^2 + \left(\frac{1}{r} \frac{d\theta}{dr} \right)^2 \right] rdr \, , \tag{4.383}$$

$$L_{II} = \int_0^a \frac{1}{r} \frac{d\theta}{dr} \frac{d^2\theta}{dr^2} rdr = \frac{1}{2} \left(\frac{d\theta}{dr} \right)^2 \Bigg|_0^a \, . \tag{4.384}$$

Obviously, $L(\sigma) = L_I + 2\sigma L_{II}$. Thus, the relative error due to replacement $L(\sigma)$ by $L(0.3)$ is

$$\frac{L(\sigma) - L(0.3)}{L(0.3)} = 2(\sigma - 0.3) \frac{L_{II}}{L_I + 0.6 L_{II}} \, . \tag{4.385}$$

For evaluating the possible error for the case of simply supported boundary we will use expression for the mode shape by formula (2.150)

$$\theta(r/a) = (1 - r^2 / a^2)(1 - r^2 / 4a^2) \, . \tag{4.386}$$

Substituting this mode shape in formulas (4.383) and (4.384) results in values $L_I = 2.92 / a^2$ and $L_{II} = 1.12 / a^2$. Thus

$$\frac{L(\sigma) - L(0.3)}{L(0.3)} \approx 0.6(\sigma - 0.3) \tag{4.387}$$

and for values of Poisson's ratio $0.25 < \sigma < 0.35$ variation of modulus of this quantity is less than 0.03. In the case of clamped boundary $L_{II} = 0$, thus $L(\sigma) = L_I$ is independent of σ.

For the free boundary substituting the mode shape by formula (4.200) into integrals (4.383) and (4.384) results in

$$\frac{L(\sigma) - L(0.3)}{L(0.3)} \approx 0.2(\sigma - 0.3) \, , \tag{4.388}$$

and the variation is less than 1%.

After this discussion the expression (4.379) for the potential energy of the radial uniform circular plate can be represented as

$$W_{pot}^{E} = \frac{1}{2}\xi_0^2\left[\int_0^t Y_\sigma(z)(z-z_0)^2 dz\right]\cdot L(0.3)_{bc} = \frac{1}{2}\xi_0^2\frac{t^3 Y_{\sigma eqv}}{12}\cdot L(0.3)_{bc} = \frac{1}{2}\xi_0^2 K_{eqv}^{E}, \quad (4.389)$$

where subscript *bc* indicates that the integral must be calculated with mode shape that correspond to a certain boundary condition, and it is defined that

$$Y_{\sigma eqv} = \frac{12}{t^3}\int_0^t Y_\sigma(z)(z-z_0)^2 dz \quad (4.390)$$

is the equivalent Young's modulus of the plate with nonuniform through the thickness elastic properties. This is the common definition for all the mechanical systems that experience flexural deformation (see examples of beams and rings).

Thus, from relation (4.389) the equivalent rigidity of the plate at particular combination of elastic moduli of materials used, and under a certain boundary conditions (*bc*), to which $L(0.3)_{bc}$ corresponds, is

$$K_{eqv}^{E} = \frac{t^3 Y_{\sigma eqv}}{12}\cdot L(0.3)_{bc}\,. \quad (4.391)$$

For the simply supported boundary $L(0.3)_{ss} = 7.2\pi / a^2$, for the clamped boundary $L(0.3)_{cl} = 21.4\pi / a^2$, and for the free boundary $L(0.3)_{free} = 22\pi / a^2$.

Expression (4.378) for the kinetic energy in case of radial uniform plate can be represented as

$$W_{kin} = \frac{1}{2}\dot{\xi}_0^2 S_{eff} t \rho_{eqv} = \frac{1}{2}\dot{\xi}_0^2 M_{eqv} \quad (4.392)$$

where

$$\rho_{eqv} = \frac{1}{t}\int_0^t \rho(z)dz \quad (4.393)$$

is the equivalent density of nonuniform through the thickness plate, and

$$S_{eff} = 2\pi\int_0^a \theta^2(r/a)rdr \quad (4.394)$$

is the effective surface area of the plate. Thus, from (4.392)

$$M_{eqv} = \rho_{eqv} t S_{eff} \,. \tag{4.395}$$

All the results for the equivalent Young's moduli and densities of the circular plates nonuniform through the thickness are the same as presented in Sections 4.5.6.1 and 4.5.6.3 for the beams having nonuniformity of the same kind. Namely, for the trilaminar plates Y_{eqvt} and ρ_{eqvt} are given by Eqs. (4.319) and (4.321); for the bilaminar plates $Y_{eqvb}(z_{0m})$ and ρ_{eqvb} are given by Eqs. (4.355) and (4.357).

4.5.7.2 Flexural Vibration of the Radially Nonuniform Circular Plates

Two modifications of the radially nonuniform plates have practical sense: the one with active material partially replaced by passive material, as shown by the cross shaded parts in Figure 4.21 (b)- (c), in which the coordinate x, l and l_a must be changed to r, a and r_a, respectively) that is labeled A, and the case with these parts of material removed labeled B. The trilaminar plates are the most widely used in A modification. The bilaminar plates are often used in B modification as well. In terms of demonstrating the approach to calculating parameters of the radially nonuniform plates the most representative are the bilaminar plates of B modification, as they have the most pronounced nonuniformity. Besides transducer design of this type is especially widely used for in air applications. Therefore, just this modification will be considered in detail for brevity. This is moreover reasonable that calculating procedures for other modifications may be the same, as it will be seen from the following analysis. From the same consideration of brevity all the results of calculations will be presented in this section for the most important case of simply supported boundary conditions, although expressions for all the transducer parameters are general. The same assumption will be made as in the case of nonuniform beams that positions of the neutral planes in the fully passive part and in the bilaminar part of the plate may be determined in the same way as for the corresponding radially uniform plates. The most pronounced is nonuniformity of the bilaminar plate with active laminate having the maximum reasonable thickness for a given combination of active and passive materials, i.e., in the case that $\delta = z_{0m}$. Therefore, estimation of the strongest influence of nonuniformity will be made for the value of $y = z_{0m} / t$. As calculating the equivalent parameters becomes straightforward after the mode of vibration is known, the modes of vibration of the radially nonuniform bilaminar plates have to be determined. At first the assumption can be made that the mode of vibration remains the same as for uniform by radius plate, i.e., that $\xi(r) = \xi_1\theta_1$. But this assumption can be considered valid only to the first approximation and for limited ranges of relative radii and thicknesses of the active and passive layers, and these limits may be different for different boundary conditions. In order to estimate the level of accuracy of the first approximation, the second approximation must be considered by representing the displacement distribution over the surface of the plate as superposition of two normal modes of vibration, namely,

$$\xi(r) = \xi_1\theta_1 + \xi_2\theta_2\,, \tag{4.396}$$

where it is denoted for brevity $\theta_1 = \theta(k_1 r)$ and $\theta_2 = \theta(k_2 r)$, as the normal modes of vibration for the corresponding boundary conditions. Finally, the relative contribution of the second mode has to be determined for the range of reasonable relative dimensions of the active and passive parts of the mechanical system.

4.5.7.3 Equivalent Parameters of the Radially Nonuniform Bilaminar Plates to the First Approximation

Using the results obtained in Section 4.5.6.4 for the equivalent Young's modulus of bilaminar beams (plates) and assuming that $\sigma(z,r)$ is replaced by $\sigma = 0.3$, the expression for the potential energy of the radially nonuniform bilaminar plate can be represented as

$$\begin{gathered} W_{pot} = \frac{\xi_1^2}{2} 2\pi \left\{ \frac{t^3}{12} Y_{eqvb} \int_0^a F(\theta_1) r dr - \right. \\ \left. - \frac{t^3}{12} Y_{eqvb} \int_{r_a}^a F(\theta_1) r dr + \frac{t_p^3}{12} Y_{\sigma p} \int_{r_a}^a F(\theta_1) r dr \right\} = \frac{\xi_1^2}{2} K_{eqvb1} \end{gathered}, \tag{4.397}$$

where it is denoted for brevity

$$\left(\frac{d^2\theta}{dr^2}\right)^2 + 2\cdot 0.3 \frac{1}{r}\frac{d\theta}{dr}\frac{d^2\theta}{dr^2} + \left(\frac{1}{r}\frac{d\theta}{dr}\right)^2 = F(\theta)\,. \tag{4.398}$$

The first term in the brace of expression (4.397) forms the equivalent rigidity of uniform bilaminar plate, K_{eqvb}. The second term compensates for the addition made to the rigidity of the active–passive part of the plate that results in forming the equivalent rigidity of the uniform plate; the third term is the equivalent rigidity of the passive part of the plate that for the case B has thickness t_p. Further in this section we will omit "*eqv*" in subscripts of the notations for the equivalent rigidities and masses for brevity. Thus, for example, K_{eqvb} will read as K_b. After obvious manipulations the expression for the equivalent rigidity of the radially nonuniform bilaminar plate will be obtained as

$$K_{b1} = K_b\left[1-\left(1-\frac{(1-y)^3 Y_{\sigma p}}{Y_{eqvb}(z_0)}\right)\frac{F_{r1}}{L_1(0.3)}\right], \tag{4.399}$$

where in the general case $Y_{eqvb}(z_0)$ is determined by formula (4.354). For the case that $\delta = z_{0m}$,

$$K_{b1} = K_b\left[1-\left(1-\frac{\gamma_Y[1-(z_{0m}/t)]^3}{4(z_{0m}/t)^2}\right)\frac{F_{r1}}{L_1(0.3)}\right]. \tag{4.400}$$

Here in addition to expressions (4.398) the notations are introduced

$$F_{ri} = 2\pi\int_{r_a}^{a} F(\theta_i)rdr, \quad 2\pi\int_0^a F(\theta_i)rdr = L_i(\sigma) \quad (i=1,2,...). \tag{4.401}$$

The kinetic energy of the nonuniform plate is determined by expression

$$W_{kin} = \frac{1}{2}2\pi\left[t\rho_{eqvb}\int_0^a(\dot{\xi}_1\theta_1)^2 rdr - \delta\rho_a\int_{r_a}^a(\dot{\xi}_1\theta_1)^2 rdr\right] = \frac{1}{2}\dot{\xi}_1^2 M_{b1}, \tag{4.402}$$

where the first term is the equivalent mass of uniform bilaminar plate for the first mode of vibration, M_b, and the second term is the correction that takes account for nonuniformity of the plate. It follows from this relation after replacing ρ_{eqvb} by its expression (4.357) that

$$M_{b1} = M_b\left[1-\frac{y}{\gamma_\rho-(\gamma_\rho-1)y}\frac{S_{eff\,r1}}{S_{eff\,1}}\right]. \tag{4.403}$$

Here the notation for the effective partial area is introduced

$$S_{eff\,ri} = 2\pi\int_{r_a}^a \theta_i^2(r/a)rdr, \quad S_{eff\,i} = 2\pi\int_0^a \theta_i^2(r/a)rdr \quad (i=1,2,...) \tag{4.404}$$

4.5.7.4 Equivalent Parameters of Radially Nonuniform Plate to the Second Approximation

When considering the equivalent parameters to the second approximation, we must substitute the displacement $\xi(r)$ by formula (4.396) into expressions for the potential and kinetic energies of the bilaminar plate. This will result in

$$W_{pot}=\frac{1}{2}2\pi\left\{\frac{t^3}{12}Y_{eqvb}\int_0^a F(\xi_1\theta_1+\xi_2\theta_2)rdr-\frac{t^3}{12}Y_{eqvb}\int_{r_a}^a F(\xi_1\theta_1+\xi_2\theta_2)rdr\right.$$
$$\left.-\frac{t_p^3}{12}Y_{\sigma p}\int_{r_a}^a F(\xi_1\theta_1+\xi_2\theta_2)rdr\right\}=\frac{1}{2}(\xi_1^2K_{b1}+2\xi_1\xi_2+\xi_2^2K_{b2}). \tag{4.405}$$

Note that following expressions (4.400) and (4.401)

$$F(\xi_1\theta_1+\xi_2\theta_2)=\xi_1^2F(\theta_1)+2\xi_1\xi_2F(\theta_1,\theta_2)+\xi_2^2F(\theta_2), \tag{4.406}$$

$$F(\theta_1,\theta_2)=\frac{d^2\theta_1}{dr^2}\frac{d^2\theta_2}{dr^2}+0.3\frac{1}{r}\left(\frac{d\theta_1}{dr}\frac{d^2\theta_2}{dr^2}+\frac{d\theta_2}{dr}\frac{d^2\theta_1}{dr^2}\right)+\frac{1}{r^2}\frac{d\theta_1}{dr}\frac{d\theta_2}{dr}. \tag{4.407}$$

After performing manipulations similar to those made to the first approximation, in course of which it is taken into consideration that the functions θ_1 and θ_2 are orthogonal in the interval $[0\le r\le a]$, the following results for the rigidities will be obtained:

K_{b1} is determined by formula (4.399),

$$K_{b2}=K_b\frac{L_2(0.3)}{L_1(0.3)}\left\{1-\left[1-\frac{\gamma_Y(1-y)^3}{4(z_{0m}/t)^2}\right]\frac{F_{r2}}{L_2(0.3)}\right\}, \tag{4.408}$$

$$K_{b12}=K_b\frac{F_{r12}}{L_1(0.3)}\frac{\gamma_Y(1-y)^3}{4(z_{0m}/t)^2}, \tag{4.409}$$

where

$$F_{r12}=2\pi\int_{r_a}^a F(\theta_1,\theta_2)rdr\ . \tag{4.410}$$

The kinetic energy to the second approximation is

$$W_{kin}=\frac{1}{2}2\pi\left[t\rho_{eqvb}\int_0^a(\dot{\xi}_1\theta_1+\dot{\xi}_2\theta_2)^2rdr-\delta\rho_a\int_{r_a}^a(\dot{\xi}_1\theta_1+\dot{\xi}_2\theta_2)^2rdr\right]=$$
$$=\frac{1}{2}(\xi_1^2M_{b1}+2\xi_1\xi_2M_{b12}+\xi_2^2M_{b2}), \tag{4.411}$$

where from the following expressions for the equivalent masses are obtained. The mass M_{b1} is determined by formula (4.403),

$$M_{b2}=M_b\frac{S_{eff\,2}}{S_{eff\,1}}\left[1-\frac{y}{\gamma_\rho-(\gamma_\rho-1)y}\frac{S_{eff\,r2}}{S_{eff\,2}}\right], \tag{4.412}$$

$$M_{b12} = -M_b \frac{yS_{eff\,r12}}{[\gamma_\rho - (\gamma_\rho - 1)y]S_{eff1}}, \tag{4.413}$$

where it is denoted

$$S_{eff\,r12} = 2\pi \int_{r_a}^{a} \theta_1 \theta_2 r dr . \tag{4.414}$$

4.5.7.5 Free Vibration of the Radially Nonuniform Plate as Two Degree of Freedom System

As far as expressions for the potential and kinetic energies of a radially nonuniform vibrating plate are known, the Lagrange's equations of free vibration of the plate can be derived and the lowest resonance frequency and the mode of vibration to the second approximation can be determined. Namely,

$$\frac{d}{dt}\left(\frac{\partial W_{kin}}{\partial \dot{\xi}_i}\right) - \frac{\partial W_{pot}}{\partial \xi_i} = 0, \quad (i = 1,2). \tag{4.415}$$

After substituting expressions (4.405) and (4.411) for the potential and kinetic energies these equations become

$$(K_{b1} - \omega^2 M_{b1})\xi_1 + (K_{b12} - \omega^2 M_{b12})\xi_2 = 0, \tag{4.416}$$

$$(K_{b12} - \omega^2 M_{b12})\xi_1 + (K_{b2} - \omega^2 M_{b2})\xi_2 = 0. \tag{4.417}$$

The frequency equation, from which the lower resonance frequency of the plate to the second approximation may be found, is

$$(1-\Omega)\left[\left(\frac{\omega_{b2}^2}{\omega_{b1}^2}\right) - \Omega\right] - \left(\frac{M_{b12}^2}{M_{b1}M_{b2}}\right)\left[\left(\frac{K_{b12}}{M_{b12}\omega_{b1}^2}\right) - \Omega\right]^2 = 0 \tag{4.418}$$

where $\omega_{b1}^2 = (K_{b1} / M_{b1})$, $\omega_{b2}^2 = (K_{b2} / M_{b2})$, and $\Omega = (f / f_{b1})^2$ is the relative frequency square. After the lowest value of Ω is obtained (we denote this quantity as $\Omega_I = (f_I / f_{b1})^2$, where f_I is the lower resonance frequency of the plate to the second approximation), the ratio of displacements ξ_2 and ξ_1 at this frequency (or the mode shape coefficient, ms_I) will be determined from Eq. (4.116), as

$$ms_I = \left.\frac{\xi_2}{\xi_1}\right|_{at\ f=f_I} = -\frac{M_{b1}}{M_{b12}} \frac{1-\Omega_I}{(K_{b12} / K_{b1}\omega_{b1}^2) - \Omega_I}. \tag{4.419}$$

The distribution of vibration at the first resonance frequency is. $\xi(r) = \xi_1[\theta_1(r) + ms_I\theta_2(r)]$. The mode shape of vibration to the second approximation can be presented now as

$$\theta(r) = \frac{\xi(r)}{\xi(0)} = \frac{1}{1+ms_I}[\theta_1(r) + ms_I\theta_2(r)]. \tag{4.420}$$

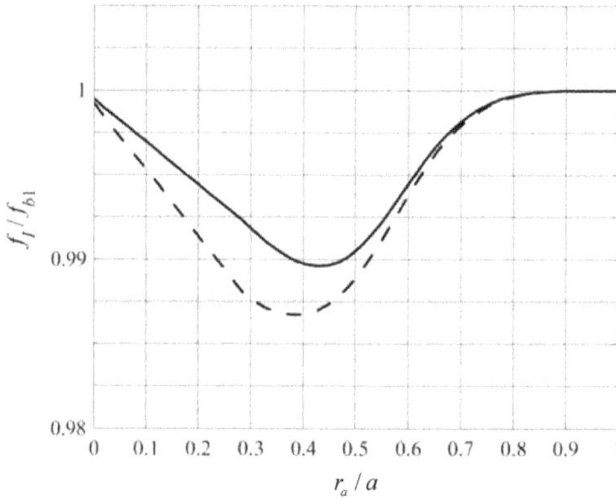

Figure 4.27: Ratio f_I / f_{b1} of the resonance frequencies of nonuniform bilaminar circular plates vs. r_a / a calculated to the second (f_I) and to the first (f_{b1}) approximations for combination of PZT-4 with aluminum (solid line) and steel (dashed line) at $y = z_{0m} / t$.

Results of calculating the first resonance frequencies and the mode shape coefficients vs. ratios r_a / a for nonuniform plates to the second approximation are presented in Figure 4.27 and Figure 4.28 for different passive materials used in combination with PZT-4 ceramics under the condition that $y = z_{0m} / t$, in which case the effect of nonuniformity is the most pronounced.

In the extreme cases at $r_a = a$ and $r_a = 0$ the mode shape coefficient $ms_I = 0$, as the plates are uniform in both these cases. The maximum deviation of the mode shapes from those for the uniform plate can be expected at $r_a / a \approx 0.5$. Plots of the mode shapes calculated at ratio $r_a / a = 0.6$ for different combination of materials used are presented in Figure 4.29. It follows from the plots that even in this case deviation from the mode shape for uniform plate is negligible. The probable reasonable limit of reducing the radius of the active laminates can be estimated as approximately $r_a / a \geq 0.6$, because the layers and for the most critical case *B* that the parts of active material are removed. As the result, the resonance frequencies of the plates calculated to the second approximation almost don't deviate from those determined to the first approximation, as it is seen from Figure 4.27.

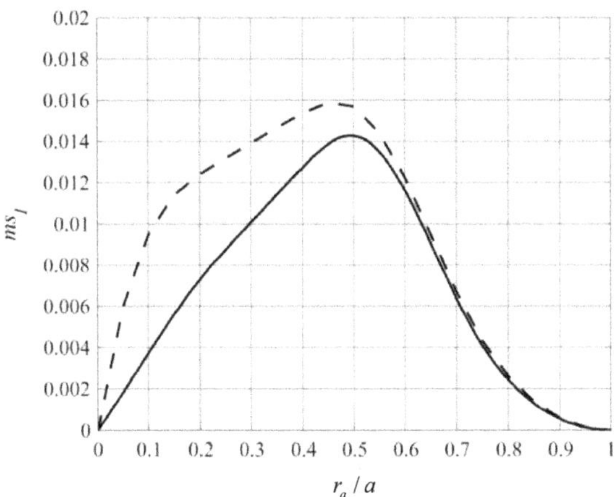

Figure 4.28: The mode shape coefficients $ms_I = \xi_2 / \xi_1$ vs. r_a / a at lowest resonance frequencies of nonuniform bilaminar circular plates for combination of PZT-4 with aluminum (solid line) and steel (dashed line) at $y = z_{0m} / t$.

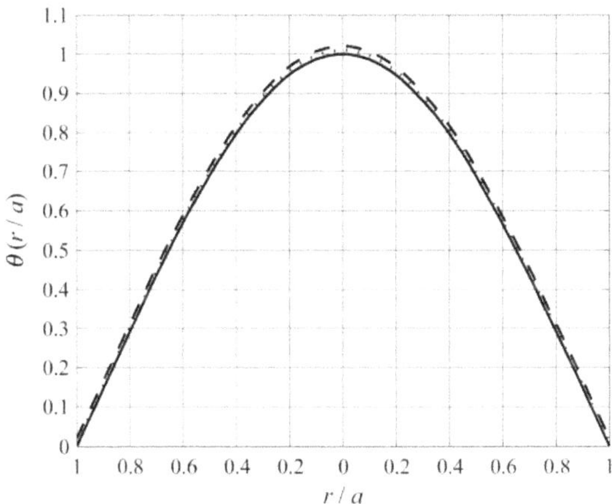

Figure 4.29: The mode shapes $\theta(r / a)$ determined to the second approximation at ratio $r_a / a = 0.5$ (solid line for aluminum–PZT, dashed line for steel–PZT) in comparison with the mode shape $\theta_1(r / a)$ for the uniform plate (dotted line) that is used in calculations to the first approximation.

Thus, the conclusion can be made that the parameters of the nonuniform bilaminar plates can be calculated using the mode shape of vibration of uniform plate with sufficient accuracy at least up to ratios $r_a / a \geq 0.6$ even for the maximum reasonable thicknesses of the active

Note, that as it follows from Figure 4.28 the maximum contribution of the second mode to actual mode of vibration takes place at $r_a / a = 0.5$. Therefore, the first mode of vibration dominates in all the range $0 \le r_a / a \le 1$, and calculating of equivalent parameters using the first mode only is possible with the same degree of accuracy at all the values of r_a / a. But, as it will be shown in Chapter 9, the effective coupling coefficient of the plate drops at $(r_a / a) < 0.5$, and further reducing of relative size of the active laminate does not make sense.

After it is proven that the first mode solution is sufficiently accurate for calculating parameters of the plate designs with the most pronounced nonuniformity (bilaminar of *B* modification), other variants of the transducers (bilaminar and trilaminar of *A* modification) can be considered using the same approximation and analogous calculating procedures. Thus, for example, the equivalent rigidities and masses for bilaminar design of *A* modification can be obtained from the expressions (4.397) for the potential and (4.402) for the kinetic energies slightly changed. Namely, in the last term in the brace of expression (4.397) quantity t_p has to be replaced by the full size thickness t. And in the last term in the brackets of expression (4.402) the density ρ_a has to be replaced by ($\rho_a - \rho_p$). This will result in the following expressions for the relative rigidity and mass. For the rigidity

$$\frac{K_{b1}}{K_b} = 1 - \left(1 - \frac{Y_{\sigma p}}{Y_{eqv\,b}(z_0)}\right)\frac{F_{r1}}{L_1(0.3)} \tag{4.421}$$

in the general case, and

$$\frac{K_{b1}}{K_b} = 1 - \left(1 - \frac{\gamma_Y}{4(z_{0m} / t)^2}\right)\frac{F_{r1}}{L_1(0.3)} \tag{4.422}$$

in the case that $\delta = z_{0m}$. And for the mass,

$$\frac{M_b}{M_{b1}} = 1 + \frac{y(\gamma_\rho - 1)}{\gamma_\rho - (\gamma_\rho - 1)y}\frac{S_{eff\,r1}}{S_{eff\,1}}. \tag{4.423}$$

4.5.8 Approximate Methods of Solving Vibration Problems

So far, the supporting functions have been represented by the system of normal modes for a real elastic body under consideration, or for a body of the same configuration under the same boundary conditions. But in general, for bodies not having suitable configuration or/and vibrating under real boundary conditions determining the normal modes itself is equivalent to solving vibration problem that hardly may be achieved analytically. In this case an attempt can be made to solve such vibration problem approximately by means of direct methods with *a priori* limitations placed on a number of considered degrees of freedom, or by using Finite Element Analysis (FEA). We will consider examples of employing the most usable for calculating mechanical systems of transducers direct methods based on considering energies of the systems: the one employing Rayleigh's principle that is applicable to one degree of freedom approximation, and another that can be used for calculating systems with several degrees of freedom, the Ritz's method of successive approximations.

4.5.8.1 Flexural Vibrations of a Center Supported Circular Plate

Vibration of the center supported plate, as potential mechanical system for electromechanical transducer, was considered analytically in Ref. 10 based on application of the Rayleigh's principle. Distribution of normal displacement $\xi(r)$ of the neutral surface of the plate supported at the center must meet conditions of zero displacement and zero slope at the center,

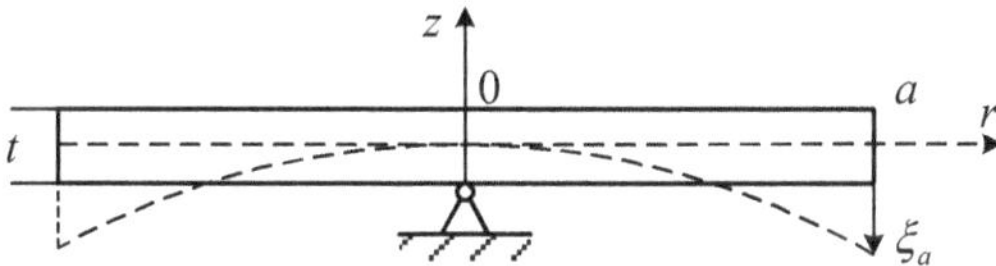

Figure 4.30: Schematic view of the center supported plate.

$$\xi(r)\big|_{r=o} = 0\,, \tag{4.424}$$

$$\left.\frac{d\xi}{dr}\right|_{r=0} = 0\,, \tag{4.425}$$

and zero moment at the edge

$$\left.\frac{d\xi^2}{dr^2} + \frac{\sigma}{r}\frac{d\xi}{dr}\right|_{r=a} = 0 \tag{4.426}$$

(the same conditions are for the plate clamped at the center).

As result of solving this problem in Ref. 10, the wave number for the lowest resonance mode of vibration was determined as $ka = 1.94$ and the corresponding resonance frequency as

$$f_{res1} = 0.172\frac{t}{a^2}\sqrt{\frac{Y}{\rho(1-\sigma^2)}}\,. \tag{4.427}$$

We will consider the example of center supported plate in detail, because it is typical and convenient for illustrating application of the Rayleigh's principle in general. In the simple words the principle states that the resonance frequency of an assumed mode of a system vibration cannot be smaller, then the resonance frequency of the true lowest normal mode of the system. Its application falls into the following steps.

Step 1: Determining of an Assumed Mode (Deflection Curve) of Vibration of the System.

The expression for the assumed deflection curve must satisfy the boundary conditions for the system and include an adjustable parameter. In our case it can be the power series

$$\xi(r/a) = c_0 + c_1\frac{r}{a} + c_2\frac{r^2}{a^2} + c_3\frac{r^3}{a^3} + c_4\frac{r^4}{a^4} + \ldots . \tag{4.428}$$

To satisfy the conditions (4.424) and (4.425) it should be $c_0 = c_1 = 0$. From condition of normalizing the function one of coefficients can be unity, so let $c_3 = 1$. One equation for determining correlation between unknown coefficients will be obtained using boundary condition (4.426). One more equation may be obtained by employing the Rayleigh's principle. Thus, it should be sufficient to retain three terms in the series (4.403) and to present the assumed deflection curve as function

$$\xi(r/a) = \frac{r^2}{a^2} + c_3\frac{r^3}{a^3} + c_4\frac{r^4}{a^4} . \tag{4.429}$$

Using the condition (4.426) (with $\sigma = 0.3$, as sufficiently accurate for all the usable PZT ceramic compositions) will be obtained

$$c_4 = -0.20 - 0.52c_3 . \tag{4.430}$$

As the result, expression for the assumed deflection curve that satisfies all the boundary conditions,

$$\xi(r/a) = \frac{r^2}{a^2} + (c_3 - 0.52)\frac{r^3}{a^3} - 0.20\frac{r^4}{a^4} , \tag{4.431}$$

has the only parameter that may be determined by application of Rayleigh's principle.

Step 2: Calculating the Kinetic and Potential Energies of the Vibrating System.

The next step is to calculate the kinetic and potential energies of the vibrating system according to the expressions

$$W_{kin} = \frac{1}{2} 2\pi \rho t \int_0^a \dot{\xi}(r/a)^2 r dr = \frac{1}{2} \dot{\xi}(1)^2 M_{eqv} = \frac{1}{2} \xi(1)^2 \omega^2 M_{eqv}, \tag{4.432}$$

$$W_{pot} = \pi D \int_0^a \left[\left(\frac{\partial^2 \xi}{\partial r^2} + \frac{1}{r} \frac{\partial \xi}{\partial r} \right)^2 - 2(1-0.3) \frac{\partial^2 \xi}{\partial r^2} \frac{1}{r} \frac{\partial \xi}{\partial r} \right] r dr = \frac{1}{2} \xi(1)^2 K_{eqv}. \tag{4.433}$$

Here $\xi(1)$ is displacement of the reference point on the edge of the plate. After substituting expression (4.431) for $\xi(r/a)$ and integrating we arrive at the following expressions for M_{eqv} and K_{eqv}

$$M_{eqv} = 7.5M \cdot (0.07c_3^2 + 0.26c_3 + 0.24), \tag{4.434}$$

$$K_{eqv} = \frac{132\pi t^3 Y}{12(1-\sigma^2)} \cdot (0.15c_3^2 + 0.40c_3 + 0.29). \tag{4.435}$$

Step 3: Application of Rayleigh's Principle.

Equating maximum values of the kinetic and potential energies we obtain

$$\omega^2(c_3) = \frac{K_{eqv}}{M_{eqv}} = 1.47 \frac{t^2}{a^4} \frac{Y}{\rho(1-\sigma^2)} \frac{0.15c_3^2 + 0.40c_3 + 0.29}{0.07c_3^2 + 0.26c_3 + 0.24}. \tag{4.436}$$

According to Rayleigh's principle the closest approximation to real normal mode of deflection will be achieved by value of coefficient c_3 that gives minimum to the frequency, which is found from equation

$$[\omega^2(c_3)]' = 0. \tag{4.437}$$

Differentiating expression (4.436) with respect to c_3 will result in equation

$$c_3^2 + 2.8c_3 + 1.7 = 0. \tag{4.438}$$

The root of the equation that gives minimum to the frequency is $c_3 = -0.91$. After determining $c_4 = 0.28$ from Eq. (4.436), expression for the assumed deflection curve will be presented as

$$\xi(r/a) = \frac{r^2}{a^2} - 0.91\frac{r^3}{a^3} + 0.28\frac{r^4}{a^4}. \tag{4.439}$$

The resonance frequency determined by using this expression for deflection curve with help of relation (4.436) at $c_3 = 0.91$ is

$$f_{res1} = 0.171\frac{t}{a^2}\sqrt{\frac{Y}{\rho(1-\sigma^2)}}, \tag{4.440}$$

i.e., differs from those by Eq. (4.427) by less than 1%.

The equivalent mass and rigidity calculated from expressions (4.434) and (4.435) are

$$M_{eqv} = 0.48\pi a^2 \rho t = 0.48M, \quad K_{eqv} = \frac{6.6\pi t^3 Y}{12a^2(1-\sigma^2)}. \tag{4.441}$$

4.5.8.2 Flexural Vibrations of a Circular Plate Supported by the Post of a Finite Radius

The center supported (clamped) plate is the ideal model. In practical applications, such as mechanical system of a hydrophone (accelerometer), real supporting element has a finite diameter. We will assume that the clamped conditions are in place on the contour of the supporting post at radius $r = b$ (Figure 4.31).

The Rayleigh-Ritz method will be used for solving the problem with an admissible function obtained in Ref. 12 for annular plate clamped on the inner and free on the outer boundary. Namely,

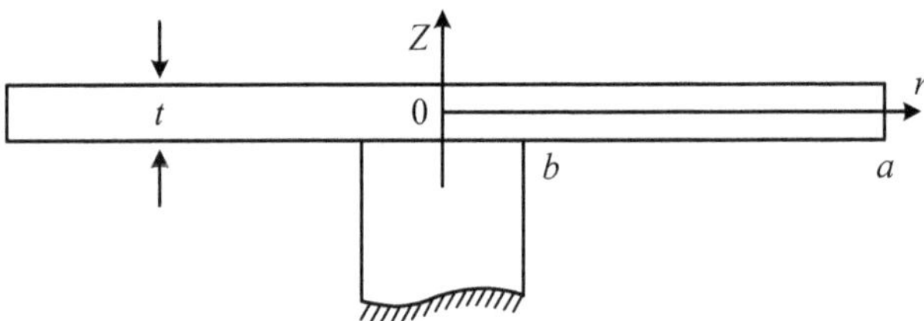

Figure 4.31: The schematic view of the disk supported by the post of finite diameter.

$$\xi(r,t) = \xi_1(t)\cdot\left(1-\frac{r^2}{b^2}\right)\ln\frac{r}{b} + \xi_2(t)\cdot\left(1-\frac{r^2}{b^2}\right)^2. \tag{4.442}$$

For brevity we introduce notation

$$f(r/b) = 1-\frac{r^2}{b^2}. \tag{4.443}$$

Thus,

$$\xi(r,t) = \xi_1(t)\cdot f(r/b)\cdot\ln\frac{r}{b} + \xi_2(t)\cdot f^2(r/b). \tag{4.444}$$

The expression contains generalized displacements ξ_1 and ξ_2. To make the mode of deflection of the plate certain, the ratio ξ_2 / ξ_1 must be determined. As the first step we must represent the kinetic and potential energies associated with vibration of the plate. The kinetic energy is

$$W_{kin} = \frac{1}{2}\rho t \cdot 2\pi \int_b^a \dot{\xi}^2 r dr = \frac{1}{2}\left(\dot{\xi}_1^2 M_1 + 2\dot{\xi}_1\dot{\xi}_2 M_{12} + \dot{\xi}_2^2 M_2\right), \tag{4.445}$$

where

$$M_1 = 2\pi\rho t \int_b^a [f(r/b)\cdot \ln(r/b)]^2 r dr, \tag{4.446}$$

$$M_2 = 2\pi\rho t \int_b^a f(r/b)^4 r dr, \tag{4.447}$$

$$M_{12} = 2\pi\rho t \int_b^a f(r/b)^3 \ln(r/b) r dr. \tag{4.448}$$

The potential energy is

$$\begin{aligned} W_{pot} &= \frac{1}{2}\left\{ D\cdot 2\pi \int_b^a \left[\left(\frac{\partial^2 \xi}{\partial r^2}\right)^2 + 2\sigma \frac{1}{r}\frac{\partial \xi}{\partial r}\frac{\partial^2 \xi}{\partial r^2} + \left(\frac{1}{r}\frac{\partial \xi}{\partial r}\right)^2 \right] r dr \right\} \\ &= \frac{1}{2}\left(\xi_1^2 K_1 + 2\xi_1\xi_2 K_{12} + \xi_2^2 K_2\right), \end{aligned} \tag{4.449}$$

where

$$\begin{aligned} K_1 = 2\pi D \int_b^a \Bigg\{ &\left[\frac{\partial^2}{\partial r^2} f \ln(r/b)\right]^2 + 2\sigma \frac{1}{r}\left[\frac{\partial^2}{\partial r^2} f \ln(r/b)\right]\left[\frac{\partial}{\partial r} f \ln(r/b)\right] + \\ &+ \left[\frac{1}{r}\frac{\partial}{\partial r} f \ln(r/b)\right]^2 \Bigg\} r dr, \end{aligned} \tag{4.450}$$

$$K_2 = 2\pi D \int_b^a \left\{ \left[\frac{\partial^2}{\partial r^2} f^2\right]^2 + 2\sigma \frac{1}{r}\left[\frac{\partial^2}{\partial r^2} f^2\right]\left[\frac{\partial}{\partial r} f^2\right] + \left[\frac{1}{r}\frac{\partial}{\partial r} f^2\right]^2 \right\} r dr, \tag{4.451}$$

$$\begin{aligned} K_{12} = 2\pi D \int_b^a &\left\{ \left[\frac{\partial^2}{\partial r^2} f \ln(r/b)\right]\left[\frac{\partial^2}{\partial r^2}(f^2)\right] + \sigma \frac{1}{r}\left[\frac{\partial^2}{\partial r^2} f \ln(r/b)\right]\frac{\partial}{\partial r}(f^2) \right\} r dr + \\ + 2\pi D \int_b^a &\left\{ \upsilon \frac{1}{r}\left[\frac{\partial}{\partial r} f \ln(r/b)\right]\frac{\partial^2}{\partial r^2}(f^2) + \frac{1}{r^2}\left[\frac{\partial}{\partial r} f \ln(r/b)\right]\left[\frac{\partial}{\partial r}(f^2)\right] \right\} r dr. \end{aligned} \tag{4.452}$$

Here

$$D = \frac{Yt^3}{12(1-\sigma^2)} . \tag{4.453}$$

The next step is in obtaining Lagrange's equations regarding the generalized displacements ξ_1 and ξ_2. They are

$$\frac{d}{dt}\left(\frac{\partial W_{kin}}{\partial \dot{\xi}_i}\right) + \frac{\partial W_{pot}}{\partial \xi_i} = 0, \quad (i = 1, 2) \tag{4.454}$$

or

$$(j\omega M_{11} + K_1 / j\omega)\dot{\xi}_1 + (j\omega M_{12} + K_{12} / j\omega)\dot{\xi}_2 = 0 , \tag{4.455}$$

$$(j\omega M_{12} + K_{12} / j\omega)\dot{\xi}_1 + (j\omega M_2 + K_2 / j\omega)\dot{\xi}_2 = 0 . \tag{4.456}$$

After obvious manipulations we obtain equations

$$(\omega^2 - \omega_1^2)\xi_1 + \frac{M_{12}}{M_1}(\omega^2 - \omega_{12}^2)\xi_2 = 0 , \tag{4.457}$$

$$\frac{M_{12}}{M_2}(\omega^2 - \omega_{12}^2)\xi_1 + (\omega^2 - \omega_2^2)\xi_2 = 0 . \tag{4.458}$$

where

$$\omega_1^2 = K_1 / M_1, \quad \omega_2^2 = K_2 / M_2, \quad \omega_{12}^2 = K_{12} / M_{12} . \tag{4.459}$$

The frequency equation of this set of equations is

$$(\omega^2 - \omega_1^2)(\omega^2 - \omega_2^2) - \frac{M_{12}^2}{M_1 M_2}(\omega^2 - \omega_{12}^2)^2 = 0 . \tag{4.460}$$

After introducing notation

$$\frac{\omega^2}{\omega_1^2} = \Omega , \tag{4.461}$$

Eq. (4.460) becomes

$$(\Omega - 1)\left(\frac{\omega_1^2}{\omega_2^2}\Omega - 1\right) - \frac{K_{12}^2}{K_1 K_2}\left(\frac{\omega_1^2}{\omega_{12}^2}\Omega - 1\right)^2 = 0 . \tag{4.462}$$

Denote solutions of this equation Ω_1 and Ω_2 assuming that $\Omega_1 < \Omega_2$. With value of Ω_1

the lower resonance frequency, ω_r, of the plate will be obtained from relation (4.461) as

$$\omega_r^2 = \omega_1^2 \Omega_1 . \tag{4.463}$$

Upon substituting this value of the resonance frequency into one of Eqs. (4.457) or (4.458) , the ratio of the generalized displacements that we denote A will be found, e.g.,

$$A = \frac{\xi_2}{\xi_1} = \frac{M_{12}}{M_2} \frac{\omega_r^2 - \omega_{12}^2}{\omega_2^2 - \omega_r^2} . \tag{4.464}$$

By combining expressions (4.442) and (4.464) we obtain the deflection curve

$$\xi(r,t) = \xi_1(t) \left[\left(1 - \frac{r^2}{b^2}\right) \ln\frac{r}{b} + A\left(1 - \frac{r^2}{b^2}\right)^2 \right] \tag{4.465}$$

and the resonance mode shape of vibration of the plate

$$\theta(r/b) = \frac{\xi(r,t)}{\xi(a,t)} = \left[\left(1 - \frac{r^2}{b^2}\right) \ln\frac{r}{b} + A\left(1 - \frac{r^2}{b^2}\right)^2 \right] \cdot \left[\left(1 - \frac{a^2}{b^2}\right) \ln\frac{a}{b} + A\left(1 - \frac{a^2}{b^2}\right)^2 \right]^{-1} . \tag{4.466}$$

With expression for the mode shape of vibration known all the equivalent parameters of the plate and related electromechanical transducer can be determined. Results of calculations made for various relations of inner and outer radiuses b/a are as follows.

Resonance Frequencies

Following notations accepted in Ref. 173 we denote

$$\omega_{r\,b/a} = \lambda^2 \frac{t}{a^2\sqrt{12}} \sqrt{\frac{Y}{\rho(1-\sigma^2)}} . \tag{4.467}$$

Plot of coefficient λ vs. ratio b/a is presented in Figure 4.32 together with the plot that corresponds to data obtained in Ref. 14 and 15, where exact solution for vibration of annular thin plate clamped on the inner radius was presented. Values of coefficients λ vs. ratios b/a are also given in Table 4.5.

Mode Shapes of Vibration vs. Ratio b/a

Coefficients A at different values of b/a are presented in Table 4.5, and the corresponding mode shapes are shown in Figure 4.33.

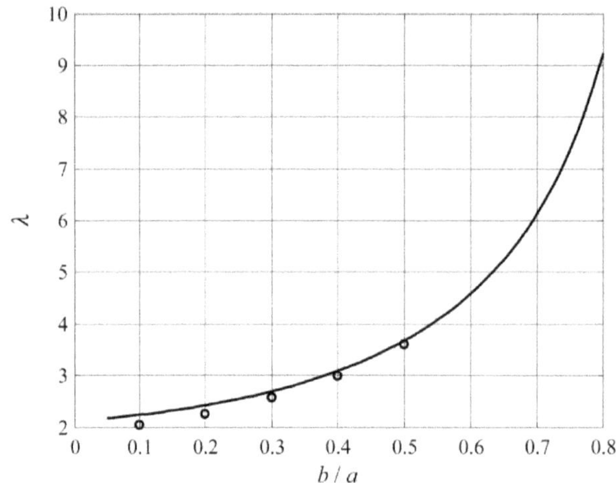

Figure 4.32: Plot of coefficient λ vs. radii ratios of b/a. Results of calculations made in Refs. 14 and 15 are presented by the circles.

Table 4.5: Calculated parameters for flexural disk supported by a post of finite diameter.

	b/a	0.7	0.6	0.5	0.1	0.05	0 (center supported)
	A	0.23	0.18	0.13	0.01	0.003	-
λ	admiss. function	6.13	4.59	3.68	2.25	2.18	1.95
λ	Static approx.	6.17	4.61	3.69	2.19	2.13	
λ	"exact" Refs. 14 & 15	6.08	4.53	3.61	2.06		
	$\frac{S_{eff\,b/a}}{S_{b/a}}$	0.28	0.28	0.30	0.36	0.37	0.48
	static	0.26	0.28	0.29	0.36	0.37	
	$\frac{S_{av\,b/a}}{S_{b/a}}$	0.41	0.44	0.45	0.52	0.53	0.64
	static	0.41	0.42	0.44	0.52	0.53	
	κ	200	79	41	9.13	8.35	6.9

Equivalent Parameters of the Plate for Various Values of b/a

The equivalent mass of the plate is determined by formula

$$M_{eqv\,b/a} = \rho t S_{eff\,b/a}, \tag{4.468}$$

where

$$S_{eff} = 2\pi \int_b^a \theta^2(r) r dr \ . \tag{4.469}$$

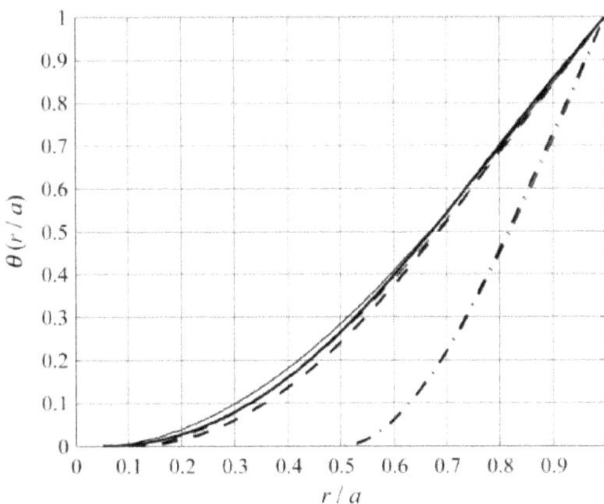

Figure 4.33: Plots of the mode shapes of vibration of the plates at different ratios b/a: 0.05 (solid line), 0.1 (dashed line), o.5 (dash-dotted line). The static approximation is shown in thin solid line.

Values of $S_{eff\,b/a}$ normalized to $S_{b/a} = \pi a^2(1 - b^2/a^2)$ are given in Table 4.5.

The equivalent rigidity $K_{eqv\,b/a}$ can be determined from formula

$$K_{eqv\,b/a} = \omega_{r\,b/a}^2 M_{eqv\,b/a} \ . \tag{4.470}$$

Following formulas (4.457) and (4.458) we will obtain that

$$K_{eqv\,b/a} = \lambda^4 \frac{S_{eff\,b/a}(1 - b^2/a^2)}{S_{b/a}} \frac{\pi t^3}{12a^2} \frac{Y}{(1-\sigma^2)} \ . \tag{4.471}$$

The coefficients

$$\kappa = \lambda^4 S_{eff\,b/a}(1 - b^2/a^2)/S_{b/a} \tag{4.472}$$

are given in the Table 4.5.

Deformation of the plate by uniformly distributed load q (hydrostatic pressure, sound pressure at low frequencies, inertia forces under action of acceleration) is proportional to the equivalent force, $F_{eqv\,b/a}$, that is determined by expression

$$F_{eqv\,b/a} = qS_{av\,b/a} \tag{4.473}$$

where

$$S_{av\,b/a} = 2\pi \int_b^a \theta(r) r dr \,. \tag{4.474}$$

Values of $S_{av\,b/a} / S_{b/a}$ are given in the Table 4.5.

The resonance frequencies of vibration, $\omega_r = \sqrt{K_{eqv} / M_{eqv}}$, are proportional to λ^2. Their approximate values are somewhat higher than those known from classical theory of thin plates that are shown in Table 4.5 following results presented in Refs. 14 and 15.

The results obtained may have practical applications in two areas. At small ratios *b*/*a* they can be used for mechanical systems of electromechanical transducers. In this capacity they will be considered in Chapter 9. At large ratios *b*/*a* the circular disks can be used as passive parts of the length expander transducers that serve for matching with acoustic field. Such application will be considered in Chapter 10.

It must be remembered that all the results are obtained under the assumptions of applicability of the thin plates theory and clamped boundary conditions on the inner radius at all the ratios *b*/*a*. In case the thickness to radius ratio of a plate becomes significant, the corrections for rotary inertia and shear deformations must be introduced.

4.5.8.3 Application of Ritz's Method of Successive Approximations

As it has been noted before, any complete system of functions satisfying boundary conditions can be taken as a supporting system. As an example, consider the problem of axial symmetric free flexural vibrations of a circular plate with clamped edge. The precise analytical solution to the problem is given in Section 4.4.3. Let us represent the displacement $\xi_z(r,t)$ in the form of series

$$\xi_z(r,t) = \sum_{i=1}^{\infty} \xi_i(t)(1 - r^2 / a^2)^{i+1} \,. \tag{4.475}$$

It is easy to verify that the functions $\theta_i(r/a) = (1 - r^2 / a^2)^{i+1}$ satisfy conditions (4.190) on the clamped edge of the plate. They form a complete system of functions within interval [0, *a*] (the proof of this is provided in Ref. 5). Let us represent the kinetic and potential energies of the plate, determining W_{pot} by integrating expression (4.186) for w_{pot} over the volume of the plate,

$$W_{kin} = \pi\rho t \int_0^a \dot{\xi}_z^2 r dr \,, \tag{4.476}$$

$$W_{pot} = \pi D \int_0^a \left[\left(\frac{\partial^2 \xi_z}{\partial r^2} + \frac{1}{r}\frac{\partial \xi_z}{\partial r} \right)^2 - 2(1-\sigma)\frac{\partial^2 \xi_z}{\partial r^2}\frac{1}{r}\frac{\partial \xi_z}{\partial r} \right] r dr \,. \tag{4.477}$$

Upon substituting the displacements $\xi_z(r,t)$ in the form of series (4.475) into expressions (4.476) and (4.477), the coefficients M_{il} and K_{il} of the system of equations (4.224) may be found. Following Ritz's method, we substitute only one (first) term of series (4.475) into the energy expressions, as the first step. The Lagrange equation in this case will be

$$K_1 - \omega^2 M_1 = 0 \,, \tag{4.478}$$

where

$$K_1 = 64\pi D / 3a^2, \quad M_1 = M/5, \quad \omega_1^{(1)} = (10.33/a^2)\sqrt{D/\rho t} \,. \tag{4.479}$$

This solution must be considered as the first approximation. The obtained value of natural frequency is somewhat higher than the exact value,

$$\omega_1 = (10.2/a^2)\sqrt{D/\rho t} \,, \tag{4.480}$$

determined by formula (4.202). The exact normal mode of vibration, which is described by function (4.197) at $k_1 a = 3.2$ and the mode of vibration to the first approximation,

$$\theta_1(r/a) = (1 - r^2/a^2)^2 \,, \tag{4.481}$$

are shown in Figure 4.34.

The second approximation will be obtained, if to take into consideration two terms of series (4.475). Upon their substituting into expressions for the energies, the following Lagrange equations will be obtained

$$(64\pi D/3a^2 - \omega^2 M/5)\xi_1 + (16\pi D/a^2 - \omega^2 M/6)\xi_2 = 0 \,, \tag{4.482}$$

$$(16\pi D/a^2 - \omega^2 M/6)\xi_1 + (96\pi D/5a^2 - \omega^2 M/7)\xi_2 = 0 \,. \tag{4.483}$$

More accurate expressions derived from these equations for the first resonance frequency,

$$\omega_1^{(2)} = (10.21/a^2)\sqrt{D/\rho t} \,, \tag{4.484}$$

and for the mode of vibration almost coincide with the exact values (for the mode shape this can be concluded from Figure 4.34). At the same time the expressions for the second resonance frequency,

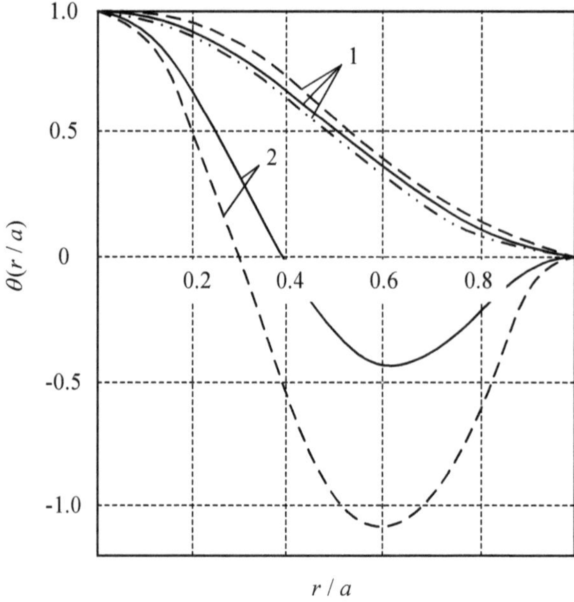

Figure 4.34: Successive approximations of the modes of plate vibration. 1 – $\theta_1(r/a)$; 2 – $\theta_2(r/a)$. Solid lines - exact values, dashed lines – the first approximation, dashed dotted lines – the second approximation.

$$\omega_2^{(1)} = (43.04/a^2)\sqrt{D/\rho t}\,, \tag{4.485}$$

and for the second mode of vibration are obtained. The resonance frequency presents a rough approximation to its exact value,

$$\omega_2 = (39.6/a^2)\sqrt{D/\rho t}\,, \tag{4.486}$$

and comparison of the modes of vibration made in Figure 4.34 shows significant difference. To obtain more accurate expressions for the subsequent values of the resonance frequencies and modes of vibration, the procedure of successive approximations must be continued.

4.5.9 Employing the Static Approximation to the First Mode of Vibration

The selection of supporting (assumed) functions and of a number of retained degrees of freedom for practical calculations, which always prove to be approximate to a certain extent, depend substantially on the objectives of the calculations. With respect to transducers for underwater applications the most typical problems are those of computing their characteristics either for the frequency range around one or two adjacent lower resonance frequencies (mainly, for transducers – projectors), or for a broad frequency range below the first resonance (mainly, for the transducers – receivers). The frequency responses of displacements in the elastic bodies can be presented in generalized coordinates as a result of superposition of the frequency responses of the partial systems, which correspond to the particular degrees of freedom of the body.

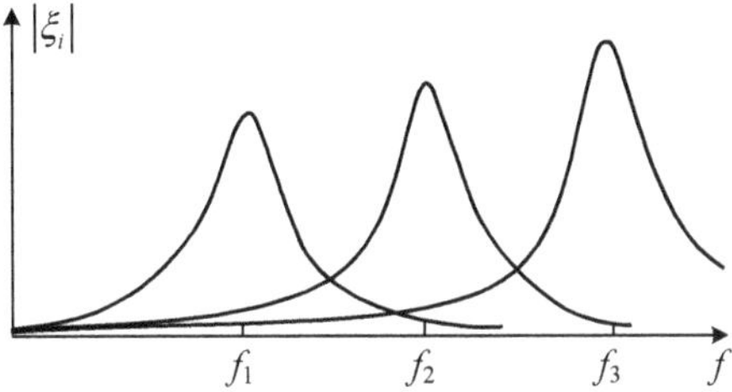

Figure 4.35: Frequency responses of displacements at the normal modes of vibrations.

In the simplest case, when the supporting functions are normal modes, this can be qualitatively illustrated by means of Figure 4.35, in which the frequency responses of moduli of displacements, $|\xi_i|$, that correspond to the resonance modes of vibration are presented. The total displacement at each frequency is the sum of displacements in all vibration modes (under a real summation the phases of displacements should be taken into account). Considering responses in frequency ranges around the certain resonance frequencies, it is practically possible to ignore contributions of other vibration modes compared to the resonance ones, and to limit a number of considered degrees of freedom. This helps to greatly simplify the solution for the vibration problem. In contradistinction to this, results of solving the same problem in geometric coordinates contain a mix of contributions of all the resonance modes including insignificant contributions of the modes that correspond to resonance frequencies remote far away from the range under consideration. This makes such solution less physically transparent, than those in the form of expansion into series in terms of the normal modes.

In calculations within the frequency range below the first resonance, the mode of vibration pertaining to this resonance frequency contributes the most to the overall displacement. However, the total contribution of the higher resonance modes also exists, in principle. In this case, instead of taking into calculation many resonance modes of vibration it may be expedient to use the static (at $\omega \to 0$) modes of displacement as the supporting functions.

A qualitative explanation of the fact that the static deflection under action of uniformly distributed force ΔF_m (for example, under the hydrostatic pressure) gives a good approximation to the first resonance mode of vibration can be given with example of a simply supported beam. The first mode of vibration and uniform distribution of the pressure over the surface of the beam are shown in Figure 4.36 (a).

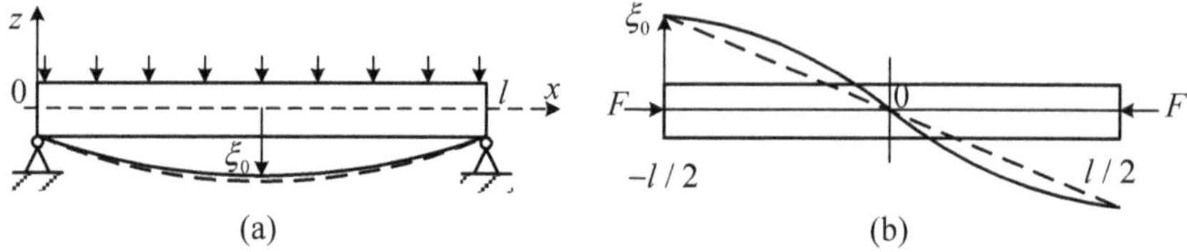

Figure 4.36: (a) Beam under bending. The first mode of vibration - solid line, and the static deflection under uniformly distributed pressure - dashed line; (b) Extensional deformation of a bar. Distribution of deformation in the first resonance mode – solid line, static deformation – dashed line.

If the acting forces were distributed in accordance with the mode shape at the resonance, then it would be the only mode exited (see formula (4.239)). By the way, if the beam vibrated in the first mode, the inertia forces loading the beam would be distributed according to this mode. Although the uniform distribution of force differs from the normal mode, contribution to the equivalent force of the pressure acting on the parts of surface that are close to the ends is small in accordance with formula (see (4.246))

$$F_{mi} = \Delta F_m \int_{\Sigma} \theta_i(\boldsymbol{r}_\Sigma) d\Sigma = \Delta F_m S_{avi} \,. \tag{4.487}$$

Accuracy of the static deflection approximation can be estimated qualitatively with the same example of simply supported beam, for which the normal modes are $\theta_i(x) = \cos i\pi x / l$. Vibration of the beam can be considered as rigidity controlled in all the modes in the frequency range below the first resonance frequency, i.e., it may be considered that $\omega M_{eqvi} \ll K_{eqvi} / \omega$. Expression for the rigidities can be presented as $K_{eqvi} = K_{eqv1} i^4$ (see formulas (4.134)). The equivalent forces can be expressed as $F_{eqvi} = \Delta F_m S_{avi}$, where $S_{avi} = 2l / i\pi$ at $i = 2n-1$, $n = 1, 2, \ldots$ and $S_{avi} = 0$ at $i = 2n$. Thus, the magnitude of displacement in the mode θ_i is

$$\bar{\xi}_i \approx j\omega \frac{2l\Delta F_{mm}}{K_{eqv1}(2n-1)^5} = \bar{\xi}_1 \frac{1}{2n-1}. \tag{4.488}$$

Thus, the maximum displacement is

$$\bar{\xi}(0) = \sum_{i=1}^{\infty} \bar{\xi}_i = \bar{\xi}_1 \sum_{n=1}^{\infty} \frac{1}{(2n-1)^5}. \tag{4.489}$$

It is known[20] that

$$\sum_{n=1}^{\infty} \frac{1}{(2n-1)^4} = \frac{\pi^4}{96} \approx 1.01. \tag{4.490}$$

Obviously,

$$\sum_{n=1}^{\infty} \frac{1}{(2n-1)^5} < \sum_{n=1}^{\infty} \frac{1}{(2n-1)^4}, \tag{4.491}$$

thus, the contribution of all the higher modes to total static displacement is less than 1% in comparison with displacement in the first mode. Similar estimations can be made for the circular plates vibrating in flexure. The modes of static deflection under the action of forces uniformly distributed over surface of the beams and circular plates are known from Ref. 2.

They are for the rectangular beams:

with simply supported ends,

$$\theta(x) = (16/5l)(x - 2x^3/l^2 + x^4/l^3); \tag{4.492}$$

with clamped ends,

$$\theta(x) = (16x^2/l^2)(1 - 2x/l + x^2/l^2); \tag{4.493}$$

with one end clamped and with a force acting on the free end,

$$\theta(x) = (x^2/2l^2)(3 - x/l). \tag{4.494}$$

For the circular plates (at $\sigma = 0.3$):

with simply supported edge,

$$\theta(r/a) = (1 - r^2/a^2)(1 - r^2/4a^2); \tag{4.495}$$

with clamped edge,

$$\theta(r/a) = (1 - r^2/a^2)^2. \tag{4.496}$$

The above considered approximating procedure for determining displacements of the circular plate with clamped edge illustrated with Figure 4.34 converged quickly to the exact solution (the first approximation was practically sufficient), because the mode of the total static displacement $\theta(r/a) = (1 - r^2/a^2)^2$ is close to the vibration mode at the first resonance frequency. By contrast, in calculations in the range of the high resonance frequencies such a selection of supporting functions would necessitate considering many degrees of freedom to get an acceptable approximation (as can be seen from Figure 4.34, the second approximation is still far from the exact form of the second mode).

It is instructive to consider using the static displacement of the circular plate center supported by the post of finite diameter (Figure 4.31) that can be obtained, as solution for deformation of the clamped-free annular plate under uniformly distributed static load with density q, which is[2]

$$\xi_{st}(r/b) = \frac{qb^4}{64D}\left[\left(\frac{r^2}{b^2} - 1\right)^2 - 2A\left(\frac{r^2}{b^2} - 1 - 2\ln\frac{r}{b}\right)\right], \tag{4.497}$$

where

$$A = \frac{(3+\sigma)(a/b)^4 - (1+\sigma)(a/b)^2}{(1+\sigma)(a/b)^2 + (1-\sigma)}. \tag{4.498}$$

Plot of the static mode shape $\theta_{st}(r/b) = \xi_{st}(r/b)/\xi_{st}(a/b)$ for ratios $b/a = 0.05$ is presented in Figure 4.33 in comparison with those obtained by using the admissible functions (4.442) for illustrating application of the Rayleigh-Ritz method of solving the problem. The coincidence of the mode shapes is almost complete. This leads to very close results of calculating equivalent parameters of the plates that are shown in Table 4.5 for various values of b/a, though they are obtained in much simple way.

It is noteworthy that for the mechanical systems vibrating in the extensional modes the static approximation to the first resonance mode does not hold in general. This can be illustrated using the same procedure as for the beam in flexure with example of longitudinally vibrating uniform bar that is shown in Figure 4.36 (b). In this case $K_{eqvi} = K_{eqv1} i^2$ (see formula (4.133)), and expression analogous to (4.489) will be

$$\bar{\xi}(l/2) \approx \bar{\xi}_1 \sum_{n=1}^{\infty} \frac{1}{(2n-1)^2} = \frac{\pi^2}{8} \approx 1.26 . \tag{4.499}$$

Thus, the contribution of the higher modes is significant.

The general conclusion can be made that the static approximation to the first resonance mode of vibration is applicable to the mechanical systems that have the next active resonance frequency far away from the first (the plates, beams, symmetrically mass loaded bars).

4.5.10 Flexural Vibration of a Slotted Ring

Free flexural vibration of the uniform slotted ring in its plane will be considered as an example of solving vibration problem for a mechanical system in static approximation regarding the mode shape of its vibration. The geometry of a slotted ring is shown in Figure 4.37. The ring is assumed to be short. The thickness and height of the ring are denoted *t* and *h*, respectively.

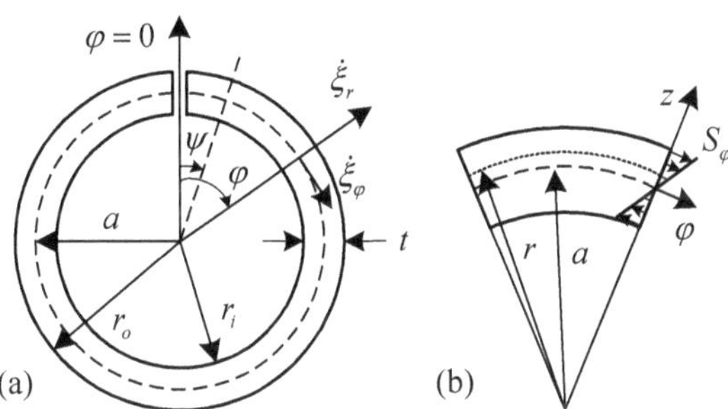

Figure 4.37: (a) Slotted ring geometry, (b) distribution of strain through the thickness of the ring.

The condition (4.224) for existing of the neutral (not stretched) circumference surface in the ring,

$$\frac{d\xi_\varphi}{d\varphi} + \xi_r = 0 , \tag{4.500}$$

is assumed to be fulfilled. In a thin ($t \ll a$) and short ($h \ll a$) ring the only not zero stress is the stress in the circumferential direction, $T_\varphi = YS_\varphi$, and the strain S_φ in circumferential direction is

$$S_\varphi = -\frac{r-a}{a^2}\left(\frac{\partial^2 \xi_r}{\partial \varphi^2} + \xi_r\right) \tag{4.501}$$

according to formula (4.220). The minus sign means that the positive moment (acting in anticlockwise direction at the right end of the segment and in the clockwise direction on the left end) produces negative strain (contraction) in the upper layer (at $r > a$). At low frequency approximation, i.e., at frequencies around and below the first resonance frequency, the assumed mode of the flexural vibration of the ring in its plane may be taken the same as the mode of its static deformation under action of hydrostatic pressure p uniform over surface. Following Ref. 17, the equation of static deformation of a thin ring may be presented in the form of

$$\xi_r + \xi_r'' = -\frac{Ma^2}{YJ}, \tag{4.502}$$

where $M(\varphi)$ is the bending moment, $J = t^3 h/12$ is the inertia moment of the rectangular cross section, and the superscript $('')$ denotes the second derivative with respect to coordinate φ. From the geometry considerations

$$M(\varphi) = pha^2 \int_0^{\varphi} \sin(\varphi - \theta) d\theta = pha^2(1 - \cos\varphi). \tag{4.503}$$

The following solution is after Ref. 18. After substituting the moment (4.503) into Eq. (4.502) and denoting $ha^2 p / YJ = N$ we obtain

$$\xi_r + \xi_r'' = -N(1 - \cos\varphi). \tag{4.504}$$

The assumed solution for Eq. (4.504) is

$$\xi_r = A\cos\varphi + B\sin\varphi + C\varphi\sin\varphi + D. \tag{4.505}$$

This can be verified by substituting ξ_r in Eq. (4.503). After substituting ξ_r into Eq. (4.504) we arrive at

$$D + 2C\cos\varphi = -N(1 - \cos\varphi). \tag{4.506}$$

For satisfying this equality at arbitrary φ it should be

$$D = -N, \quad C = N/2. \tag{4.507}$$

From the symmetry considerations $\xi_r'|_{\varphi=\pi} = 0$. Thus, $B = -N\pi/2$ and the radial displacement is

$$\xi_r = A\cos\varphi - \frac{N\pi}{2}\sin\varphi + \frac{N}{2}\varphi\sin\varphi - N. \tag{4.508}$$

Following condition (4.224)

$$\xi_\varphi = -\left(A + \frac{N}{2}\right)\sin\varphi - \frac{N}{2}(\pi - \varphi)\cos\varphi - N(\pi - \varphi). \tag{4.509}$$

Assuming that the mode of static deformation coincides to the fist approximation with the resonance mode of vibration, the total amount of motion of a ring should be zero, i.e.,

$$\rho tha \int_0^{2\pi} u_y d\varphi = \rho tha \int_0^{2\pi} (\dot{\xi}_r \cos\varphi - \dot{\xi}_\varphi \sin\varphi) d\varphi = 0, \tag{4.510}$$

where $u_y = \dot{\xi}_r \cos\varphi - \dot{\xi}_\varphi \sin\varphi$ is the projection of velocity of vibration on the axis y. Integral of the horizontal component of velocity, u_x, is equal to zero due to symmetry. After taking in consideration the condition (4.510), we obtain $A = -1.25N$ and finally the distributions of displacements will be presented in the form

$$\xi_r = -1.25N\cos\varphi - \frac{N\pi}{2}\sin\varphi + \frac{N}{2}\varphi\sin\varphi - N, \tag{4.511}$$

$$\xi_\varphi = 0.75N\sin\varphi - \frac{N}{2}(\pi - \varphi)\cos\varphi - N(\pi - \varphi). \tag{4.512}$$

Let the reference point be at $\varphi = 0$ (the slot is supposed to be thin enough). The radial displacement of this point is

$$\xi_r(0) = -2.25N = \xi_{ro}. \tag{4.513}$$

Then the expressions for the radial and tangential displacements (mode shapes) will be represented in normalized form as

$$\frac{\xi_r(\varphi)}{\xi_{ro}} = \theta_r(\varphi) = 0.55\cos\varphi + 0.70\sin\varphi - 0.22\varphi\sin\varphi + 0.44, \tag{4.514}$$

$$\frac{\xi_\varphi(\varphi)}{\xi_{ro}} = \theta_\tau(\varphi) = -0.33\sin\varphi + 0.22(\pi - \varphi)\cos\varphi + 0.44(\pi - \varphi). \tag{4.515}$$

The mode shapes are shown in Figure 4.38 together with results of experimental verification of the radial displacement[18]. The fact that at $\varphi = \pm 150°$ both the radial and tangential displacements are close to zero can be exploited for mounting the rings to structural elements of a transducer design.

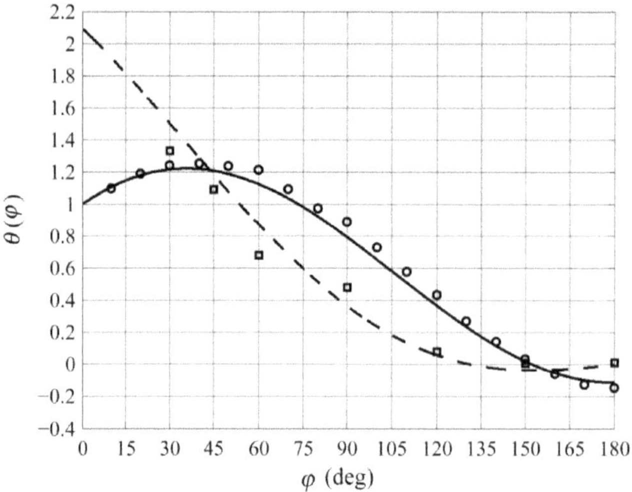

Figure 4.38: Mode shapes of the radial (solid line) and tangential (dashed line) displacements. The experimental data are shown by the circles and squares.

Qualitative comparison between the mode shapes of radial displacements for the complete ring, slotted ring and for the "arms" of the tuning fork comprised of two bars clamped at one end is made in Figure 4.39. The comparison is useful in terms of a qualitative prediction of effectiveness of low frequency radiation by these systems. Whereas $S_{av} = 0$ in the case (a) and baffling of the areas vibrating in anti-phase is required, in the cases (b) and (c) $S_{av} \neq 0$ and the radiation resistance has a fairly good value without baffles. It will be obtained after substituting expression for $\theta_r(\varphi)$ into formula (4.247) that for the slotted ring $S_{av} = 0.66S$, where S is the total surface area of the ring.

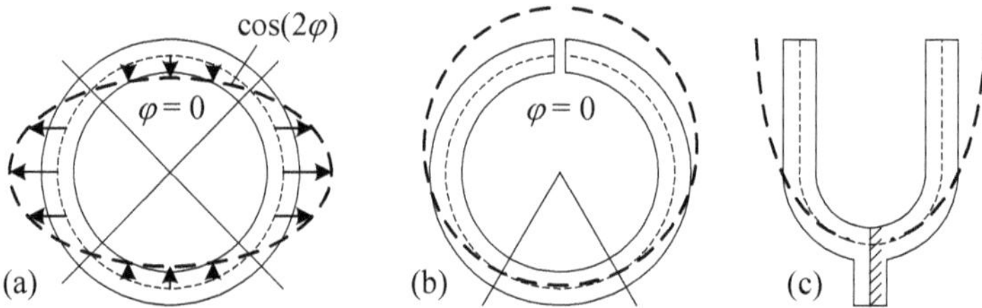

Figure 4.39: Geometries of mode shapes of the radial displacements: (a) complete ring in flexure, (b) slotted ring, (c) tuning fork.

To determine the equivalent parameters of the ring, expressions for the potential and kinetic energies must be considered. For the potential energy it is

$$W_{pot} = \frac{1}{2}\int_{\tilde{V}} S_\varphi T_\varphi d\tilde{V} = \frac{1}{2}\frac{S_{cs}t^2 Y}{12a^4}\int_0^{2\pi}\left(\frac{\partial^2 \xi_r}{\partial \varphi^2} + \xi_r\right)^2 a d\varphi = \frac{1}{2}K_{eqv}\xi_{ro}^2. \tag{4.516}$$

For the kinetic energy

$$W_{kin} = \frac{1}{2}\int_{\tilde{V}} \rho u^2 d\tilde{V} = \frac{1}{2}\rho S_{cs}\int_0^{2\pi}\left(\dot{\xi}_r^2 + \dot{\xi}_\varphi^2\right) a d\varphi = \frac{1}{2}M_{eqv}\dot{\xi}_{ro}^2. \tag{4.517}$$

Here $S_{cs} = th$ is the cross section area of a ring, M_{eqv} and K_{eqv} are the equivalent mass and rigidity of the ring. After substituting values of ξ_r and ξ_φ from formulas (4.514) and (4.515) into expressions for the energies we obtain

$$M_{eqv} = 1.57M, \quad K_{eqv} = \frac{1}{C_{eqv}} = \frac{20}{\pi}\frac{a^3}{t^3 hY}. \tag{4.518}$$

The resonance frequency of the ring is

$$f_r = 1/2\pi\sqrt{M_{eqv}C_{eqvm}} = 0.02\frac{t}{a^2}\sqrt{\frac{Y}{\rho}}. \tag{4.519}$$

4.6 Coupled Vibrations in the Mechanical Systems

So far, we have considered one-dimensional vibrations of elastic bodies. In this case the degrees of freedom of the elastic body were related to different distributions of vibration in one dimension. Though in the most cases such an approximated analysis of vibration of elastic bodies used in transducer designs is justified, sometimes it is necessary to consider simultaneous vibrations that occur in two dimensions. An exact solution to the problems of two-dimensional vibrations of elastic bodies having commensurable dimensions presents considerable difficulties. At the same time, a decent for practical purposes approximations can be obtained, if to consider the two-dimensional vibrations as coupled vibrations of the properly chosen one-dimensional partial elastic systems.

4.6.1 The General Outline of the Theory of Coupled Vibrations

Consider some general information on the theory of coupled vibrations, assuming that the vibrations of an elastic body take place in two dimensions and that each partial system has one degree of freedom (i.e., has one resonance frequency in the operating range). The state of the body is described by two generalized coordinates, ξ_{1p} and ξ_{2p}, so that distribution of displacements in its volume can be represented as

$$\xi(x, y) = \xi_{1p}\theta_1(x) + \xi_{2p}\theta_2(y), \tag{4.520}$$

where $\theta_1(x)$ and $\theta_2(y)$ are vibration modes of the partial systems. (Partial is the system that remains, if to put one of the generalized coordinates to zero). In the general case the expressions for the kinetic and potential energies of the system are of the form

$$W_{kin} = M_1\dot{\xi}_{1p}^2/2 + M_{12}\dot{\xi}_{1p}\dot{\xi}_{2p} + M_2\dot{\xi}_{2p}^2/2, \tag{4.521}$$

$$W_{pot} = K_1\xi_{1p}^2/2 + K_{12}\xi_{1p}\xi_{2p} + K_2\xi_{2p}^2/2. \tag{4.522}$$

Here M_1, M_2 and K_1, K_2 are the equivalent masses and equivalent rigidities of the partial systems; M_{12} and K_{12} are the mutual masses and rigidities, which quantitatively characterize inertial and elastic interaction between the partial systems.

Generalized forces f_{m1} and f_{m2} that generate vibrations of the partial systems are independent, and the energy supplied by these forces is

$$W_e = f_{m1}\xi_{1p} + f_{m2}\xi_{2p}. \tag{4.523}$$

Expressions (4.521) and (4.522) can describe the coupled vibrations not only in an elastic body, but also in mechanical systems composed of two separate bodies, between which a constructive connection exists. For illustration consider the typical example of the two degree of freedom mechanical system shown in Figure 4.40 that has lumped parameters. Masses M_1' and M_2' vibrate in the vertical direction on the springs having the stiffness constants K_1' and K_2'. (The primes are introduced to distinguish the example from the general case). Configuration of the system is completely defined by displacements ξ_1 and ξ_2 of the masses.

The potential and kinetic energies of the system are

$$W_{pot} = \frac{1}{2}K_1'\xi_1^2 + \frac{1}{2}K_2'(\xi_2 - \xi_1)^2 = \frac{1}{2}(K_1' + K_2')\xi_1^2 - K_2'\xi_1\xi_2 + \frac{1}{2}K_2'\xi_2^2, \tag{4.524}$$

$$W_{kin} = \frac{1}{2}M_1'\dot{\xi}_1^2 + \frac{1}{2}M_2'\dot{\xi}_2^2. \tag{4.525}$$

The partial systems are as follows: at $\xi_2 = 0$ the mass M_1' vibrating between springs K_1' and K_2' having a combined stiffness $K_1' + K_2'$; at $\xi_1 = 0$ the mass M_2' vibrating on the spring

K_2'. The resonance frequencies of the partial systems are $f_{1p} = (1/2\pi)\sqrt{(K_1' + K_2')/M_1'}$ and $f_{2p} = (1/2\pi)\sqrt{K_2'/M_2'}$, respectively. The coupling between the partial systems is elastic (as $M_{12}' = 0$) with mutual rigidity $K_{12}' = K_2'$.

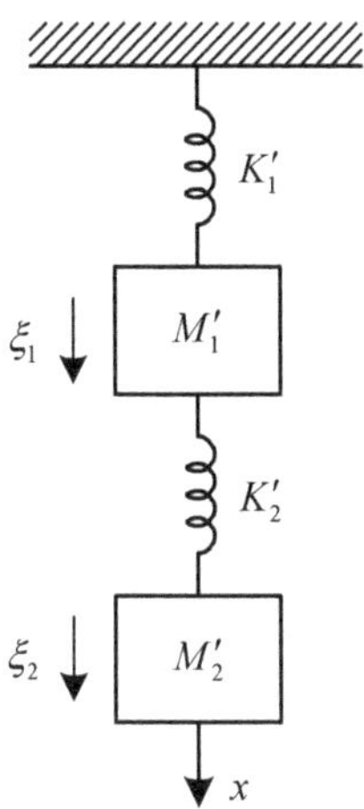

Figure 4.40: Coupled mechanical system with lumped parameters.

Using expressions (4.521)-(4.523), the Lagrange's equations for a system with two coupled degrees of freedom can be obtained as

$$(K_1 - \omega^2 M_1)\xi_{1p} + (K_{12} - \omega^2 M_{12})\xi_{2p} = F_1, \tag{4.526}$$

$$(K_{12} - \omega^2 M_{12})\xi_{1p} + (K_2 - \omega^2 M_2)\xi_{2p} = F_2. \tag{4.527}$$

The following notations will be introduced: for the natural frequencies of partial systems (partial frequencies),

$$f_{1p} = (1/2\pi)\sqrt{K_1/M_1} \text{ and } f_{2p} = (1/2\pi)\sqrt{K_2/M_2}; \tag{4.528}$$

and for the coefficients of inertial and elastic coupling between the systems,

$$\gamma_i = M_{12}/\sqrt{M_1 M_2} \text{ and } \gamma_e = K_{12}/\sqrt{K_1 K_2}. \tag{4.529}$$

In the example of mechanical system shown in Figure 4.40 the coupling is elastic, therefore, $\gamma_i = 0$ and $\gamma_e = K_{12}'/\sqrt{K_1' K_2'} = -\sqrt{K_2'/(K_1' + K_2')}$.

Further in the general analysis of the coupled vibrations in elastic bodies only the inertial coupling will be considered for brevity assuming that $K_{12} = 0$, and notation γ for the coefficient of the inertia coupling will be used. The variant of elastic coupling can be considered in analogous way. This will be illustrated with examples of particular mechanical systems.

4.6.1.1 Free Vibrations in Coupled Systems

With notations (4.528), (4.529) introduced and at condition that $K_{12}=0$, Eqs. (4.526) and (4.527) for free vibrations become

$$M_1(f_{1p}^2-f^2)\xi_1-M_{12}f^2\xi_2=0, \tag{4.530}$$

$$-M_{12}f^2\xi_1+M_2(f_{2p}^2-f^2)\xi_2=0. \tag{4.531}$$

The frequency equation for the system will be obtained from the condition that its determinant is equal to zero

$$(f_{1p}^2-f^2)(f_{2p}^2-f^2)-\gamma^2 f^4=0, \tag{4.532}$$

where from the values of the natural frequencies of vibration in a coupled system f_1 and f_2 can be determined as follows

$$f_{1,2}^2=\frac{1}{2(1-\gamma^2)}\left[f_{1p}^2-f_{2p}^2\pm(f_{1p}^2-f_{2p}^2)\sqrt{1+\chi^2}\right], \tag{4.533}$$

where $\chi=2\gamma f_{1p}f_{2p}/|f_{2p}^2-f_{1p}^2|$. This quantity is called the coefficient of connectivity.

From the system of equations (4.530) and (4.531) relations between the amplitudes of displacements in the partial systems at natural frequencies, the mode shape coefficients ms_i, may be obtained as

$$ms_1=\left.\frac{\xi_2}{\xi_1}\right|_{f=f_1}=\frac{M_1(f_{1p}^2-f_1^2)}{f_1^2 M_{12}},\quad ms_2=\left.\frac{\xi_2}{\xi_1}\right|_{f=f_2}=\frac{M_2(f_{2p}^2-f_2^2)}{f_2^2 M_{12}}. \tag{4.534}$$

Analysis of expression (4.533) shows that values of the partial frequencies are situated between the natural frequencies of the system. If we assume that $f_1<f_2$ and $f_{1p}<f_{2p}$, then relation $f_1<f_{1p}<f_{2p}<f_2$ is fulfilled. Thus, from expressions (4.534) follows that $ms_1>0$ and $ms_2<0$.

Quantitative characterization of interaction between partial systems and of a shift of natural frequencies relative to the partial ones is determined by the value of the coefficient of connectivity χ. At small values of γ the connectivity of partial systems significantly increases at $f_{1p}\to f_{2p}$, and conversely, with great difference between partial frequencies the partial systems become independent, even if the coefficient of coupling γ is relatively large.

From formulas (4.533) and (4.534) follows that at $f_{1p} = f_{2p} = f_p$

$$f_1^2 = f_p^2 / (1+\gamma), \quad f_2^2 = f_p^2 / (1-\gamma), \tag{4.535}$$

$$ms_1 = 1, \quad ms_2 = -1, \tag{4.536}$$

i.e., vibrations of the partial systems occur with the same amplitude, but at the lower natural frequency they are in phase, while at the higher natural frequency in anti-phase.

It is interesting to consider behavior of natural frequencies at constant coefficient γ in dependence from detuning between the partial systems, which is characterized by value of the detuning factor

$$\beta = f_{1p} / f_{2p}. \tag{4.537}$$

Denoting $\Omega = f^2 / f_{1p}^2$, as the normalized coefficient of natural frequencies of a coupled system, we bring Eq. (4.532) to the form

$$(1-\gamma^2)\Omega^2 - \left(1+\frac{1}{\beta}\right)\Omega + \frac{1}{\beta^2} = 0. \tag{4.538}$$

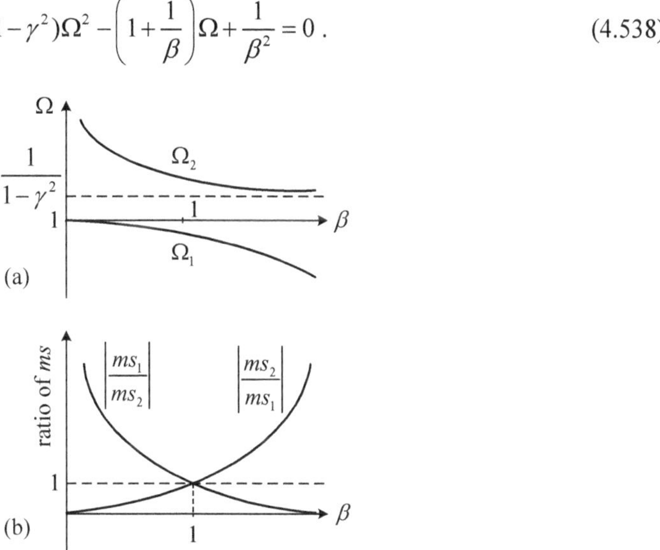

Figure 4.41: Dependences of (a) natural frequencies and (b) ratios of the mode shape coefficients from the detuning factor β.

The resulting dependence of Ω on β is qualitatively presented in Figure 4.41 (a).

At $\beta = 0$, we obtain $\Omega_1 = 1$ and $\Omega_2 \to \infty$, so that

$$f_1 = f_{1p} \text{ and } f_2 \to \infty . \tag{4.539}$$

At $\beta \to \infty$, $\Omega_1 \to 0$ and $\Omega_2 \to (1-\gamma^2)^{-1}$, so that

$$f_1 \to 0 \text{ and } f_2 \to f_{1p} / \sqrt{1-\gamma^2} . \tag{4.540}$$

The mode shape coefficients (4.534) in the new notations become

$$ms_1 = \frac{M_1}{M_{12}} \frac{1-\Omega_1}{\Omega_1}, \quad ms_2 = \frac{M_2}{M_{12}} \frac{1-\Omega_2}{\Omega_2}, \quad \frac{ms_1}{ms_2} = \frac{\Omega_2(1-\Omega_1)}{\Omega_1(1-\Omega_2)} . \tag{4.541}$$

At $\beta \to 0$, $\Omega_1 = 1$ and $ms_1 = 0$; $\Omega_2 \to \infty$ and $ms_2 \to -M_2 / M_{12}$.

At $\beta \to \infty$, $\Omega_1 \to 0$, $ms_1 \to \infty$; $\Omega_2 \to 1/(1-\gamma^2)$, $ms_2 \to -M_{12} / M_1$.

Dependences of ratios of the mode shape coefficients from the detuning factor β is illustrated in Figure 4.41 (b). The further away from the value $\beta = 1$, at which the connectivity between the partial systems is the greatest and $ms_1 = -ms_2$, the less is the contribution of the partial systems into vibration of each other. If the admissible value of the ratio $|ms_1 / ms_2|$ is set, the values of the detuning factor can be determined, at which interaction between the partial systems becomes insignificant.

4.6.1.2 Forced Vibrations in the Coupled Systems

Let us present Eqs. (4.526) and (4.527) replacing the displacements by the velocities of vibration $U_{1p} = \dot{\xi}_{1p}$ and $U_{2p} = \dot{\xi}_{2p}$. Besides, the following notations will be introduced

$$\begin{gathered} j(\omega^2 - \omega_{1p}^2) M_1 / \omega = Z_1, \\ j(\omega^2 - \omega_{1p}^2) M_2 / \omega = Z_2, \\ j\omega \, M_{12} = Z_c, \end{gathered} \tag{4.542}$$

where Z_1 and Z_2 are the self-impedances of the partial systems in the absence of coupling, Z_c is the coupling impedance of the partial systems (in our case the inertial coupling is assumed). As a result, the equations describing forced vibrations in a coupled system can be represented as

$$Z_1 U_{1p} + Z_c U_{2p} = F_1 , \tag{4.543}$$

$$Z_c U_{1p} + Z_2 U_{2p} = F_2 . \tag{4.544}$$

Solution to this system of equations is

$$U_{1p} = \frac{F_1 Z_2}{Z_1 Z_2 - Z_c^2} - \frac{F_2 Z_c}{Z_1 Z_2 - Z_c^2}, \tag{4.545}$$

$$U_{2p} = \frac{F_2 Z_1}{Z_1 Z_2 - Z_c^2} - \frac{F_1 Z_c}{Z_1 Z_2 - Z_c^2}. \tag{4.546}$$

The distribution of displacements in the mechanical system can be found in a general case using these equations. The concepts of input, Z_{11} and Z_{22}, and transfer, Z_{12}, impedances of the coupled system are commonly introduced as follows

$$\begin{gathered} Z_{11} = \frac{F_1}{U_{1p}}\bigg|_{F_2=0} = \frac{Z_1 Z_2 - Z_c^2}{Z_2}, \\ Z_{22} = \frac{F_2}{U_{2p}}\bigg|_{F_1=0} = \frac{Z_1 Z_2 - Z_c^2}{Z_1}, \\ Z_{12} = -\frac{F_1}{U_{2p}}\bigg|_{F_2=0} = -\frac{F_2}{U_{1p}}\bigg|_{F_1=0} = \frac{Z_1 Z_2 - Z_c^2}{Z_c}. \end{gathered} \tag{4.547}$$

The last relation presents formulation of the reciprocity principle for the mechanical systems.

Using expressions for the input and transfer impedances, Eqs. (4.543) and (4.544) can be rewritten in the form

$$U_{1p} = \frac{F_1}{Z_{11}} - \frac{F_2}{Z_{12}}, \quad U_{2p} = \frac{F_2}{Z_{22}} - \frac{F_1}{Z_{12}}. \tag{4.548}$$

The coupled system has two resonance frequencies, close to which the amplitudes of forced vibrations will increase in both partial systems. At small losses the frequencies, at which the maximum magnitudes of vibration are reached, are close to natural frequencies of the system, because they are determined from the condition of the minimum value of $|Z_1 Z_2 - Z_c^2|$, whereas the natural frequencies correspond to condition $\mathrm{Im}\{Z_1 Z_2 - Z_c^2\} = 0$.

If only one of the forces, for example F_1, is acting on the system, then the frequency responses of velocities of vibration in partial systems qualitatively look, as it is shown in Figure 4.42 for velocity U_{1p}. The peculiarity of this case is that U_{1p} goes to zero at partial frequency f_{2p}. This property can be utilized in designing vibration absorbers and rejecting filters. In the transducer (predominantly receiver) designs this can result in appearance of a trough in the frequency response provided the partial frequency of a coupled mechanical system (it can be an enclosing case or/and structural elements, to which the transducer is attached) gets into its operating range.

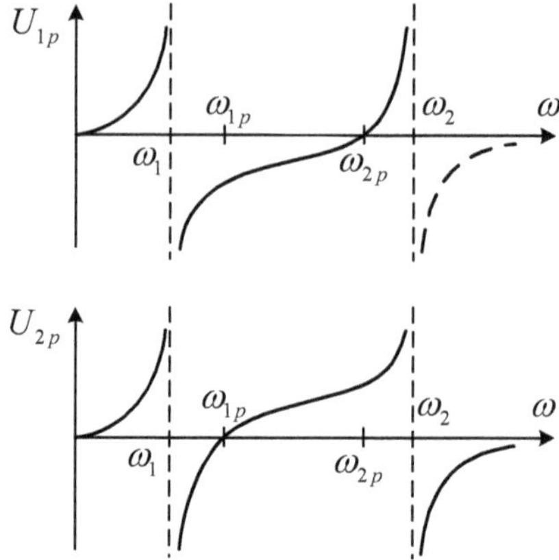

Figure 4.42: The frequency responses of the coupled system.

4.6.2 Examples of Coupled Vibrations in Mechanical Systems

4.6.2.1 Double Plate Symmetrical Mechanical System

Consider the mechanical system that is comprised of two identical circular plates hinged on a common base in the form of a rigid circular ring, shown in Figure 4.43. We assume that the magnitudes of vibration of centers of the plates are ξ_{1p} and ξ_{2p}. The mass and displacement of the base are denoted M_b and ξ_b. Distribution of displacements over surfaces of the plates can be represented as

$$\xi_1(r/a) = \xi_b + \xi_{1p}\theta(r/a), \quad \xi_2(r/a) = \xi_b + \xi_{2p}\theta(r/a). \tag{4.549}$$

Determine the kinetic and potential energies of the system taking into consideration the kinetic

energy of the base $M_b\dot{\xi}_b^2/2$. As a result, we obtain

$$W_{kin} = \frac{1}{2}2\pi\rho t\int_0^a(\dot{\xi}_b + \dot{\xi}_{1p}\theta)^2 rdr + \frac{1}{2}2\pi\rho t\int_0^a(\dot{\xi}_b + \dot{\xi}_{2p}\theta)^2 rdr + M_b\dot{\xi}_b^2/2, \tag{4.550}$$

$$W_{pot} = \frac{1}{2}(K_{eqv}\xi_{1p}^2 + K_{eqv}\xi_{2p}^2), \tag{4.551}$$

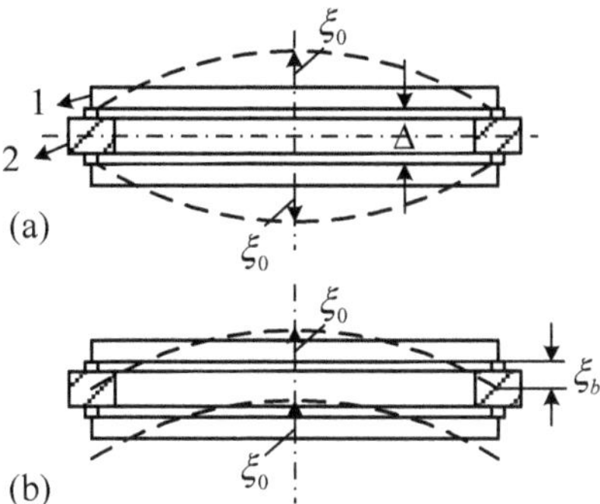

Figure 4.43: Coupled vibrations of double plate mechanical system: 1- plate; 2 – base (support). Modes of vibration of the double plate structure: (a) symmetrical vibration of plates in opposite phase, $\dot{\xi}_b = 0$; (b) in phase vibration of plates, $\dot{\xi}_b \neq 0$.

where K_{eqv} is the equivalent rigidity of the simply supported plate. After performing integration in expression for W_{kin}, and introducing notation $\rho t S_{av} = M_{av}$ for the average mass of a plate, we obtain

$$W_{kin} = \frac{2M_{pl} + M_b}{2}\dot{\xi}_b^2 + \dot{\xi}_b(\dot{\xi}_{1p} + \dot{\xi}_{2p})M_{av} + M_{eqv}\frac{\dot{\xi}_{1p}^2 + \dot{\xi}_{2p}^2}{2}. \tag{4.552}$$

Expressions for the equivalent rigidity and mass, and average surface S_{av} are presented by formulas (2.151). Namely, $M_{eff} = 0.29M_{pl}$ and $S_{av} = 0.46\pi a^2$. From Lagrange's equation for the generalized coordinate ξ_b

$$\frac{d}{dt}\left(\frac{\partial W_{kin}}{\partial \dot{\xi}_b}\right) = 0, \tag{4.553}$$

follows that

$$\dot{\xi}_b = -\frac{(\dot{\xi}_{1p} + \dot{\xi}_{2p})M_{av}}{M_b + 2M_{pl}}. \tag{4.554}$$

Upon substituting the velocity $\dot{\xi}_b$ into Eq. (4.552) it will be obtained

$$W_{kin} = \frac{1}{2}M_1\dot{\xi}_{1p}^2 + M_{12}\dot{\xi}_{1p}\dot{\xi}_{2p} + \frac{1}{2}M_2\dot{\xi}_{2p}^2, \tag{4.555}$$

where

$$M_1 = M_2 = M_{eqw}\left[1 - \frac{1}{3(1 + M_b / 2M_{pl})}\right], \tag{4.556}$$

$$M_{12} = -\frac{0.1M_{pl}}{1+M_b/2M_{pl}}. \tag{4.557}$$

Thus, according to formula (4.529)

$$\gamma = \frac{M_{12}}{\sqrt{M_1M_2}} = -\frac{1}{3(1+M_b/2M_{pl})-1}, \tag{4.558}$$

and

$$M_1 = M_{eqv}\frac{1}{1+|\gamma|}. \tag{4.559}$$

As the coefficient of coupling, γ, is negative, the absolute value is taken to avoid confusion regarding the sign of this quantity in formulas, where it is used.

The partial frequencies are

$$f_{1p} = f_{2p} = f_p = \sqrt{\frac{K_{eqv}}{M_1}} = f_0\sqrt{1+|\gamma|}, \tag{4.560}$$

where f_0 is the first resonance frequency of the simply supported circular plate according to formula (4.184). Since the partial frequencies are equal, the natural frequencies of the coupled system can be found by formulas (4.535), namely

$$f_1^2 = \frac{f_p^2}{1-|\gamma|} = \frac{f_0^2(1+|\gamma|)}{1-|\gamma|}, \quad f_2^2 = \frac{f_p^2}{1+|\gamma|} = f_0^2. \tag{4.561}$$

In this case $f_2 < f_1$.

According to relations (4.536) at frequency f_2, $ms_1 = -1$, i.e., $\xi_{2p} = -\xi_{1p}$. The plates vibrate in opposite phase (Figure 4.43 (a)), and displacement of the base $\xi_b = 0$. The mechanical system vibrating in this mode is ideal for designing pressure hydrophones. Being fixed to a platform by the base, the hydrophone should not be sensitive to a structural vibration. At frequency f_1 we obtain $\xi_{1p} = \xi_{2p} = \xi_p$. The plates vibrate in phase (Figure 4.43 (b)), and displacement of the base may be determined by formula (4.554), as

$$\dot{\xi}_b = -\frac{2\dot{\xi}_p M_{av}}{M_b + 2M_{pl}} = \frac{0.45}{1+M_b/2M_{pl}}. \tag{4.562}$$

The mechanical system vibrating in this mode can be used for designing the pressure gradient hydrophones.

Note that at $M_b = 0$ $\gamma = -0.5$ and $f_1 = 1.73 f_0$ from relation (4.561). At $(1 + M_b / 2M_{pl}) \to \infty$ we obtain $\gamma \to 0$ and $f_1 \to f_2 = f_0$. In this case the plates may vibrate independently, and the mode of vibration depends on the phase correlation of forces that generate vibration. A close effect can be achieved at finite values of the ratio $M_b / 2M_{pl}$. This can be estimated from formula

$$f_1 / f_o = \sqrt{[(M_b / 2M_{pl}) + 1] / [(M_b / 2M_{pl}) + 0.34]} \tag{4.563}$$

that can be obtained from relation (4.561) using expression (4.558) for the coefficient γ.

4.6.2.2 Coupled Vibrations in Rectangular Plates and Bars

In general, the mechanical systems of piezoelectric ceramic transducers are elastic bodies of a size, for which it is hard to accurately determine or predict the distribution of displacements over the volume under vibration. Approximate approach to solving of the problem was suggested in Ref. 19 and is known as "hypothesis of Giebe and Blechshmidt". The approach to solving the problem is based on the assumptions that: (a) vibration of a real elastic body may be represented as the coupled vibration of the partial one-dimensional systems, to which the real body approaches at the extreme values of its aspect ratios, and (b) the coupling factors between the partial systems may be selected in such a way as to yield the known resonance frequencies for their extreme one-dimensional configurations.

Thus, for example, it was suggested to consider the extensional vibration of a rectangular plate as a coupled vibration of two bars, as shown in Figure 4.44, with inertia coupling between them. For this case the first resonance frequencies of the partial systems are $f_{1p} = (1/2L_1)\sqrt{Y/\rho}$ and $f_{2p} = (1/2L_2)\sqrt{Y/\rho}$, and the coefficient of coupling was chosen as $\gamma_m = \sigma$. In Ref. 20 such approach was used for determining the resonance frequencies of piezoelectric bodies of several configurations, and the results obtained where sufficiently accurate.

However, the technique based on the Gibbe and Blechschmidt approach cannot be directly applied to calculating the electromechanical parameters of transducers. Moreover, it cannot be applied to treatment of transducers as electroacoustic, i.e., under the action of acoustic loads, because this requires knowing how the modes of vibration of piezoelements change vs. their aspect ratios. For these applications the approach needs to be modified.

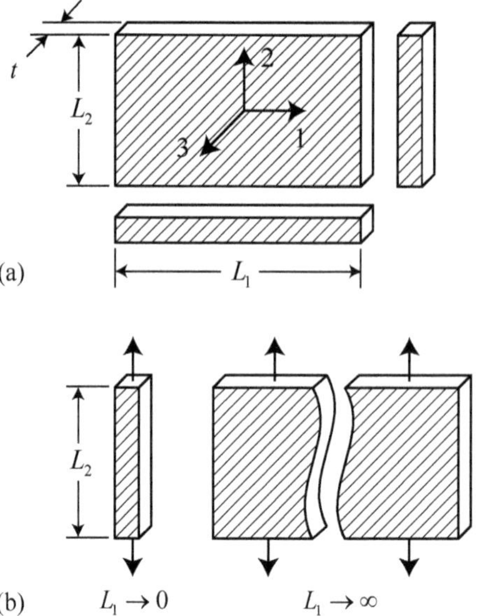

Figure 4.44: Drawing of a rectangular plate considered as a coupled system of two bars (a), the extreme one-dimensional systems.

This can be done by employing the general coupling theory, if to assume that the partial modes of displacement in the mechanical system are chosen as suggested according to part (a) of the Gibbe and Blechschmidt hypothesis. In this section the modification will be illustrated with example of the mechanical systems in the shape of thin rectangular plates (Figure 4.44 (a)) and long bars (Figure 4.45), vibration of which can be regarded as two-dimensional. Here we assume that mechanical

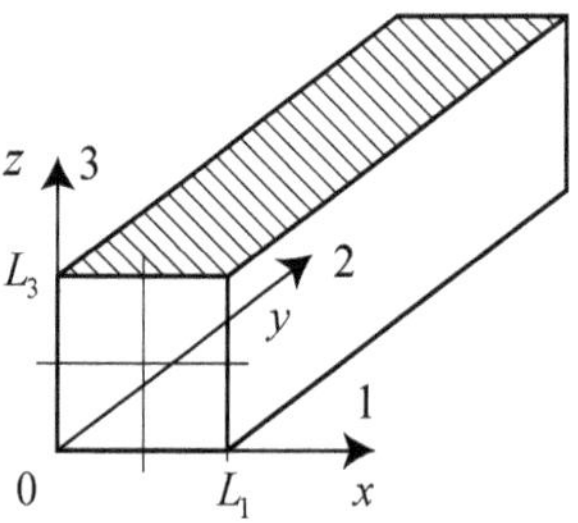

Figure 4.45: Geometry of the long bar and coordinate systems. It is assumed that deformation $S_2 = 0$.

systems are passive isotropic bodies. In the case that they are made of piezoelectric ceramics, their properties depend on the directions of their polarization and may become anisotropic. The differences that arise from these effects will be considered when the transducers employing these mechanical systems will be treated. To make referencing the results of this section easier, the surfaces that will presumably be covered by electrodes are marked with dashed lines, and the crystallographic coordinate systems corresponding to the directions of polarization are used.

6.1.3.1.1. Coupled Vibrations in the Rectangular Plates

Consider vibration of a thin rectangular isotropic plate in its plane. The geometry of the plate is shown in Figure 4.44 (a). We assume that the thickness of the plate, *t*, is much smaller than its lateral dimensions L_1 and L_2, and therefore the boundary condition can be imposed that the stress $T_3 = 0$. We are interested in the dependence of the plate's parameters on its aspect ratio, L_1 / L_2. The partial distributions of displacements in the plate will be represented as the fundamental modes of vibration of thin bars having lengths L_1 and L_2,

$$\xi_x = \xi_{1p} \cos(\pi x / L_1), \quad \xi_y = \xi_{2p} \cos(\pi y / L_2)\,. \tag{4.564}$$

The corresponding distribution of displacement and strain in the plate will be

$$\xi(x,y) = \xi_{1p} \cos \pi x / L_1) + \xi_{2p} \cos(\pi y / L_2)\,, \tag{4.565}$$

$$S_1 = \xi_{1p}(\pi / L_1)\sin(\pi x / L_1), \quad S_2 = \xi_{2p}(\pi / L_2)\sin(\pi y / L_2)\,. \tag{4.566}$$

(The minus sign is omitted here and further for brevity. Its meaning is that at positive directions of displacements ξ_{1p} and ξ_{3p} the strains are compressive, i.e., negative by adopted sign convention. But this is not important in context of this section.)

For the stress follows from Eqs. (4.10)

$$T_1 = \frac{Y}{1-\sigma^2}(S_1 + \sigma S_2), \quad T_2 = \frac{Y}{1-\sigma^2}(\sigma S_1 + S_2)\,. \tag{4.567}$$

For the strain in the direction of z axis we obtain

$$S_3 = -\frac{\sigma}{1-\sigma}(S_1 + S_2)\,. \tag{4.568}$$

The energies associated with vibration of the plate are determined as follows. The potential energy is

$$W_{pot} = \frac{1}{2}\int_{\tilde{V}}(T_1S_1 + T_2S_2)d\tilde{V} = \frac{1}{2}\frac{Y}{1-\sigma^2}\int_{\tilde{V}}(S_1^2 + 2\sigma S_1S_2 + S_2^2)d\tilde{V} . \qquad (4.569)$$

Here is the volume of the mechanical system. After substituting expressions (4.566) for the strain and integrating over the volume of the plate we obtain

$$W_{pot} = \frac{1}{2}(K_1\xi_{1p}^2 + 2K_{12}\xi_{1p}\xi_{2p} + K_2\xi_{2p}^2), \qquad (4.570)$$

where

$$K_1 = \frac{\pi^2 Yt}{2(1-\sigma^2)}\frac{L_2}{L_1}, \quad K_2 = \frac{\pi^2 Yt}{2(1-\sigma^2)}\frac{L_1}{L_2}, \quad K_{12} = \frac{4Y\sigma t}{1-\sigma^2} . \qquad (4.571)$$

The kinetic energy of the vibrating plate is

$\tilde{V}$

$$W_{kin} = \frac{\rho}{2}\int_{\tilde{V}}(\dot{\xi}_x^2 + \dot{\xi}_y^2 + \dot{\xi}_{z\Sigma}^2)d\tilde{V} . \qquad (4.572)$$

Here the velocity of the surface vibration, $\dot{\xi}_z$, is denoted as $\dot{\xi}_{z\Sigma}$. It can be found from Eq. (4.568) that $\dot{\xi}_{z\Sigma} = \dot{\xi}_3 = \dot{S}_3 z$. Thus,

$$\dot{\xi}_{z\Sigma} = -\frac{\sigma t}{2(1-\sigma)}(\dot{S}_1 + \dot{S}_2) . \qquad (4.573)$$

Note that velocity $\dot{\xi}_{z\Sigma}$, being relatively small compared with velocities in lateral directions, is important for measuring the mode shapes of the plate vibration. This is the quantity that determines the acoustic radiation in the case that a laterally vibrating plate is used for this purpose. After integrating over the volume of the plate, the kinetic energy can be represented as

$$W_{kin} = \frac{1}{2}M_1\dot{\xi}_{1p}^2 + M_{12}\dot{\xi}_{1p}\dot{\xi}_{2p} + \frac{1}{2}M_2\dot{\xi}_{2p}^2, \qquad (4.574)$$

where

$$M_1 = \frac{M}{2}\left[1 + \frac{\pi^2\sigma^2t^2}{12(1-\sigma^2)L_1^2}\right], \quad M_2 = \frac{M}{2}\left[1 + \frac{\pi^2\sigma^2t^2}{12(1-\sigma^2)L_2^2}\right],$$
$$M_{12} = M\frac{\sigma^2t^2}{3(1-\sigma^2)L_1L_2} . \qquad (4.575)$$

In these relations, $M = L_1L_2t\rho$ is the mass of the plate. So far as $t \ll L_1, L_2$, the contribution of the terms with factors t^2/L^2 is negligible and

$$M_1 = M_2 = M / 2, \quad M_{12} \approx 0 . \tag{4.576}$$

Thus, the coupling between the partial systems may be regarded as pure elastic.

Considering expressions for the potential (4.570) and kinetic (4.574) energies together with expressions for the equivalent rigidities (4.571) and masses (4.575), we can arrive at the following conclusions. In the case under consideration the partial system that is determined by the condition $\xi_{1p} = 0$ or $S_1 = 0$, at $L_1 \to \infty$, constitutes a strip that is infinite in the x direction and vibrates along its width L_2. Similarly, another partial system (at $\xi_{2p} = 0$) is the strip that is infinite in the y direction and vibrates along dimension L_1. Thus, the partial resonance frequencies are

$$f_{ip} = \frac{1}{2\pi}\sqrt{\frac{K_i}{M_i}} = \frac{1}{2L_i}\sqrt{\frac{Y}{\rho(1-\sigma^2)}} \quad (i = 1,2) . \tag{4.577}$$

The coupling between the partial systems is elastic with the coupling factor

$$\gamma_e = \frac{K_{12}}{\sqrt{K_1 K_2}} = \frac{8\sigma}{\pi^2} . \tag{4.578}$$

(The subscript will be further omitted.) For a plate made of PZT-4 ceramic polled in z direction ($\sigma = \sigma_1^E = 0.33$) $\gamma = 0.27$.

After the energies associated with free vibration of the plate (without taking into account the energy losses and external loads) are determined, the Lagrange's equations can be represented in the following form analogous to Eqs. (4.530) and (4.531)

$$(f_{1p}^2 - f^2)U_{1p} + (K_{12} / K_1) f_{1p}^2 U_{2p} = 0 , \tag{4.579}$$

$$(K_{12} / K_2) f_{2p}^2 U_{1p} + (f_{2p}^2 - f^2) U_{2p} = 0 . \tag{4.580}$$

The frequency equation follows from this set of equations in the form

$$(f_{1p}^2 - f^2)(f_{2p}^2 - f^2) - \gamma^2 f_{1p}^2 f_{2p}^2 = 0 . \tag{4.581}$$

We assume further for definiteness that the dimension L_2 (i.e., f_{2p}) is kept constant and L_1 changes. After denoting $f^2 / f_{2p}^2 = \Omega$ as the normalized non-dimensional resonance frequency factor and

$$\frac{f_{2p}}{f_{1p}} = \frac{L_1}{L_2} = \beta , \tag{4.582}$$

as ratio of the partial frequencies, Eq. (4.581) may be transformed to

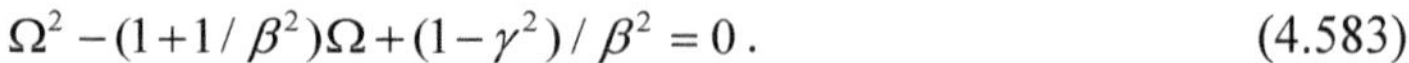

$$\Omega^2 - (1 + 1/\beta^2)\Omega + (1 - \gamma^2)/\beta^2 = 0 . \tag{4.583}$$

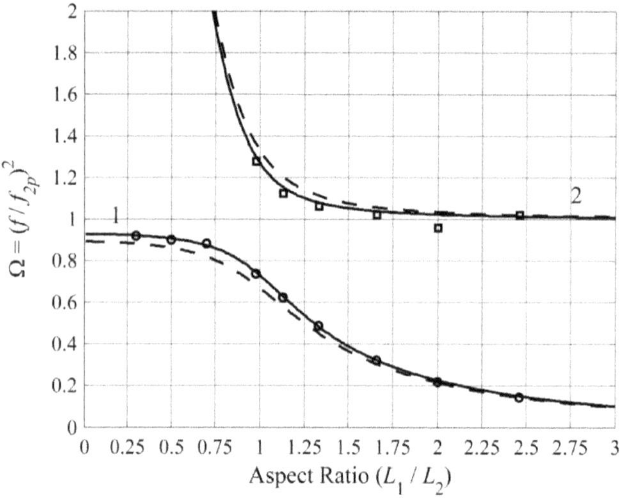

Figure 4.46: Resonance frequency dependence on aspect ratio for lower (1) and upper (2) frequency branches (solid – calculated, dashed - Ref. 20, markers – measured, Ref. 21).

From this equation two branches of resonance frequencies that correspond to solutions Ω_1 and Ω_2 may be found as functions of the aspect ratio. We will use the convention that $\Omega_1(\beta)$ forms the lower and $\Omega_2(\beta)$ the upper frequency branch. The frequency dependencies calculated at value of the coupling factor $\gamma = (8/\pi^2)\sigma_1 = 0.27$ are shown in Figure 4.46 by solid lines. It follows from Eq. (4.583) that at $\beta = 1$

$$\Omega_1 = 1 - \gamma, \quad \Omega_2 = 1 + \gamma, \quad \frac{f_2^2}{f_1^2} = \frac{\Omega_2}{\Omega_1} = \frac{1+\gamma}{1-\gamma} . \tag{4.584}$$

After the resonance frequencies are determined, the corresponding mode shape factors, which will be defined as

$$ms_i = \left. \frac{U_{1p}}{U_{2p}} \right|_{\Omega = \Omega_i} \quad (i = 1,2) \tag{4.585}$$

may be found from the set of Eqs. (4.579) and (4.580). Namely,

$$ms_i = -\frac{K_{12}}{K_1} \frac{1}{1 - \Omega_i \beta^2} . \tag{4.586}$$

It follows from expressions (4.578) and (4.571) that $(K_{12} / K_1) = \gamma\beta$. Thus, Eq. (4.586) can be represented as

$$ms_i = -\frac{\gamma\beta}{1-\Omega_i\beta^2}. \tag{4.587}$$

The dependence of the mode shape factors on the aspect ratio β is shown in Figure 4.47. It follows from Eq. (4.587) that at frequencies pertaining to the lower branch the velocities U_1 and U_2 are in anti-phase. In particular, at $\beta = 1$ $ms_1 = -1$ and $ms_2 = 1$. This means that at higher and lower resonance frequencies the velocities have the same magnitude but are in-phase at the higher frequency and in anti-phase at the lower frequency.

It is interesting to estimate the relation of the obtained solutions for the resonance frequencies with known results for the limiting one-dimensional configurations of a plate, which are a long bar (at $\beta \to 0$ and L_2 constant) and a long thin strip (at $\beta \to \infty$ and L_2 constant). It follows from Eq. (4.583) that

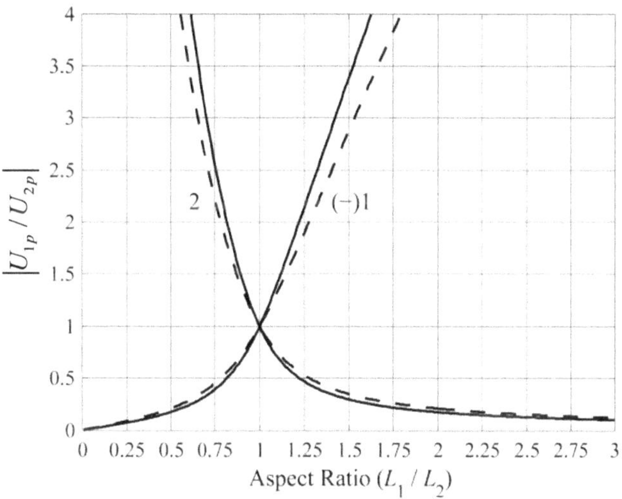

Figure 4.47: Dependence of the mode shape factors on aspect ratio for lower (1) and upper (2) frequency branches (solid - calculated, dashed – calculated using $\gamma = \sigma_1^E$ following Ref. 20).

$$\Omega_{1,2} = \frac{1+\beta^2}{2\beta^2}\left[1 \pm \sqrt{1-\frac{4\beta^2(1-\gamma^2)}{(1+\beta^2)^2}}\right]. \tag{4.588}$$

At $\beta \to 0$

$$\Omega_{1,2} \approx \frac{1+\beta^2}{2\beta^2}\{1 \pm [1-2\beta^2(1-\gamma^2)]\}, \tag{4.589}$$

and

$$\Omega_1 \approx (1-\gamma^2), \quad \Omega_2 \to 1/\beta^2 . \tag{4.590}$$

Thus,

$$f_1 \to f_{2p}\sqrt{1-\gamma^2}, \quad f_2 \to (f_{2p}/\beta) = f_{1p} . \tag{4.591}$$

Taking into account expression (4.571) for the partial frequencies f_{ip}, we arrive at the resonance frequency for a thin bar of the length L_2

$$f_1 = \frac{1}{2L_2}\sqrt{\frac{Y(1-\gamma^2)}{\rho(1-\sigma^2)}}, \tag{4.592}$$

and for an infinitely long strip of the width L_1

$$f_2 = \frac{1}{2L_1}\sqrt{\frac{Y}{\rho(1-\sigma^2)}}. \tag{4.593}$$

Given that the exact value for the resonance frequency of the thin bar must be $f_{bar} = (1/2L_2)\sqrt{Y/\rho}$ and that $\gamma = 0.27$ we obtain $f_1 = 1.02 f_{bar}$. Thus, the error of the current approach can be considered as negligible for the limiting case at $\beta \to 0$. For another extreme case (long strip) the value of resonance frequency obtained by formula (4.593) is exact.

6.1.3.1.2. Coupled Vibrations in the Long Rectangular Bars

The geometry and the electrode location for a long bar are shown in Figure 4.45. The condition $S_2 = 0$ is held. Distribution of displacements is assumed to be

$$\xi_x = \xi_{1p}\cos(\pi x/L_1), \quad \xi_z = \xi_{3p}\cos(\pi z/L_3), \tag{4.594}$$

where from the strains are

$$S_1 = \xi_{1p}(\pi/L_1)\sin(\pi x/L_1), \quad S_3 = \xi_{3p}(\pi/L_3)\sin(\pi z/L_3). \tag{4.595}$$

(Regarding the omitted sign minus, see the note under Eq. (4.566).)

From Eqs. (4.9) follows that

$$T_1 = c_{11}S_1 + c_{13}S_3, \quad T_3 = c_{13}S_1 + c_{33}S_3, \tag{4.596}$$

where for an isotropic material

$$c_{11} = c_{33} = c = \frac{Y(1-\sigma)}{(1+\sigma)(1-2\sigma)}, \quad c_{13} = \frac{Y\sigma}{(1+\sigma)(1-2\sigma)}. \tag{4.597}$$

The same technique as was used for treating the plate can be applied to this case in the straightforward way. The partial systems in this case are the infinite in lateral dimensions plates of thickness L_1 (in which case $S_3 = 0$) and of thickness L_3 (in which case $S_1 = 0$). From considering expressions for the potential and kinetic energies analogous to (4.570) and (4.574) we obtain that

$$K_1 = \frac{c\pi^2 L_2}{2}\frac{L_3}{L_1}, \quad K_3 = \frac{c\pi^2 L_2}{2}\frac{L_1}{L_3}, \quad K_{13} = 4c_{13}L_2, \tag{4.598}$$

and

$$M_1 = M_3 = \frac{L_1 L_2 l_3}{2}\rho = \frac{M_{bar}}{2}, \quad M_{13} \approx 0. \tag{4.599}$$

Thus, the partial frequencies are

$$f_{1p} = \frac{1}{2L_1}\sqrt{\frac{c}{\rho}}, \quad f_{3p} = \frac{1}{2L_3}\sqrt{\frac{c}{\rho}}, \tag{4.600}$$

and the expressions for the ratio of the partial frequencies is $\beta = (L_1 / L_3)$. The coupling is elastic with the coupling factor

$$\gamma = \frac{K_{13}}{\sqrt{K_1 K_3}} = \frac{8}{\pi^2}\frac{c_{13}}{c} = \frac{8}{\pi^2}\frac{\sigma}{1-\sigma}. \tag{4.601}$$

For example, for a bar made of isotropic material having Poisson's ratio $\sigma = 0.33$ $\gamma = 0.4$.

In case that the bars are made of isotropic material, dependencies of the modal resonance frequencies and mode shape factors vs. coefficient β can be found from Eqs. (4.583) and (4.587), respectively, at corresponding value of the coupling factor γ, if to replace subscript 3 by 2.

For the piezoceramic bars all the calculations complicate because of anisotropy of elastic properties. This case will be considered in Chapter 10.

It is noteworthy that the results presented in this section are restricted to the fundamental modes of vibration of the partial systems, and as such are applicable for frequency range near and below the resonance frequencies of the two lowest modes of their vibration. The analysis is not intended for a much broader frequency range that could be of interest for a general treatment of the vibration of piezoelectric plates and bars per se. The overtones of the partial systems are neglected because the corresponding modes of vibration are typically not suitable for practical or effective electromechanical transduction.

4.6.2.3 Coupled Vibrations of Cylindrical Discs and Solid Rods

Piezoelectric elements in the shape of thick discs or solid circular cylinders (Figure 4.48) that undergo longitudinal vibrations in the axial direction are used for many applications in a broad frequency range. The height to diameter aspect ratio of these piezoelements, $h/2a$, may change from very small (at $h/2a \ll 1$, Figure 4.48 (b)), to rather large (at $h/2a \gg 1$, Figure 4.48 (c)). In these limits of extreme aspect ratios, the piezoelements are commonly called "discs" and "rods," respectively, and can be considered as one-dimensional systems in terms of calculating their vibration in the axial direction. In the intermediate cases that the height of the piezoelements is comparable with their diameter, the piezoelements may be called "finite-height cylinders." This definition will be used further for all the geometries in general so far as the dependence is concerned of their parameters on the aspect ratios.

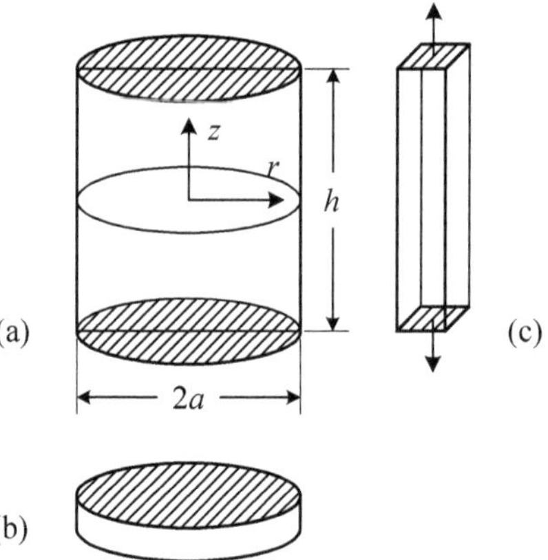

Figure 4.48: (a) Geometry of the finite size cylinder, (b) the extreme case at $h/2a \ll 1$, (c) the extreme case at $h/2a \gg 1$.

There have been many results of investigations related to the vibration of the circular discs having various height to diameter aspect ratios published over the span of several decades

(Ref.21-24). The main goal of these experimental works was in investigating and classifying the axisymmetric vibrational modes of the circular discs and their dependence on the height to diameter aspect ratio. Piezoceramic discs were used in these investigations with fully electroded end surfaces as a convenient way to generate axisymmetric vibrations. These works showed complexity of vibrations especially for thin discs. Authors of the referred works performed very careful and detailed measurements of the modal distributions of vibration over the discs surface in the axial direction that proved to be essentially nonuniform in the most cases, which makes them virtually unusable for effective sound radiation. This information can be used for qualitatively predicting the ranges of aspect ratios that have to be avoided in practical designing of the piezoelements for electroacoustic transducers.

The following qualitative description can be made from the point of view of the coupled vibration concept, of how parameters of piezoelements change as the aspect ratio progresses from extremely small to larger values (or as the resonance frequency of axial mode of vibration, f_{ac}, changes from high to lower values, while the diameter of the discs remains constant).

When the resonance frequency of the axial vibration approaches the resonance frequencies of the radial modes of vibration, $f_{rad.i}$, the effect of coupling between these modes takes place that results in nonuniform distribution of axial displacement over the surface due to contribution of the radial modes. So far as the aspect ratio remains small, the radial modes are of high order (about $i > 5-6$), their effective coupling coefficients are small, and contribution of these modes to the axial displacement can be practically neglected. In this range of aspect ratios the axial (thickness) vibration of a disc can be approximately considered as one dimensional

As the aspect ratio increases, the radial modes of lower order having higher effective coupling coefficients come into action and the effects of their coupling with the axial mode become more pronounced, including raising nonuniformity of the distribution of axial vibration that may result in corrupting the directional factor of a corresponding transducer. These effects are especially significant in the ranges of aspect ratios around values that correspond to the condition $f_{ax} \approx f_{rad.i}$, at which the coupling between modes is strongest. After the aspect ratio reaches the values, at which interaction must be taken into account with the first (the lowest, at $f_{rad.1}$) mode of radial vibration only, the cylinder can be considered as a two degree-of-freedom system, and the coupled vibration technique can be applied to calculating its parameters. With a further increase of the aspect ratio the longitudinal vibration of relatively tall cylinder can be considered isolated and one-dimensional theory of calculating the corresponding transducers is applicable.

In this section we will analyze the dependence of the resonance frequencies and the mode shape coefficients of the finite-size cylindrical piezoelements for the range of aspect ratios, for which coupled vibration technique can be applied to provide means for calculating the operating characteristics of the transducers that employ vibration of such piezoelements, and to determine approximate value of aspect ratio, after which the longitudinal vibration of the cylinder can be considered as one-dimensional. For this purpose the coupled vibration analysis technique can be used that is described in Section 4.6.1. In order to be qualified as a two degree of freedom system, to which the coupled technique approach can be applied, the corresponding partial systems must each have a single degree of freedom in all the intended range of changing the aspect ratio of the body. For specifying the partial systems the assumed distribution of displacements in the body must be suggested. Following the Gibbe and Blechshmidt hypothesis the assumed partial distribution of the displacements will be chosen, as the modes of vibration of the one-dimensional bodies at the extreme values of the height to diameter aspect ratio *h*/2*a*. In our case they can be: the radial vibrating thin disc and axial vibrating thin bar at $h \to 0$ and at $a \to 0$, respectively, as shown in Figure 4.48. The partial displacements in the radial and axial directions will be denoted as ξ_a and ξ_h. Characterization of the partial systems will be obtained with reference to Eqs. (4.521) and (4.522), in which the partial displacements have to be replaced by ξ_a and ξ_h. The pair of the partial systems that comply with these equations represent radially vibrating infinitely long cylinder (in limit $h \to \infty$, $\xi_h \to 0$) and the thickness vibrating disc of a large diameter (in limit $a \to \infty$, $\xi_a \to 0$), because at $\xi_h = 0$ the remaining equations characterize the radially vibrating infinitely long cylinder, and at $\xi_a = 0$ they characterize the axially vibrating disc of infinite radius.

If to consider the axial vibration of a finite-size cylinder around its lowest resonance frequency only, which is practically the case, then distribution of displacement in the axial direction is

$$\xi(z) = \xi_h \sin(\pi z / h), \tag{4.602}$$

and it remains the same in all the range of aspect ratios.

The radial vibration of the thin disc is considered in Section 4.4.2. It cannot be regarded as one degree of freedom systems in the entire frequency range of interest, because it has multiple resonances within this range. The axisymmetric distribution of displacement in the radial direction being represented as expansion in the series by normal modes of vibration (4.163) is

$$\xi(r) = \sum_{i=1}^{\infty} \xi_i J_1(k_i r) = \sum_{i=1}^{\infty} \xi_{ai} J_1(k_i r) / J_1(k_i a). \tag{4.603}$$

Here ξ_{ai} are the displacements at $r = a$, $k_i = \lambda_i / a$, and λ_i are the eigenvalues for the vibration problem (Section 4.4.2). In order to reveal the range of aspect ratios, in which application of the two-dimensional coupling technique is possible, the first term of the series has to be retained only,

$$\xi(r) = \xi_a J_1(k_1 r) / J_1(k_1 a), \tag{4.604}$$

where $k_1 a = 2.05$. The corresponding resonance frequency is (4.168)

$$f_a = \frac{2.05}{2\pi a} \sqrt{\frac{Y}{\rho(1-\sigma^2)}}. \tag{4.605}$$

The resonance frequency of the next radial mode is at $k_2 a = 5.83$. It can be considered that the first resonance mode of vibration remains dominant up to approximately $ka = 3$.

Consider as the partial resonance frequencies the resonance frequency of the bar vibrating in its fundamental mode,

$$f_h = \frac{1}{2h} \sqrt{Y / \rho}, \tag{4.606}$$

and the resonance frequency of the radially vibrating thin disc by formula (4.605) corrected to $ka = 3$. Then the aspect ratios that correspond to condition $\beta = (f_a / f_h) = 1$ of the strongest coupling for the resonance modes of the disc can be determined from relation

$$\beta = 6h / 2\pi a \sqrt{(1-\sigma^2)} = 1 \text{ or } (h / 2a) > 0.5 \text{ at } \sigma = 0.33. \tag{4.607}$$

With increase of the aspect ratio (increase the height with $2a$ constant) above approximately $h / 2a \approx 1.5$ the third mode of the longitudinal vibration cannot be ignored in terms of applicability of the coupled vibration technique due to its coupling with the first radial mode.

(The strongest coupling between third mode of longitudinal vibration and first radial mode of a tall cylinder takes place at $h/2a \approx 1.8$). But the range of aspect ratios above this value does not present problems in terms of transducer designing, because at these aspect ratios the longitudinal vibration of the cylinder can be treated as one-dimensional, as will be shown below, and the radial modes of vibration of such tall rod do not present a practical interest.

Thus, the entire range of aspect ratios can be divided conditionally into several regions that may differ qualitatively in terms of contribution of radial vibration of a disc into its thickness vibration and in terms of applicability of treating the vibration as a coupled two-dimensional problem. At aspect ratios $h/2a < 0.1$, the contribution of effects of vibration in radial direction can be practically neglected due to negligible values of the modal coupling coefficients. The thickness vibration can be considered as the extreme case of one-dimensional vibration of a plate with infinitely large lateral dimensions.

The range of aspect ratios approximately in the interval $0.5 > h/2a > 0.1$, is characterized by coupling the thickness (axial) mode of vibration with multiple radial modes having relatively small separation between the resonance frequencies. This makes the vibration problem rather complicated, and results in highly nonuniform distributions of displacements on the top cylinder surfaces. This range of the aspect ratios is hardly usable for transducers designing. Information on the displacement distributions for this range of the aspect ratios can be found in Ref. 21. In the range of aspect ratios approximately $1.5 > h/2a > 0.5$ the first radial mode is dominant, and the radially vibrating disc can be considered as a single degree of freedom system. In this range of aspect ratios the vibration problem for the finite-size cylindrical piezoelement can be treated as a two degree of freedom system by applying the coupled theory approach. Dependences of the resonance frequencies and mode shape coefficients vs aspect ratio are further considered analytically for this range of the aspect ratios.

The strains in the body of a cylinder according to the assumed distributions of displacements (4.602) and (4.603) are

$$S_1 = \frac{\partial \xi_r}{\partial r} = \xi_a \frac{1}{J_1(k_1 a)} \frac{\partial J_1(k_1 r)}{\partial r} = \xi_a \frac{k_1}{J_1(k_1 a)} \left[J_0(k_1 r) - \frac{J_1(k_1 r)}{k_1 r} \right], \tag{4.608}$$

$$S_2 = \frac{\xi_r}{r} = \xi_a \frac{k_1}{J_1(k_1 a)} \frac{J_1(k_1 r)}{k_1 r}, \tag{4.609}$$

$$S_3 = \frac{\partial \xi_z}{\partial z} = \xi_h \frac{\pi}{h} \cos\frac{\pi z}{h}. \tag{4.610}$$

The corresponding expressions for stresses must be obtained from Eqs. (4.9) under the conditions that for the partial system at $\xi_a = 0$ (axially vibrating disc of a large diameter) $S_1 = S_2 = 0$,and at $\xi_h = 0$ (radially vibrating cylinder of a large height) $S_3 = 0$. Thus,

$$T_1 = \frac{Y}{(1+\sigma)(1-2\sigma)}[(1-\sigma)S_1 + \sigma S_2], \tag{4.611}$$

$$T_2 = \frac{Y}{(1+\sigma)(1-2\sigma)}[\sigma S_1 + (1-\sigma)S_2], \tag{4.612}$$

$$T_3 = \frac{Y}{(1+\sigma)(1-2\sigma)}(1-\sigma)S_3. \tag{4.613}$$

Our final goal is in treating coupled vibration of piezoelements of the same configuration.

This will be done in Chapter 10. Elastic properties of the piezoelectric ceramics are anisotropic, and equations analogous to (4.9) look like

$$\begin{aligned} T_1 &= c_{11}^E S_1 + c_{12}^E S_2 + c_{13}^E S_3, \\ T_2 &= c_{12}^E S_1 + c_{11}^E S_2 + c_{13}^E S_3, \\ T_3 &= c_{13}^E S_1 + c_{13}^E S_2 + c_{33}^E S_3, \end{aligned} \tag{4.614}$$

where c_{ik}^E are the elastic moduli of the ceramics at electric field $E_3 = 0$. In order to make expressions for the equivalent parameters of an isotropic passive body like those of the piezoelements of the same configuration, and for sake of brevity it is convenient to introduce the notations

$$\frac{Y(1-\sigma)}{(1+\sigma)(1-2\sigma)} = c_a = c_h, \tag{4.615}$$

where c_a is analogous to c_{11}^E and c_h is analogous to c_{33}^E,

$$\frac{Y\sigma}{(1+\sigma)(1-2\sigma)} = c_{ah}, \tag{4.616}$$

where c_{ah} is analogous to c_{13}^E and c_{12}^E.

Consider the expressions for the potential and kinetic energies of the coupled system.

$$W_{pot} = \frac{1}{2} 2\pi \int_{-h/2}^{h/2} \int_{0}^{a} (S_1 T_1 + S_2 T_2 + S_3 T_3) r dr dz = \frac{1}{2}(K_{eqva}\xi_a^2 + 2K_{ah}\xi_a \xi_h + K_{eqvh}\xi_h^2), \quad (4.617)$$

$$W_{kin} = \frac{1}{2} 2\pi \int_{-h/2}^{h/2} \int_{0}^{a} \rho[\dot{\xi}_r^2 + \dot{\xi}_z^2] r dr dz = \frac{1}{2}(M_{eqva}\dot{\xi}_a^2 + 2M_{ah}\dot{\xi}_a \dot{\xi}_h + M_{eqvh}\dot{\xi}_h^2). \quad (4.618)$$

After substituting the above expressions for the stress, strain, and displacements, the following expressions will be obtained for the equivalent rigidities and masses that belong to the partial systems (their "self" parameters) and for parameters K_{ah}, M_{ah} that characterize coupling between the partial systems

$$K_{eqva} = 13.4 h c_a, \quad K_{eqvh} = \frac{\pi^3 a^2}{2h} c_h, \quad K_{ah} = 4.2\pi a c_{ah}; \quad (4.619)$$

$$M_{eqva} = 0.86M, \quad M_{eqvh} = 0.5M, \quad M_{ah} = 0. \quad (4.620)$$

The coefficient of coupling between the partial systems is elastic,

$$\gamma = \left(K_{ah} / \sqrt{K_{eqva} K_{eqvh}}\right) \approx \frac{c_{ah}}{\sqrt{c_a c_h}} = \frac{\sigma}{1-\sigma}. \quad (4.621)$$

For example, consider a cylinder made of aluminum ($\sigma = 0.33$) with $\gamma = 0.49$.

The partial resonance frequencies are

$$f_a = \frac{1}{2\pi}\sqrt{\frac{K_{eqva}}{M_{eqva}}} = \frac{0.43}{a}\sqrt{\frac{Y}{\rho}}, \quad f_h = \frac{1}{2\pi}\sqrt{\frac{K_{eqvh}}{M_{eqvh}}} = \frac{0.61}{h}\sqrt{\frac{Y}{\rho}}. \quad (4.622)$$

For the ratio of the partial resonance frequencies, we obtain

$$\beta = \frac{f_a}{f_h} \approx 1.4 \frac{h}{2a}. \quad (4.623)$$

After the coefficients γ and β are determined, calculating dependencies of the resonance frequencies and mode shape factors can be produced in the way, as it was demonstrated in the previous section (see Eqs. (4.583) and (4.58).

4.6.2.4 Coupled Vibration of Thin-Walled Tubes

Cylindrical piezoelectric ceramic transducers are widely used for underwater applications. Calculation of their parameters is well known in the case that the transducers are built from the thin-walled short rings, for which the assumption of one-dimensional nature of vibrations in the circumferential direction is valid (see Section 4.4.4). With increasing the height to diameter aspect ratio of the comprising piezoelements the one-dimensional approximation fails, and vibrations of the cylindrical piezoelements must be considered as two-dimensional coupled vibrations in the circumferential and axial directions.

It has to be noted that a vagueness of terminology exists regarding naming of the thin-walled cylindrical piezoelements, if to consider their two-dimensional vibration. So far as the passive elastic cylindrical elements are concerned, it is common to refer to elements as "rings" at aspect ratios $h/2a \ll 1$ and as the "tubes" at $h/2a \gg 1$. But it is not clear when a ring transitions into a tube. At the same time, a convention exists among suppliers of piezoceramic parts to specify the thin-walled cylindrical parts with electrodes applied to the side surfaces as tubes regardless of their height to diameter ratio. Following this convention, we will name the objects of this treatment as tubes because of their intended application as the piezoelements.

The first treatment of vibration of piezoelectric ceramic thickness poled tubes in the two-dimensional approximation was carried out in Ref. 25 in the framework of the "membrane" theory of shells following the treatment of vibration of the passive thin isotropic elastic tubes described by A. E. Love [1]. An alternative to the partial differential equations of motion treatment of the problem that was introduced by Love was suggested by Gibbe and Blechshmidt [19]. They considered vibration of the tubes as dynamical interaction of two coupled partial mechanical systems, namely, of a thin ring undergoing radial vibrations and of a thin longitudinally vibrating bar, as shown in Figure 4.49. We will adhere to this latter approach.

A common prediction of both analyses was the existence of a so called "dead zone," i.e., some frequency range, in which no resonance vibrations of a tube may occur. This physically improbable result is rooted in the membrane theory approximation, in which case the thickness of a tube is assumed to be small to the extent that the energy of flexural deformations can be neglected in comparison with the energy of extensional deformations at any aspect ratio. It was shown in Ref. 26 that such an assumption is especially wrong for the range of aspect ratios around the point of strongest coupling. To correct this shortcoming, the energy related to flexural vibration of the bar as one of the partial systems must be taken into account, when considering vibration of the tubes. This involves introducing one more generalized coordinate, makes the problem three degree of freedom, and requires using technique that is somewhat different from those used in the previous sections for treating the coupled vibrations. This will be done following Ref. 27.

6.1.3.1.3. Assumptions on the Distribution of Deformations in the Thin-Walled Tubes

Consider extensional axial symmetric vibrations of a thin-walled isotropic elastic tube shown in Figure 4.49 (a) as the coupled vibrations of the two partial one-dimensional systems, namely, of the radially vibrating ring (Figure 4.49 (b)) and of the longitudinally vibrating thin bar (Figure 4.49 (c)).

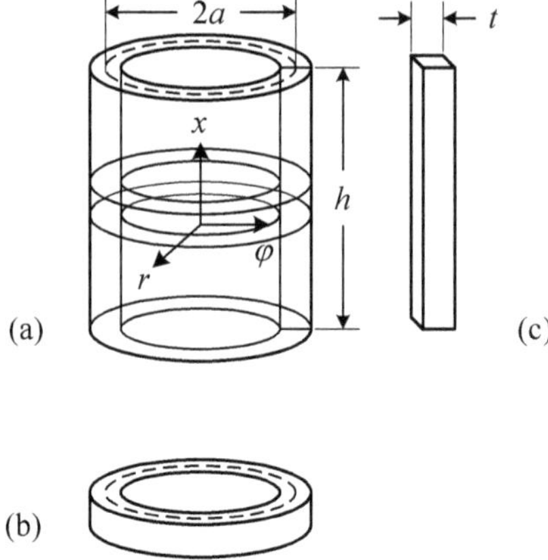

Figure 4.49: A thin-walled elastic tube (a) and its partial subsystems: (b) the radially vibrating short ring and (c) the longitudinally vibrating thin bar.

The common assumption for the thickness, t, of the "thin-walled" tube is that $t \ll 2a$. In other words, this means that the resonance frequency of vibration of the ring in the axial direction and the resonance frequency of the radial vibration of the ring, $f_r = c / 2\pi a$, are very far apart. Thus, these vibrations can be considered as independent, and at frequencies close to the radial resonance of the ring the thickness is very small compared with the extensional wavelength. Therefore, the stresses in radial direction, T_r, are practically constant, and being zero on the ring surfaces they remain negligible inside of its volume, i.e., $T_r = 0$, which allows the problem to be treated as two-dimensional.

Consider the extreme cases for the height of a tube, in which $h / 2a \ll 1$ and $h / 2a \to \infty$. In the case that $h / 2a \ll 1$, the tube reduces to short ring, for which the first resonance frequency of the axial vibrations, $f_{h1} \doteq c / 2h$, is much higher than the resonance frequency of the radial vibration, f_r. By the reasons discussed for the thickness of a ring it can be assumed that the axial stress $T_x = 0$ in the volume of the ring. Thus, the problem becomes one-dimensional with a well-known solution. In the case that $h/2a \to \infty$, the tube becomes very long, the resonance frequency of the axial vibrations becomes much lower than for the radial vibration, and the vibrations in the vicinity of the radial resonance frequency may be considered as one-dimensional with the mechanical conditions $T_r = 0$ and $S_x = 0$, where S_x is the strain in the axial direction. The latter condition is valid, strictly speaking, for an infinitely long tube because of the symmetry considerations, but it may be assumed to be applicable to a tube long in comparison with the extensional wavelength in a frequency range of interest. Both one-dimensional approximations are valid practically to a broad extent of aspect ratios so far as the separation between the resonance frequencies of vibration in the axial and radial directions is sufficiently large. But it remains to be estimated, what is large enough in terms of an acceptable accuracy of calculations based on these approximations. It follows from the general theory of coupled vibrations that the strongest coupling between the partial systems takes place if the resonance frequencies of the partial systems are equal. In our case this condition fulfills at first in the vicinity of the aspect ratio $\beta = h/2a = \pi/2$, at which point $f_r = f_{h1}\xi_{ri} = -i\pi\sigma a/h$, and then repeatedly at the aspect ratios related to the harmonics of the axial resonance frequency. Thus, it can be expected that the one-dimensional ring approximation may be valid for the tube with the aspect ratios $h/2a \ll \pi/2$. It is not clear, what the lowest acceptable value of the aspect ratio is for the one-dimensional long tube approximation to be valid. This must be determined.

To employ the coupled vibration technique to analysis of a tube vibration, an assumption must be made regarding distribution of displacements over its surface. At first consider deformations of a ring (Figure 4.49 (b)). Denote the radial displacement of the ring surface as ξ_0, then the strain S_φ in the circumferential direction may be presented as

$$S_\varphi = \frac{2\pi(a+\xi_0)-2\pi a}{2\pi a} = \frac{\xi_0}{a}, \tag{4.624}$$

and the strain S_x in the axial direction will be determined as

$$S_x = -\sigma S_\varphi = -\sigma\xi_0 / a . \tag{4.625}$$

Correspondingly, the displacement in the axial direction ξ_{xr} generated in the tube by the radial displacement will be

$$\xi_{xr} = -\xi_0 \frac{\sigma x}{a}. \tag{4.626}$$

Consider now deformations of a thin bar (Figure 4.49 (c)). Displacement in the axial direction, ξ_{xx}, may be represented as an expansion in the series

$$\xi_{xx} = \sum_{i=1}^{2n-1} \xi_{xi} \sin(i\pi x / h), \tag{4.627}$$

where n is a number of modes taken into account for an approximation, that may be considered as acceptable. Expression for the strain in the axial direction is

$$S_x = \partial \xi_{xx} / \partial x, \tag{4.628}$$

and the strain in the lateral direction, S_φ, will be found as

$$S_\varphi = -\sigma S_x = -\sigma \partial \xi_{xx} / \partial x. \tag{4.629}$$

The deformation corresponding to this strain, which is produced in the circumferential direction of the ring, cause a displacement of the ring surface, ξ_{rx}. This displacement can be determined as

$$\xi_{rx} = a S_\varphi \tag{4.630}$$

according to formula (4.624). Thus, the deformation of the bar in the axial direction generates the radial displacement of the tube surface

$$\xi_{rx} = -\sigma a (\partial \xi_{xx} / \partial x), \tag{4.631}$$

which can be expressed with reference to Eq.(4.627) as

$$\xi_{rx} = \sum_{i=1}^{2n-1} \xi_{ri} \cos(i\pi x / h), \tag{4.632}$$

where, $i = 0, ..., 4$

Summarizing the assumptions made above, the distribution of displacements in an axial symmetrically vibrating thin-walled tube can be represented as follows

$$\xi_x = \xi_{xx} + \xi_{xr} = (-\sigma x / a)\xi_0 + \sum_{i=1}^{2n-1} \xi_{xi} \sin(i\pi x / h), \tag{4.633}$$

$$\xi_r = \xi_0 + \xi_{rx} = \xi_0 + \sum_{i=1}^{2n-1} \xi_{ri} \cos(i\pi x / h). \tag{4.634}$$

The distribution of displacements in a tube is shown qualitatively in Figure 4.50 (a), as a superposition of displacements generated by vibration of the partial systems (only the fundamental mode of a bar vibration is illustrated for simplicity).

So far as the distribution of displacements is defined by Eqs. (4.629) and (4.630), the axial symmetric strain distribution in the body of the tube can be represented in the cylindrical coordinates x, φ as follows

$$S_x = \frac{\partial \xi_x}{\partial x} - z\frac{\partial^2 \xi_r}{\partial x^2} = -\frac{\sigma}{a}\xi_0 + \sum_{i=1}^{2n-1} \xi_{xi} \frac{i\pi}{h} \cos(i\pi x / h) - z\sum_{i=1}^{2n-1} \xi_{ri} \left(\frac{i\pi}{h}\right)^2 \cos(i\pi x / h), \tag{4.635}$$

$$S_\varphi = \frac{\xi_r}{a} = \frac{\xi_0}{a} + \frac{1}{a}\sum_{i=1}^{2n-1} \xi_{ri} \cos(i\pi x / h). \tag{4.636}$$

The term $(-z\,\partial^2 \xi_r / \partial x^2)$ in Eq. (4.635) accounts for the strains due to the flexural deformations of the wall of the tube. The coordinate axis z goes in the radial direction and has its origin on the mean circumferential surface of the tube, as it is shown in Figure 4.50 (b). This term is a matter of principle in this treatment. It takes into consideration the energy of the flexural deformations of the wall of a tube having a finite thickness and makes the proposed approach to the problem different from that which was used in the framework of the "membrane" theory.

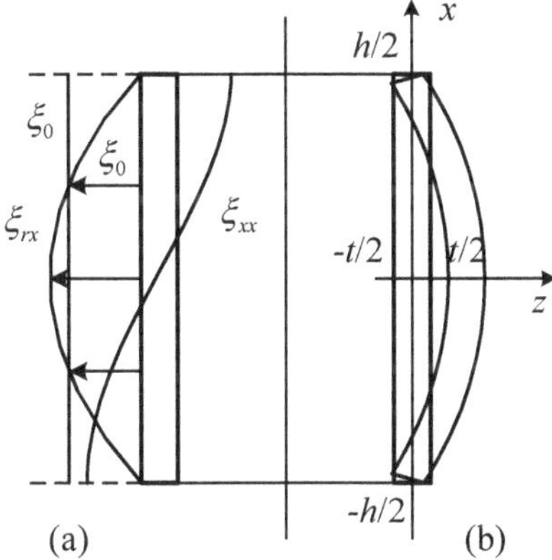

Figure 4.50: Distribution of displacements in a tube: (a) in the extensional vibrations, (b) in the flexural vibrations.

The stresses in the tube will be found as follows from Eqs. (4.10) (remember that $T_r = 0$),

$$T_x = \frac{Y}{1-\sigma^2}(S_x + \sigma S_\varphi), \tag{4.637}$$

$$T_\varphi = \frac{Y}{1-\sigma^2}(\sigma S_x + S_\varphi). \tag{4.638}$$

By substituting the strains S_x and S_φ, defined by Eqs. (4.635) and (4.636), into Eq. (4.637) it is easy to make certain that the boundary conditions $T_x(\pm h/2) = 0$ on the free ends of a tube are met. With distribution of strain in the tube known, the equations of motion of an isotropic passive tube can be derived as the Lagrange's equations.

A note must be made regarding a number of terms in the series (4.627) to be taken into account for practical calculations in order to achieve an acceptable level of accuracy for the results. This depends on the range of aspect ratios under consideration. For the range below and around the first region of strong coupling, namely, $0 < h/2a < 3$, it should be sufficient to retain only the first term of the series, which corresponds to the fundamental mode of the axial vibration. For the higher range of aspect ratios, but below and around the second region of a strong coupling, that is for $0 < h/2a < 6$, the next term must be included, which corresponds to the third harmonic of the axial vibration. One more note has to be made regarding representation of the flexural term in Eq. (4.635). The flexural term is represented based on the elementary theory of bending. This can be considered as sufficiently accurate until ratio of the half wave of flexure to thickness for a bar as a partial system is much larger than unity, i.e., $h/it \gg 1$, where i is a number of half waves of flexure on the length of the bar. Otherwise, corrections accounted for the effects of rotary inertia and shearing deformations on the kinetic and potential energies related to the flexural vibrations must be taken into consideration, as it is illustrated in Section 4.3.5.

6.1.3.1.4. Results of Solving Equations of Coupled Vibration of Thin-Walled Tubes

The equations of free vibration of isotropic elastic tubes can be obtained as Lagrange's equations. The second approximation will be considered, in which case $n = 2$ in Eqs. (4.635) and (4.636). The new notations will be introduced for the generalized coordinates as follows: $\xi_{r1} \to \xi_1$, $\xi_{r3} \to \xi_3$, $\xi_{x1} \to \xi_2$, $\xi_{x3} \to \xi_4$. Thus, the five generalized coordinates ξ_i will be used to describe a solution to the problem. To obtain the Lagrange's equations the kinetic and potential energies of a tube must be considered. Under the assumption that for flexural deformations elementary theory of bending is applicable the energies will be determined as follows.

The kinetic energy of a tube is

$$W_{kin} = \frac{1}{2}\int_{\tilde{V}} \rho(\dot{\xi}_r^2 + \dot{\xi}_x^2)d\tilde{V} = \frac{1}{2}2\pi a t \rho \int_{-h/2}^{h/2} (\dot{\xi}_r^2 + \dot{\xi}_x^2)dx . \tag{4.639}$$

Here $\dot{\xi}$ is the time derivative of the displacement or the velocity, and $\tilde{V}$ is the volume of the tube. The following expression for the kinetic energy will be obtained after substituting the displacements ξ_r and ξ_x from Eqs. (4.633) and (4.634) and after integrating over the height

$$W_{kin} = \frac{1}{2}\left(\sum_{i=0}^{4} M_{ii}\dot{\xi}_i^2 + 2\sum_{l=1}^{4} M_{0l}\dot{\xi}_0\dot{\xi}_l \right), \tag{4.640}$$

where

$$\begin{gathered} M_{00} = M\left(1 + \frac{\sigma^2 h^2}{12a^2}\right), \quad M_{ii} = \frac{M}{2} \quad (at\ i = 1,2,3,4), \quad M_{01} = \frac{2}{\pi}M, \\ M_{02} = -\frac{2\sigma h}{\pi^2 a}M, \quad M_{03} = -\frac{2}{3\pi}M, \quad M_{04} = \frac{2\sigma h}{9a\pi^2}M. \end{gathered} \tag{4.641}$$

Here $M = 2\pi a h t \rho$ is the mass of the tube.

The potential energy of a tube is

$$W_{pot} = \frac{1}{2}\int_{\tilde{V}} (S_x T_x + S_\varphi T_\varphi) d\tilde{V} , \tag{4.642}$$

Substituting expressions (4.635) and (4.636) for the strain and expressions (4.637) and (4.638) for the stress under integral (4.642), integrating over the volume of the tube and a little manipulation yields

$$W_{pot} = \frac{1}{2}\left[\sum_{i=0}^{4} K_{ii}\xi_i^2 + 2(K_{01}\xi_0\xi_1 + K_{12}\xi_1\xi_2 + K_{03}\xi_0\xi_3 + K_{34}\xi_3\xi_4) \right], \tag{4.643}$$

where it is denoted

$$K_{00} = \frac{2\pi thY}{a}, \quad K_{ii} = \frac{\pi htY}{a(1-\sigma^2)}\left[1+\frac{(i\pi)^4}{48}\left(\frac{t}{h}\right)^2\left(\frac{2a}{h}\right)^2\right] \quad (i=1,3),$$
$$K_{22} = \frac{\pi^3 atY}{h(1-\sigma^2)}, \quad K_{44} = \frac{9\pi^3 atY}{h(1-\sigma^2)}, \quad K_{01} = \frac{4thY}{a}, \tag{4.644}$$
$$K_{03} = \frac{4thY}{3a}, \quad K_{12} = \frac{\pi^2 t\sigma Y}{1-\sigma^2}, \quad K_{34} = \frac{3\pi^2 t\sigma Y}{1-\sigma^2}.$$

As it was noted, for small aspect ratios corrections to the flexure related parameters of the masses and rigidities must be introduced. Following formulas (4.140) for the masses and rigidities at $i = 1, 3$ with the corrections accounted for the rotary inertia and shearing deformations will be obtained

$$M_{ii} = \frac{M}{2}\left[1+\frac{(i\pi t)^2}{12h^2}\right], \tag{4.645}$$

$$K_{ii} = \frac{\pi htY}{a(1-\sigma^2)}\left[1+\frac{(i\pi)^4}{48}\left(\frac{t}{h}\right)^2\left(\frac{2a}{h}\right)^2\right]\left[1-\frac{(i\pi)^2}{20}\frac{Y}{\mu}\left(\frac{t}{h}\right)^2\right], \tag{4.646}$$

where μ is the shear modulus. Calculations made with these corrections are accurate in the range of aspect ratios $h/2a > 0.5$ at $i = 1$ and $h/2a > 1.5$ at $i = 3$.

The Lagrange's equations of free vibration of a tube,

$$\frac{d}{dt}\left(\frac{\partial W_{kin}}{\partial \dot{\xi}_i}\right)+\frac{\partial W_{pot}}{\partial \dot{\xi}_i} = 0 \quad (i = 0,1,2,3,4), \tag{4.647}$$

after substituting expressions (4.640) and (4.643) for the kinetic and potential energies, differentiating and converting to the complex quantities yield

$$(j\omega M_{eqvi} + K_{eqvi} / j\omega)U_i = 0 \quad (i = 0,1,2,3,4). \tag{4.648}$$

Here U_i is the complex amplitude of velocity $\dot{\xi}_i$, and M_{eqvi} and K_{eqvi} are the equivalent parameters defined as follows

$$M_{eqv0} = M_{00} + \sum_{i=1}^{4} M_{0i}(U_i/U_0), \quad M_{eqv1} = M_{11} + M_{01}(U_0/U_1),$$
$$M_{eqv2} = M_{22} + M_{02}(U_0/U_2), \quad M_{eqv3} = M_{33} + M_{03}(U_0/U_3), \tag{4.649}$$
$$M_{eqv4} = M_{44} + M_{04}(U_0/U_4),$$

$$K_{eqv0} = K_{00} + K_{01}(U_1 / U_0) + K_{03}(U_3 / U_0), \quad K_{eqv2} = K_{22} + K_{12}(U_1 / U_2),$$
$$K_{eqv4} = K_{44} + K_{34}(U_3 / U_4), \quad K_{eqv1} = K_{11} + K_{01}(U_0 / U_1) + K_{12}(U_2 / U_1), \tag{4.650}$$
$$K_{eqv3} = K_{33} + K_{03}(U_0 / U_3) + K_{34}(U_4 / U_3).$$

After substituting the parameters thus determined into Eqs. (4.648), these equations finally may be represented as

$$Z_{ii}U_i + \sum_{l \neq i} z_{il}U_l = 0 \quad (i,l = 0,1,2,3,4)\,, \tag{4.651}$$

where

$$z_{il} = z_{li}, \quad z_{13} = z_{23} = z_{14} = z_{24} = 0, \quad Z_{ii} = (K_{ii} / j\omega)(1 - \omega^2 / \omega_{ii}^2),$$
$$z_{o1} = (K_{01} / j\omega)(1 - \omega^2 / \omega_{01}^2), \quad z_{02} = j\omega M_{02}, \quad z_{03} = (K_{03} / j\omega)(1 - \omega^2 / \omega_{03}^2), \tag{4.652}$$
$$z_{04} = j\omega M_{04}, \quad z_{12} = K_{12} / j\omega, \quad z_{34} = K_{34} / j\omega.$$

When representing the impedances (4.652), relations (4.650) and (4.651) are taken into consideration and it is denoted

$$\omega_{ii}^2 = K_{ii} / M_{ii}, \quad \omega_{01}^2 = K_{01} / M_{01}, \quad \omega_{03}^2 = K_{03} / M_{03}\,. \tag{4.653}$$

The frequencies ω_{ii} can be interpreted in the following way

$$\omega_{00}^2 = K_{00} / M_{00} = \frac{Y}{a^2\rho[1 + \sigma^2 h^2 / (12a^2)]} = \omega_{ring}^2\,, \tag{4.654}$$

$$\omega_{11}^2 = K_{11} / M_{11} = \frac{Y}{a^2\rho(1-\sigma^2)} + \frac{\pi^4 t^2 Y}{12h^4\rho(1-\sigma^2)} = \omega_{tube}^2 + \omega_{fh1}^2\,, \tag{4.655}$$

$$\omega_{22}^2 = K_{22} / M_{22} = \frac{\pi^2 Y}{h^2\rho(1-\sigma^2)} = \omega_{h1}^2\,, \tag{4.656}$$

$$\omega_{33}^2 = K_{33} / M_{33} = \omega_{tube}^2 + \omega_{fh3}^2 = \omega_{tube}^2 + 9\omega_{fh1}^2\,, \tag{4.657}$$

$$\omega_{44}^2 = K_{44} / M_{44} = \omega_{h3}^2 = 9\omega_{h1}^2\,. \tag{4.658}$$

The expressions for ω_{11} and ω_{33} are given for the aspect ratios, at which elementary theory of bending is applicable, i. e., when the corrections in Eqs. (4.645) and (4.646) can be neglected.

In Eqs. (4.654)-(4.658) the following notations are introduced: ω_{ring} for the resonance frequency of the radial vibration of a thin ring of the height h small compared with its radius; ω_{h1} and ω_{h3} for the fundamental resonance frequency and the third harmonics of vibration of infinite thin strip in direction of its width h; ω_{tube} for the resonance frequency of the radial vibration of a thin-walled tube of infinite height; ω_{fh1} and ω_{fh3} for the first and third resonance frequencies of the flexural vibrations in the "width" mode of an infinite strip of thickness t simply supported on the edges. The expressions for the frequencies in Eqs. (4.654)-(4.656) may be found in Ref. 1. Equation (4.655) can be transformed to the form

$$\omega_{11}^2 = \omega_{tube}^2 \left[1 + \frac{\pi^4}{48} (t/h)^2 (2a/h)^2 \right], \tag{4.659}$$

where from it follows that frequency ω_{11} depends on both the aspect ratio and the thickness to height ratio of a tube. Further these frequencies will be labeled for brevity as ω_i at $i = 0, ..., 4$.

The set of Eqs. (4.651) provides a solution to the problem of free coupled vibrations in a passive isotropic elastic tube. Results of calculating the resonance frequencies as functions of aspect ratio $h/2a$ are represented in Figure 4.51. The calculations were carried out for PZT-4 ceramic tubes with the outer diameter D_o = 38.2 mm and thickness t = 3.2 mm, which correspond to the dimensions of tubes used in the experimental study[28]. Parameters of PZT-4 were used as presented in Ref. 29 (see Appendix B, Table B.1). Results of calculations showed that at least up to the aspect ratios $h/2a \doteq 3$ the second approximation doesn't contribute noticeably to values of frequency branches 0, 1 and2 calculated using just the first approximation.

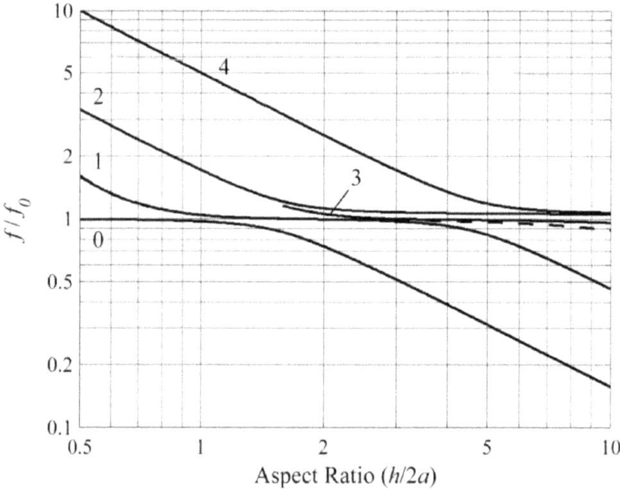

Figure 4.51: The resonance frequencies ω_i of the thin-walled tube (PZT-4, $2a$ = 35 mm, t = 3.2 mm) normalized by the resonance frequency of a short ring, f_0 = 30 kHz, as a function of aspect ratio $h/2a$. The curves are labeled by number $i = 0, ..., 4$.

The only deviation due to the second approximation is shown in Figure 4.51 by dashed line and is related to curve 1. Thus, given that the dependencies 0, 1 and 2 are of the most practical interest, the generalized velocities U_3, U_4 and all the related impedances in Eqs. (4.651) can be set to zero. Therefore, the further analysis will be restricted by the first approximation the more so because the range of aspect ratios $h/2a < 3$ is the most interesting for practical applications and the restriction simplifies analysis for this range without loss of accuracy. Besides, the same technique can be used to analyze the solution at larger values of the aspect ratios in case this is needed.

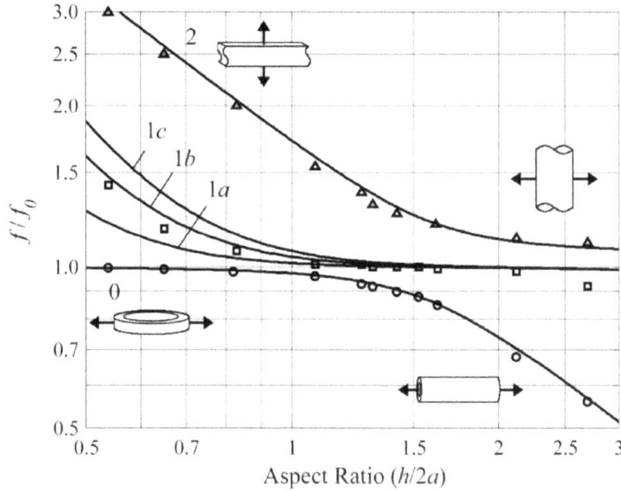

Figure 4.52: Normalized resonance frequencies of the tube (PZT-4, $2a = 35$ mm) calculated to the first approximation at different wall thickness t: 2.0 mm (1a), 3.2 mm (1b), 4.0 mm (1c). Experimental data from Ref. 28 are shown by circles for branch 0, by squares for branch 1 and by diamonds for branch 2.

The part of the plot in Figure 4.51 related to the first approximation only is displayed in Figure 4.52 up to values $h/2a = 3$ in a larger scale together with results of calculations made for different thickness of the wall of the tube and with experimental data included. The notable difference between the frequency plots displayed in Figure 4.51 and those presented in Ref. 25 is in existence of the flexure related branches 1 and 3, which could not be predicted by the "membrane" theory and which cross the so called "dead zone." It is of note that for these branches the normalized numerical values are valid for the thickness $t = 3.2$ mm only, whereas the extensional vibration related frequency branches do not depend on the thickness.

The results of calculations made for the tubes with different thicknesses presented in Figure 4.52 show significant dependence of the "intermediate" frequency branch 1 from the thickness at aspect ratios up to values in vicinity of the point of the strongest coupling. This is indicative of a flexural nature of the corresponding modes of vibration in this range of aspect ratios. Above the point of the strongest coupling this frequency branch becomes independent of thickness as the extensional branches do. This is in accord with results of observations made in Ref. 28.

Effects of the coupled vibration on performance of the piezoceramic finite-size cylindrical transducers will be considered in Chapter 7.

4.7 Input Impedances of Mechanical Systems

4.7.1 Input Impedance of a Homogeneous Uniform Bar

The input impedance of a bar of the uniform cross section S_{cs} that is loaded at the opposite end with impedance Z_L, is determined by the relation (4.101), or from the two-port circuit in Figure 4.6 (a) as

$$Z_{in} = \frac{j\rho c S_{c.s} \tan kl + Z_L}{1 + j(Z_L / \rho c S_{c.s}) \tan kl}. \tag{4.660}$$

Besides, the ratio of vibration velocity, U_L, of the load to velocity of vibration at the bar input, U, in this case is as follows

$$U_L / U = 1/\cos kl[1 + j(Z_L / \rho c S_{cs}) \tan kl]. \tag{4.661}$$

Consider the input impedances of a bar at different loading conditions.

The input impedance of a bar with free end ($Z_L = 0$) is

$$Z_{in} = j\rho c S_{cs} \tan kl. \tag{4.662}$$

This function is shown in Figure 4.53 by the solid line. For the case that $l \to \lambda/4$ (the quarter wavelength bar), which corresponds to the parallel resonance frequency ω_{pr}, the input impedance approaches infinity $Z_{in} \to \infty$, and close to this frequency the equivalent circuit of the bar can be represented by the parallel contour. Thus, attaching the quarter wavelength unloaded bar to some surface may result in its clamping. When the length of the bar decreases to

$l \to \lambda / 6$, then within 10% accuracy the input impedance becomes

$$Z_{in} \approx j\rho c S_{cs} kl = j\omega M , \tag{4.663}$$

and the bar behaves as a mass $M = \rho S_{cs} l$.

The input impedance of a bar loaded by active resistance $Z_L = r_L$, where $r_L / \rho c S_{cs} < 1$, may be obtained from formula (4.660) as

$$Z_{in} \approx j\rho c S_{cs} \tan kl + r_L / \cos^2 kl . \tag{4.664}$$

Thus, a bar can be used for transforming the load resistance to a great extent by changing its *kl*.

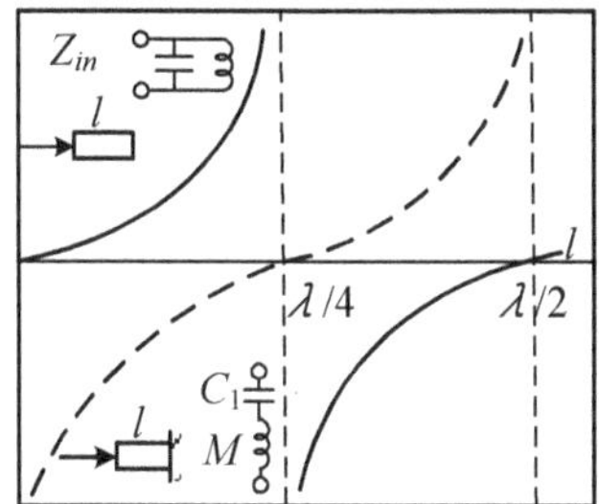

Figure 4.53: Input impedance of a bar: bar with free end-solid line, bar with the clamped end – dashed line.

The input impedance of a bar with the clamped end ($Z_L \to \infty$) is

$$Z_{in} = j\rho c S_{cs} / \tan kl . \tag{4.665}$$

This function is shown in Figure 4.53 by the dashed line.

At the length $l = \lambda / 4$, we have $Z_{in} = 0$, which corresponds to the resonance frequency of the bar. In the frequency range close to this frequency the equivalent circuit of the bar can be represented as the series contour with

$$M_l = M / 2, \quad C_l = 8 / \pi^2 Y S_{cs} . \tag{4.666}$$

At $l < \lambda / 6$

$$Z_{in} \approx -j\rho c S_{cs} / kl = \rho c^2 S_{cs} = K_l / j\omega , \tag{4.667}$$

and the bar behaves as the rigidity

$$K_l = l / \rho c^2 S_{cs} = l / Y S_{cs} . \tag{4.668}$$

Thus, the reactive and active components of the input impedance of a homogenous bar of constant cross section with a load applied to its end, can be changed significantly depending on the wave resistance $\rho c S_{cs}$ and the bar wave size *kl*. Because of this bar can be used for matching transducers with the external loads, in particular, for compensating the reactive component of the transducer internal impedance at a particular operating frequency (resonance tuning) and for transforming of the load impedances and the external forces in order to approach the optimal operating conditions for the transducer. To expand the matching capabilities, along with homogenous bars of constant cross section the systems with variable area of cross section are used, such as systems composed of bars having different cross sections and built from different materials.

It must be remembered that all the considerations of this article are valid under the condition that cross sections of a bar vibrate in piston like fashion. In practice this approximation takes place at significant values of the bar height to diameter aspect ratio ($h/2a > 1.5$). At smaller aspect ratios effects of coupled vibration in the axial and radial directions may become significant, and eventually the flexural vibrations, as of the thick disk, interfere. An accurate analytical solution for the vibration problem becomes hardly possible and using FEA for calculations becomes the appropriate option. In the extreme case of small aspect ratio (at $h/2a \to 0$) the input impedances of the thin disk vibrating in axial direction under action of forces applied to its surface are considered in the next section.

4.7.2 Input Impedances of a Circular Disk

Circular disks are widely used elements of transducer designs. The caps of a cylindrical transducer can be regarded as the circular disks. Calculating the cylindrical transducer in this case requires determining input impedances on the edge of the disk in radial (Z_{inr}) and perpendicular to its surface (Z_{inx}) directions. A circular disk may be considered also as an extreme case of a matching element of a Tonpilz like transducer. In this case the bar transducer is often made as a hollow cylinder, and its calculation requires knowing the input impedance of a disk in the direction perpendicular to its plane on the circumference with a radius that may change from considerations of optimizing the matching conditions. In both the cases it is expedient to use disks with such dimensions that their resonance frequencies were above an operating frequency range.

4.7.2.1 Input Impedance of a Circular Disk in the Radial Direction

Assuming that in the frequency range below the first resonance frequency the resonance mode of vibration holds, the disk with the free edge can be considered as one degree of freedom system having the mode of vibration

$$\theta_1(r/a) = J_1(2.05r/a)/J_1(2.05). \tag{4.669}$$

The total system of Eqs. (4.254) for a disk in this approximation is reduced to equation

$$Z_{m11}U_1 = F_{m1}, \tag{4.670}$$

where F_{m1} is the total radial force acting on the edge of the disk, U_1 is vibration velocity of the edge in the radial direction, and Z_{m11} is determined by relation from (4.253). Thus, the input impedance of the disk will be found as

$$Z_{inr} = F_{m1}/U_1 = Z_{m11}, \tag{4.671}$$

that is,

$$Z_{inr} = F_{mi}/U_1 = j\omega M_{11} + (K_{11}/j\omega) + r_{mL} = -jK_{11}[1-(\omega/\omega_1)^2]/\omega + r_{mL}. \tag{4.672}$$

The equivalent mass M_{11} and rigidity K_{11} have to be determined by formulas (4.236). For this case

$$M_{11} = 2\pi t\rho\int_0^a \theta_1^2(r/a)rdr, \quad K_{11} = \frac{2\pi tY}{1-\sigma^2}\int_0^a \theta_1'^2(r/a)rdr. \tag{4.673}$$

After substituting the mode shape (4.84) and integrating we obtain

$$M_{11} = 0.78M, \quad K_{11} = \frac{10.3tY}{1-\sigma^2}, \quad \omega_1 = \frac{2.05}{a}\sqrt{\frac{Y}{\rho(1-\sigma^2)}}. \tag{4.674}$$

4.7.2.2 Input Impedance at the Edge of a Disk in the Transverse Direction

We represent displacements of the disk as

$$\xi(r) = \xi_a + \xi_o\theta(r/a), \tag{4.675}$$

where

$$\theta_1(r/a) = (1-r^2/a^2)(1-r^2/4a^2) \tag{4.676}$$

is the mode of static deflection of simply supported disk; ξ_a is the displacement of its edge (Figure 4.54 (a)). In this approximation the disk represents a system with two coupled degrees of freedom, and the general system of equations of vibration (4.254) is reduced to equations of type (4.526) and (4.527). The partial systems represent a rigid disk that vibrates as a piston with displacements ξ_a, and a simply supported disk that performs vibration $\xi(r) = \xi_o \theta(r/a)$. Therefore, in Eqs. (4.526) and (4.527) we have $\xi_{1p} = \xi_a$, $\xi_{2p} = \xi_o$, $K_1 = 0$, $M_1 = M$.

The equivalent parameters for the second partial system must be determined considering that the thickness of the disk can be significant to ensure its static strength and needed value of the resonance frequency. Therefore, the equivalent parameters must be determined with corrections for the rotary inertia and transverse shear that are introduced in Section 9.4.3 by formulas (9.266). Namely,

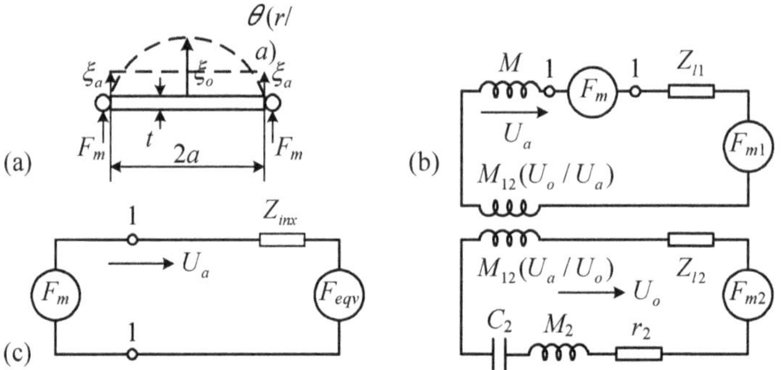

Figure 4.54: To the input impedance on the edge of a circular disk: (a) general view of the disk, (b) equivalent circuit for calculating input impedance between points 1, 1; (c) simplified equivalent circuit.

$$M_2 = 0.3M\left[1 + 0.5(t/a)^2\right], \tag{4.677}$$

$$K_2 \approx \frac{2}{a^2}\frac{t^3 Y}{(1-\sigma^2)}\left[1 - 0.25\frac{Y}{(1-\sigma^2)\mu}\frac{t^2}{a^2}\right]. \tag{4.678}$$

The mutual mass M_{12} and rigidity K_{12} must be calculated by formulae (4.236) assuming that $\theta_1(r) = 1$ (piston-like vibrations) and $\theta_2(r) = \theta(r/a)$. Thus, will be obtained that $K_{12} = 0$ and

$$M_{12} = 2\pi\rho t\int_0^a \theta(r/a) r dr = \rho t S_{av} = 0.45M, \tag{4.679}$$

(remember that for a circular simply supported plate $S_{av} = 0.45S_{pl}$).

When calculating the input impedance of a disk at the condition that its surface is not loaded, it should be taken $F_2 = 0$. Thus, Eqs. (4.526) and (4.527) of vibration of the disk under the force applied to the edge can be represented in the generalized velocities as

$$j\omega MU_a + j0.45\omega MU_o = F_m, \tag{4.680}$$

$$j0.45\omega MU_a + j\omega M_2[1-(\omega_{pl}^2/\omega)^2]U_o = 0. \tag{4.681}$$

From these equations follows that

$$Z_{inx} = \frac{F_m}{U_a} = j\omega M\left(1+\frac{0.7\omega^2}{\omega_{pl}^2-\omega^2}\right)\left(1+0.5\frac{t^2}{a^2}\right), \tag{4.682}$$

where $\omega_{pl}^2 = K_2/M_2$. If the disk represents a transducer cap, then it may experience an action of external forces and impedances, which in its turn will result in the change of the input impedances. In order to consider the external actions in estimating the dependence of vibration velocity U_o from the force F_m, the equivalent circuit shown in Figure 4.54 (b) can be used that results from Eqs. (4.680) and (4.681).

The equivalent forces F_{m1}, F_{m2} and impedances of loads Z_{l1} and Z_{l2} must be determined from expressions (4.240) and (4.242) provided the external actions are known. If the transducer cap experiences the acoustic pressure P_o, which is uniform over its surface, then

$$F_{m1} = P_o S_{pl} = P_o \pi a^2, \quad F_{m2} = P_o S_{av}. \tag{4.683}$$

This corresponds to the case that dimensions of a transducer are small compared to the length of acoustic wave. It is natural to assume that impedances of the loads can be neglected in comparison with the mechanical self-impedance of the cap (remember that its resonance frequency is supposed to be above an operating frequency range), therefore in this case $Z_{l1} \approx Z_{l2} \approx 0$. By the Thevenin's theorem applied regarding the points 1, 1 in the circuit Figure 4.54 (b), this circuit can be transformed to those shown in Figure 4.54 (c). In this circuit Z_{inx} is determined by formula (4.682). The equivalent force, F_{eqv}, accounts for the combined action of forces F_{m1} and F_{m2} reduced to the open output of the circuit between points 1, 1 (see Figure 4.54 (b)). Considering that $U_a = 0$,

$$U_o = -F_{m2}\omega / jM_2(\omega_{pl}^2 - \omega^2), \tag{4.684}$$

and

$$F_{eqv} = F_{m1} + jM_{12}U_o = F_{m1}\left[1 - \frac{M_{12}F_{m2}\omega^2}{M_2F_{m1}(\omega_{pl}^2 - \omega^2)}\right]. \tag{4.685}$$

After substituting expressions (4.683) for F_{m2} and F_{m2}, and expressions (4.677) and (4.679)for M_2 and M_{12} into this relation will be obtained

$$F_{eqv} = P_o\pi a^2[1 - 0.67(1 + 0.5t^2 / a^2)\omega^2 / (\omega_{pl}^2 - \omega^2)]. \tag{4.686}$$

4.7.2.3 Input Impedance of the Free Disk on the Circle of Radius $b < a$

We will represent the distribution of displacements over the disk surface as

$$\xi(r) = \xi_o + \xi_1\theta_1(r / a), \tag{4.687}$$

where $\theta_1(r / a)$ is the natural mode of vibration of a circular plate with the free edge. This function is expressed by formula (4.200) at the eigenvalue $k_1a = 3.0$, i.e.,

$$\theta_1(r / a) = 1.1J_0(3r / a) - 0.1I_0(3r / a). \tag{4.688}$$

Peculiarity of the displacement distribution (4.687) in comparison with the previous case is that for the free plate the normal mode $\theta_0(r / a) = 1$ exists that formally corresponds to natural frequency $\omega_0 = 1$ (the free plate may vibrate as a piston). At $\omega_0 = 0$ $Z_{in} = j\omega M_{pl}$, and formula (4.687) represents sum of the first two terms of expansion of displacement $\xi(r)$ into series in terms of the normal modes. The equations of vibration in the form of Eqs. (4.526) and (4.527) in a similar to the previous case manner. In this case $M_{10} = 0$, since the modes $\theta_0(r / a) = 1$ and $\theta_1(r / a)$ are orthogonal, and the partial systems (a rigid piston and a plate vibrating in the mode $\theta_1(r / a)$) are independent. The equivalent mass and rigidity for the free circular plate that account for the inertia of rotation and shear (see Section 9.4.3) are

$$M_1 = 0.26M[1 + 1.5\cdot(t^2 / a^2)], \tag{4.689}$$

$$K_1 = \left[6t^3Y / a^2(1 - \sigma^2)\right][1 - 0.7\cdot Y\mu(t^2 / a^2) / (1 - \sigma^2)]. \tag{4.690}$$

The resonance frequency of the partial system, which is the free plate, is $\omega_{1p}^2 = \omega_{pl}^2 = K_1 / M_1$.

The generalized forces acting on the partial systems can be obtained from expression (4.523) converted into the complex form as

$$\bar{\dot{W}}_e = F_m \bar{\dot{\xi}}^*(b) = F_m \bar{\dot{\xi}}_o^* + F_m \bar{\dot{\xi}}_1^* \theta_1(b/a) . \tag{4.691}$$

Thus,

$$F_{m0} = F_m, \quad F_{m1} = F_m \theta_1(b/a) , \tag{4.692}$$

and equations of the forced vibration of the disk are

$$j\omega M U_o = F_m , \tag{4.693}$$

$$j\omega M_1 (1 - \omega_{pl}^2 / \omega^2) U_1 = F_m \theta_1(b/a) . \tag{4.694}$$

The distribution of velocity over surface of the disk will be represented as

$$U(\omega, r/a) = U_0(\omega) + U_1(\omega)\theta_1(r/a) = \frac{F_m}{j\omega M}\left[1 - \frac{M}{M_1}\frac{\theta_1(b/a)\theta_1(r/a)}{\omega_{pl}^2/\omega^2 - 1}\right]. \tag{4.695}$$

Determined from this relation input impedance of the disk is

$$Z_{in}(\omega, b/a) = \frac{F_m}{U(\omega, b/a)} = j\omega M\left[1 - \frac{M}{M_1}\frac{\theta_1^2(b/a)}{\omega_{pl}^2/\omega^2 - 1}\right]^{-1} . \tag{4.696}$$

Behavior of the input impedance vs. ratio *b/a* is qualitatively shown in Figure 4.55 (a).

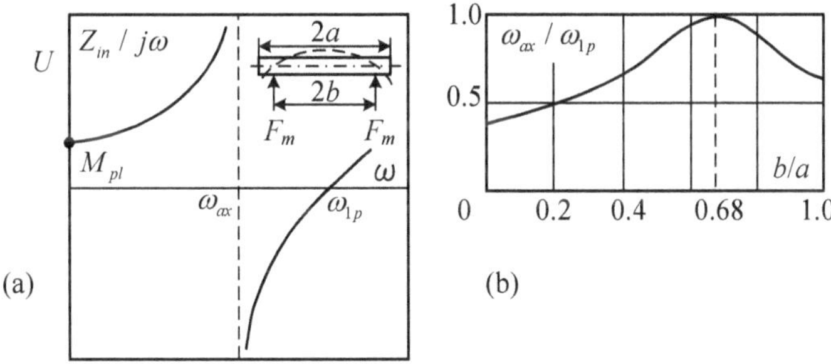

Figure 4.55: Input impedance of a circular disk on circumference of a variable radius.

At some frequency ω_{ar} (anti-resonance frequency) that depends on the radius of the circle, at which generating force is applied, the input impedance becomes infinitely big, $Z_{in} \to \infty$. It is not desirable to have this frequency in the operating frequency range. In Figure 4.55 (b) dependence of $\omega_{ar} / \omega_{pl}$ on the ratio *b*/*a* is shown. It can be concluded from the figure that optimal in terms of removing frequency ω_{ar} from the operating frequency range is value of *b*/*a* – 0.68 that corresponds to the radius of the nodal circle at which $\theta_1(r/a) = 0$. Expression (4.695) shows that in the case that acting force is applied on the nodal circle the distribution of vibration over the surface must be uniform. A violation of vibration uniformity may be caused by the fact that in reality the force is applied not exactly on the line, but on a ring area of finite width. Besides, in the absence of the first vibrational mode contribution of the next vibrational mode $\theta_2(r/a)$, which so far was neglected, must be considered. However, using expression (4.200) at $ka = 6.29$ for the mode of vibration $\theta_2(r/a)$, it is easy to verify that at $b/a = 0.68$ in the frequency range $\omega < \omega_{pl}$

$$|U_2/U_1| < 10\theta_2(0.68)\frac{\omega_{pl}^2}{\omega_{pl2}^2} < 0.15\frac{\omega^2}{\omega_{pl}^2}, \tag{4.697}$$

where ω_{pl2} is the corresponding resonance frequency (note that $\omega_{pl}/\omega_{pl2} = 0.23$). Given that $\omega/\omega_{1p} < 1$, contribution of the second mode of vibration is relatively small. Thus, by appropriate application of the acting force vibration of even relatively thin disk can be made close to uniform. More details on this issue will be presented in Chapter 10.

4.8 References

1. A. E. Love, *Mathematical Theory of Elasticity* (Dover, New York 1935).
2. S. P. Timoshenko, *Strength of Materials*, Part I, Elementary Theory and Problems, 2nd Ed. (Van Nostrand, New York, 1940).
3. *Handbook of Mathematical Functions*, edited by M. Abramowitz and I. A. Stegun (National Bureau of Standards Applied Mathematical Series, Washington, DC, 1964).
4. L. Ya. Gutin, "To the theory of the magnetostrictive transducer", Zhurnal Tekhnicheskoi Fiziki, Vol. XV, No. 4-5, 1945. In Selected works (Sudostroenie, Leningrad, 1977), p. 135 (in Russian).
5. L. V. Kantorovich and V. I. Krylov, *Approximate Methods of Higher Analysis*, translated from the 3rd Russian edition by Curtis D. Benster (Interscience, New York, 1959).
6. A. W. Leissa, *Vibration of Shells* (Reprinted by the Acoustical Society of America, New York, 1993).
7. W. E. Baker, "Axisymmetric Modes of Vibration of the Thin Spherical Shell," J. Acoust. Soc. Am. **33**, 1749-1758 (1961).
8. A. Kalnins, "Effect of Bending on Vibrations of Spherical Shells," J. Acoust. Soc. Am. **36,** 74-81 (1964).
9. B. Aronov, D. A. Brown, X. Yan, and C. L. Bachand, "Modal analysis of the electromechanical conversion in piezoelectric ceramic spherical shells", J. Acoust. Soc. Am. **130**(2), 753-763 (2011).
10. R. S. Woollett, *Theory of the Piezoelectric Flexural Disk Transducer with Applications to Underwater Sound*, USL Research Report No. 490, Naval Undersea Warfare Center, Newport (1960).
11. R. V. Southwell, "On the free transverse vibrations of a uniform circular disk clamped at its center and on the effect of rotation," Proc. Roy. Soc. (London), Ser. A, **101**, 133-153 (1922).
12. G. V. Joga Rao and K. Vijayakumar, "On Admissible Functions for Flexural Vibration and Buckling of Annular Plates", Journal of Aeronautical Society of India, Vol. 15, No. 1, pp. 1- 5 (1963).
13. A. W. Leissa, *Vibration of Plates* (Reprinted by the Acoustical Society of America, New York, 1993).
14. S. M. Vogel and D. W. Skinner, "Natural Frequency of Transversely Vibrating Uniform Annular Plates," J. Appl. Mech. Trans. ASME, **32**, 926-931 (1965).
15. T. B. Gabrielson, "Frequency Constants for Transverse Vibration of Annular disks," J. Acoust. Soc. Am., **105**, 3311-3317 (1999).
16. I. S. Gradshtein, and I. M. Ryzhik, *Table of Integrals, Series, and Products*, (Academic, New York, 1963).

17. T. Boussinesq, "Résistance d'un Anneau à la flexion, quand sa surface exlérieure supporte une pression normale" ("Resistance of a ring to bending under external constant pressure"), Comptes Rendus **97**, 843 (1883).
18. B. S. Aronov, "Piezoelectric slotted ring transducer", J. Acoust. Soc. Am. **133**(6), 3875-3884 (2013).
19. E. Giebe and E. Blechschmidt, "Experimental and Theoretical Studies of Extensional Vibrations of Rods and Tubes," Ann. Physik **18**, 417-485 (1933).
20. M. Onoe and H. F. Tiersten, "Resonant Frequencies of Finite Piezoelectric Vibrators with High Electromechanical Coupling," IEEE Trans. Ultrasonic Engineering **10**, 32-39 (1963).
21. B. S. Aronov, C. L. Bachand, and D. A. Brown "Analytical modeling of piezoelectric ceramic transducers based on coupled vibration analysis with application to rectangular thickness poled plates," J. Acoust. Soc. Am. **126**(6), 2983–2990 (2009).
22. E. A. G. Shaw, "On the resonant vibration of thick barium titanate discs," J. Acoust. Soc. Am. **28**, 38–50 (1956).
23. S. Ikegami, I. Ueda, and S. Kobayashi, "Frequency spectra of resonant vibration in disc plates of PbTiO3 piezoelectric ceramic," J. Acoust. Soc. Am. **55**, 339-344 (1974).
24. S. Ueha, S. Sakuma, and E. Mori, "Measurement of vibration velocity distribution and mode analysis in thick discs of Pb(ZrTi)O3," J. Acoust. Soc. Am. **73**(5), 1842-1847 (1983).
25. J. F. Haskins and J. L. Walsh, "Vibrations of ferroelectric cylindrical shells with transverse isotropy," J. Acoust. Soc. Am. **29**, 729-734 (1957).
26. M. C. Junger and F. G. Rosato, "The propagation of elastic waves in thin-walled cylindrical shells," J. Acoust. Soc. Am. **26**, 709–713 (1954).
27. B. S. Aronov, "Coupled vibration analysis of the thin-walled cylindrical piezoelectric ceramic transducers," J. Acoust. Soc. Am. **125**(2), 803-818 (2009).
28. B. S. Aronov, D. A. Brown and S. Regmi, "Experimental investigation of coupled vibrations in piezoelectric cylindrical shells," J. Acoust. Soc. Am., **120**(3), 1374-1380 (2006).
29. D. A. Berlincourt, D. R. Curran, and H. Jaffe, *Piezoelectric and Piezomagnetic Materials and their Function in Transducers*, in Physical Acoustics, Vol. I, Part A, edited by W. P. Mason (Academic, New York, 1964).

CHAPTER 5

ELECTROMECHANICAL CONVERSION

5.1 Equations of State for Piezoceramic Medium

According to the energy approach that is accepted in this treatment and formulated in Chapter 1, all the equations describing a transducer operation are derived from variational principle. The main characteristic feature of using variational principle to deriving equations of vibration for piezoceramic bodies is that the state of a piezoceramic body is defined not only by mechanical but also by electrical generalized coordinates. Therefore, in the expression (1.94) for the Lagrangian L of an electromechanical system that a piezoceramic body presents, a suitable thermodynamic function characterizing its energy state should be used instead of the density of the potential energy, w_{pot}. Since the processes of vibrations of elastic bodies are considered to be adiabatic (proceed under conditions of thermal insulation at constant entropy), a suitable thermodynamic function may be the internal energy, w_{int},

$$\delta w_{int} = T_i \delta S_i + E_m \delta D_m + T \delta S, \tag{5.1}$$

where $T_i \delta S_i$ and $E_m \delta D_m$ are independent mechanical and electric energies applied to an element of a body (thermodynamic functions are related to the unit volume), T is the temperature, S is the entropy; $T\delta S$ is the thermal energy which vanishes in adiabatic process ($\delta S = 0$). Variations of the thermodynamic functions are meaningful only, as a measure of work performed in process of change of state of a body. For thermodynamics of piezoceramic media see, for example, Ref. 1. For a piezoceramic body that is under an action of external forces, or for a unit volume of this body, we will present Lagrangian in the form analogous to that accepted by relations (1.94) and (1.92)), replacing the potential energy with the internal energy. Namely, for a unit volume

$$L = w_{kin} - w_{int} + w_e \tag{5.2}$$

and for a body

$$L = W_{kin} - W_{int} + W_e, \quad W_{int} = \int_{\tilde{V}} w_{int} dV. \tag{5.3}$$

Equations of motion for piezoceramic bodies can be obtained in the same way, as they were derived for elastic bodies made from passive materials, in the form of Euler's equations (4.1) in the generalized coordinates, or (4.2) in the geometric coordinates. The only difference being that electric coordinate must be included in the generalized coordinates. For deriving a particular Euler's equation, it is necessary to obtain the explicit expressions for the internal energy of a volume element and for the entire piezoceramic body via generalized coordinates. To this end, the equations of state that describe relationship between variables, which are involved in determining the internal energy of a body, must be considered.

The equations of state for piezoceramic medium are derived based on the thermodynamic functions chosen depending on what variables are used as independent. For the adiabatic process it is convenient along with the internal energy function to use enthalpy H

$$\delta H = -S_i \delta T_i - D_m \delta E_m + T \delta S , \tag{5.4}$$

where the stress, T_i, and the electric field, E_m, are independent variables, and the electric enthalpy H_2

$$\delta H_2 = T_i \delta S_i - D_m \delta E_m + T \delta S , \tag{5.5}$$

in which case the independent variables are the strain, S_i, and electric field E_m.

For a general transducer operation analysis, the function H_2 is preferable with S_i and E_m as independent variables. Although the choice of a particular initial function is not crucial, since all the thermodynamic functions are interrelated. The convenience of using strain S_i as an independent variable is because in a general analysis the mode of vibration of a body and the strain distribution therein are often assumed to be known. In addition, energy transfer into a load is generally caused by the transducer surface displacements, which are directly related to the strains. The equations of state derived from the enthalpy H_2 have the following form

$$T_i = \left(\frac{\partial H_2}{\partial S_i} \right)_{E,S} = T_i(S_i, E_m), \quad D_m = -\left(\frac{\partial H_2}{\partial E_m} \right)_{S_i,S} = D_m(S_{i,} E_m) . \tag{5.6}$$

For particular calculations in many cases it is convenient to use stress T_i, and E_m as independent variables. The respective equations of state which are derived from the function H are

$$S_i = -\left(\frac{\partial H}{\partial T_i}\right)_{E,S} = S_i(T_i, E_m), \quad D_m = -\left(\frac{\partial H}{\partial E_m}\right)_{T,S} = D_m(T_i, E_m). \tag{5.7}$$

The subscript indices for the derivatives of the thermodynamic functions indicate that the values of the corresponding variables remain constant. If to consider transducers at small deviations of independent variables from their position of equilibrium and under a linear approximation, which is justified for almost all practical modes of operating the transducers built from the modern piezoceramic materials, then equations of state (5.6) can be represented, as follows:

$$T_i = c_{ik}^E S_k - e_{im} E_m, \tag{5.8}$$

$$D_m = e_{mi} S_i + \varepsilon_{mk}^S E_k, \tag{5.9}$$

where $c_{ij}^E = \left(\partial T_i / \partial S_j\right)_E$ are the moduli of elasticity at constant electric field, $\varepsilon_{mk}^S = \left(\partial D_m / \partial E_k\right)_S$ are dielectric constants at constant strain, and

$$e_{im} = \left(\partial T_i / \partial E_m\right)_S = -\left(\partial D_m / \partial S_i\right)_E \tag{5.10}$$

are piezoelectric constants. The latter expression is a relationship of electromechanical reciprocity for the piezoelectric transduction. In analogous way for equations (5.7) we obtain

$$S_i = s_{ik}^E T_k + d_{im} E_m, \tag{5.11}$$

$$D_m = d_{mi} T_i + \varepsilon_{mk}^T E_k, \tag{5.12}$$

where $s_{ik}^E = \left(\partial S_i / \partial T_j\right)_E$ are the elastic compliances at constant electric field, $\varepsilon_{mk}^T = \left(\partial D_m / \partial E_k\right)_T$ are the dielectric constants at constant mechanical stress, and $d_{im} = \left(\partial S_i / \partial E_m\right)_T = \left(\partial D_m / \partial T_i\right)_E$ are the piezoelectric moduli In equations (5.8), (5.9) and (5.10) summation is supposed to be performed with respect to repeating indices, in accordance with the rule accepted for tensor quantities. These equations are called local piezoelectric equations (or the constitutive equations) since they describe situation in a small volume element, within which the values of independent variables do not change. In the case that deformation is uniform and electric field is independent of coordinates, the volume, in which piezoelectric equations are valid, may not be small. In the absence of piezoelectric effect equations (5.8) and (5.9) become independent equations of mechanical state (Hooke's law $T_i = c_{ik} S_k$) and equations of dielectric state $D_m = \varepsilon_{mk} E_k$. These equations should be used with respect to those portions of the transducer volume where the ceramics is not polarized. This is a widespread case in the transducers designing, and in addition to the values of constants c_{ik}^{E}, s_{ik}^{E}, ε_{ik}^{T}, ε_{mk}^{S} it is necessary to know the values of elastic and dielectric constants of not polarized ceramics as well.

The elastic constants c_{ik}^{E} and s_{ik}^{E} relate two second-order tensors, and they form the fourth-order tensors. (It is more accurate to call these constants electro elastic, because their values depend on the electrical state of a piezoelement.) Piezoelectric constants relate second-order tensors and vector and form a third-order tensor. Dielectric constants form a second-order tensor. All the above-listed constants are determined experimentally. The number of non-zero and independent constants depends on the symmetry of the material structure. For the polarized ceramics, which belongs to the ∞ *mm* class of symmetry, the matrices of elastic, dielectric and piezoelectric constants that correspond to equations (5.11) and (5.12) are presented in Table 5.1. Note that the volume element of piezoceramic is related to the orthogonal coordinate system (Figure 5.1), where the direction of axis 3 (unit vector $\boldsymbol{q}_3$) coincides with the direction of poling vector $\boldsymbol{P}$. Directions of the unit vectors $\boldsymbol{q}_1$ and $\boldsymbol{q}_2$ are arbitrary, but such that all the three form a right-hand coordinate system (rotation of $\boldsymbol{q}_1$ to the coincidence with $\boldsymbol{q}_2$ must be seen as counter-clockwise from the end of vector $\boldsymbol{q}_3$). For simplicity we will consider the direction of $\boldsymbol{q}_1$ being such that vector $\boldsymbol{E}$ is always in the plane 2, 3, whereby $E_1 = 0$, and $\boldsymbol{E} = E_3\boldsymbol{q}_3 + E_2\boldsymbol{q}_2$.

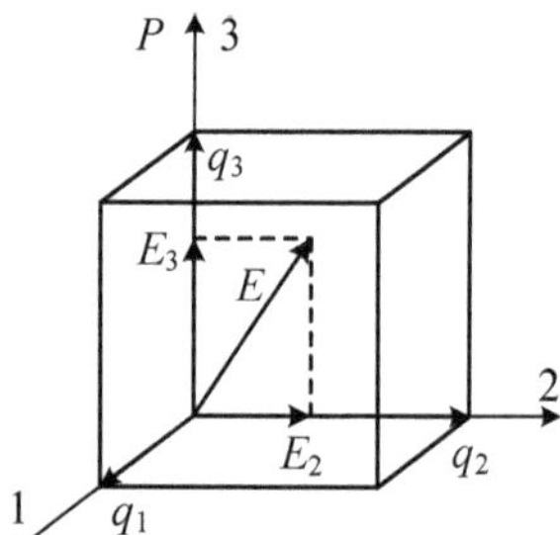

Figure 5.1: Volume element in the crystallographic coordinate system.

The following relation holds between the elastic constants,

$$s_{ij}^{E} = (-1)^{i+j}\,\Delta_{ij} / \Delta, \tag{5.13}$$

where Δ is the determinant of matrix c_{ij}^{E}, Δ_{ij} is a minor formed by deletion of the line *i* and column *j* (similarly for c_{ij}^{E}). Namely, (we will omit superscripts *E* of elastic constants in the relations between them for brevity)

$$\Delta = (s_{11} - s_{12})[s_{33}(s_{11} + s_{12}) - 2s_{13}^2], \quad \Delta_{11} = s_{11}s_{33} - s_{13}^2, \quad \Delta_{33} = s_{11}^2 - s_{12}^2,$$
$$\Delta_{13} = s_{12}s_{13} - s_{11}s_{13}, \quad \Delta_{12} = s_{13}^2 - s_{12}s_{33}. \tag{5.14}$$

Piezoelectric constants are related to each other through the equations

$$d_{mi} = e_{mj}s_{ji}^E, \quad e_{mi} = d_{mj}c_{ji}^E. \tag{5.15}$$

Since in process of manipulations the constants of different kind may be used interchangeably, it is useful to present some relations between them that follow from expressions (5.13)-(5.15).

$$c_{33} - c_{13}^2 / c_{11} = s_{11} / \left(s_{11}s_{33} - s_{13}^2\right), \quad c_{11} - c_{13}^2 / c_{33} = s_{11} / \left(s_{11}^2 - s_{12}^2\right),$$
$$c_{11} - c_{12}^2 / c_{11} = s_{33} / \left(s_{11}s_{33} - s_{13}^2\right), \quad c_{12} - c_{13}^2 / c_{33} = -s_{12} / \left(s_{11}^2 - s_{12}^2\right); \tag{5.16}$$

$$e_{33} - e_{31}c_{13} / c_{11} = (d_{33}s_{11} - d_{31}s_{13}) / (s_{11}s_{33} - s_{13}^2), \quad e_{31} - e_{33}c_{13} / c_{33} = d_{31} / \left(s_{11} + s_{12}\right),$$
$$e_{31}(1 - c_{12} / c_{11}) = [d_{33}(s_{11} + s_{12s_{13}}) + d_{31}s_{13}] / (s_{11}s_{33} - s_{13}^2). \tag{5.17}$$

Relationships are also valid obtained from (5.16) and (5.17) by replacing c_{ij} with s_{ij}, and d_{mi} with e_{mi}, and vice versa.

Table 5.1: Matrices of constants of piezoelectric ceramics.

-	T_1	T_2	T_3	T_4	T_5	T_6	E_1	E_2	E_3
S_1	s_{11}^E	s_{12}^E	s_{13}^E	0	0	0	0	0	d_{31}
S_2	s_{12}^E	s_{11}^E	s_{13}^E	0	0	0	0	0	d_{31}
S_3	s_{13}^E	s_{13}^E	s_{33}^E	0	0	0	0	0	d_{33}
S_4	0	0	0	s_{44}^E	0	0	0	d_{15}	0
S_5	0	0	0	0	s_{44}^E	0	d_{15}	0	0
S_6	0	0	0	0	0	$s_{66}^{E\ 1)}$	0	0	0
D_1	0	0	0	0	d_{15}	0	ε_{11}^T	0	0
D_2	0	0	0	d_{15}	0	0	0	ε_{11}^T	0
D_3	d_{31}	d_{31}	d_{33}	0	0	0	0	0	ε_{33}^T

In course of calculating electromechanical transducers, it is often necessary to refer to properties of analogous mechanical systems made of isotropic passive materials, or to consider mechanical systems of the transducers as comprised of active and passive materials. Therefore, it is convenient to use analogous notations for the elastic constants of passive and piezoceramic materials. Such notations for the piezoceramic materials may be introduced by comparing the equations of state (4.10) and (5.11) at $m = 3$ and $E_3 = 0$, the latter being rewritten in the form

$$S_1 = s_{11}^E[T_1^E + (s_{12}^E / s_{11}^E)T_2^E + (s_{13}^E / s_{11}^E)T_3^E], \tag{5.18}$$

$$S_2 = s_{11}^E[(s_{12}^E / s_{11}^E)T_1^E + T_2^E + (s_{13}^E / s_{11}^E)T_3^E], \tag{5.19}$$

$$S_3 = s_{33}^E[(s_{13}^E / s_{33}^E)T_1^E + (s_{13}^E / s_{33}^E)T_2^E + T_3^E]. \tag{5.20}$$

Further we introduce the following notations

$$1 / s_{11}^E = Y_1^E, \quad 1 / s_{33}^E = Y_3^E, \tag{5.21}$$

analogous to Young's moduli, and

$$-s_{12}^E / s_{11}^E = \sigma_1^E, \quad -s_{13}^E / s_{11}^E = \sigma_3^E, \quad -s_{13}^E / s_{33}^E = \sigma_{13}^E, \tag{5.22}$$

analogous to Poisson's ratios. In these notations (that are also useful in terms of brevity) the above equations will look like

$$S_1 = \frac{1}{Y_1^E}[T_1^E - \sigma_1^E T_2^E - \sigma_3^E T_3^E], \tag{5.23}$$

$$S_2 = \frac{1}{Y_1^E}[-\sigma_1^E T_1^E + T_2^E - \sigma_3^E T_3^E], \tag{5.24}$$

$$S_3 = \frac{1}{Y_3^E}[-\sigma_{13}^E T_1^E - \sigma_{13}^E T_2^E + T_3^E]. \tag{5.25}$$

Values of the introduced elastic constants are presented in Table 5.2 for several piezoceramic compositions following the original data from Ref. 2 (see Appendix B, Table B.1.)

Table 5.2: Values of the elastic constants of piezoceramic compositions.

	Y_1^E GPa	Y_3^E GPa	σ_1^E	σ_3^E	σ_{13}^E
PZT-4	81	64	0.33	0.43	0.34
PZT-5	61	53	0.35	0.44	0.38
PZT-8	87	74	0.32	0.42	0.35

Consider the piezoelectric equations (5.8), (5.9) and (5.11), (5.12) from the point of view of the rule of signs accepted in Section 1.5.2. For a visual demonstration we will identify behavior of the domain areas in piezoceramic with behavior of a dipole (two charged balls on a compliant bar, as shown in Figure 5.2), the electric moment of which is parallel to the direction of the dominant orientation of the electric moments of domains. We will also assume that the strains in the piezoelectric element and the changes of charges on the electrodes (changes of charge density D) caused by mechanical actions and by the external electric field are due to deformations (rotations and tensions/compressions) of domain areas in our model of equivalent dipoles. Figure 5.2 shows the cross section of a piezoelectric element with two dipoles, the moments of which are equally inclined relative to direction of the poling vector since the directions of domain moments are also symmetric relative to the polar axis.

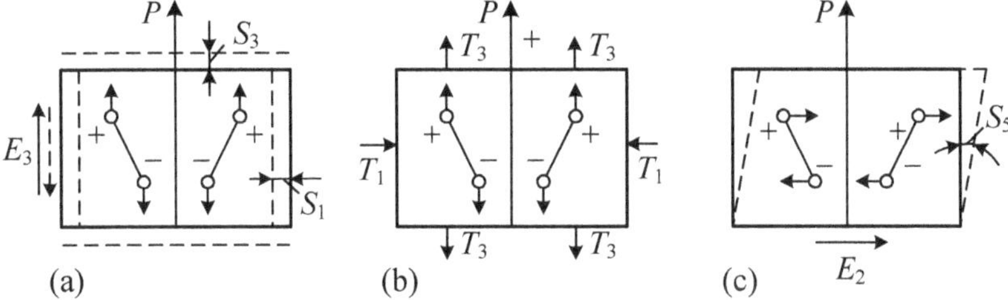

Figure 5.2: On the rule of signs in the piezoelectric equations for the tensile (a, b) and shear (c) deformations.

In a free piezoelectric element (at $T_k = 0$) $S_i = d_{3i}E_3$. The action of electric field E_3 in direction of the polar axis should result in a positive tensile strain S_3 and negative compressive strains S_1 and S_2 in the transverse direction, since in this case the dipoles elongate and rotate towards the polar axis (Figure 5.2 (a)). Since $d_{33} > 0$, $d_{31} < 0$, it is this direction of $\boldsymbol{q}_3$, which is conventionally positive. With changes of direction of vector $\boldsymbol{q}_3$ (dashed line in (a)), dipoles shorten and turn away from the polar axis, which leads to a negative compressive strain S_3 and positive tensile strains S_1 and S_2. It is easy to verify that for the conventionally positive direction of $\boldsymbol{q}_2$ $\boldsymbol{E} \times \boldsymbol{P} > 0$.

In a short-circuited piezoelectric element (at $E_3 = 0$) $S_i = s_{ik}T_k$. Tensile mechanical stresses cause longitudinal tensile strains ($s_{ii} > 0$) and transverse compressive strains ($s_{ik} < 0$). Therefore, they are conventionally positive. It follows from equation (5.12) that $D_m = d_{mi}T_i$, and the tensile mechanical stress T_3 must cause positive polarization (increase of charge density on the electrodes) since $d_{33} > 0$, while T_1 and T_2 must cause negative polarization (decrease of charge density on the electrodes) since $d_{31} < 0$. Indeed, as can be seen from Figure 5.2 (b), the stress T_3 results in dipole elongation and rotation towards the polar axis, and hence increase of the charge density on the electrodes. By contrast, the tensile stresses T_1 and T_2 cause withdrawal of charges from the electrodes and, consequently, a decrease in the charge density. Positive shear stresses lead to dipole rotation towards the electrodes (Figure 5.2 (c)), which corresponds to the value of $d_{15} > 0$.

From the equations (5.8) and (5.9) it follows (and, similarly to the above case, can be illustrated by means of Figure 5.2) that the electric field E_m in the assumed conventionally positive direction generates in a clamped piezoelectric element ($S_i = 0$, $T_i = -e_{im}E_m$) mechanical compressive stresses T_3 and negative shear stress T_5 ($e_{33} > 0$, $e_{15} > 0$) as well as positive tensile stresses T_1 and T_2 ($e_{i3} < 0$). In a short circuited piezoelectric element ($E_3 = 0$, $D_3 = e_{3i}S_i$) the tensile strains S_3 and positive shear strains S_5 cause positive polarization, whereas tensile strains S_1 and S_2 cause the negative one.

Elastic and dielectric constants characterize piezoceramics as an ideal elastic electric medium. For a proper transducer designing the mechanical and dielectric losses of energy in process of operation must be taken into consideration. The quality of piezoceramics in terms of internal energy losses are characterized by its mechanical (Q_m) and electric (Q_e) quality factors or by the more convenient in some cases angles of losses δ_m and δ_e, which are related to the quality factors by formulas $\tan\delta_m = 1/Q_m$ and $\tan\delta_e = 1/Q_e$.

The values of the piezoceramic constants depend on the initial equilibrium state of ceramics, namely, on the static mechanical stresses and temperature, at which the transducers operate. Under strong dynamic mechanical stress and electric field non-linearity of properties of piezoceramics may become noticeable, which must be taken into consideration in designing powerful transducers. Information on dependencies of piezoceramic parameters from the strong static and dynamic actions can be found in Ref. 2 and in Chapter 11.

5.2 Energy State of a Volume Element

Consider the energy state of a piezoceramic volume element in the rectangular coordinate system with unit vectors $\boldsymbol{q}_1$, $\boldsymbol{q}_2$, $\boldsymbol{q}_3$ (Figure 5.1). As it was noted, the possible variants of mutual direction of vectors $\boldsymbol{P}$ and $\boldsymbol{E}$ for piezoceramics are reduced to two, namely, $\boldsymbol{E} = E_3\boldsymbol{q}_3$ in the case that the same electrodes are used for piezoceramic poling and transducer operation, and $\boldsymbol{E} = E_2\boldsymbol{q}_2$ in the case that working electrodes are placed on the faces of a piezoelectric element, which are parallel to the polar axis, upon removing the electrodes used for poling, in order to realize electromechanical conversion under the shear deformation.

Some mechanical and electrical boundary conditions exist on the faces of a volume element in each particular case. If no energy exchange between the volume element and environment takes place, they are ideal boundary conditions. Thus, the faces may be free of mechanical stresses ($T = 0$) or clamped ($S = 0$), and the electrodes may be short circuited ($E = 0$) or open ($D = 0$).

Variation of the internal energy of a unit volume element,

$$\delta w_{int} = T_i \delta S_i + E_m \delta D_m, \tag{5.26}$$

may be caused by independent mechanical and/or independent electrical energy entering the volume. If only mechanical energy is applied, then independent electric energy is absent and

$$\delta w_{\text{int}} = \delta w_{mch} = T_i \delta S_i. \tag{5.27}$$

If only electrical energy is applied, then independent mechanical energy is absent and

$$\delta w_{\text{int}} = \delta w_{el} = E_m \delta D_m. \tag{5.28}$$

Certainly, it does not mean that in the first case no strains or mechanical stresses arise and in the second case no charges or electric fields are generated, as it would be without electromechanical conversion performed by piezoelectric material. As for piezoceramics, here in the first case the electric polarization energy depending on the mechanical actions appears by virtue of the direct piezoelectric effect, which is a component of the independent mechanical energy δw_{mch}. In the second case the strain energy depending on an electrical action appears by virtue of the reverse piezoelectric effect, which is a component of the independent electric energy δw_{el}. For distinguishing between energies due to independent electric and mechanical actions in general case that both may be applied, we will underline the terms pertaining to independent mechanical actions. Thus, the expression for the internal energy in this general case will be presented as

$$\delta w_{int} = \underline{T_i \delta S_i} + E_m \delta D_m \,. \tag{5.29}$$

First, we assume that only electric energy is supplied to a volume element. Then the electric field E_m is generated in the element, and strains S_i allowed by boundary conditions arise due to reverse piezoelectric effect. The electric field E_m is the independent variable. Taking into account that in this case independent mechanical energy vanishes $\underline{\delta w_{mech}} = \underline{T_i \delta S_i} = 0$, and referring to Eq. (5.8) we arrive at the following expression for Eq. (5.28),

$$\delta w_{int} = \delta w_{el} = E_m \delta D_m = \varepsilon^S_{mm} E_m \delta E_m + e_{mi} E_m \delta S_i \,. \tag{5.30}$$

The first term in Eq. (5.30) is the energy that would be supplied by an external electric source with the field strength changing by δE_m to the volume element, if it was completely clamped ($\delta S_i = 0$). This energy will be designated as δw^S_{el}. The second term represents the energy, which would be supplied to the volume element by the electric source, if the electric field E_m was kept constant and the strain changed by δS_i. This quantity will be called the electromechanical energy and designated δw_{em}. The concept of electromechanical energy was introduced in Chapter 1 (see Eq. (1.51)) for the case of one- dimensional deformation. Here it will be considered for the general case. At m = 3 we have $\delta w_{em\,3} = e_{31} E_3 \delta S_i$, at m = 2 $\delta w_{em\,2} = e_{24} E_2 \delta S_4$. Multiplying both parts of Eq. (5.8) by δS_i and taking into account that $T_i \delta S_i = 0$ due to ideal boundary conditions, we obtain

$$\delta w_{em\,3} = e_{3i} E_3 \delta S_i = c^E_{ii} S_i \delta S_i + c^E_{ik} S_i \delta S_k = \delta w^E_{mch\,3} \quad (i,k = 1,2,3)\,, \tag{5.31}$$

$$\delta w_{em\,2} = e_{24} E_2 \delta S_4 = c^E_{44} S_4 \delta S_4 = \delta w^E_{mch2} \,. \tag{5.32}$$

The expressions for δw^E_{mech} describe increments of the volume strain energies calculated under the condition that the values of elastic moduli are determined at $\delta E_m = 0$. Relations (5.31) and (5.32) demonstrate that electromechanical energy δw_{em} can be considered as that part of energy supplied to the volume element, which is converted into the strain energy determined at values of elastic moduli at E_m constant.

Since all the subsequent manipulations concerning the two possible variants of mutual orientation of vectors ***P*** and ***E*** at m = 3 and at m = 2 are analogous, we will perform them for the most common case of m = 3, omitting subscript m in designations of energies. The relations for the case of m = 2 will be presented in their final form, when it will be needed. Thus, summarizing expressions (5.30)-(5.32) we arrive at the following relations, which will be used in further calculations

$$\delta w_{int} = \delta w_{el} = \delta w_{el}^S + \delta w_{em} , \tag{5.33}$$

$$\delta w_{int} = \delta w_{el}^S + \delta w_{mch}^E , \tag{5.34}$$

where

$$\delta w_{el} = \varepsilon_{33}^S E_3 \delta E_3 , \quad \delta w_{em} = e_{3i} E_3 \delta S_i . \tag{5.35}$$

If the volume element is considered in the mode of electromechanical conversion under action of an external mechanical load (which is the case), then the energy balance instead of (5.30) should be presented in the form

$$\delta w_{el} = \delta w_{el}^S + \delta w_{em} = \delta w_{int} + \delta w_e , \tag{5.36}$$

where δw_e is the mechanical energy generated because of electromechanical conversion and propagating into the load. Thus, in this case

$$\delta w_{int} = \delta w_{el}^S + \delta w_{em} - \delta w_e . \tag{5.37}$$

Using equation (5.8), we obtain

$$e_{mi} E_m \delta S_i = c_{ik}^E S_k \delta S_i - T_{ir} \delta S_i , \tag{5.38}$$

where T_{ir} represents the mechanical stresses that are generated on the surface of the volume element as reaction of the load. This expression may be transformed as follows.

Since the energy δw_e flows into the load, i.e., out of the volume element, $T_{ir} \delta S_i = -\delta w_e$ (the situation is like that illustrated in Figure 1.14). In presence of reaction of the load an electric field may be generated in the volume element, which is directed perpendicular to the faces free of electrodes. Since there is no free charges on these faces, the corresponding charge densities are zero, and from equation (5.8) we obtain: $E_1 = -e_{15} S_5 / \varepsilon_{11}^S$ and $E_2 = -e_{24} S_4 / \varepsilon_{11}^S$ at $m = 3$; $E_3 = -e_{3i} S_i / \varepsilon_{33}^S$ and $E_1 = -e_{15} S_5 / \varepsilon_{11}^S$ at $m = 2$. When the external field E_m is applied, the strains S_4, S_5 at $m = 3$ and S_1, S_2, S_3 at $m = 2$ are not generated directly, but they may arise as a response of the environment to deformation of the volume element. As result of manipulation of expression (5.38) upon substituting thus obtained values of electric field and considering that $T_{ir} \delta S_i = -\delta w_e$, we obtain for $m = 3$ and $m = 2$, respectively,

$$\begin{aligned}
\delta w_{em} &= e_{3i}E_3\delta S_i = c_{ii}^E S_i\delta S_i + c_{ik}^E S_i\delta S_k + c_{ll}^D S_l\delta S_l + \delta w_e = \\
&= \delta w_{mch3}^E + \delta w_e \quad (i,k = 1,2,3;\ l = 4), \\
\delta w_{em} &= e_{24}E_2\delta S_4 = c_{44}^E S_4\delta S_4 + c_{ii}^D S_i\delta S_i + c_{ik}^D S_i\delta S_k + \delta w_e = \\
&= \delta w_{mch2}^E + \delta w_e \quad (i,k \neq 4).
\end{aligned} \tag{5.39}$$

In relations (5.39) it is taken into account the correlation between piezoceramic constants $c_{ik}^E + e_{mk}^2 / \varepsilon_{mm}^S = c_{ik}^D$, which can be obtained from equations (5.8) and (5.9) at $D_m = 0$. We can see that in the general case, when the volume element is located inside of a body that experiences deformation, the expressions for δw_{mch3}^E and δw_{mch2}^E resulting from the relations (5.39) have to be used. Relations (5.39) show that in this case δw_{em} is the total energy that is converted into the mechanical form as the result of electromechanical conversion as well, however, now a part of the mechanical energy propagates into a load. Substituting (5.39) into (5.37) results in

$$\delta w_{int} = \delta w_{el}^S + \delta w_{em} - \delta w_e = \delta w_{el}^S + \delta w_{mch}^E . \tag{5.40}$$

Thus, while expression (5.37) for internal energy is always valid, the expression (5.34) holds for the ideal boundary conditions only, and in the general case, when reaction of the environment is present, one should use expression (5.37) instead.

Determine the total energy imparted to the volume element as its state changes from the initial stage at $E_3 = 0$, $S_i = 0$ to that characterized by values E_3, S_i. Evidently, it will be

$$w_{el}^S = \frac{1}{2}\varepsilon_{33}^S E_3^2, \quad w_{mch}^E = \frac{1}{2}\left(c_{ii}^E S_i^2 + c_{ik}^E S_i S_k + c_{ll}^D S_l^2\right) \quad (i,k \neq 4,5;\ l = 4,5) . \tag{5.41}$$

To determine the total change in w_{em}, imagine that the entire interval of change of E_m and S_i is divided into N equal parts. Assume that within each part $\delta S_i = S_i / N$ and the electrical field remains constant and undergoes a sudden change by E_m / N at its end. The electromechanical energy w_{em} can be found as the limit of the integral sum

$$w_{em} = \lim_{N\to\infty} \sum_{n=1}^{N-1} e_{im} \frac{S_i}{N} \cdot \frac{E_m}{N} \cdot n = \frac{1}{2} e_{im} E_m S_i . \tag{5.42}$$

Now, the value of w_{int} can be represented based on the relation (5.33) as

$$w_{int} = w_{el} = w_{el}^S + |w_{em}| . \tag{5.43}$$

Here w_{em} is taken by modulus because of the following reasons. As can be seen from relation (5.32), w_{em} is the essentially positive quantity (the energy that this quantity represents flows into the volume element). This complies also with expression (5.35), if to take a proper account for the signs of piezoelectric constants and strains arising from the conventionally positive direction of E_3. However, in practical calculations it is difficult to keep track of the signs of the strains, and one may erroneously obtain a negative value of w_{em}. Taking this quantity by modulus eliminates the possibility of such an error.

Expression for the internal energy can also be represented in a form other than that in Eq. (5.30). Considering that the values of E_m and S_i characterize the state of the volume element with faces free of stress (except for those fixed by virtue of boundary conditions), the internal energy can be represented as

$$\delta w_{int} = E_m \delta D_m = \delta w_{el}^{T_i} = \varepsilon_{mm}^{T_i} E_m \delta E_m \,. \tag{5.44}$$

Here $w_{el}^{T_i} = \varepsilon_{mm}^{T_i} E_m^2 / 2$ can be regarded as the electric energy of the volume element determined at the value of dielectric constant $\varepsilon_{mm}^{T_i}$, which corresponds to the existing boundary conditions (superscript T_i indicates, which mechanical stress are equal to zero). Comparing the new expression for δw_{em} with formula (5.23) leads to the relation

$$w_{em} = w_{el}^{T_i} - w_{el}^{S} \,. \tag{5.45}$$

When the mechanical state of the volume element changes from the state corresponding to a completely clamped volume to the state corresponding to the free (to the extent that is allowed by the pre-set boundary conditions), the energy equal to electromechanical energy can be converted into mechanical work performed in the external medium, provided that the electric field is kept constant. Therefore, the electromechanical energy w_{em} also can be called convertible, as it is done in Ref. 7. Finally, in order to emphasize once again the continuity of the connection between the electric and mechanical states of piezoceramics, this energy can be called mutual, as it is done in Ref. 2.

Consider now the internal energy, δw_{int}, in the case that independent flow of mechanical energy is supplied to the volume element and the electrodes are open ($D_m = 0$, electrical open circuit (no-load) condition), i.e.,

$$\underline{\delta w_{mch}} = \underline{T_i \delta S_i} \,. \tag{5.46}$$

Using equation (5.8) we obtain (the case that $m = 3$ will be considered only)

$$\delta w_{int} = \underline{\delta w_{mch}} = \underline{c_{ik}^{E} S_k \delta S_i} - \underline{e_{3i} E_{3oc} \delta S_i} = \underline{\delta w_{mch}^{D}} \ . \tag{5.47}$$

From formula (5.8) at $D_3 = 0$ we obtain $E_{3\,oc} = -e_{3k} S_k / \varepsilon_{33}^{S}$. Thus,

$$\underline{-e_{3i} E_{3oc} \delta S_i} = \underline{\varepsilon_{33}^{S} E_{3oc} \delta E_{3oc}} \ . \tag{5.48}$$

Let us designate

$$\underline{-e_{3i} E_{3oc} \delta S_i} = \underline{\delta w_{me}}, \quad \underline{\varepsilon_{33}^{S} E_{3oc} \delta E_{3oc}} = \underline{\delta w_{el}^{S}} \ . \tag{5.49}$$

As we can see from (5.48), $\underline{\delta w_{me}}$ is a part of mechanical energy supplied to the volume element that can be considered as converted to electric form $\underline{\delta w_{el}^{S}}$. Since (see Eq. (5.31))

$$\underline{c_{ik}^{E} S_i \delta S_k} = \underline{\delta w_{mch}^{E}} , \tag{5.50}$$

we can obtain from formula (5.47) with regard to introduced designations that

$$\underline{\delta w_{me}} = \underline{\delta w_{mch}^{D}} - \underline{\delta w_{mch}^{E}} , \tag{5.51}$$

$$\delta w_{int} = \underline{\delta w_{mch}} = \underline{\delta w_{mch}^{E}} + \underline{\delta w_{me}} = \underline{\delta w_{mch}^{E}} + \underline{\delta w_{el}^{S}} \ . \tag{5.52}$$

Comparing Eqs. (5.49) and (5.35), we can see that, if to replace E_{3oc} by E_3, the expressions for δw_{em} and $\underline{\delta w_{me}}$ will differ only by sign. The opposite signs for $\underline{\delta w_{em}}$ and $\underline{\delta w_{me}}$ reflect the fact that, while in the case of electromechanical conversion the electric field E_3 of conventionally positive direction (coinciding with direction of the poling vector) causes positive strain ($e_{33} > 0$), in the case of the mechanoelectric conversion the positive tensile strain causes generating of the field E_{3oc} of the opposite direction. This corresponds with relation (5.10) of reciprocity of piezoelectric conversion.

The expressions for energies (5.41) and (5.42) have to be used in the general case, when the mechanical boundary conditions are complicated. In the most of cases the number of independent components of tensors of strain or stress is restricted, and it is expedient to simplify these expressions in advance. This will be done for the same kinds of boundary conditions, as those considered in Chapter 4 with respect to a volume element of a passive isotropic material.

Note that a brief and less general summary of results of this Section that are useful for understanding the essence of the subject is presented in Ref. 3.

5.3 Expressions for Energy Densities

The energy densities will be considered for various mutual directions of acting deformations and polarization vector that are illustrated in Table 5.3. In the table configurations of the piezoelectric elements combined under numbers I-VI are shown, in which the respective boundary conditions are realized, and various orientations of the polar axis in these piezoelectric elements (designated by numbers 1, 2, 3). The coordinate axes, direction of which coincides with that of acting deformation, are designated by dashed lines.

At first, we consider variants of deformations that are not accompanied by shear, that is, at $S_4 = S_5 = S_6 = 0$. The numbers of variants correspond to the numbers of their images in Table 5.3.

I. Deformation through the thickness of a plate, other dimensions of which are large. The strains in the plane of the plate are absent from considerations of symmetry. Two directions of axis 3 can be considered: 1) perpendicular to the plate, whereby $S_1 = S_2 = 0$; 2) parallel to the plane of the plate, whereby $S_2 = S_3 = 0$. By substituting of S_i into expressions (5.41) and (5.42) for these two directions of axis 3, in the first of which the longitudinal and in the second one the transverse piezoelectric effect is realized, we obtain

$$1)\quad 2w_{el}^S = \varepsilon_{33}^S E_3^2, \quad 2w_{mch}^E = c_{33}^E S_3^2, \quad 2w_{em} = e_{33} E_3 S_3, \tag{5.53}$$

$$2)\quad 2w_{el}^S = \varepsilon_{33}^S E_3^2, \quad 2w_{mch}^E = c_{11}^E S_1^2, \quad 2w_{em} = e_{31} E_3 S_1. \tag{5.54}$$

II. Deformation through the width of a thin plate, one dimension of which is large, or in the direction of the circumference of a long thin cylinder. Since the plate (cylinder) has a small thickness and mechanical stresses on its side surfaces are zero, they can be considered as zero also throughout the thickness of the plate (cylinder). The strains in the direction of large length are absent from the symmetry considerations. In this case the strains that differ by the direction of axis 3 are possible: 1) axis 3 is coincident with the strain direction, the longitudinal piezoelectric effect, $T_1 = 0$, $S_2 = 0$; 2) axis 3 is perpendicular to the plane of the plate, the transverse piezoelectric effect, $T_3 = 0$, $S_2 = 0$; 3) axis 3 is parallel to the length of the plate, the transverse piezoelectric effect, $T_2 = 0$, $S_3 = 0$.

For the variant 1 (longitudinal piezoelectric effect, $T_1 = 0$, $S_2 = 0$) from equation (5.8) at $i = 1$ we obtain $S_1 = \left(e_{31} E_3 - c_{13}^E S_3\right) / c_{11}^E$. Substituting the values of S_1 and $S_2 = 0$ into expressions (5.41) and (5.42) we find by formula (5.43) that

Table 5.3: Coefficients characterizing piezoeffect under various boundary conditions (1/3)

Mechanical System	I		II		
Boundary Conditions Parameter	1 $S_1 = S_2 = 0,\ S_3$	2 $S_2 = S_3 = 0,\ S_1$	1 $S_1 = 0,\ T_2 = 0, S_3$	2 $S_2 = 0, T_3 = 0, S_1$	3 $S_3 = 0, T_2 = 0,\ S_1$
$C_{e\Delta}^{S_i}$	ε_{33}^{S}	ε_{33}^{S}	$\varepsilon_{33}^{S_{3,1}} = \varepsilon_{33}^{S} + e_{31}^{2} / c_{11}^{E}$	$\varepsilon_{33}^{S_{1,2}} = \varepsilon_{33}^{S} + e_{33}^{2} / c_{33}^{E}$ $= \varepsilon_{33}^{T}(1 - k_p^2)$	$\varepsilon_{33}^{S_{1,3}} = \varepsilon_{33}^{S} + e_{31}^{2} / c_{11}^{E}$
K_{Δ}^{E}	c_{33}^{E}	c_{11}^{E}	$c_{33}^{E} - (c_{13}^{E})^2 / c_{11}^{E}$	$c_{11}^{E} - (c_{13}^{E})^2 c_{33}^{E}$ $= s_{11}^{E} / [(s_{11}^{E})^2 - (s_{12}^{E})^2)]$	$c_{11}^{E} - (c_{12}^{E})^2 / c_{11}^{E}$
n_{Δ}	e_{33}	e_{31}	$e_{33} - e_{31} c_{13}^{E} / c_{11}^{E}$	$e_{31} - e_{33} c_{13}^{E} / c_{33}^{E}$ $= d_{31} / (s_{11}^{E} + s_{12}^{E})$	$e_{31}(1 - c_{12}^{E} / c_{11}^{E})$
k_c	$\dfrac{e_{33}}{\sqrt{\varepsilon_{33}^{S} c_{33}^{D}}}$	$\dfrac{e_{31}}{\sqrt{\varepsilon_{33}^{S} c_{11}^{D}}}$	-	-	-
Designation of k_c in Ref. 2	k_t	$k_{31}^{''}$	$k_{33}^{'}$	$k_{31}^{'}$	$k_{31}^{'}$
K_{Δ}	$\dfrac{(1-\sigma)Y}{(1+\sigma)(1-2\sigma)}$	-	-	$\dfrac{Y}{1-\sigma^2}$	-

Table 5.3: Coefficients characterizing piezoeffect under various boundary conditions (2/3)

Mechanical System	III		IV		
Boundary Conditions Parameter	1 $T_2 = T_3 = 0, S_1$	2 $T_1 = T_2 = 0, S_3$	1 $S_1 \neq S_2, T_3 = 0$	2 $T_3 = 0, S_1 = S_2$	3 $T_2 = 0, S_1 \neq S_3$
$C_{e\Delta}^{S_i}$	$\varepsilon_{33}^{S_1} = \varepsilon_{33}^{T} - d_{31}^2 / s_{11}^{E}$ $= \varepsilon_{33}^{T}(1 - k_{31}^2)$	$\varepsilon_{33}^{S_3} = \varepsilon_{33}^{T} - d_{33}^2 / s_{33}^{E}$ $= \varepsilon_{33}^{T}(1 - k_{33}^2)$	$\varepsilon_{33}^{S_{1,2}}$	$\varepsilon_{33}^{S_{12}} = \varepsilon_{33}^{T} - 2d_{31}^2 / (s_{11}^{E} + s_{12}^{E})$ $= \varepsilon_{33}^{T}(1 - k_p^2)$	$\varepsilon_{33}^{S_{1,3}}$
K_{Δ}^{E}	$\frac{1}{s_{11}^{E}}$	$\frac{1}{s_{33}^{E}}$	w_{mch}^{E} from (5.69)	$\frac{2}{s_{11}^{E} + s_{12}^{E}}$	w_{mch}^{E} from (5.73)
n_{Δ}	$\frac{1}{s_{11}^{E}}$	$\frac{d_{33}}{s_{33}^{E}}$	w_{em} from (5.70)	$\frac{2d_{31}}{s_{11}^{E} + s_{12}^{E}}$	w_{em} from (5.74)
k_c	$\frac{d_{31}}{\sqrt{\varepsilon_{33}^{T} s_{11}^{E}}}$	$\frac{d_{33}}{\sqrt{\varepsilon_{33}^{T} s_{33}^{E}}}$	-	$k_{31}\sqrt{\frac{2}{1 + s_{12}^{E} / s_{11}^{E}}}$	-
Designation of k_c in Ref. 2	k_{31}	k_{33}	-	k_p	-
K_{Δ}	Y		w_{pot} from (4.16)	$\frac{2Y}{1 - \sigma}$	w_{pot} from (4.16)

Table 5.3: Coefficients characterizing piezoeffect under various boundary conditions (3/3)

Mechanical System	V S_1 S_2		VI S_4
Boundary Conditions Parameter	1 3 2 1 $S_3 = 0, S_1 \neq S_2$	$S_1 = S_2$	1 P E_2 S_4
$\tilde{N}_{e\Delta}^{S_i}$	ε_{33}^{S}	ε_{33}^{S}	ε_{11}^{S}
K_{Δ}^{E}	w_{mch}^{E} from (5.76)	$2(c_{11}^{E} + c_{12}^{E})$	$c_{44}^{E} = 1 / s_{44}^{E}$
n_{Δ}	w_{em} from (5.77)	$2e_{31}$	$e_{15} = d_{15} / s_{44}^{E}$
k_c	-	-	$\dfrac{d_{15}}{\sqrt{\varepsilon_{11}^{T} s_{44}^{E}}}$
Designation of k_c in Ref. 2	-	$k_p^{'}$	k_{15}
K_{Δ}	-	-	μ

$$w_{int} = \frac{1}{2}(\varepsilon_{33}^{S} + e_{31}^{2} / c_{11}^{E})E_3^2 + \frac{1}{2}(e_{33} - e_{31}c_{13}^{E} / c_{11}^{E})E_3 S_3 = w_{el}^{S_{3,1}} + \frac{1}{2}(c_{33}^{E} - c_{13}^{E2} / c_{11}^{E})S_3^2 . \quad (5.55)$$

Considering relations (5.16) and (5.17) between the constants of piezoceramics, the expressions for components of the internal energy can be represented as follows

$$2w_{el}^{S_{3,1}} = \varepsilon_{33}^{S_{3,1}} E_3^2, \quad \varepsilon_{33}^{S_{3,1}} = \varepsilon_{33}^{S} + e_{31}^{2} / c_{11}^{E}, \quad (5.56)$$

$$2w_{mch}^{E} = (c_{33}^{E} - c_{13}^{E2} / c_{11}^{E})S_3^2 = \frac{s_{11}^{E}}{s_{11}^{E} s_{33}^{E} - s_{13}^{E2}} S_3^2 , \quad (5.57)$$

$$2w_{em} = (e_{33} - e_{31}c_{13}^{E} / c_{11}^{E})E_3 S_3 = \frac{d_{33}s_{11}^{E} - d_{31}s_{13}^{E}}{s_{11}^{E} s_{33}^{E} - s_{13}^{E2}} E_3 S_3 . \quad (5.58)$$

Here designation $\varepsilon_{33}^{S_{3,1}} = \varepsilon_{33}^{S} + e_{31}^{2} / c_{11}^{E}$ is introduced for the dielectric constant determined under the condition that $S_3 = 0$ and $S_1 = 0$,the index of acting strain coming first in the subscript $S_{3,1}$. Designations with superscripts of the analogous kind will be used further in general.

For the variant 2 (transverse piezoeffect, $T_3 = 0$, $S_2 = 0$):

$$2w_{el}^{S_{1,2}} = \varepsilon_{33}^{S_{1,2}} E_3^2, \quad \varepsilon_{33}^{S_{1,2}} = \varepsilon_{33}^{S} + e_{33}^{2} / c_{33}^{E} , \quad (5.59)$$

$$2w_{mch}^{E} = (c_{11}^{E} - c_{13}^{E2} / c_{33}^{E})S_1^2 = \frac{s_{11}^{E}}{s_{11}^{E2} - s_{12}^{E2}} S_1^2 , \tag{5.60}$$

$$2w_{em} = (e_{31} - e_{33}c_{13}^{E} / c_{33}^{E})E_3S_1 = \frac{d_{31}}{s_{11}^{E} + s_{12}^{E}} E_3S_1 . \tag{5.61}$$

For the variant 3 ($T_2 = 0$, $S_3 = 0$):

$$2w_{el}^{S_{1,3}} = \varepsilon_{33}^{S_{1,3}} E_3^2 , \quad \varepsilon_{33}^{S_{1,3}} = \varepsilon_{33}^{S_{3,1}} , \tag{5.62}$$

$$2w_{mch}^{E} = (c_{11}^{E} - c_{12}^{E2} / c_{11}^{E})S_1^2 = \frac{s_{33}^{E}}{s_{11}^{E}s_{33}^{E} - s_{13}^{E2}} S_1 , \tag{5.63}$$

$$2w_{em} = e_{31}(1 - c_{12}^{E} / c_{11}^{E})E_3S_1 = \frac{d_{33}(s_{11}^{E} + s_{12}^{E}) + d_{31}s_{33}^{E}}{s_{11}^{E}s_{33}^{E} - s_{13}^{E2}} E_3S_1 . \tag{5.64}$$

III. Deformations of a thin bar in the direction of its length, or of a thin short ring in the direction of its circumference. This variant is considered in Chapter 1 (Eqs. (1.47)–(1.50)) for the case of the transverse piezoeffect ($T_2 = T_3 = 0$). The variant of the longitudinal piezoeffect can be considered in the analogous way. The final results for the energy densities are

$$2w_{el}^{S_i} = \varepsilon_{33}^{S_i} E_3^2 , \quad \varepsilon_{33}^{S_i} = \varepsilon_{33}^{T} - d_{3i}^2 / s_{ii}^{E} , \tag{5.65}$$

$$2w_{mch}^{E} = s_{ii}^{E} T_i^2 = S_i^2 / s_{ii}^{E} , \tag{5.66}$$

$$2w_{em} = d_{3i}E_3T_i = (d_{3i} / s_{ii}^{E})E_3S_i . \tag{5.67}$$

Here the subscripts $i = 1$ and $i = 3$ correspond to the transverse and longitudinal piezoeffects, respectively.

IV. Two-dimensional deformation in the plane of a thin plate.

In the variant of the transverse piezoelectric effect, $T_3 = 0$, $S_1 \neq 0$, $S_2 \neq 0$. The same conditions apply for deformations that take place in a thin-walled spherical shell, and along the axis of a thin-walled cylinder. In the general case (variant 1), in which $S_1 \neq S_2$, we obtain using equations (5.8), (5.9) and relations (5.41), (5.42) that:

$$2w_{el}^{S_{1,2}} = \varepsilon_{33}^{S_{1,2}} E_3^2 , \tag{5.68}$$

$$\begin{aligned} 2w_{mch}^{E} &= (c_{11}^{E} - c_{13}^{E2} / c_{33}^{E})(S_1^2 + S_2^2) + 2(c_{12}^{E} - c_{13}^{E2} / c_{33}^{E})S_1S_2 = \\ &= \frac{s_{11}^{E}}{s_{11}^{E2} - s_{12}^{E2}} [S_1^2 - 2(s_{12}^{E} / s_{11}^{E})S_1S_2 + S_2^2], \end{aligned} \tag{5.69}$$

$$2w_{em} = (e_{31} - e_{33}c_{13}^E / c_{33}^E)\mathrm{E}_3(S_1 + S_2) = \frac{d_{31}\mathrm{E}_3}{s_{11}^E + s_{12}^E}(S_1 + S_2). \tag{5.70}$$

Here relations (5.16) and (5.17) between the constants of the piezoceramics are used. For a pulsating sphere (variant 2) $S_1 = S_2 = S$ and

$$2w_{mch}^E = 2S^2 / (s_{11}^E + s_{12}^E), \quad 2w_{em} = 2d_{31}\mathrm{E}_3 S / (s_{11}^E + s_{12}^E). \tag{5.71}$$

In the case that direction of polarization is parallel to the plane of the plate (variant 3) $T_2 = 0$, $S_1 \neq 0$, $S_3 \neq 0$, and we obtain

$$2w_{el}^{S_{3,1}} = \varepsilon_{33}^{S_{3,1}} E_3^2, \tag{5.72}$$

$$2w_{mch}^E = \frac{1}{s_{11}^E s_{33}^E - s_{13}^E}(s_{11}^E S_3^2 - 2s_{13}^E S_1 S_3 + s_{33}^E S_1^2), \tag{5.73}$$

$$2w_{em} = \frac{\mathrm{E}_3}{s_{11}^E s_{33}^E - s_{13}^{E2}}[(d_{31}s_{33}^E - d_{33}s_{13}^E)S_1 + (d_{33}s_{11}^E - d_{31}s_{13}^E)S_3]. \tag{5.74}$$

V. Two-dimensional deformation in the plane of cross section of a long cylindrical bar. Unlike the above example, here under the transverse piezoeffect $S_3 = 0$, $T_3 \neq 0$, and

$$2w_{el}^S = \varepsilon_{33}^S E_3^2, \tag{5.75}$$

$$2w_{mch}^E = (c_{11}^E S_1^2 + 2c_{12}^E S_1 S_2 + c_{11}^E S_2^2), \tag{5.76}$$

$$2w_{em} = e_{31}\mathrm{E}_3(S_1 + S_2). \tag{5.77}$$

VI. Shear deformation of an elemental volume.

If no mechanical transformation of shear into strains of other kinds is assumed, then $S_i = 0$ at $i \neq 4$, and from equations of type (5.31) and (5.35) at $m = 2$ we obtain

$$2w_{el}^S = \varepsilon_{11}^S E_2^2, \tag{5.78}$$

$$2w_{mch}^{\mathrm{E}} = c_{44}^E S_4^2 = T_4^2 / s_{44}^E, \tag{5.79}$$

$$2w_{em} = e_{24}S_4E_2 = d_{24}T_4E_2. \tag{5.80}$$

Returning to expression (5.34) for the internal energy of a piezoceramic volume element we can conclude that the electric (E) and mechanical (S_i) variables are would be separated. The quantity $w_{el}^{S_i}$ is determined as the energy of the volume made from ceramics with dielectric constant known for each particular case of deforming. The quantity w_{mech}^E is determined as the potential energy of the element of passive ceramic material (with elastic constants at $E = 0$)

considering anisotropy of elastic properties of piezoceramics.

Essentially new is the quantity w_{em}, which characterizes electromechanical conversion of energy in piezoceramics. For evaluating the quality of piezoceramics as an active material, electromechanical coupling coefficients are widely used. Their values are directly related to values of the electromechanical energy w_{em} and depend on conditions of a piezoelement deforming. Before considering methods of determining the electromechanical coupling coefficients and calculating their values for the most common conditions of deforming of piezoceramic volume elements, we note that in all the above considered cases of deformation, except for examples with two-dimensional deformation at $S_1 \neq S_2$, density of the internal energy is expressed through one electric (E) and one mechanical (S_i) variable. For all of these cases analogy can be drawn between energy conversion in elemental volume of piezoceramics and in electromechanical two-port network. Indeed, if one considers a unit volume element ($\Delta x_1 = \Delta x_2 = \Delta x_3 = \Delta x = 1$), then the value of E is numerically equal to the voltage between electrodes V ($E = V / \Delta x$), S_i is numerically equal to the displacement of the respective face, ($S_i = \xi_i / \Delta x$), and $\varepsilon_{33}^{S_i}$ is equal to the capacitance of the volume element, $C_{e\Delta}^{S_i}$ $(C_{e\Delta}^{S_i} = \varepsilon_{33}^{S_i} \Delta x^2 / \Delta x)$. Taking into account these considerations and relations (5.33), expressions for the internal energy and its components, $w_{\text{int}} = w_{el} = w_{el}^{S_i} + w_{em}$, $w_{em} = w_{mch}^{\text{E}}$, can be represented as follows

$$w_{el}^{S_i} = C_{e\Delta}^{S_i} V^2 / 2, \tag{5.81}$$

$$w_{mch}^{\text{E}} = K_{\Delta}^{\text{E}} S_i^2 / 2, \tag{5.82}$$

$$w_{em} = n_{\Delta} V S_i / 2. \tag{5.83}$$

Here $C_{e\Delta}^{S_i}$, K_{Δ}^{E}, n_{Δ} are the coefficients, which stand for specific capacitance, specific rigidity ($C_{\Delta}^{E} = 1 / K_{\Delta}^{\text{E}}$ is specific flexibility) and electromechanical transformation coefficient for different variants of one-dimensional deformation of the volume element. Using these coefficients, we can represent piezoelectric equations in the following general form,

$$T = K_{\Delta}^{\text{E}} S - n_{\Delta} E_3, \tag{5.84}$$

$$D_3 = n_{\Delta} S + C_{e\Delta}^{S} E_3, \tag{5.85}$$

where S is the only active strain. Similar representation can be used for the case of shear deformation.

Expressions for $C_{e\Delta}^{S_i}$, K_{Δ}^{E} and n_{Δ} for different variants of one-dimensional deformations that follow from above expressions for the energy densities are presented in Table 5.3. The table also contains expressions for K_{Δ} related to the case that a passive isotropic material is used at the same boundary conditions. For the two-dimensional deformations the references are given to the formulae that must be used for determining the respective energy densities.

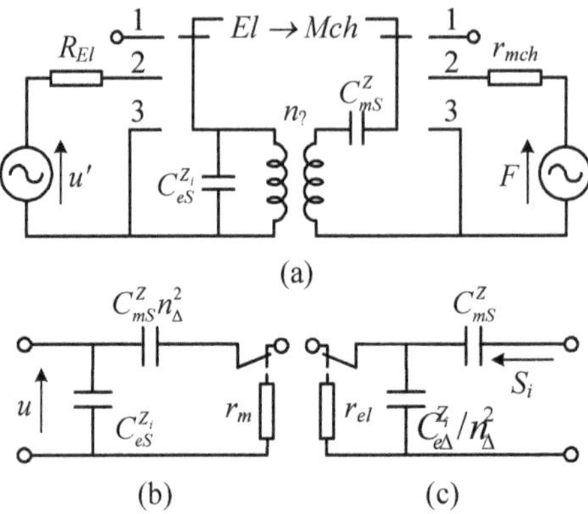

Figure 5.3: Electromechanical two-port network.

It is easy to verify that the energy relations (5.81)-(5.83) are valid for the electric two-port network shown in Figure 5.3 (a), if to use generally accepted rules for calculating electric circuits and to assume that the fictitious transformer introduced in the circuit maintains the transformation ratio n_{Δ} down to the frequency $\omega = 0$ (corresponds to the static deformation).

In the mode of the electromechanical conversion the energy is supplied to the electric input (position 2 of switch *El*) and the mechanical output is under the conditions of open circuit ($S_i = 0$) or short circuit ($T_i = 0$) in positions 1 or 3 of switch *Mch*, respectively.

In the mode of mechanoelectrical conversion, in which case energy is supplied to the mechanical input (position 2 of switch *Mch*), one can obtain expressions for internal energy of the volume element, which correspond to conditions of short circuit ($E = 0$) or open circuit ($D = 0$) of the electric side in positions 3 or 1 of switch *El*, respectively. The same results follow from the expressions (5.47)-(5.52).

5.4 Coupling Coefficients

5.4.1 About the Definitions for the Electromechanical Coupling Coefficients

Electromechanical coupling coefficients (further just coupling coefficients, k_c) are important parameters of piezoelectric material and piezoelectric bodies subjected to deformation. They are introduced for evaluating quality of piezoceramics as piezo active material for various conditions of deformation. The basic energy expression for k_c is[4]

$$k_c^2 = \frac{\text{energy stored in the mechanical form}}{\text{total input energy}}. \tag{5.86}$$

For the receive mode of operation this definition can be rephrased as

$$k_c^2 = \frac{\text{energy stored in the electrical form}}{\text{total input mechanical energy}}. \tag{5.87}$$

In a piezoelectric body of finite size the quality of electromechanical conversion depends on distribution of strains over its volume. For evaluating effects of energy conversion in piezoelectric bodies under nonuniform strain distribution the concept of effective coupling coefficients, k_{eff}, is introduced. For this case expressions (5.84) and (5.85) can be modified, as it was done in Chapter 2 (formula (2.88)), to

$$k_{eff}^2 = \frac{\text{energy stored in mechanical form in the considered mode of vibration}}{\text{total input energy}} \text{ at } \omega \to 0. \tag{5.88}$$

The effective coupling coefficient is related to a static strain distribution in the body that corresponds to a particular mode of vibration. Several examples of the application of this formula were considered in Chapter 2.

It is difficult in some cases to calculate and especially to analyze the possible ways of optimizing the effective coupling coefficients using the above expressions directly. Therefore, one more formula for coupling coefficient was introduced in Ref. 2

$$k_m = \frac{w_{em}}{\sqrt{w_e w_{mech}}}. \tag{5.89}$$

The extension of this formula to the case of the effective coupling coefficient of a piezoelectric body,

$$k_{eff}^2 = \frac{(\int_{\tilde{V}} w_{em} d\tilde{V})^2}{(\int_{\tilde{V}} w_e d\tilde{V})(\int_{\tilde{V}} w_{mech} d\tilde{V})}, \tag{5.90}$$

was reported in Ref. 5. In formulas (5.89) and (5.90) w_{em}, w_e and w_{mech} are the densities of the mutual, dielectric and elastic energies in our notations (the original notations in Ref. 2 are U_m, U_e and U_d, respectively). Applications of these formulas are not straightforward and require additional explanations. A systematic review of the definitions for the coupling coefficients that includes explanation of possible shortcomings in application formulas (5.89) and (5.90) is presented in Ref. 6. Results of this work will be used for further analysis of issues related to the concept of effective coupling coefficients.

It is noteworthy that sometimes (e.g., in Ref. 2) the coupling coefficients of piezoceramics, k_m, are referred to as static coupling coefficients, and the effective coupling coefficients, k_{eff}, as dynamic coupling coefficients. Such definitions do not follow from the essence of the matter since the notion of strain nonuniformity is not necessarily related to motion. The dynamic strains may be uniform throughout the volume, as, for example, in pulsating vibration of a ring or of a spherical shell, while the static strains may be nonuniform, as in the case of bending of plates and beams.

At first, we will turn to the coupling coefficients of piezoceramic material at various mechanical boundary conditions. The effective coupling coefficients for piezoceramic bodies under nonuniform deformation distributions will be considered in Section 5.6 after the general analysis of the energy state of a deformed body will be done.

5.4.2 Coupling Coefficients of Piezoceramic Material

The energy stored in mechanical (electrical) form in the expressions (5.86) and (5.87) is the electromechanical, w_{em}, or the mechanoelectrical, w_{me}, energy (depending on direction of the energy conversion). The coupling coefficient will be denoted as k_m until particular boundary conditions are specified. Using relation (5.43) the coupling coefficient can be represented in the following equivalent forms

$$k_m^2 = \frac{w_m^E}{w_{el}^S + w_m^E} = \frac{|w_{em}^E|}{w_{el}^S + w_m^E} \tag{5.91}$$

$$k_m^2 = \frac{|w_{em}|}{w_{el}^T} = \frac{\underline{|w_{me}|}}{\underline{w_{mch}^D}}, \tag{5.92}$$

and, after using relation (5.33),

$$k_m^2 = \frac{|w_{em}|}{w_{el}^T} = \frac{|w_{em}|}{w_{el}^S + |w_{em}|}, \tag{5.93}$$

where from

$$\frac{k_m^2}{1-k_m^2} = \frac{|w_{em}|}{w_{el}^S}. \tag{5.94}$$

The reason, by which w_{em} and $\underline{w_{me}}$ in relations (5.91)-(5.94) are taken by modulus, was explained when discussing formula (5.53). Convenience of application of the alternative expressions (5.92) and (5.94) for the coupling coefficients depends on what quantities are more appropriate to use as independent variables in a particular variant of the boundary conditions.

Since the expressions for w_{em} and w_{el} under various boundary conditions are determined in Section 5.3, formulas (5.91)-(5.94) can be readily used for calculating the corresponding coupling coefficients. To this end, we will manipulate formulas (5.91)-(5.94) in such a way that it explicitly includes specific quantities $C_{e\Delta}^{S_i}$, K_Δ^E, n_Δ that were introduced for a unit volume. Using relations (5.81)-(5.83) we obtain

$$k_m^2 = \frac{1}{1 + C_{e\Delta}^{S_i} / \left(n_\Delta^2 C_\Delta^E\right)}. \tag{5.95}$$

Expressions for the coupling coefficients k_m at the various boundary conditions considered in Section 5.3 are presented in Table 5.3

Using formula (5.91), we can obtain relations between the values of elastic and dielectric constants of piezoceramics under various mechanical and electrical boundary conditions that involve the respective coupling coefficients. Since it follows from expressions (5.44), (5.45) and (5.51), (5.52) that $\delta w_{em} = \delta w_{el}^{T_i} - \delta w_{el}^{S_i}$, $\delta w_{el} = \delta w_{el}^{T_i}$ and, similarly, $\underline{\delta w_{me}} = \underline{\delta w_{mch}^D} - \underline{\delta w_{mch}^E}$, $\underline{\delta w_{mch}} = \underline{\delta w_{mch}^D}$, the formula (5.91) can be transformed, as follows:

$$k_m^2 = \frac{w_{el}^{T_i} - w_{el}^{S_i}}{w_{el}^{T_i}} = \frac{\underline{w_{mch}^D} - \underline{w_{mch}^E}}{\underline{w_{mch}^D}}. \tag{5.96}$$

These relations can also be obtained by means of the circuit shown in Figure 5.3. In the case that independent electric energy is supplied, after transforming the mechanical impedance to the electric side (Figure 5.3 (b)) we obtain

$$w_{el} = \frac{1}{2}V^2\left(C_{e\Delta}^{S_i} + n_\Delta^2 C_\Delta^E\right) = \frac{1}{2}V^2 C_{e\Delta}^{T_i} = w_{el}^{T_i}, \tag{5.97}$$

$$w_{el}^{S_i} = \frac{1}{2}V^2 C_{e\Delta}^{S_i}, \quad \frac{w_{el}^{T_i} - w_{el}^{S_i}}{w_{el}^{T_i}} = \frac{1}{1 + C_{e\Delta}^{S_i} / n_\Delta^2 C_\Delta^E} = k_m^2 . \tag{5.98}$$

From these expressions follows that

$$C_{e\Delta}^{S_i} = C_{e\Delta}^{T_i}(1 - k_m^2), \quad \varepsilon_{33}^{S_i} = \varepsilon_{33}^{T_i}(1 - k_m^2). \tag{5.99}$$

In the case that independent mechanical energy is supplied, after transforming the electrical impedance to the mechanical side (Figure 5.3 (c)) in analogous way will be obtained that

$$K_\Delta^D\left(1 - k_m^2\right) = K_\Delta^{\mathrm{E}} . \tag{5.100}$$

Thus, using relationships (5.99) and (5.100), in variant I.1 of the boundary conditions (see Table 5.3) we obtain

$$\varepsilon_{33}^S = \varepsilon_{33}^{T_3}(1 - k_t^2), \quad c_{33}^D(1 - k_t^2) = c_{33}^{\mathrm{E}}, \tag{5.101}$$

and in the variants III.1 and III.2

$$\varepsilon_{33}^{S_i} = \varepsilon_{33}^T\left(1 - k_{3i}^2\right), \quad s_{ii}^{\mathrm{E}}\left(1 - k_{3i}^2\right) = s_{ii}^D, \tag{5.102}$$

where $i = 1$ and $i = 3$ for the transverse and longitudinal piezoelectric effect, respectively.

Since the values of the coupling coefficients depend on a type of boundary conditions, a question arises, what their maximum values are and under what mechanical action they can be realized. In the case that $m = 2$ the only and hence the maximum coupling coefficient is k_{15}. In addition to the relevant expression presented in Table 5.3 the expression

$$k_{15}^2 = 1 - \varepsilon_{11}^S / \varepsilon_{11}^T \tag{5.103}$$

that follows from relation analogous to formula (5.99) is another option. In the case that $m = 3$, the maximum value of the coupling coefficient $k_{\max 3}$ (it is called “invariant” in Ref. 2) can be obtained by generating in a volume element stresses T_3 and $T_1 = T_2$ of opposite signs, e.g., tension along the polar axis and compression in the perpendicular plane, whereby

$$k_{\max 3}^2 = 1 - \varepsilon_{33}^S / \varepsilon_{33}^T \tag{5.104}$$

5.4.3 Cycles of Energy Conversion by a Piezoelement

For better understanding the physical meaning of concepts of the electromechanical energy and of the coupling coefficient, consider the cycles of conversion of the electrical energy into mechanical energy, and the reverse conversion of energy of a mechanical source into the electric energy that are performed by a piezoelement. The piezoelement will be schematically represented by the two-port system (short bar) having electrical and mechanical inputs. The electromechanical conversion is illustrated with diagrams in Figure 5.4.

The process of the energy conversion will be assumed as proceeding in several stages. We assume that at the first stage (it will be labeled by the superscript *I*) an external force *F* is applied that clamps the piezoelement in the direction of axis 3 ($S_3^I = 0$). The external electric source produces electric field E_3^I inside the element. As follows from equations (5.11)and (5.12), in the piezoelement arise the mechanical stress and charge density

$$T_3^I = -d_{33}\mathrm{E}_3^I / s_{33}^{\mathrm{E}}, \tag{5.105}$$

$$D_3^I = \left(\varepsilon_{33}^T - d_{33}^2 / s_{33}^{\mathrm{E}}\right)\mathrm{E}_3^I = \varepsilon_{33}^{S_3}\mathrm{E}_3^I. \tag{5.106}$$

The change of state of the piezoelement at this stage is represented in Figure 5.4 by segments 0, *I*. The electric source has supplied energy

$$w_{el}^I = \varepsilon_{33}^{S_3}\left(E_3^I\right)^2 / 2. \tag{5.107}$$

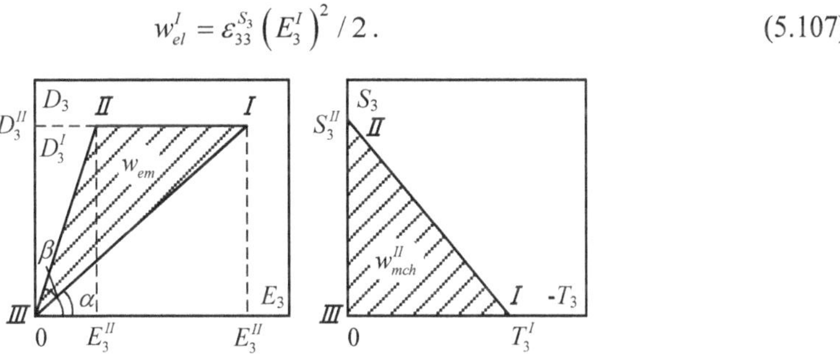

Figure 5.4: Cycle of the electromechanical conversion: $\tan\alpha = \varepsilon_{33}^{S_3}$, $\tan\beta = \varepsilon_{33}^{T_3}$.

At the second stage the piezoelement is disconnected from the electric source with its electrodes remaining open (thus $D_3^{II} = D_3^I$). The clamping force is removed, which is equivalent to connecting a small resistance of mechanical load, r_{ml} , to mechanical terminals of the piezoelement. In the ideal cycle $r_{ml} \to 0$, and at the end of the second stage $T_3^{II} = 0$. In the reality r_{ml} , however small, still has a finite value, so that $T_3^{II} \approx 0$. Using equations (5.11) and (5.12) we obtain by the end of the second stage for an ideal cycle

$$D_3^{II} = \varepsilon_{33}^{T} E_3^{II} = D_3^{I} = \varepsilon_{33}^{S_3} E_3^{I},$$
$$E_3^{II} = \varepsilon_{33}^{S_3} E_3^{I} / \varepsilon_{33}^{T}, \quad (5.108)$$
$$S_3^{II} = d_{33} E_3^{II} = d_{33} \varepsilon_{33}^{S_3} E_3^{I} / \varepsilon_{33}^{T}.$$

(Note that $T_1 = T_2$ all the time, and condition $T_3^{II} = 0$ is equivalent to condition of mechanically completely free piezoelement. Therefore, use of ε_{33}^{T} is justified.) The change of state of the piezoelement at this stage is represented by segments *I, II*. The energy

$$\left| w_{mch}^{II} \right| = T_3^{I} S_3^{II} / 2 = d_{33}^2 \varepsilon_{33}^{S_3} \left(E_3^{I} \right)^2 / 2 s_{33}^{E} \varepsilon_3^{T} = w_{el}^{I} k_{33}^2 \quad (5.109)$$

passes into the mechanical load. Energy w_{mch}^{II} is taken by modulus, because formally it has sign minus ($\delta w_{mch}^{II} = \delta T_3^{I} \cdot S_3^{II}$ and $\delta T_3^{I} < 0$), which means by accepted rule of signs that the energy flux flows out of volume of the piezoelement.

At the final stage, the electrodes are connected to a small internal resistance of the electric energy source, r_{in}. Ideally $r_{in} \to 0$, $E_3^{III} = 0$ and $S_3^{III} = 0$. In reality $\mathrm{E}_3^{III} \approx 0$. In the ideal cycle by the end of this stage the piezoelectric element returns to the initial state of the piezoelement, in which $E_3 = 0$, $S_3 = 0$, and $T_3 = 0$. The change of state is shown by segments *II, III*. On this path the piezoelement returns to the electric source the energy

$$\left| w_{el}^{III} \right| = D_3^{II} \mathrm{E}_3^{II} / 2 = \varepsilon_{33}^{S_3} \left(\mathrm{E}_3^{I} \right)^2 / 2 \varepsilon_{33}^{T} = w_{el}^{I} \varepsilon_{33}^{S_3} / \varepsilon_{33}^{T}, \quad (5.110)$$

while it does not perform any mechanical work ($T_3^{II} = T_3^{III} = 0$). (The modulus of the energy, w_{el}^{III} is used since the energy flux flows out of the piezoelement and formally should have sign minus. Indeed, $\delta w_{el}^{III} = D_3^{II} \delta \mathrm{E}_3^{III}$, where $\delta \mathrm{E}_3^{III} < 0$.)

The difference between the energy w_{el}^{I}, supplied to the piezoelement, and $\left| w_{el}^{III} \right|$, returned by the element to the electric source, is equal to the mechanical work performed by the piezoelement, i.e., to the energy transferred into mechanical load. Thus,

$$w_{el}^{I} - \left| w_{el}^{III} \right| = w_{el}^{I} \left(1 - \varepsilon_{33}^{S_3} / \varepsilon_{33}^{T} \right) = \left| w_{mch}^{II} \right| = w_{el}^{1} k_{33}^2. \quad (5.111)$$

From the graphic representations of the ideal cycle in Figure 5.4 follows that the energy supplied by electric source to the piezoelement is proportional to area of the triangles $0, I, \mathring{A}_3^{I}$ or $0, I, D^{I}$, i.e., $S_{0ID^{I}} \to w_{el}^{T}$. The energy returned to the source in the end of the cycle is proportional to area of the triangle $0, II, D_3^{II}$. Area of the triangle $0, I, II$

$$S_{0,I,D_3^{II}} - S_{0,II,D_3^{II}} = S_{0,I,II} \rightarrow w_{em} \tag{5.112}$$

represents electromechanical energy, as a part of supplied electrical energy that can be converted into mechanical energy in the ideal cycle.

For the case of the mechanoelectrical conversion of energy of a mechanical source into the energy transferred to an electric load the graphic representation of the ideal cycle can be obtained in the analogous way. Therefore, we present only basic relations relevant to each of the conversion stages that are illustrated in Figure 5.5.

Stage I. The source of mechanical energy produces strain S_3^I (let it be tension, $S_3^I > 0$). Electrodes of the piezoelectric element are open, $D_3^I = 0$. It follows from equations (5.11) and (5.12) that

$$E_3^I = -d_{33} T_3^I / \varepsilon_{33}^T, \tag{5.113}$$

$$S_3^I = \left(s_{33}^E - d_{33}^2 / \varepsilon_{33}^T\right) T_3^I = s_{33}^D T_3^I. \tag{5.114}$$

The mechanical energy supplied is

$$w_{mch}^I = T_3^I S_3^I / 2 = s_{33}^D \left(T_3^I\right)^2 / 2. \tag{5.115}$$

The change of state of the piezoelement is characterized by segments 0, *I*.

Stage II. Having fixed the face of the piezoelement (maintaining $S_3^{II} = S_3^I$), we terminate its electrodes by a small resistance of electric load, $r_{el} \rightarrow 0$. In the ideal cycle, $r_{el} = 0$ and $E_3^{II} = 0$. The mechanical stress in the piezoelement changes to a value of T_3^{II} that can be found from the condition

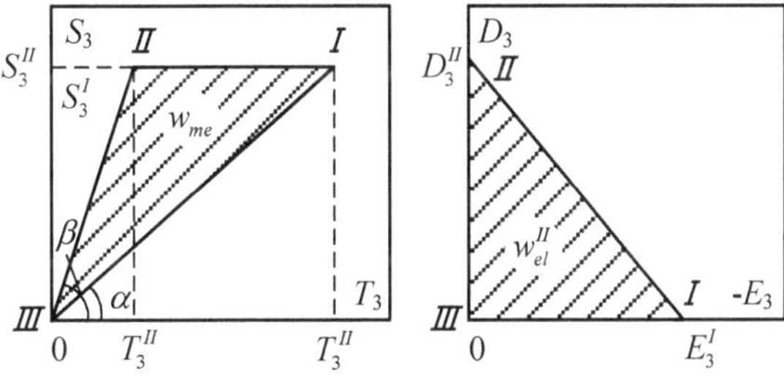

Figure 5.5: Cycle of the mechanoelectrical conversion: $\tan\alpha = s_{33}^D$, $\tan\beta = s_{33}^E$.

$$S_3^{II} = s_{33}^E T_3^{II} = S_3^I = s_{33}^D T_3^I, \tag{5.116}$$

where from

$$T_3^{II} = s_{33}^D T_3^I / s_{33}^E, \quad D_3^{II} = d_{33} T_3^{II} = d_{33} s_{33}^D T_3^I / s_{33}^E . \tag{5.117}$$

The electric energy that enters the load is

$$\left| w_{el}^{II} \right| = D_3'' E_3' / 2 = d_{33}^2 s_{33}^D (T_3')^2 / 2\varepsilon_{33}^T s_{33}^E = w_{mch}^I k_{33}^2 . \tag{5.118}$$

This energy is taken by modulus because it escapes the piezoelement, and formally is negative by the rule of signs, as this was considered regarding the mechanical energy that flows into load in the previous case. In Figure 5.5 the change of state is shown by segment *I, II.*

Stage III. The force that maintained the strain constant is reduced to zero in the ideal cycle, and $T_3^{III} = S_3^{III} = 0$. (In the real cycle they drop to small values that are determined by the internal resistance of the mechanical source, $r_{mch} \to 0$). The electrodes remain short circuited, thus, $E_3^{III} = E_3^{II} = 0$ and no change of electrical energy takes place. Not spent part of mechanical energy,

$$\left| w_{mch}^{III} \right| = T_3^{II} S_3^I / 2 = \left(s_{33}^D T_3^I \right)^2 / 2 s_{33}^E = w_{mch}^I s_{33}^D / s_{33}^E , \tag{5.119}$$

on this stage returns to the mechanical source. (The energy is taken by modulus because it flows out of volume of the piezoelement.) Change of state of the piezoelement to the initial one (in the ideal cycle to $E_3^{III} = D_3^{III} = 0$, $S_3^{III} = T_3^{III} = 0$) is represented by segment *II, III*. As the result,

$$\left| w_{el}^{II} \right| = w_{mch}^I - \left| w_{mch}^{III} \right| = w_{mch}^I \left(1 - s_{33}^D / s_{33}^E \right) = w_{mch}^I k_{33}^2 . \tag{5.120}$$

This part of supplied mechanical energy that may be converted into electrical energy in the ideal cycle (mechanoelectrical energy) is represented in Figure 5.5 by area of triangle 0, *I, II.*

Both electromechanical and mechanoelectrical energies are also called convertible or mutual energies[7]. The ratio of the convertible energy to the total energy supplied to the piezoelement is equal to the coupling coefficient square in the ideal cycle. In a real cycle the convertible part of the energy is smaller.

5.4.4 To comparing the definitions of the coupling coefficients

Clarification must be made regarding using formula (5.89) (and hence formula (5.90)) for calculating the coupling coefficients. This expression differs from the physical clear definition (5.86). It is introduced in a formal way and has shortcomings. Consider the way, how it is derived in Ref. 2. Starting from the general expression for the internal energy

$$w_{int} = \frac{1}{2} S_i T_i + \frac{1}{2} D_m E_m , \tag{5.121}$$

and using the piezoelectric equations (5.30), the authors obtained

$$w_{int} = \frac{1}{2} T_i s_{ik}^E T_k + \frac{1}{2} T_i d_{im} E_m + \frac{1}{2} E_m d_{im} T_i + \frac{1}{2} E_3^2 \varepsilon_{33}^T = U_e + 2U_m + U_d , \tag{5.122}$$

where $U_e = T_i\, s_{ik}^E\, T_k / 2$, $U_m = E_m d_{im} T_i / 2$ and $U_d = \varepsilon_{33}^T E_3^2 / 2$ are the elastic, mutual and dielectric energies, respectively. Afterwards, the definition (5.89) for k_m is introduced by analogy with the correlation coefficient between two actions. However, it must be remembered that $E_m D_m / 2$ and $T_i S_i / 2$ in relation (5.121) are *independent electrical* and *independent mechanical* energies. Therefore, the relations (5.121) and (5.122) should be represented in our designations in the form of

$$\begin{aligned} w_{int} &= \underline{\frac{1}{2} S_i T_i} + \frac{1}{2} E_m D_m = \underline{\frac{1}{2} T_i s_{ik}^E T_k} + \underline{\frac{1}{2} T_i d_{im} E_m} + \frac{1}{2} T_i d_{im} E_m + \frac{1}{2} \varepsilon_{33}^T E_3^2 \\ &= \underline{w_{mech}^E} + \underline{w_{me}} + w_{em} + w_{el}^T . \end{aligned} \tag{5.123}$$

Now it is clear that the middle terms in relation (5.122) being outwardly similar are not equal in general and could not be doubled. They may be equal only in the case that a certain relation exists between otherwise independent electrical and mechanical actions. Being obtained from equation $w_{em} = \underline{w_{me}}$, this relation is

$$\frac{E_3}{\underline{T_i}} = -\frac{d_{3i}}{\varepsilon_{33}^T} . \tag{5.124}$$

Because of the above mentioned inaccuracy the piezoelectric equations (5.8) and (5.9) with strains as independent variables cannot be used for deriving expression (5.89). In fact, the same procedure, as described by expressions (5.121) and (5.122), being applied in this case leads to

$$w_{int} = \frac{1}{2} S_i c_{ik}^E S_k + \frac{1}{2} S_i e_{im} E_m - \frac{1}{2} E_m e_{im} S_i + \frac{1}{2} \varepsilon_{33}^S E_3^2 = U_e + U_d , \tag{5.125}$$

and the mutual term disappears.

Despite the inaccurate derivation of formula (5.89) the expression of this kind can be used for determining k_m, because it can be obtained from the original definition (5.86) in the form of expressions (5.91). Indeed, if to take into account that $|w_{em}| = w_{mech}^E$, in the case that the stresses are independent variables expression (5.92) for k_m^2 can be represented as

$$k_m^2 = \frac{|w_{em}|}{w_{el}^T} = \frac{|w_{em}|}{w_{el}^T} \cdot \frac{|w_{em}|}{w_{mech}^E}, \text{ i.e., } k_m = \frac{|w_{em}|}{\sqrt{w_{el}^T w_{mech}^E}} . \tag{5.126}$$

In the case that the strains are independent variables, it can be found from expression (5.94) that

$$\frac{k_m^2}{1-k_m^2} = \frac{|w_{em}|}{w_{el}^S} = \frac{|w_{em}|}{w_{el}^S} \cdot \frac{|w_{em}|}{w_{mech}^E}, \text{ i.e., } \frac{k_m^2}{\sqrt{1-k_m^2}} = \frac{|w_{em}|}{\sqrt{w_{el}^S w_{mech}^E}} . \tag{5.127}$$

Expressions (5.126) and (5.127) can be used for calculating k_m, as well as formulas (5.91).

5.5 Internal Energy of Piezoceramic Body

The main content of this section was presented in Ref. 3.

5.5.1 Basic Considerations

We define the internal energy of a piezoceramic body as an integral of density of internal energy, w_{int}, over its volume. Taking into consideration that a volume element inside the body experiences reaction of the surrounding parts of the body in course of deformation (denote it T_{ri}), the expression for w_{int} have to be used that follows from relation (5.37). Given that $w_e = T_{ri} S_i / 2$,

$$W_{int} = \int_{\tilde{V}} w_{int} d\tilde{V} = \int_{\tilde{V}} w_{el}^S d\tilde{V} + \int_{\tilde{V}} w_{em} d\tilde{V} - \frac{1}{2} \int_{\tilde{V}} T_{ri} S_i d\tilde{V} . \tag{5.128}$$

Applying Green's transformation to the last integral and keeping in mind that no external volume forces are present, we obtain

$$\int_{\tilde{V}} T_{ri} S_i d\tilde{V} = \int_{\Sigma} \boldsymbol{f} \cdot \boldsymbol{\xi} \, d\Sigma = W_\Sigma = 0 . \tag{5.129}$$

Here $\boldsymbol{f}$ is the density of the forces acting on the surface Σ of a body, $\boldsymbol{\xi}$ is the displacement of the surface points, W_Σ is the work of external forces, i.e., the mechanical energy that flows through the surface of the body. Since deformations of the body are supposed to occur under the ideal boundary conditions, $W_\Sigma = 0$. The following information must be available for calculating the integrals in (5.128) in addition to the energy densities that are already considered for various mechanical boundary conditions.

Configuration of the piezoceramic poling electric field must be known since it determines the crystallographic coordinate system q_1, q_2, q_3, for which the tensors of piezoceramic constants presented in Table 5.1 are valid. It is expedient to perform integration in this coordinate system, otherwise the tensors of the constants must be transformed to a chosen coordinate system.

The configuration of the operating electric field must be known. According to our convention, we have to consider only two variants: $\boldsymbol{E} \,||\, \boldsymbol{q}_3$ ($m = 3$, $E_3 \neq 0$, $E_2 = 0$) and $\boldsymbol{E} \perp \boldsymbol{q}_3$ ($m = 2$, $E_2 \neq 0$, $E_3 = 0$). The first variant corresponds to the most common case, in which the same electrodes connected in the same way are used both for ceramics polarization and for operation. If in the operating mode some parts of the electrodes are connected in antiphase, then field component E_2 may develop in the parts of the volume around the gap between these parts of the electrodes. For implementing the second variant, the operating electrodes should be applied in such a way, as to insure the condition $\boldsymbol{E} \perp \boldsymbol{q}_3$. Usually this involves removing the electrodes used for polarization and applying new electrodes. If necessary, the general case of arbitrary mutual direction of vectors $\boldsymbol{E}$ and $\boldsymbol{P}$ can be considered as superposition of these two variants.

Distribution of strain, S_i, over the volume must be known. Since for solving vibration problems in generalized coordinates the systems of supporting functions are used that characterize distribution of displacements over the volume, in the general analysis of the energy state of a body the strain distributions can be considered as *a priori* known.

Prior to determining the particular configurations of polarization field, which depend on the shape of the piezoelectric elements and layout of electrodes on their surfaces, we will carry out analysis of integral (5.128) in the curvilinear coordinate system of the general form. Configuration of piezoelement of a general form is qualitatively shown in Figure 5.6. We assume that the unit vector $\boldsymbol{q}_3$ is tangential to the lines of force of the polarization field, and the unit vectors $\boldsymbol{q}_1$, $\boldsymbol{q}_2$ are tangential to the equipotential surfaces of this field. The elemental volume $d\tilde{V} = H_i dq_i$ is limited by the side surface of a tube of current formed by lines of force of operating electric field and equipotential surfaces of this field (Figure 5.6).

At first, we will assume that the entire volume of the piezoelement is confined between two electrodes. Since both variants of mutual directions of the poling vector $\boldsymbol{P}$ and of the operating electric field $\boldsymbol{E}$ can be considered in analogous way, a detailed analysis will be made for the variant of $\boldsymbol{E} \,||\, \boldsymbol{q}_3$ and the final result will be given only for the variant of $\boldsymbol{E} \perp \boldsymbol{q}_2$. All the analysis will be performed for the mode of the electromechanical conversion.

The components of internal energy densities $w_{el}^{S_i}$ and w_{em} in formula (5.128) are defined by relations (5.41) and (5.42). Peculiarity of the energy conversion by the elemental volume inside the piezoelectric body may occur due to possible mechanical and electrical interactions between this element and the neighboring parts of the body. The mechanical interactions don't produce effect on the result of integrating over the volume under the ideal mechanical boundary conditions, as it follows from expression (5.129).

The electrical interactions between elements inside the body can result from the fact that distribution of the electric field in a deformed body (it will be denoted by E_3') may differ from distribution E_3 in a clamped body under the condition that the same voltage is applied, since for an elemental volume the field generated by the remaining part of the body as a result of the piezoelectric effect turns out to be external as well. Therefore, E_3 should be replaced by E_3' in Eqs. (5.8), (5.9) and (5.11), (5.12), if the volume element under consideration is inside the body. Thus, one must know the distribution of E_3' over the volume of the body with respect to E_3, when integrating density of the internal energy, w_{int}. Thus, relation between values of E_3' and E_3 must be established.

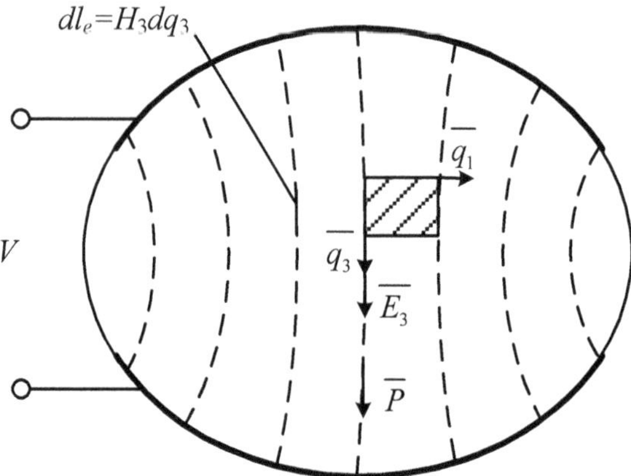

Figure 5.6: Piezoelectric body represented in the crystallographic coordinate system.

Since piezoceramics is dielectric material, and there are no electrodes inside the volume upon which free charges could form,

$$\operatorname{div}\boldsymbol{D} = \frac{1}{H_1H_2H_3}\left[\frac{\partial(D_1H_2H_3)}{\partial q_1} + \frac{\partial(D_2H_1H_3)}{\partial q_2} + \frac{\partial(D_3H_1H_2)}{\partial q_3}\right] = 0 . \qquad (5.130)$$

If $\boldsymbol{E} \,||\, \boldsymbol{q}_3$, then $D_1 = D_2 = 0$, because there are no electrodes on the respective surfaces of the body, and

$$\partial(D_3H_1H_2)/\partial q_3 = 0, \quad D_3H_1H_2 = \text{constant} . \qquad (5.131)$$

After applying Eq. (5.9) to the elemental volume of the body in Figure 5.6 with substitution of E_3' for E_3, multiplying both sides of the equation by H_1H_2, and integrating along the line of electric field l_e ($dl_e = dl_3 = H_3dq_3$), we obtain

$$D_3 = \frac{\varepsilon_{33}^S}{l_eH_1H_2}\int_0^{l_e} E_3'H_1H_2H_3dq_3 + \frac{e_{3i}}{l_eH_1H_2}\int_0^{l_e} S_iH_1H_2H_3dq_3 . \qquad (5.132)$$

Note that the length l_e in general may depend on the coordinates q_1, q_2. The first term in this relation represents the charge density $\varepsilon_{33}^S E_3$ due to the electric field E_3 in the clamped body under the applied voltage. The second term represents the charge density generated by the deformation of the body due to the piezoelectric effect. Substituting D_3 expressed by Eq. (5.132) into Eq. (5.9), we obtain the electric field inside the vibrating body E_3' in the form

$$E_3' = E_3 + \frac{e_{3i}}{\varepsilon_{33}^S l_eH_1H_2}\int_0^{l_e} S_iH_1H_2H_3dq_3 - \frac{e_{3i}S_i}{\varepsilon_{33}^S} . \qquad (5.133)$$

Note that in relations (5.132) and (5.133) summing by the repeating index is assumed. Namely, in general it may be $e_{3i}S_i = e_{31}(S_1 + S_2) + e_{33}S_3$, but here S_i are the working strain only. For particular boundary conditions, which result in one-dimensional deformation, $e_{3i}S_i / \varepsilon_{33}^{S_i} = (n_\Delta / C_{y\Delta}^{S_i})S_i$, where S_i is the working strain, $n_\Delta, C_{e\Delta}^{S_i}$ are the constants presented in Table 5.3. With regard to relation (5.95), $e_{3i} / \varepsilon_{33}^S$ can be replaced by $K_\Delta^E k_c^2 / (1 - k_c^2)$, where k_c is the respective coupling coefficient.

Under nonuniform deformation of a body the equation (5.132) for D_3 and the following equation for stress T_i,

$$T_i = c_{ik}^E S_k - e_{3i}E_3 - \frac{e_{3i}^2}{\varepsilon_{33}^S H_1H_2l_e}\int_0^{l_e} S_iH_1H_2H_3dq_3 + \frac{e_{3i}^2}{\varepsilon_{33}^S} S_i , \qquad (5.134)$$

must be used for its internal points instead of the local Eqs. (5.8) and (5.9). It is obtained from

Eq. (5.8) after replacing E_3 by expression (5.133) for E_3', The expression (5.128) that includes components of the internal energy have to be rewritten correspondingly, as

$$W_{int} = \int_{\tilde{V}} w_{el}^{S_i} d\tilde{V} + \int_{\tilde{V}} w_{em} d\tilde{V} = \frac{1}{2}\int_{\tilde{V}} \varepsilon_{33}^S \left(E_3'\right)^2 d\tilde{V} + \frac{1}{2}\int_{\tilde{V}} e_{3i} E_3' S_i d\tilde{V} . \tag{5.135}$$

When integrating after substituting E_3' by its expression (5.133), the assumption that $H_1 H_2$ is approximately independent of coordinate q_3 can be adopted for practically all the piezoelectric ceramic transducer designs. (Otherwise, the poling electric field in the piezoelectric elements could be nonuniform that would result in a poor quality of polarization). Under this assumption and considering that

$$\int_0^{l_e} H_3 dq_3 = l_e \tag{5.136}$$

the integration in Eq. (5.135) becomes straightforward and leads to the following results.

$$\frac{1}{2}\int_V \varepsilon_{33}^S \left(E_3'\right)^2 dV =$$

$$= \frac{1}{2}\int_V \varepsilon_{33}^S E_3^2 dV + \frac{1}{2}\frac{e_{3i}^2}{\varepsilon_{33}^S} \int_{q_1,q_2} \left[\int_0^{\ell_e} S_i^2 H_3 dq_3 - \frac{1}{l_e}\left(\int_0^{l_e} S_i H_3 dq_3 \right)^2 \right] H_1 H_2 dq_1 dq_2, \tag{5.137}$$

$$\frac{1}{2}\int_V e_{3i} S_i E_3' dV =$$

$$= \frac{1}{2}\int_V e_{3i} S_i E_3 dV - \frac{e_{3i}^2}{2\varepsilon_{33}^S} \int_{q_1,q_2} \left[\int_0^{l_e} S_i^2 H_3 dq_3 - \frac{1}{l_e}\left(\int_0^{l_e} S_i H_3 dq_3 \right)^2 \right] H_1 H_2 dq_1 dq_2. \tag{5.138}$$

Integrating in the second term is supposed over the equipotential surface perpendicular to the direction of polarization, $\boldsymbol{q}_3$.

It should be noted that by virtue of relations (5.38), in which E_3 must be replaced by E_3', and (5.129)

$$\frac{1}{2}\int_{\tilde{V}} e_{3i} E_3' S_i d\tilde{V} = \int_{\tilde{V}} w_{mch}^E d\tilde{V} + \frac{1}{2}\int_{\tilde{V}} T_{ri} S_i d\tilde{V} = \int_{\tilde{V}} w_{mch}^E d\tilde{V} = W_m^E . \tag{5.139}$$

Here W_m^E is the strain energy of piezoceramic body, calculated under the assumption that the electric field is kept constant during deformation.

Several designations will be introduced for brevity:

$$\frac{1}{2}\int_V \varepsilon_{33}^S E_3^2 dV = W_{el}^S \tag{5.140}$$

for the electric energy supplied to the clamped body by the source generating the electric field of strength E_3;

$$\frac{1}{2}\int_V e_{3i} E_3 S_i d\tilde{V} = W_{em} \tag{5.141}$$

for the electromechanical energy, and

$$\frac{e_{3i}^2}{2\varepsilon_{33}^S}\int_{q_1,q_2}\left[\int_0^{l_e} S_i^2 H_3 dq_3 - \frac{1}{l_e}\left(\int_0^{l_e} S_i H_3 dq_3\right)^2\right] H_1 H_2 dq_1 dq_2 = \Delta W \tag{5.142}$$

for the additional term, which depends both on the strain distribution and on the configuration of electric field in the clamped body. This quantity accounts for influence exerted on the internal energy by possible differences in conditions of electromechanical conversion over the volume of the body, which may arise under deformation. In other words, this term characterizes electrical interaction between elements in a deformed body.

For the boundary conditions, which result in one-dimensional deformation, $e_{3i}S_i / \varepsilon_{33}^{S_i} = (n_\Delta / C_{e\Delta}^{S_i})S_i$, where S_i is the working strain, n_Δ, $C_{e\Delta}^{S_i}$ are the constants presented in Table 5.3. With regard to relation (5.95), $e_{3i} / \varepsilon_{33}^S$ can be replaced by $K_\Delta^E k_c^2 / (1-k_c^2)$, where k_c is the respective coupling coefficient.

Summarizing expressions (5.138) through (5.142), we find that

$$W_{em} = W_m^E + \Delta W . \tag{5.143}$$

After substituting expressions (5.137) and (5.138) into formula (5.135), and taking into consideration the introduced designations, we obtain

$$W_{int} = W_{el}^{S_i} + W_{em} = W_{el}^{S_i} + W_m^E + \Delta W . \tag{5.144}$$

In the case that the strains do not change along the direction of electrical lines of force (i. e., $\partial S_i / \partial q_3 = 0$), $\Delta W = 0$, and relations (5.143) and (5.144) become

$$W_{int} = W_{el}^S + W_m^E, \quad W_{em} = W_m^E . \tag{5.145}$$

That is, under the condition that $\Delta W = 0$ the electric and mechanical variables would be separated. This means that the internal energy of a deformed body can be calculated as a sum of the electric energy of the body being clamped and of the mechanical energy determined at the constant electric field. This statement can be qualified as formulation of the *Theorem of Separation* of the electrical and mechanical variables in the deformed piezoceramic bodies.

In the variant that $\boldsymbol{E} \perp \boldsymbol{q}_3$ ($E_3 = 0, E_2 \neq 0$), relations (5.143) through (5.145) remain valid, and for the energies involved therein the following expressions hold:

$$W_{el}^S = \frac{1}{2}\int_{\tilde{V}} \varepsilon_{11}^S E_2^2 d\tilde{V}, \qquad (5.146)$$

$$W_{em} = \int_{\tilde{V}} e_{24} E_2 S_4 d\tilde{V} = \int_{\tilde{V}} d_{24} E_2 T_4 d\tilde{V}, \qquad (5.147)$$

$$W_m^E = \frac{1}{2}\int_{\tilde{V}} w_{mch2}^E d\tilde{V}, \qquad (5.148)$$

$$\Delta W = \frac{e_{24}^2}{2\varepsilon_{11}^S}\int_{q_1 q_3}\left[\int_0^{l_e} S_4^2 H_1 H_2 H_3 dq_2 - \frac{1}{l_e}\int_0^{l_e} S_4 H_2 dq_2 \int_0^{l_e} S_4 H_1 H_2 H_3 dq_2\right] dq_1 dq_3. \qquad (5.149)$$

The condition, under which $\Delta W = 0$, is $\partial S_4 / \partial q_2 = 0$.

When calculating energies by formulas (5.139) through (5.141) for the particular bodies under certain boundary conditions, the expressions for energy densities $w_{el}^{S_i}$, w_{mch}^E and w_{em} must be used that are given in Table 5.3.

Let us assume that the piezoelectric body has one mechanical degree of freedom. The distribution of displacements inside the body can be represented as $\xi(\boldsymbol{r},t) = \xi_o(t)\theta(\boldsymbol{r})$, where $\xi_o(t)$ is the displacement of a reference point on the surface of the body and $\theta(\boldsymbol{r})$ is a nondimensional function of the geometrical coordinates, which does not change in the frequency range under consideration. Then all the components of the W_{int} may be expressed by ξ_0 as the generalized mechanical coordinate and by the voltage V as the generalized electrical coordinate, namely,

$$W_{el}^S = \frac{C_{el}^S V^2}{2}, \quad W_m^E = \frac{K_m^E \xi_o^2}{2} = \frac{\xi_o^2}{2C_m^E}, \quad W_{em} = \frac{1}{2}V\xi_o n, \quad \Delta W = \frac{\Delta K \xi_o^2}{2} = \frac{\xi_o^2}{2\Delta C}. \qquad (5.150)$$

Here C_{el}^S is the electrical capacitance of a clamped body, K_m^E and C_m^E are the equivalent rigidity and compliance ($K_m^E = 1/C_m^E$), n is the electromechanical transformation coefficient and

$\Delta K = 1/\Delta C$ is the additional rigidity of vibrating body, which is associated with energy ΔW. Generally, the equivalent electromechanical circuit of such a transducer with one mechanical degree of freedom can be represented as shown in Figure 5.7 (a). In the case that electrical and mechanical variables are separable ($\Delta W = 0$), the circuit element ΔC must be excluded.

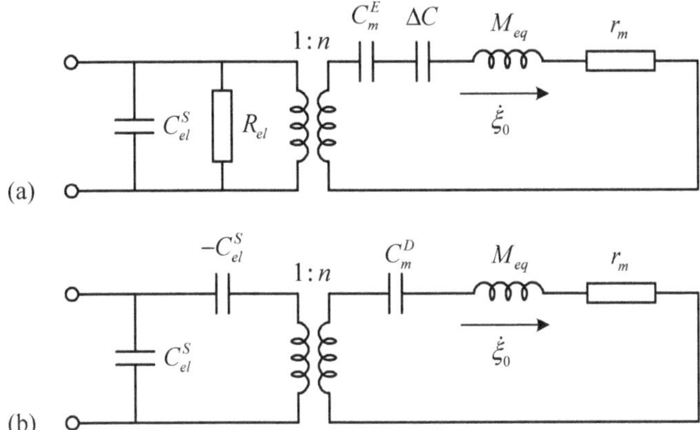

Figure 5.7: Equivalent circuit of a transducer with one mechanical degree of freedom: (a) the general representation (in the case that electrical and mechanical coordinates are separable, i.e. $\Delta W = 0$, the compliance ΔC must be excluded), (b) the Mason's equivalent circuit for an end-electroded bar vibrating in the fundamental mode.

5.5.2 About the Physical Meaning of Quantity ΔW

The energy term ΔW that is defined by expression (5.142) is the matter of principle in terms of separation of the electrical and mechanical variables. Consider the physical meaning of this term and quantitative estimate of its magnitude in comparison with the elastic energy W_m^E. It is convenient to examine these issues with typical examples of the longitudinally vibrating bars having different electrical boundary conditions, as illustrated in Figure 5.8, and with rectangular beams in flexural vibration, shown in Figure 5.11.

5.5.2.1 Longitudinally Vibrating Bars

The internal energy densities of the longitudinally vibrating bar are given by the expressions

$$w_{el}^S = w_{el}^{S_i} = \varepsilon_{33}^{S_i} \frac{E_3^2}{2}, \quad w_m^E = \frac{S_i^2}{2s_{ii}^E}, \quad w_{em} = \frac{d_{3i}}{2s_{ii}^E} S_i E_3 . \qquad (5.151)$$

Distribution of displacements and strain in the bars under consideration are

$\xi(x) = \xi_0 \cos(\pi x / l)$ and $S_i = -\xi_0 (\pi / l)\sin(\pi x / l)$, respectively. The equivalent parameters for the bars must be calculated using expressions (5.139) through (5.142) and energy densities by formulas (5.151). The coordinate system is rectangular. Thus, when calculating ΔW by formula (5.149) it must be taken $H_1 = H_2 = H_3 = 1$, $dq_3 = dx_3$, $dq_1 = dx_1$, $dq_2 = dx_2$. Besides for working strain along direction of polarization

$$e_{3i}^2 S_i^2 / \varepsilon_{33}^{S_i} = d_{33}^2 S_3^2 / s_{33}^{E2} \varepsilon_{33}^{S_3} = k_{33}^2 S_3^2 / \left(1 - k_{33}^2\right) s_{33}^E, \tag{5.152}$$

and for the working strain in the transverse direction

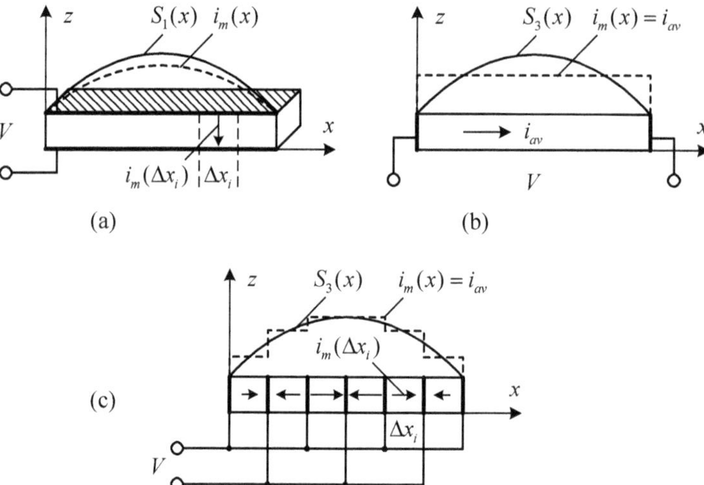

Figure 5.8: Piezoelectric bars vibrating in the longitudinal fundamental mode: (a) side-electroded, transverse piezoeffect; (b) end-electroded, longitudinal piezoeffect; (c) segmented axially polarized, longitudinal piezoeffect.

$$e_{3i}^2 S_i^2 / \varepsilon_{33}^{S_i} = d_{31}^2 S_1^2 / s_{11}^{E2} \varepsilon_{33}^{S_1} = k_{31}^2 S_1^2 / \left(1 - k_{31}^2\right) s_{11}^E. \tag{5.153}$$

The following results of calculating the equivalent parameters will be obtained.

For the side-electroded bar (transverse polarization) the working strain is $S_1(x)$, $dS_1 / dq_3 = dS_1(x) / dz = 0$ and the equivalent parameters are

$$C_{el}^{S_1} = \varepsilon_{33}^T \left(1 - k_{31}^2\right) \frac{wl}{t}, \quad K_m^E = \frac{1}{C_m^E} = \frac{\pi^2 tw}{2 s_{11}^E l}, \quad n = \frac{2 w d_{31}}{s_{11}^E}, \quad \Delta K = 0. \tag{5.154}$$

The side-electroded bar is a typical case that electrical and mechanical variables are separable. The equivalent circuit for the transducer looks like it is shown in Figure 5.7 (a) without

ΔC and obviously coincides with the common equivalent circuit for a side-electroded bar vibrating in the fundamental mode.

For the end-electroded bar (axial polarization) the working strain is $S_3(x)$, $dS_3 / dq_3 = dS_3(x)/dx = -\xi_o(\pi x/l)^2 \cos(\pi x/l)$. Thus,

$$\Delta W = \frac{1}{2}\frac{k_{33}^2}{1-k_{33}^2}\frac{S_{cs}}{s_{33}}\left[\int_0^l S_3^2 dx - \frac{1}{l}\left(\int_0^l S_3 dx\right)^2\right] = \frac{\Delta K_3 \xi_o^2}{2}. \tag{5.155}$$

The equivalent parameters are

$$C_{el}^{S_3} = \varepsilon_{33}^T\left(1-k_{33}^2\right)\frac{wt}{l},\quad K_m^E = \frac{1}{C_m^E} = \frac{\pi^2 tw}{2s_{33}^E l},\quad n = \frac{2d_{33}wt}{s_{33}^E l},$$
$$\Delta K_3 = \frac{1}{\Delta C} = K_m^E\left(1-\frac{8}{\pi^2}\right)\frac{k_{33}^2}{(1-k_{33}^2)}. \tag{5.156}$$

The end-electroded bar is the most typical representation of a transducer with nonuniform strain distribution along the electrical field, which results in the additional rigidity ΔK associated with the energy ΔW. In the case that PZT-4 is used with $k_{33}^2 \doteq 0.5$, $\Delta K_3 = 0.2K_m^E$. This effect is represented by the term ΔC in the equivalent circuit shown in Figure 5.7 (a). The Mason's equivalent circuit (Ref. 2) shown in Figure 5.7 (b) represents the same end-electroded bar vibrating in the fundamental mode. The parameters C_m^D and $(-C_{el}^S)$ in Figure 5.7 (b) are responsible for the same effect of the internal energy component $W_m^E + \Delta W$, as parameters C_m^E and ΔC in Figure 5.7 (a).

For the segmented axially polarized bar at the parallel electric connection of the segments, as it is shown in the Figure 5.8 (c), the poling directions and directions of the working electric field coincide in each segment. In terms of deformation nothing has changed in the bar. When computing W_{el}^S, W_{em} and ΔW the integration must be performed over each segment and the results must be added up. Finally for energy ΔW, which in this case will be denoted as ΔW_N (N is the number of segments), we obtain

$$\Delta W_N = \frac{1}{2}\frac{k_{33}^2}{1-k_{33}^2}\frac{S_{cs}}{s_{33}^E}\left[\int_0^l S_3^2 dx - \frac{1}{\Delta x}\sum_{n=1}^{N}\left(\int_{\Delta x(n-1)}^{\Delta xn} S_3 dx\right)^2\right] = \frac{\Delta K_{3N}\xi_o^2}{2}. \tag{5.157}$$

After substituting expression for S_3 into the integrals will be found that

$$\Delta K_{3N} = K_m^E \frac{k_{33}^2}{1-k_{33}^2}\left[1-\beta\left(\sin^2\frac{\pi}{2N}\right)\left(\frac{\pi}{2N}\right)^{-2}\right]. \tag{5.158}$$

Here $\beta=1$ at $N \geq 2$ and $\beta=2$ at $N=1$, which correspond to the preceding case of the solid axially polarized bar. All the other equivalent parameters are

$$C_{el}^{S_3} = \varepsilon_{33}^T\left(1-k_{33}^2\right)\frac{wtN^2}{l}, \quad K_m^E = \frac{\pi^2 tw}{2s_{33}^E l}, \quad n = \frac{2d_{33}wtN}{s_{33}^E l}. \tag{5.159}$$

The ratio $\Delta K_{3N}/\Delta K_{31}$ vs. the number of segments N is presented in Figure 5.9.

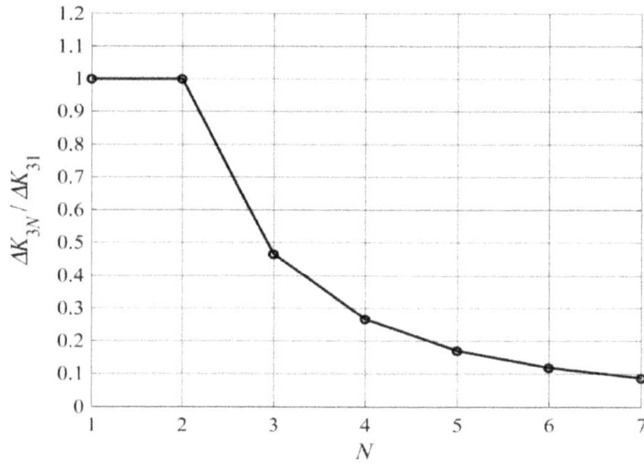

Figure 5.9: The ratio $\Delta K_{3N} / \Delta K_{31}$ as a function of the number of segments N.

At number of segments $N > 6$ on a half wavelength of deformation $\Delta K_6 < 0.1\Delta K_1$, and in practice is negligible. For instance, in the case of PZT-4, $\Delta K_6 < 0.02K_m^E$. With $\Delta K_N \ll K_m^E$ (conventionally, at $N \geq 6$, if the modern PZT ceramics is used), the equivalent circuit for the axially poled segmented bar is qualitatively the same as the equivalent circuit for the transverse polarized bar, as it would look like circuit in Figure 5.8 (a) with ΔC removed, because $1/\omega\Delta C \ll 1/\omega C_m^E$ and $1/\omega\Delta C$ can be neglected. This reflects the fact that the conversion of energy in these two cases occurs qualitatively in the same manner, and it differs from the case of the axially polarized solid bar.

In order to explain the physical difference in the quality of energy conversion between the transverse polarized, axially polarized solid and axially polarized segmented bars, let us assume that the bar is divided into small elements Δx as shown in Figure 5.8 (a), and consider these elements as the individual elemental energy converters. The electrical energy, which is utilized by the elements, may be represented as $w_{el}(x)=\left[w_{el}^S(x)+w_{em}(x)\right]\sim\left[i_s(x)+i_m(x)\right]$ (the "~" sign indicates the proportionality). The terms $w_{el}^S(x)$ and $i_s(x)$ are the electrical energy and the current through the element in the case that the bar is clamped, $w_{em}(x)$ and $i_m(x)$ are the motional part of electrical energy utilized by the element and the motional current through the element due to the deformation of the whole bar. The motional part of the electrical energy consumed by the element is $w_{em}(x)\sim i_m(x)\cdot E_3(x)$. The part of the electrical energy, which is converted into the mechanical energy in the course of the element deformation, is $w_{em}(x)=w_{mch}^E$, and according to Eq. (5.42) it is proportional to the strain, i.e., $w_m^E(x)\sim S(x)\cdot E_3(x)$, where $E_3(x)$ is the electric field in the element. Thus, the ratio $w_{mch}^E(x)/w_{em}(x)\sim S(x)/i_m(x)$ quantifies the part of the motional electrical energy that is converted into the mechanical form by the element of the bar with coordinate x. In the case of the transverse piezoelectric effect the electrodes of the bar distribute the motional current in such a way that $i_m(x)\sim S(x)$ (actually $i_m=d_{31}S_1/s_{11}^E$) and $w_{mch}^E(x)/w_{em}(x)=$ constant. The distribution of the motional electrical energy between the elements of the mechanical system occurs in exact accordance with their contribution to the electromechanical conversion.

In the case of the axially polarized solid bar $i_m(x)=i_{av}$, where i_{av} is the average current, which flows through all the elements, while $S(x)\sim\sin(\pi x/l)$. Accordingly, $w_{mch}^E(x)/w_{em}(x)\sim\sin(\pi x/l)$. This means that, although the elements located near the ends contribute nearly nothing to electromechanical conversion, they consume the same amount of the motional electrical energy, as the elements located in the middle part of the bar which contribute the most. The distribution of the motional electrical energy in this case is "unfair". The electrical interaction between the elements takes place in a manner that the "strong" elements of the bar feed the "weak" ones. To obtain the same amount of mechanical energy, relatively more electrical energy is needed than in the preceding case of the transverse electric field.

In the case of a segmented bar the electrodes inserted into the bar distribute the motional current between segments in accordance with the average strain $S_{3av}(\Delta x_i)$ within a segment, $i_m(\Delta x_i)\sim S_{av}(\Delta x_i)$. Therefore, we have $w_{mch}^E/w_{em}\sim S_3(x_i)/S_{3av}(\Delta x_i)$. With increasing the number of segments $\Delta x_i=l/N\to 0$ and $S_3(x_i)/S_{3av.}(\Delta x_i)\to 1$. (It can be assumed that $S_{3av.}(\Delta x)\approx S_3(x)$ at $N>6$ for a half wavelength of deformation, as it follows from Figure 5.9.) The distribution of the motional energy becomes almost as "fair" as in the case of the transverse electric field. The conversion of energy in these two cases takes place qualitatively in the same manner in a more "economical" way.

Returning to the equivalent circuits presented in Figure 5.7 it can be concluded that the representation of the circuit in Figure 5.7 (a) with term ΔC has an advantage of clarity, whereas the term ($-C_{el}^S$) in Figure 5.7 (b) does not have a clear physical meaning.

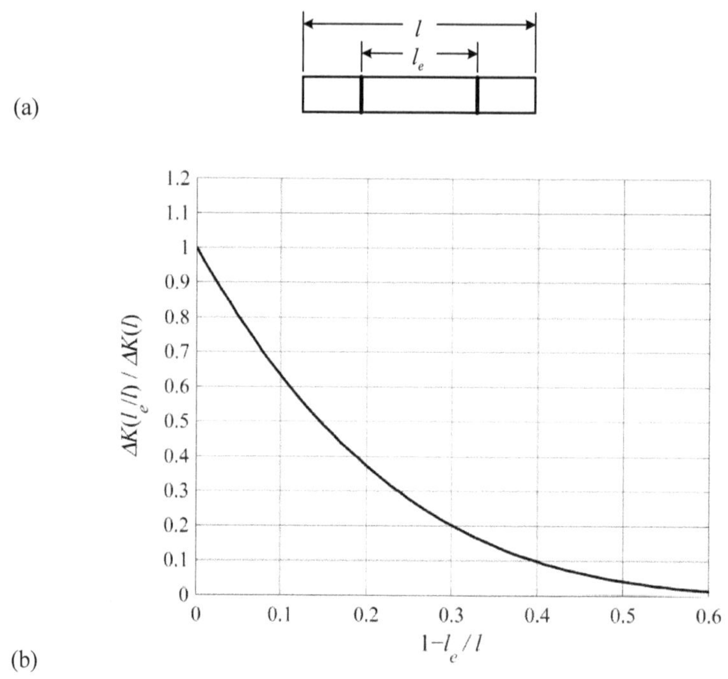

Figure 5.10: Effect of the electrodes imbedded in a piezoelectric bar: (a) location of the electrodes, (b) ratio $\Delta K(l_{el}) / \Delta K(l)$ as a function of separation between electrodes.

Due to the above-described circumstance, a solid end-electroded bar behaves, as if it was more rigid, since it requires a larger electric energy to achieve the same level of deformation than the segmented bar. Indeed, the rigidity of a solid bar is $K^E = K_3^E + \Delta K_3$ whereas that of a segmented is $K^E = K_3^E$. There exists terminology that reflects the distinction between piezo-rigid and piezo-soft forms of vibration of piezoceramic bodies (with implicit understanding that in the first case the longitudinal piezoelectric effect and in the second case the transverse effect is used). As we can see, such classification is justified to a certain extent, since with the transverse piezoelectric effect $\Delta K = 0$. However, in the context of the real transducer designs it is not indisputable, the more so as the quantities equivalent to Young's modulus for the longitudinal and transverse piezoeffect are related as $s_{11}^E / s_{33}^E < 1$.

The most inadequate consumption of the motional electrical energy in the axially polarized solid bar takes place near the ends of the bar, where strains are especially small. Therefore, it is interesting to consider the case that the electrodes are imbedded in the bar at some distance from the ends, as shown in Figure 5.10 (a). Using Eq. (5.142), where the integration is fulfilled over the length of the bar between the electrodes, yields

$$\frac{\Delta K(l_e)}{\Delta K(l)} = \frac{1}{1-8/\pi^2} \cdot \frac{l_e}{l} \left[1 + \frac{l}{\pi l_e} \cdot \sin \frac{\pi l_e}{l} - 2 \frac{\sin^2(\pi l_e / 2l)}{(\pi l_e / 2l)^2} \right]. \tag{5.160}$$

This function is depicted in Figure 5.10 (b), where from it can be concluded that $\Delta K(l_e)$ may be neglected compared to $\Delta K(l)$ at $l_e / l < 0.5$.

5.5.2.2 Rectangular Beam Vibrating in Flexure

One more example for the case that the strain changes in the direction of electric field represent transducers employing mechanical systems in the shape of beams, plates, and shells vibrating in flexure and employing the transverse piezoelectric effect. Distribution of strains in direction perpendicular to the neutral surface in these systems does not depend on their configuration in the horizontal plane and on the boundary conditions. Therefore, example can be used of the rectangular simply supported beam vibrating in the fundamental mode $\theta(x) = \sin(\pi x / l)$ that was previously considered in Section 2.6.1 and shown in Figure 5.11.

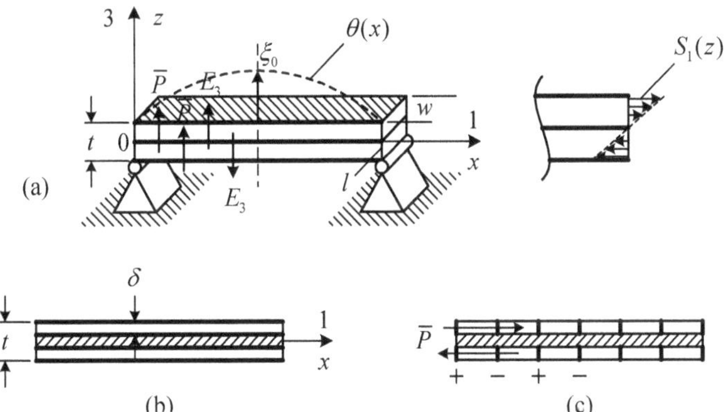

Figure 5.11: Rectangular beam under flexural deformation: (a) fully active bimorph design, (b) trilaminar design, (c) segmented transducer design, longitudinal piezoeffect.

The general expression for the strain is $S_1(z,x) = z\xi_o(\pi/l)^2 \sin(\pi x/l)$. After substituting this expression into formula (5.142), and taking into account relations (5.152) and expression $K_m^E = (\pi^4/48)(wt^3/\ell^3 s_{11}^E)$ for the equivalent rigidity of the beam we arrive at

$$\Delta W = \frac{\xi_o^2}{2}\frac{d_{31}^2}{\varepsilon_{33}^{S_1} s_{11}^E} K_m^E \left[1 - \frac{2}{t}\left(\int_0^{t/2} z\,dz\right)^2 \left(\int_0^{t/2} z^2 dz\right)^{-1}\right] = \frac{\xi_o^2}{2}\Delta K, \tag{5.161}$$

where

$$\Delta K = 0.25\frac{k_{31}^2}{1-k_{31}^2}K_m^E. \tag{5.162}$$

If the transducer is made of PZT-4 ceramic, in which case $k_{31}^2 \doteq 0.1$, then $\Delta K \doteq 0.03 K_m^E$, and this term can obviously be neglected as compared to K_m^E. In practical designing of the flexural transducers that employ the transverse piezoeffect it is more common to use trilaminar mechanical systems with the piezoelectric layers removed from the neutral plane, as shown in Figure 5.11 (b). In this design $\Delta K / K_m^E$ drops as the separation between the piezoelectric layer and the neutral plane increases exactly for the same reason as in the case of a longitudinally vibrating bar with embedded electrodes. If we assume that Young's moduli of the active and passive materials are approximately the same, then

$$\frac{\Delta K(\delta)}{\Delta K(t/2)} = \left(\frac{2\delta}{t}\right)^2 \left[3\left(1-\frac{2\delta}{t}\right) + \left(\frac{2\delta}{t}\right)^2\right]^{-1}, \tag{5.163}$$

where δ is the thickness of the piezoelectric layer and t is the total thickness of the beam. From this equation follows that at $\delta / t \leq 0.4$, which is common from considerations of optimizing the effective coupling coefficient (see Section 5.6), the term ΔK can be neglected even if the single crystal materials are used having very high coupling coefficients ($k_{31} \approx 0.5$).

Piezoelements of the rectangular beam bender transducers intended for application as projectors usually have segmented design, as shown in Figure 5.11 (c) and employ the longitudinal piezoeffect. Situation in terms of the additional rigidity in this case is exactly the same as for the segmented longitudinal bar, and formula (5.158) is valid for ΔK_N with $K_m^E = (\pi^4/48)(wt^3/\ell^3 s_{33}^E)$.

As it is clear from above discussion, for calculating components of the internal energy in each case one must know configuration of the polarization field and of the operating electric field in the clamped piezoceramic body (that is at values of dielectric constant $\varepsilon_{mm}^{S_i}$). The most widely used in the transducer designs are the piezoelements in the shape of the bars, plates, disks, and thin-walled shells with unipolar electrodes completely covering their side surfaces. The electric fields in these piezoelements form the rectangular coordinate systems. This is also true for the cases of split electrodes having different polarities, if separation between them is small compared with their linear dimensions, because at this condition an effect of leakage fields is negligible. In significantly thick ring-shaped piezoelectric elements polarized in the radial direction the poling electric field forms the cylindrical coordinate system. Piezoelements with essentially nonuniform electric fields that form a specific curvilinear coordinate system may be used relatively seldom. Such fields can be generated by electrodes of special configuration, for example, by the striped electrodes used to tangentially polarize thin-walled piezoelements. Examples of the effect of the tangential polarization will be considered in Chapter 7.

5.6 Effective Coupling Coefficients

Consider now the effective coupling coefficient, k_{eff}, for arbitrary deformed piezoelectric body[6]. Using definition (5.90) and expressions (5.143) and (5.144) for the components of internal energy of nonuniformly deformed piezoelectric body, k_{eff} can be represented as

$$k_{eff}^2 = \frac{W_{em}}{W_{el}^S + W_{em}} = \frac{W_m^E + \Delta W}{W_{el}^S + W_m^E + \Delta W}. \tag{5.164}$$

It must be noted that the concept of k_{eff} makes sense only in connection with a certain distribution of deformation in a body and is valid so long as this distribution remains invariable. In other words, each single mechanical degree of freedom of a piezoelectric body is characterized by its coupling coefficient k_{eff}. Therefore, prior to calculating k_{eff} the displacement distribution $\xi(\boldsymbol{r})$ in the body should be known. Represent this displacement distribution in the general form as

$$\xi(\boldsymbol{r}) = \xi_o \theta_\xi(\boldsymbol{r}), \tag{5.165}$$

where ξ_o is the displacement of the reference point with coordinate $\boldsymbol{r}_o$ and $\theta_\xi(\boldsymbol{r})$ is the mode shape, i.e., displacement distribution normalized in such a way that $\theta_\xi(\boldsymbol{r}_o) = 1$. Let the corresponding strain distribution be

$$S(\boldsymbol{r}) = S_o \theta_S(\boldsymbol{r}). \tag{5.166}$$

In the case that the body has one mechanical degree of freedom, the energies involved in formula (5.164) may be expressed by means of corresponding equivalent parameters defined by formulas (5.150), and the expression for k_{eff}^2 can be modified as follows

$$k_{eff}^2 = 1/(1 + C_{el}^S / n^2 C_m), \tag{5.167}$$

where $C_m = C_m^E \cdot \Delta C / (C_m^E + \Delta C)$. Formula (5.167) is especially convenient for calculating k_{eff} of the particular transducer types.

5.6.1 Optimizing the Effective Coupling Coefficient

In general, under nonuniform deformation $k_{eff} < k_m$, which means that the ability of a piezoelectric material to perform electromechanical conversion is not fully used. The question arises, whether the electromechanical conversion in nonuniformly deformed bodies can be improved. To answer this question the analytical formulation (5.164) for k_{eff} may be considered.

First, in the case that initially $\Delta W \neq 0$ k_{eff} can be increased by segmenting the mechanical system of the transducer in the direction of the electric field. As stated in Section 5.5.2, this leads to $\Delta W = 0$ at number of segments $N \geq 6$ on the have wave of deformation. Therefore, for the further analysis we will assume $\Delta W = 0$ in expression (5.164), and as the result $C_m = C_m^E$ in formula (5.167).

The expressions (5.150) for energies of a body with one mechanical degree of freedom may be represented in the general form as

$$W_{el}^S = C_{el}^S V^2 / 2 = \int_V w_{el}^S(\boldsymbol{r}) dV, \tag{5.168}$$

$$W_m^E = \xi_o^2 / 2C_m^E = \int_V w_m^E(\boldsymbol{r}) dV, \tag{5.169}$$

$$W_{em} = nV\xi_o / 2 = \int_V w_{em}(\boldsymbol{r}) dV. \tag{5.170}$$

Suppose now that the electric field is a function of coordinates

$$E_3(\boldsymbol{r}) = E_3(\boldsymbol{r}_o) \theta_E(\boldsymbol{r}), \tag{5.171}$$

where $\theta_E(\boldsymbol{r})$ is the normalized electric field distribution. Taking into account expressions (5.166) and (5.171), the energy densities w_{el}^S, w_m^E, and w_{em} as the functions of the electric field and deformation can be represented in the form

$$w_{el}^S(\boldsymbol{r}) = w_{el}^S(\boldsymbol{r_o})\theta_E^2(\boldsymbol{r}), \quad w_m^E(\boldsymbol{r}) = w_m^E(\boldsymbol{r_o})\theta_S^2(\boldsymbol{r}), \quad w_{em} = w_{em}(\boldsymbol{r_o})\theta_E(\boldsymbol{r})\theta_S(\boldsymbol{r}). \tag{5.172}$$

Upon substituting these expressions into (5.168)-(5.170) and the latter into (5.164) we arrive at

$$k_{eff}^2 = \frac{w_{em}(\boldsymbol{r_o})\int_V \theta_E(\boldsymbol{r})\theta_S(\boldsymbol{r})dV}{w_{el}^S(\boldsymbol{r_o})\int_V \theta_E^2(\boldsymbol{r})dV + w_m^E(\boldsymbol{r_o})\int_V \theta_S^2(\boldsymbol{r})dV}, \tag{5.173}$$

where from k_{eff} may be found as soon as the distribution of strain and electric field are known. The following variants of the distributions are of interest:

1. Uniform deformation, uniform electric field ($\theta_S(\boldsymbol{r}) = 1$, $\theta_E(\boldsymbol{r}) = 1$).

This is, for example, the case that thin rings and spherical shells with fully electroded inner and outer surfaces vibrate in the breathing mode. Expression (5.173) in this case becomes

$$k_{eff}^2 = \frac{w_{em}(\boldsymbol{r}_o)}{w_{el}^S(\boldsymbol{r}_o) + w_m^E(\boldsymbol{r}_o)}, \tag{5.174}$$

and comparison with formula (5.91) results in $k_{eff} = k_m$.

2. The most widespread case of an arbitrary strain distribution and uniform electric field. ($\theta_S(\boldsymbol{r}) < 1$ at $\boldsymbol{r} \neq \boldsymbol{r}_o$, $\theta_E(\boldsymbol{r}) = 1$). From expression (5.173) it can be concluded that $k_{eff} < k_m$.

3. The distribution of strain is arbitrary. The distribution of electric field matches the strain distribution ($\theta_E(\boldsymbol{r}) = \theta_S(\boldsymbol{r})$). Substitution of $\theta_E(\boldsymbol{r}) = \theta_S(\boldsymbol{r})$ into expression (5.173) results in the relation (5.174), which means that $k_{eff} = k_m$.

Thus, we arrive at the conclusion that effective coupling coefficient, k_{eff}, for a piezoelectric body under nonuniform deformation can be increased (theoretically up to the corresponding coefficient k_m), if to match the electric field and the strain distributions. To illustrate how this condition can be practically fulfilled and what the physics is behind this condition, at first refer to the examples of the length expander bars, namely, side-electroded, end-electroded and segmented axially polled vibrating in the fundamental mode. The geometry of the bars is shown in Figure 5.8, and expressions for the equivalent parameters are given in Section 5.5.2.

Substituting the equivalent parameters of the bars into formula (5.167) results in the following values for k_{eff} :

$$k_{eff}^2 = \left(1 + \frac{\pi^2}{8} \cdot \frac{1-k_{3i}^2}{k_{3i}^2}\right)^{-1} \tag{5.175}$$

for the side-electroded bar (at $i = 1$) and for the segmented bar with number of segments $N > 6$ ($i = 3$), and

$$k_{eff}^2 = 8k_{33}^2 / \pi^2 \tag{5.176}$$

for the solid end-electroded bar. Thus, k_{eff} of a solid end-electroded bar is smaller than that of a segmented bar. If PZT-4 is considered, k_{eff}^2 = 0.40 for the solid bar and k_{eff}^2 = 0.45 for the segmented bar, while $k_m^2 = 0.5$ for the material (i.e., under uniform deformation). Qualitatively this fact is explained in Section 5.5.2. Thus, the effective coupling coefficient of the longitudinal vibrating bar can be increased by rational changing the electrical field distribution.

5.6.2 Examples of Optimizing the Effective Coupling Coefficients

5.6.2.1 Length Expander Bar, Transverse Piezoeffect.

It is hardly possible to exactly fulfill the relation $\theta_E(x)=\theta_S(x)$ in a reasonable way in practical devices except for the trivial cases of uniform deformation like for the pulsating rings and spheres. Even the stepped distribution of electric field shown in Figure 5.12 (a) is very complicated and scarcely worthwhile. Fortunately, very reasonable results on k_{eff} optimization can be achieved by means of the simple approximation by the electric field to the strain distribution shown in Figure 5.12 (b), namely, just by removing electrodes from the parts of the mechanical system that undergo relatively small deformation. To make an appropriate quantitative analysis, consider expression (5.167) for k_{eff} . In this expression the term

$$\frac{n^2 C_m^E}{C_{el}^S} = \frac{n^2}{C_{el}^S K_m^E} = \alpha_c \tag{5.177}$$

depends on the shape of the electrodes. The equivalent compliance C_m^E remains the same, when the shape of electrodes changes, if the parts of mechanical system not being used as piezoactive are electroded and short circuited. If these parts are not electroded, then, strictly speaking, one must use Young's modulus of unpolarized ceramics, when evaluating the contribution of the passive parts to the total potential energy of the mechanical system.

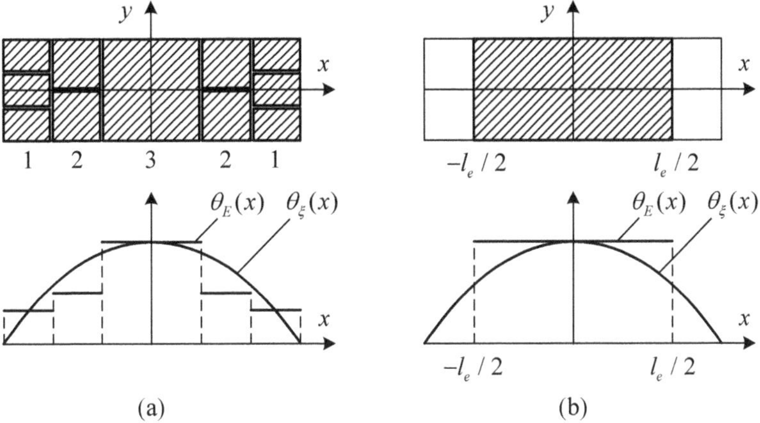

Figure 5.12: The length expander bars with different electrode shapes: (a) fragmented (stepped) electrode (parts 1 and 2 are connected in series, then all the electrodes of the same sign are connected), (b) parts of electrode are removed.

But the potential energy density w_{pot} of those parts is relatively small because of small deformations (exactly by this reason they are not used for electromechanical conversion). Thus, the values of the total W_m^E and C_m^E accordingly should not change significantly by comparison with those for a fully active mechanical system. Further in this Section we will assume that this is the case. In general, the not electroded parts of the mechanical system can be replaced by a passive material with different elastic properties. In this case the same analysis is applicable, but changes of C_m^E also must be considered.

The coefficient α_c is linked to k_{eff} by relation

$$k_{eff}^2 = \frac{\alpha_c}{1+\alpha_c}, \tag{5.178}$$

and we can judge the change of k_{eff} based on the behavior of the term α_c and specifically on the factor n^2 / C_{el}^S. Denote the value of α_c that corresponds to $k_{eff} = k_m$ by α_{cm}.

The coefficient

$$A = \frac{\alpha_c}{\alpha_{cm}} = \left(\frac{n^2}{C_{el}^S}\right) \cdot \left(\frac{n^2}{C_{el}^S}\right)_m^{-1} \tag{5.179}$$

may characterize, how close a particular electrode shape brings k_{eff} to its maximum possible value k_m. Using formula (5.167) for k_{eff} and relations (5.177)-(5.179) yields

$$k_{eff}^2 = \frac{Ak_m^2}{1-(1-A)k_m^2}. \tag{5.180}$$

The value of k_{eff} for a particular electrode shape may be determined by means of this expression given that the corresponding coefficient A is known. In the case of the transverse piezoelectric effect $k_m^2 < 0.15$ practically for all the piezoelectric ceramic compositions, and it follows from relation (5.180) that $k_{eff}^2 \approx A \cdot k_m^2$ within 5% accuracy.

Substituting $w_{el}^{S_1}$ and w_{em} for a bar from (5.65) and (5.67) into relations (5.172) and carrying out the integrals (5.168)-(5.170) over the volume of the bar produces the following general expressions for the electromechanical transformation coefficient n and capacitance $C_{el}^{S_1}$ in the case of a side-electroded length expander bar:

$$n = \frac{d_{31}w}{s_{11}^E} \int_{-l/2}^{l/2} \theta_E(x)\theta_S(x)dx, \tag{5.181}$$

$$C_{el}^{S_1} = \frac{\varepsilon_{33}^T(1-k_{31}^2)w}{t} \int_{-l/2}^{l/2} \theta_E^2(x)dx. \tag{5.182}$$

With these expressions in use, it appears that for the electric field distribution corresponding to the electrode connection shown in Figure 5.12 (a) we obtain $A = 0.95$, i.e., $k_{eff} = 0.97k_{31}$ (which is almost the maximum possible value) compared with $k_{eff} = 0.90k_{31}$ in the case of the uniform electric field. In the case of the electrode shape shown in Figure 5.12 (b)

$$\theta_E(x) = 1 \text{ at } |x| \le l_e/2 \text{ and } \theta_E(x) = 0 \text{ at } |x| > l_e/2. \tag{5.183}$$

The corresponding expressions for parameters n, $C_{el}^{S_1}$ and coefficient A will be

$$n = \frac{2wd_{31}}{s_{11}^E}\sin(\pi l_e/2l), \quad C_{el}^{S_1} = \varepsilon_{33}^T(1-k_{31}^2)wl_e/t, \tag{5.184}$$

$$A(l_e/l) = \frac{8l}{\pi^2 l_e}\sin^2(\pi l_e/2l). \tag{5.185}$$

Plot of the function $A(l_e/l)$ is depicted in Figure 5.13, where from it can be concluded that maximum value for k_{eff} can be achieved by removing about 0.13 of electrode length from each end of the bar. In this case $A \approx 0.92$, i.e., almost the same as for rather complicated electrode configuration presented in Figure 5.12 (a). Another interesting conclusion can be made that in the case that $l_e = 0.5l$ k_{eff} is the same as with the full- size electrodes.

In the variant of a solid bar with embedded electrodes the function $A(l_e / l)$ obtained by employing formulas analogous (5.181) for the case of longitudinal piezoeffect is shown in Figure 5.13 by the dashed line. If the bar is segmented on its part of length l_e, the dependence $A(l_e / l)$ has the same form, as in the case of the transverse piezoelectric effect. Comparison the dependences for the segmented and solid with embedded electrodes bars shows that at $l_e / l < 0.5$ segmenting does not lead to an increase in k_{eff}. Qualitatively this follows from the fact that the segments located close to the ends consume amount of electric energy disproportionally large to their contribution to the electromechanical conversion, and thus cause reduction of the effective coupling coefficient.

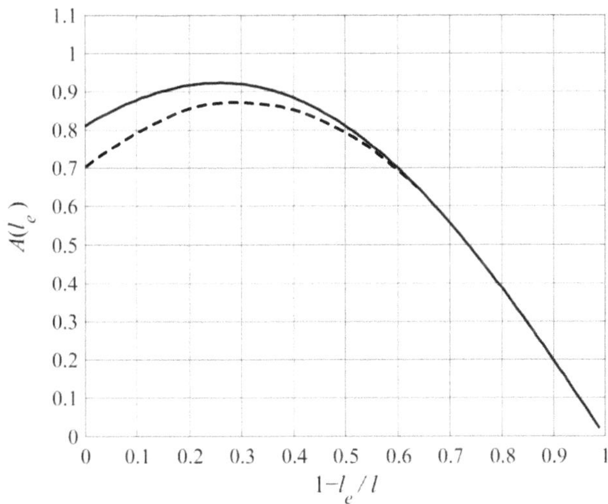

Figure 5.13: Illustration of the k_{eff} dependence on the electrode length, $A(l_e / l) \approx k_{eff}^2(l_e) / k_m^2$. Solid line for a bar at transverse piezoeffect and for a segmented bar at the longitudinal piezoeffect. Dashed line for a solid bar at the longitudinal piezoeffect with embedded electrodes.

If one considers piezoelectric transducer as a mechanical system with multiple degrees of freedom, each corresponding to a normal mode of vibration, then maximizing of k_{eff} for a particular mode makes this mode isolated, i.e., the only electromechanically active. It is worth mentioning that the opposite statement is not valid, i.e., an isolated mode does not always have maximum k_{eff}. For example, the electrode shape shown in Figure 5.14 (see Ref. 8), in which case

$$\theta_E(x,y) = 1 \text{ at } |y| \le (w/2)\sin(\pi x/l) \tag{5.186}$$

and

$$\theta_E(x) = 0 \text{ at } |y| > (w/2)\sin(\pi x/l), \tag{5.187}$$

leads to the isolated fundamental mode $\theta(x) = \cos(\pi x/l)$. But evaluating k_{eff} by applying the above considered procedure gives in this case $k_{eff} = 0.78k_{31}$ i.e., even smaller value than in the case that electrical field is uniform.

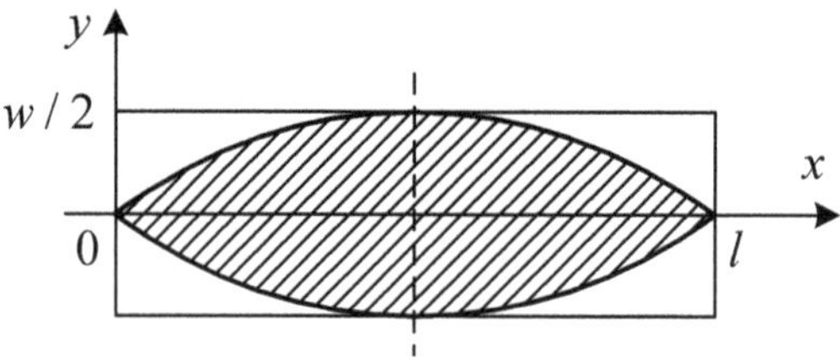

Figure 5.14: The electrodes shape that leads to the isolated fundamental mode $\theta(x) = \cos(\pi x/l)$

Sometimes it is more reasonable to consider reducing an amount of piezoelectric material in the mechanical system of a transducer without loss in k_{eff} rather than obtaining the maximum possible k_{eff}. Thus, as we have already seen, up to 0.5 of the volume of a length expander bar can be substituted by a passive material without loss in k_{eff}. Even more significant gain can be produced in the case that mechanical system undergoes flexural deformation. This will be illustrated in the next section.

The conclusions concerning the effects of the electrodes designing drawn for the piezoelement in the shape of a bar can also be extended to the cases of other piezoceramic bodies having strain distributions. Namely, for increasing k_{eff} piezoelements should be segmented in the direction of the lines of force of the electric field, if strains are nonuniform in this direction, and the electric field distribution must be approached to the strain distribution. The simplest way of increasing k_{eff} is removing the electrodes from those parts of piezoelectric body, in which strains are relatively small.

5.6.2.2 Beams and Circular Plates Under Flexure

In the case of the beam and circular plate with axisymmetric electrodes shown in Figure 5.15

the distribution of strain exists by the thickness and by the length or by the radius. Therefore, the values of k_{eff} for regular designs with uniform electrical field are relatively smaller than in the case of one-dimensional strain distribution, and the gains due to optimization may be more significant. At first consider trilaminar beam transducer with simply supported ends (Figure 5.15 (a)) vibrating in the first flexural normal mode. In this case

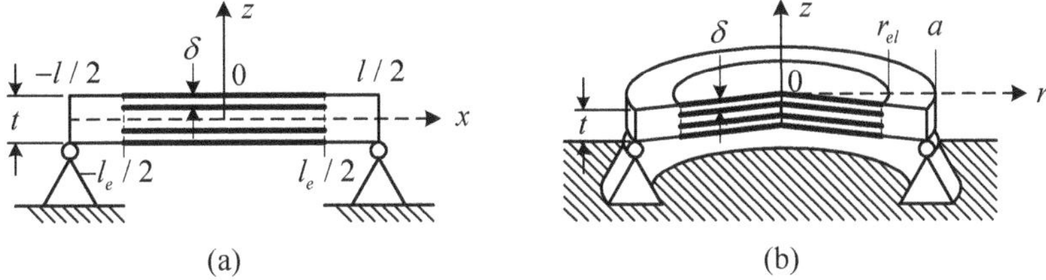

Figure 5.15: Mechanical systems under flexure: (a) trilaminar beam with simply supported ends, (b) circular plate with simply supported boundary.

$$\xi_z(x) = \xi_o \cos(\pi x / l), \tag{5.188}$$

$$S_x(x,z) = -z\left(\partial^2 \xi_z / \partial^2 x\right) = z\xi_o (\pi / l)^2 \cos(\pi x / l). \tag{5.189}$$

With electrodes inserted in mechanical system, as it is shown in Figure 5.15 (a), the electrical field in the body of the beam can be represented as follows

$$E_3(x,z) = (V/\delta)\theta_{Ex}(x)\theta_{Ez}(z), \tag{5.190}$$

where

$$\theta_{Ex}(x) = 1 \text{ at } |x| \le l_e / 2 \text{ and } \theta_{Ex} = 0 \text{ at } |x| > l_e / 2, \tag{5.191}$$

$$\theta_{Ez}(z) = 1 \text{ at } |z| \ge t/2 - \delta \text{ and } \theta_{Ez}(z) = 0 \text{ at } |z| \le t/2 - \delta. \tag{5.192}$$

Here l_e and δ are the length and the thickness of the active piezoelectric layers, which are assumed to be electrically connected in parallel, and $l_e = l$, and $\delta = t/2$ correspond to the beam fully made of active piezoelectric material. Substituting expressions for energy densities $w_{el}^{S_1}$ and w_{em} from (5.65) and (5.67) into (5.172) and carrying out the integrals (5.168)-(5.170) over the volume of the beam produces

$$n(\delta, l_e) = \frac{wd_{31}}{s_{11}^E}\frac{(t \quad \delta)\pi}{l}\sin\frac{\pi l_e}{2l}, \quad C_{el}^{S_1} = \frac{2\varepsilon_{33}^T(1-k_{31}^2)wl_e}{\delta}. \tag{5.193}$$

The coefficient $\alpha_c(\delta, l_e)$ normalized to its value at $\delta = t/2$ and $l_e = l$ may be represented in the form

$$\frac{\alpha_c(\delta,l_e)}{\alpha_c(t/2,l)} = A(l_e)A(\delta)\,, \tag{5.194}$$

where $A(l_e)$ is given by formula (5.185) and is depicted in Figure 5.13, and

$$A(\delta) = 8\frac{(1-\delta/t)^2\delta}{t}\,. \tag{5.195}$$

The coefficient $A(\delta)$ is given in as function of δ/t. This function has maximum at $\delta = t/3$, and at $\delta = t/5$ it has the same value, as at $\delta = t/2$. Summarizing the results illustrated in Figure 5.13 and Figure 5.16 one may conclude that the maximum value for k_{eff} can be achieved at $\delta = t/3$ and $l_e = 0.74l$. In this case $k_{eff} = 0.91k_{31}$, and the volume of active material is half of the total volume of the beam. Note that in the case that $\delta = t/5$ and $l_e = 0.5l$ ($\tilde{V}_{active} = 0.2\tilde{V}_{total}$), the effective coupling coefficient has the same value, as in the case that the beam is fully made of piezoelectric material.

Electromechanical conversion in the circular axially symmetric plate (Figure 5.15 (b)) vibrating in flexure is considered in Section 2.6.3. For a circular plate the expression analogous to expression (5.194) for a beam may be represented as

$$\frac{\alpha(\delta,r_{el})}{\alpha(t/2,a)} = A(\delta)A(r_{el})\,. \tag{5.196}$$

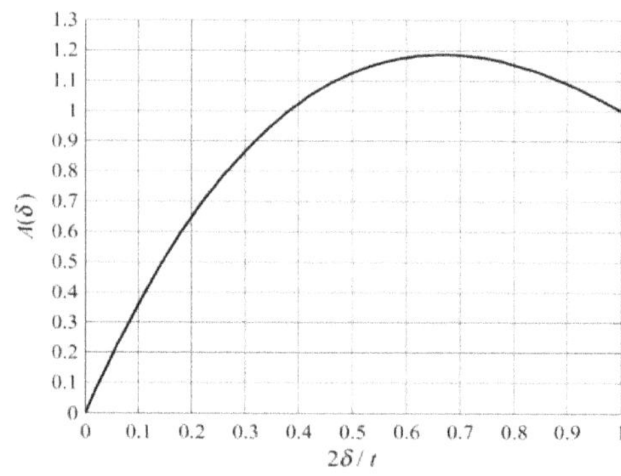

Figure 5.16: To dependence of the effective coupling coefficient on the thickness of piezoelectric layer.

The factor $A(\delta)$ is the same as given by formula (5.195) and as illustrated in Figure 5.16. The factor $A(r_{el})$ is different for different boundary conditions. For the simply supported and free plates $A(r_{el})$ as the function of r_{el} / a is given in Figure 5.17. In the case that the plates are fully made of piezoelectric material $k_{eff} = 0.9k_{31}$ for the simply supported boundary and $k_{eff} = 0.8k_{31}$ for the free boundary. The same value of k_{eff} can be obtained at $r_{el} = 0.8a$ and $\delta = t/5$ ($\tilde{V}_{activ} = 0.25\tilde{V}_{total}$) for the simply supported plate and at $r_{el} = 0.58a$ and $\delta = t/5$ ($\tilde{V}_{activ} = 0.13\tilde{V}_{total}$) for the free plate. The maximum values of k_{eff} are: $k_{eff} = 0.98k_{31}$ ($r_{el} = 0.9a$, $\delta = t/3$) for the simply supported plate, and $k_{eff} = 0.92k_{31}$ ($r_{el} = 0.75a$, $\delta = t/3$) for the free plate.

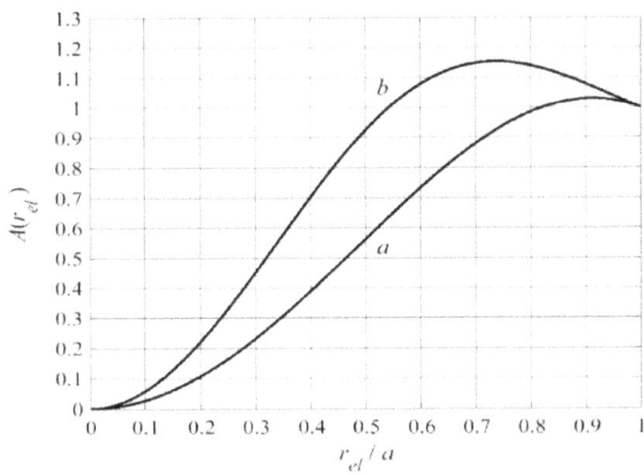

Figure 5.17: Function $A(r_{el}) \approx k_{eff}^2(r_{el}) / k_{eff}^2(a)$ as Illustration of k_{eff} dependence on the radius of electrodes: (a) with simply supported boundary, (b) with free boundary.

It must be noticed that all the numerical results related to k_{eff} optimization are obtained under the assumption that the elastic properties of the passive parts of a mechanical system are approximately the same as for the active piezoelectric parts. In the case that the elastic properties of the active and passive parts differ significantly the mechanical system must be treated as nonuniform, and the numerical results may change, although the qualitative conclusions remain valid. Related issues for the transducers with nonuniform mechanical systems will be considered in Chapter 9.

The presented analysis shows that analytical expression (5.164) for the effective coupling coefficient k_{eff} and its modification (5.167), which are based on the concept of the internal energy of a piezoelectric body, may be successfully used for optimizing the electromechanical conversion of energy under nonuniform deformation. It is shown that the absolute maximum of k_{eff}, which is equal to the corresponding coupling coefficient of the piezoelectric material, can be achieved theoretically for any mode of the strain distribution by special electrodes design, leading to the distribution of electric field that matches the strain distribution. It is illustrated with typical examples of bars, beams and plates vibrating in the longitudinal and flexural modes that very close to optimum results can be obtained just by removing electrodes from the parts of mechanical system, experiencing relatively small deformation. In practice these parts of piezoelectric material may be replaced by a passive material having approximately similar elastic properties. Another option illustrated with the same examples is to significantly reduce the amount of piezoelectric material in electromechanical transducer without loss of value of the effective coupling coefficient.

Optimizing the effective coupling coefficient is undoubtedly desirable for receivers because their specific sensitivity is proportional to this coefficient (see formula (3.181)). As to the projectors, this may be questionable, because reducing the amount of active material results in reducing the electromechanical force that generates vibration. Therefore, it can be acceptable only in the case that the transducer has sufficient reserve of the electric strength.

5.6.3 Effect of Electromechanically Passive Elements on the Effective Coupling Coefficient

Transducer designs may include electromechanically passive elements, which are essential for their operation. The typical examples in this regard are the cables (having capacitance C_c) and reinforcing mechanical elements (having equivalent rigidity K_{ad}) shown in Figure 5.18, where the reinforced segmented bar transducer is schematically depicted.

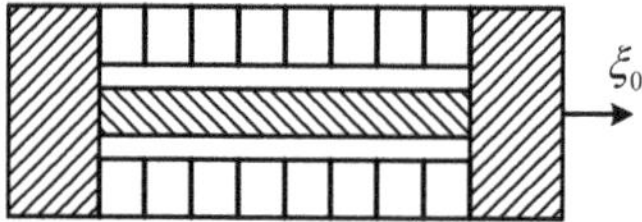

Figure 5.18: Segmented bar reinforced with a central metal bolt.

These passive elements produce effect of increase of electrical energy consumed by a transducer and of the additional energy of deformation, $W_{m\,ad}$, at the same magnitude of the transducer vibration. Namely, now the electrical energy supplied and the total mechanical energy produced are

$$W_{el} = W_{el}^S + W_{cable} = \frac{V^2}{2}(C_{el}^S + C_c), \tag{5.197}$$

where W_{el} is the electrical energy consumed by the clamped transducer, and

$$W_{mech} = W_m^E + W_{m\,ad} = \frac{\xi_o^2}{2}(K_m^E + K_{ad}), \tag{5.198}$$

where W_{mech} is the total energy of deformation and W_m^E is the useful mechanical energy that remains the same, as without the passive elements. Besides the relation holds

$$W_{em} = \frac{1}{2} V \xi_o n = W_{mech}. \tag{5.199}$$

The total electrical energy supplied to the transducer can be represented as

$$W_{e\,total} = W_{el} + W_{mech} = W_{el}^S + W_{cable} + W_m^E + W_{m\,ad}. \tag{5.200}$$

The energy balance that corresponds to the above expressions of energies leads to the equivalent circuit that is presented in Figure 5.19.

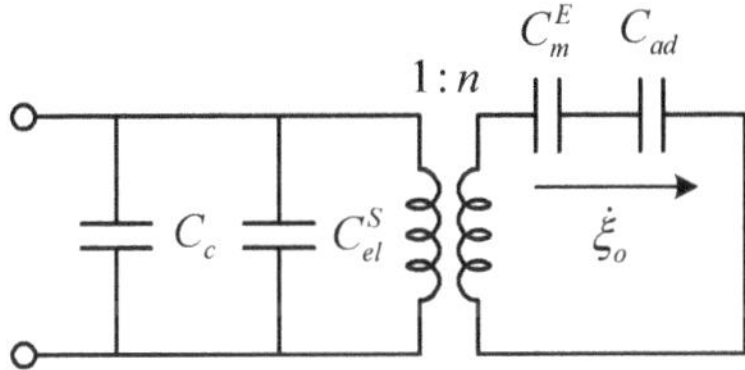

Figure 5.19: Equivalent circuit of a transducer with equivalents of the passive elements included.

After substituting expressions for the energies into the basic definition (5.164) the following result for the effective coupling coefficient, $(k_{eff})_p$, that accounts for existence of the passive elements will be obtained,

$$(k_{eff}^2)_p = \frac{W_{mech}}{W_{e\,total}} = \frac{\xi_o^2(K_m^E + K_{ad})}{V^2(C_{el}^S + C_c) + \xi_o^2(K_m^E + K_{ad})}. \tag{5.201}$$

From comparison of expressions (5.198) and (5.199) follows that

$$\frac{V}{\xi_o} = \frac{K_m^E + K_{ad}}{n}. \tag{5.202}$$

Using relations (5.201) and (5.202), we arrive at formula for effective coupling coefficient $(k_{eff})_p$ in the form

$$(k_{eff}^2)_p = \frac{\alpha_c}{\alpha_c + [1 + (K_{ad} / K_m^E)][1 + (C_c / C_{el}^S)]}, \tag{5.203}$$

where α_c is defined by expression (5.177). Remember that k_{eff} without the passive elements is determined through coefficient α_c as $k_{eff}^2 = \alpha_c / (\alpha_c + 1)$.

5.7 Equations of Vibration of Piezoceramic Bodies in Generalized Coordinates

5.7.1 Expressions for the energies involved.

Note that the basic derivation of the equations of vibration of piezoceramic bodies in the generalized coordinates was presented in Section 1.6 in a simplified form. The peculiarities of the internal energy of nonuniformly deformed piezoceramic bodies were not considered. Here more general treatment of the problem will be presented. In order to obtain equations of vibration of a piezoceramic body in generalized coordinates in the form of Euler's equations (4.1), the already known expressions for components of the internal energy W_{int} of the body have to be explicitly expressed in terms of the generalized coordinates, as it was done for a body made of a passive material in Section 4.5. It is appropriate to use for this purpose the same system of supporting functions $\theta_i(r)$ that would be selected to solve the problem of free vibration of a body of the same configuration made from a passive isotropic material. Thus, displacements in a piezoceramic body will be represented as expansion into the series (4.231)

$$\xi(\boldsymbol{r},t) = \sum_{i=1}^{\infty} \xi_i(t)\theta_i(\boldsymbol{r}). \tag{5.204}$$

The coefficients $\xi_i(t)$ of this series and the voltage $v(t)$ across the electrodes form the system of generalized coordinates. The expression for W_{kin} does not differ from the similar one in system (4.232), namely,

$$W_{kin} = \frac{1}{2}\sum_{i=1}^{\infty}\sum_{l=1}^{\infty} M_{il}\dot{\xi}_i\dot{\xi}_l, \quad M_{il} = \int_{\tilde{V}} m_\Delta \theta_i(\boldsymbol{r})\theta_l(\boldsymbol{r})d\tilde{V}, \tag{5.205}$$

where m_Δ is the specific mass of the volume element as in Eq. (4.39), M_{il} are the equivalent masses. They are the self-masses for the vibrational modes θ_i at $l = i$ and the mutual masses that characterize inertial interaction between vibrational modes θ_i and θ_l at $l \neq i$. Let us now express the components of the internal energy in terms of the generalized coordinates. We will do this for the case that $\boldsymbol{E} \uparrow\uparrow \boldsymbol{q_3}$ ($E = E_3$). For $E = E_2$ the final expressions are the same, if only to use constants of piezoceramics corresponding to this case.

The electric energy $W_{el}^{S_i}$ (5.140) is

$$W_{el}^{S_i} = \frac{1}{2}\int_{\tilde{V}} \varepsilon_{33}^{S_i} E_3^2 d\tilde{V} = \frac{1}{2} C_{el}^{S_i} v^2, \tag{5.206}$$

where

$$C_{el}^{S_i} = \left(1/v^2\right)\int_{\tilde{V}} \varepsilon_{33}^{S_i} E_3^2 d\tilde{V} \tag{5.207}$$

is capacitance of the clamped piezoelement, $\varepsilon_{33}^{S_i}$ is the value of $C_{el\Delta}^{S_i}$ that corresponds to a particular boundary condition in Table 5.3. Considering $W_{el}^{S_i}$ in a real piezoceramic body, one should take into account the energy of dielectric losses, which is characterized quantitatively by the electrical quality factor Q_e or by $\tan\delta_e = 1/Q_e$. The energy of electrical losses we determine as

$$W_{eL} = W_{el}^{S_i} / Q_e = W_{el}^{S_i} \tan\delta_e \tag{5.208}$$

(see Section 1.4). In the equations of vibration dielectric losses can be accounted for by assigning to the capacitance the meaning of a complex quantity, $\overline{C}_{el}^{S_i} = C_{el}^{S_i}\left(1 - j\tan\delta_e\right)$, or by introducing resistance of electrical loss R_{el} in accordance with formulas

$$\dot{W}_{eL} = v^2 / R_{eL} = i_{eL} v, \tag{5.209}$$

$$R_{eL} = 1/\omega C_{el}^{S_i} \tan\delta_e. \tag{5.210}$$

The electromechanical energy W_{em} (5.141) is

$$W_{em} = \int_{\tilde{V}} w_{em} d\tilde{V} = \frac{1}{2} v \sum_{i=1}^{\infty} n_i \xi_i, \tag{5.211}$$

where w_{em} is the density of electromechanical energy for the corresponding boundary conditions, and n_i is the electromechanical transformation coefficient for the vibration mode $\theta_i(r)$. It can be obtained by using relation (5.35) that

$$\delta W_{em} = v \sum_{l=1}^{\infty} n_l \delta S_l .$$

Thus, the electromechanical transformation coefficient for the vibration mode $\theta_i(r)$ can be found as

$$n_i = \frac{1}{v}\frac{\partial \dot{W}_{em}}{\partial \dot{\xi}_i}. \tag{5.212}$$

It is noteworthy that expressions for the electromechanical energy and transformation coefficient are valid, if the electrodes are divided into electrically insulated sections, which are connected in different polarities, or if only a part of the transducer volume is used as active (e.g., for optimizing the effective coupling coefficient). In these cases the signs of electric field within the sections must be changed to opposite, or the electric field in the passive parts have to be set to zero, when integrating in Eq. (5.211).

The *mechanical energy* W_m^E can be represented in the same manner as W_{pot} by formula (4.233) with the exception that the energy density for a piezoceramics body is w_{mch}^E for the corresponding boundary conditions, namely,

$$W_m^E = \int_{\tilde{V}} w_m^E d\tilde{V} = \frac{1}{2}\sum_{i=1}^{\infty}\sum_{l=1}^{\infty} K_{il}^E \xi_i \xi_l, \tag{5.213}$$

where K_{il}^E are the equivalent rigidities ($1/K_{il}^E = C_{il}^E$ are equivalent compliances): self rigidities for modes of vibration θ_i at $l = i$ and the mutual rigidities that characterize elastic interaction between the modes of vibration θ_i and θ_l at $l \neq i$. If only a part of the mechanical system is made of piezoceramics, then in calculations related to the remaining part of the volume in the expression for energy density under the sign of integral (5.213), the value of w_{mch} for passive material must be used for the same boundary conditions. However, the designations for the full equivalent rigidity and energy in this case will be retained as K_{il}^E and W_m^E in order to distinguish them from similar quantities for a body, which is made entirely of a passive material. For a real piezoceramic body the energy of mechanical losses must be accounted for by introducing into equations either the complex quantity of rigidity, $\overline{K}_{il}^E = K_{il}^E(1 + j\tan\delta_m)$, or the mechanical loss resistances r_{mLi}, which are determined by means of the relation

$$\dot{W}_{mL} = \sum_{i=1}^{\infty} r_{mLi}\dot{\xi}_i^2, \tag{5.214}$$

where (see Section 1.4)

$$\dot{W}_{mL} = \dot{W}_m^E / Q_m = \dot{W}_m^E \tan\delta_m \,. \tag{5.215}$$

In this case,

$$r_{mLi} = 1/\omega C_{ii}^E Q_m = K_{ii}^E / \omega Q_m \,. \tag{5.216}$$

The quantity ΔW is proportional to S_i^2 and hence to ξ_i^2, as can be seen from expression (5.142). We will regard this quantity as the mechanical energy (though it could equally well be classified as the electric energy because quantity $e_{3i}S_i$ has dimension of charge), and represent it in the form

$$\Delta W = \frac{1}{2}\sum_{i=1}^{\infty}\sum_{l=1}^{\infty}\Delta K_{il}\xi_i\xi_l \,, \tag{5.217}$$

where ΔK_{ii} (ΔC_{ii} = $1/\Delta K_{ii}$) correspond to the vibration modes θ_i, and ΔK_{il} characterize interaction between the vibration modes θ_i and θ_l arising due to nonuniformity of electromechanical conversion.

In the External Actions that were considered as the energy W_e of the mechanical actions in Section 4.5.1 now the action of source of the electric energy, W_{el}, must be included in the mode of electromechanical conversion. In the mode of the mechanoelectrical conversion we will consider electrodes to be open, i.e., electric load being absent. In the case that an electric load is applied, the solution obtained for the open circuited electrodes can be used by employing the Thevenin's theorem (see Section 1.5.3). The flux of electric energy supplied to transducer is $\dot{W}_{el} = vi$. Since in case of the electromechanical conversion $W_{el} = W_{int}$ and by virtue of formula (5.144) $W_{int} = W_e^{S_i} + W_{em}$, we obtain

$$\dot{W}_{el} = vi = \dot{W}_e^{S_i} + \dot{W}_{em} = C_{el}^{S_i} v\dot{v} + v\sum_{l=1}^{\infty} n_l \dot{\xi}_l \,, \tag{5.218}$$

and

$$\frac{\partial \dot{W}_{el}}{\partial \dot{\xi}_l} = v\sum_{l=1}^{\infty} n_l \,. \tag{5.219}$$

The electrical energy losses can be take into account by substituting the quantity W_{eL} according to formula (5.209) into the right-hand side of Eq. (5.218). After turning to the complex form the Eq. (5.218) becomes

$$I = j\omega C_{el}^{S_l} V + V / R_{eL} + \sum_{l=1}^{\infty} n_l U_l \, . \tag{5.220}$$

5.7.2 Derivation of equations of vibration

In the mode of electromechanical conversion the Lagrangian can be presented as

$$L = W_{kin} - W_{int} + W_{el} - W_L \, , \tag{5.221}$$

where W_L is the energy transferred into a mechanical load (in particular, the energy of acoustic radiation, W_{ac}). This energy is represented in general form by Eq. (4.240), and through the impedance of the load, Z_L, by Eq. (4.243). Euler's equations (4.1) of vibration of the piezoelectric body in the generalized coordinates is found as

$$\frac{d}{dt}\left(\frac{\partial W_{kin}}{\partial \dot{\xi}_i}\right) + \frac{\partial W_{int}}{\partial \xi_i} - \frac{\partial (W_{el} - W_L)}{\partial \xi_i} = 0 \, . \tag{5.222}$$

Upon substituting into Euler's equations W_{kin} according to formula (5.205), W_{int} from (5.144) with regard to expressions (5.212) through (5.217), W_L by Eq. (4.240), and expression (5.218) for W_{el}, we obtain equations of vibration, which are similar to equations (4.250) for the passive mechanical system. Here, like in equations (4.251), resistances that account for energy of mechanical losses are introduced

$$M_{ii}\ddot{\xi}_i + (K_{ii}^E + \Delta K_{ii})\xi_i + r_i\dot{\xi}_i + \sum_{l \neq i}^{\infty} [M_{li}\ddot{\xi}_l + (K_{li}^E + \Delta K_{li})\xi_l] = vn_i - f_{Li} \, . \tag{5.223}$$

The following distinctions from equations for a passive body exist in the case of a piezoceramic body: the forces that generate vibrations, $f_i = vn_i$, are of electromechanical origin, and specifics exists in determining the equivalent rigidities. For unifying the equations of motion with equations (4.254) for the passive bodies we will convert them to the complex form and introduce notations for the mechanical impedances of the piezoceramic bodies analogous to expressions in (4.253). Namely:

$$j\omega M_{ii} + \left(K_{ii}^E + \Delta K_{ii}\right) / j\omega + r_i = Z_{mii}^E, \quad j\omega M_{li} + \left(K_{li}^E + \Delta K_{li}\right) / j\omega = Z_{mli}^E,$$
$$Z_{mii}^E + \sum_{l \neq i} (U_l / U_i) Z_{mli}^E = Z_{mi}^E. \tag{5.224}$$

Finally, the equations (5.223) become

$$\left[Z^E_{mii} + Z_{Lii} + \sum_{l \neq i}^{\infty} (U_l / U_i)(Z^E_{mli} + Z_{Lli})\right] U_i = V n_i \quad (i = 1,2,...) . \tag{5.225}$$

Together with Eq. (5.220) for the generalized electrical coordinate these equations form the complete system of equations that describe vibrations of a piezoceramic body in the mode of electromechanical conversion (in the transmit mode).

In the mechanoelectrical conversion mode with open electrodes

$$L = W_{kin} - W_{int} + W_m , \tag{5.226}$$

where W_m is the energy supplied by an external mechanical source (in particular, energy supplied by the acoustic field) that is acting on the transducer surface by the equivalent force F_m. Both the energy and the force are expressed in general form by formulas (4.239). Based on the relations (5.52), where from $w_{int} = \underline{w^E_m} + \underline{w_{me}}$ and $\underline{w_{me}} = \underline{w^{S_i}_{el}}$, we obtain in this case that

$$W_{int} = \underline{W^E_m} + \Delta W + \underline{W_{em}}, \quad \underline{W_{em}} = \underline{W^{S_i}_e} , \tag{5.227}$$

where in expressions for $\underline{W_{em}}$ and $\underline{W^{S_i}_e}$ the quantities E_3 and V should be replaced by E_{3oc} and V_{oc}, respectively. Euler's equations for the mechanical generalized velocities in the complex form will be derived for the mechanoelectrical conversion using the same procedures, as in the previous case. As the result will be obtained

$$\left[Z^E_{mii} + Z_{Lii} + \sum_{l \neq i}^{\infty} (U_l / U_i)(Z^E_{mli} + Z_{Lli})\right] U_i + V_{oc} n_i = F_i \quad (i = 1,2,...) . \tag{5.228}$$

The equation for electrical coordinate V_{oc}, being derived directly from the relation $\dot{W}_{em} = \dot{W}^{S_i}_e + \dot{W}_{el}$ with taking into account expressions (5.210) and (5.213), is as follows

$$V_{oc} = Z^{S_i}_e \sum_{i=1}^{\infty} U_i n_i , \tag{5.229}$$

where

$$Z^{S_i}_e = 1 / Y^{S_i}_e, \quad Y^{S_i}_e = j\omega C^{S_i}_e + 1 / R_{eL} . \tag{5.230}$$

Equations (5.228) and (5.229) provide solution to the problem of calculating transducer that operates in the mechanoelectrical (receive) mode. After substituting V_{oc} by formula (5.229) into (5.228) we obtain the following equations for determining generalized velocities

$$\left[Z_{mii}^{E} + Z_{Lii} + Z_{el}^{S_i} n_i^2 + \sum_{l \neq i}^{\infty} (U_l / U_i)(Z_{mli}^{E} + Z_{Lli} + Z_{el}^{S_i} n_l n_i) \right] U_i = F_i \quad (i = 1, 2, \ldots). \qquad (5.231)$$

The set of equations (5.220) and (5.225) for the electromechanical and (5.229), (5.231) for the mechanoelectrical conversion can be obtained from the electromechanical circuit shown in Figure 5.20 with corresponding values of its components, using the common rules for calculating the electric circuits. Thus, if the energy is supplied to the electric side, the switch should be set to *EM* position. In absence of the external force it must be $F_L = 0$. If energy is supplied from the mechanical side in the absence of an electrical load (open circuited output), the switch should be set to *ME* position. The internal impedance of the source of mechanical energy is included into the circuit as Z_{Li}. In the case that transducer is electroacoustic, Z_{Li} is the radiation impedance that should be included in the equivalent circuit for both modes of operation. It represents a load in *EM* mode. In *ME* mode radiation impedance is the internal impedance of acoustic field, as source of mechanical energy with "acoustomotive" force $F_i = F_{eqvi}$, as it was discussed in Chapter 1. A detailed analysis of the acoustic field related parameters for transducers of different types will be performed in Chapter 6.

5.7.3 Equivalent Electromechanical Circuits

The equivalent electromechanical circuit of Figure 5.20 presents visualization of the equations that describe vibrations of piezoceramic body under various effects. Though the circuit provides the same information as that obtained from the equations, it may serve as a mnemonic rule, using which the equations can be recovered. One of the merits of utilizing the equivalent circuits is that this allows considering the transducers and electrical circuits, in combination with which they usually operate, in the similar manner and even without turning to equations describing the entire system.

So far the equations of vibration for a piezoceramic body in the generalized coordinates were presented in the most general form of equations (5.225) and (5.231). They include the mutual impedances $Z_{m\,mi}^{E}$, $Z_{L\,mi}$ and $Z_{el}^{S_i} n_m n_i$, which link the equations (the contours of the equivalent circuit corresponding to the vibration modes). This is due to the peculiarity of the method used, namely, that the normal modes for uniform isotropic mechanical system of the same shape and under the same boundary conditions are chosen as the system of supporting functions without regard to piezoelectric effects and mechanical loads.

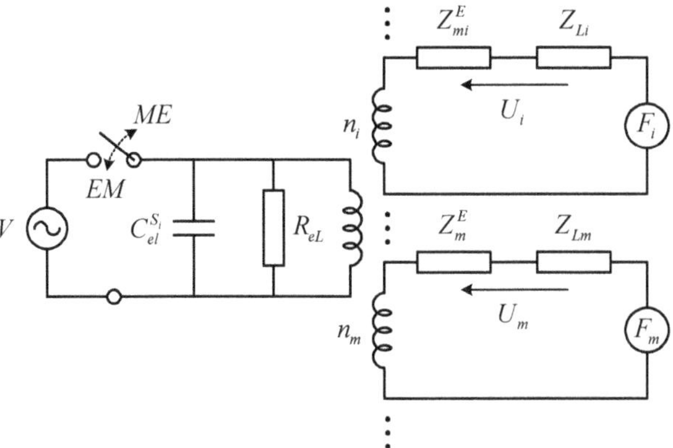

Figure 5.20: Equivalent circuit of the electromechanical transducer (general case)

Quite often interactions between the equations turn out to be insignificant, so that they can be or neglected, or easily be considered. In the first case the equations of vibration can be regarded as independent. This may be especially justified, if the transducers are supposed to operate in the frequency ranges around the resonance frequencies that correspond to the supporting vibration modes. In this case the equivalent circuit simplifies to the form that is shown in Figure 5.21, where equivalent compliances ($C^E_{ii} = 1/K^E_{ii}$) and masses (M_{ii}) are presented in the form of capacitances and inductances (as elements that in electric circuits are related to the potential and kinetic energies as well).

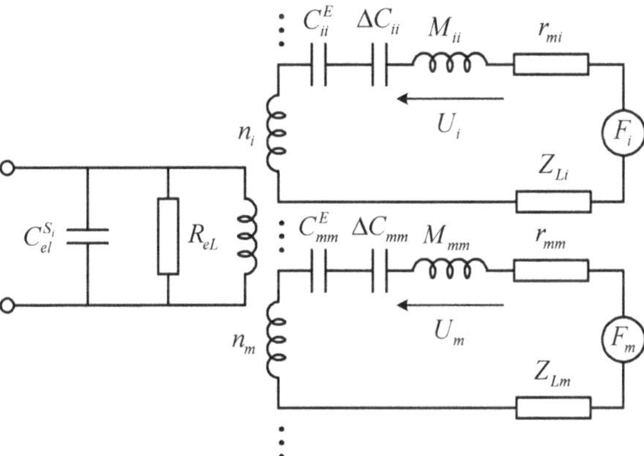

Figure 5.21: Equivalent circuit of electromechanical transducer for the case that interaction between the contours can be neglected (for the case that the generalized coordinates are normal, in particular).

Some of the electromechanical transformation coefficients n_i may vanish. This means that the respective modes of vibration (contours of the equivalent circuit) have no direct coupling with the electrical side, i.e., cannot be directly excited electrically, and being excited mechanically do not produce direct effect at the electric output (they are passive vibration modes). However, an indirect output effects may arise due to their interaction with active vibration modes. The procedure used for deriving equations of motion of system with an infinite number of degrees of freedom is also applicable without any alterations to the case, in which the number of degrees of freedom is limited. In this case the set of equations (among the contours of the equivalent circuit) only those remain, which correspond to the degrees of freedom considered. For a single mechanical degree of freedom, the result will be the same as in Section 1.6.1.

5.7.4 Examples of Application of the Equations in Generalized Coordinates

5.7.4.1 Equations of Vibration of the Piezoceramic Bars

Vibration of a bar as one degree of freedom system was considered in Section 2.4 under the assumptions that the ends of the bar are free of loads, all the volume of the bar is confined between unipolar electrodes, and in the case of longitudinal piezoeffect the bar is segmented with number of segments sufficient for considering that electric field is constant along the bar. Bars vibrating in the first normal mode were used as examples for illustrating effect of nonuniform deformation on the effective coupling coefficient in Sections 5.5.2.1 and 5.6.2.1. Here vibration of the bars utilizing the transverse and longitudinal piezoelectric effect that are presented in Figure 5.22 will be examined under more general assumptions.

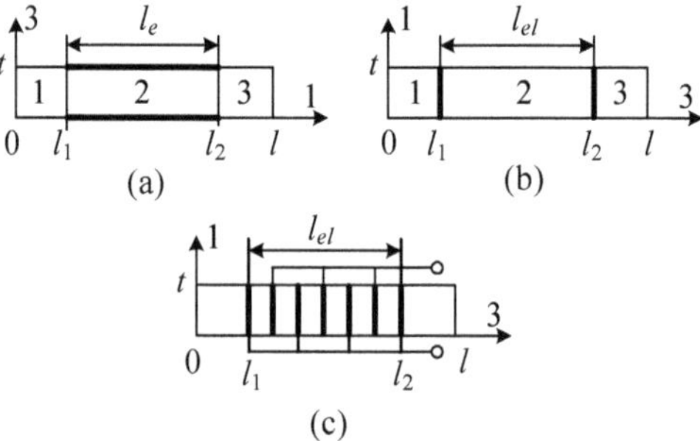

Figure 5.22: Transducers in the shape of a bar with incomplete electrodes: (a) transverse piezoeffect, (b) longitudinal piezoeffect, solid bar; (c) longitudinal piezoeffect, segmented bar.

At first, we will assume that the ends of the bars are not loaded, and active are the segments of the bars between coordinates l_1 and l_2 that have lengths l_e and are positioned symmetrically relative to the middle cross section. Choosing the system of normal modes of the problem of free vibrations of a bar, $\theta_i = \cos(i\pi x / l)$, as a system of supporting functions, we represent displacements and strains in the bar as follows:

$$\xi(x) = \sum_{i=0}^{\infty} \xi_i \theta_i(x) = \sum_{i=0}^{\infty} \xi_i \cos(i\pi x / l), \tag{5.232}$$

$$S(x) = \xi'(x) = -\sum_{i=0}^{\infty} \xi_i (i\pi / l) \sin(i\pi x / l). \tag{5.233}$$

(The minus sign in formula for strain accounts for the sign convention, according to which the strain is negative in case the displacement couses compression.)

The equivalent parameters of a bar M_{mi}, $C_e^{S_i}$, n_i, K_{mi}^E, and ΔK_{mi} have to be determined by formulas (5.205), (5.207), (5.212), (5.213), and (5.217), respectively, using the expressions (5.65)-(5.67) for the energy densities,

$$\varepsilon_{33}^{S_i} = \varepsilon_{33}^T (1 - k_{3i}^2), \quad w_m^E = S_i^2 / 2s_{ii}^E, \quad w_{em} = (d_{3i} / 2s_{ii}^E) E_3 S_i . \tag{5.234}$$

Here $i = 1$ for the transverse and $i = 3$ for the longitudinal piezoelectric effect. For the equivalent masses we obtain in all the cases $M_{00} = M$, $M_{ii} = M / 2$, where $M = \rho S_{cs} l$ is the full mass of the bar, M_{oo} is the mass corresponding to motion of the bar as a whole (without deformation). Due to uniformity of the bar $M_{mi} = 0$. In the case that $l_e \neq l$ for the transverse piezoelectric effect (Figure 5.22(a))

$$C_e^{S_1} = \varepsilon_{33}^T \left(1 - k_{31}^2\right) w l_e / t, \tag{5.235}$$

and for the longitudinal piezoelectric effect (Figure 5.22 (c))

$$C_e^{S_3} = \varepsilon_{33}^T \left(1 - k_{33}^2\right) w t / l_e = \varepsilon_{33}^T \left(1 - k_{33}^2\right) N^2 w t / l_e, \tag{5.236}$$

where N is the number of the segments within the length l_e.

In determining values of the equivalent rigidities K_{mi}^E it should be remembered that the value of the elastic constant of ceramics within segments 1 and 3, which are beyond the volume confined between the active electrodes, depends on whether the ceramics is polarized or not. If not, then it should be elastic constant of ceramic as a passive material. If they were polarized, but electrodes are removed afterwards, it should be s_{ii}^{D}, as there is no free charges on the surfaces of these segments. If the electrodes on these parts exist being electrically isolated, then the elastic properties depend on whether they are short circuited or open. In the case that the electrodes are short circuited $s_{ii} = s_{il}^{E}$, as it would be in the case that $l_e = l$. This assumption will be used in all the further treatments. The case that the isolated electrodes are open will be considered in Section 5.7.4.3. In the examples under consideration, we will be interested in how a change of the length of the active part of the bar influences electromechanical conversion, and therefore the assumption of short-circuiting the electrodes on the passive parts that does not complicate the calculations is reasonable.

To distinguish between the values of K_{mi}^{E}, ΔK_{mi} and n_i for the transverse and longitudinal piezoelectric effect letter p will be used in the subscripts, which have values $p = 1$ and $p = 3$, respectively. By formula (5.213)

$$K_{pii}^{E} = \frac{i^2\pi^2 S_{cs}}{2s_{pp}^{E} l}, \quad K_{pmi}^{E} = 0. \tag{5.237}$$

Note, that if differences in values of s_{pp} along the length would be taken into account, then in the general case (at $l_e \neq l$), $K_{pmi}^{E} \neq 0$, i.e., an elastic interaction between the modes of vibration would exist.

Quantity ΔK_{mi} may be obtained from expression (5.142) after it is represented as

$$\Delta W = \frac{1}{2}\sum_{m=1}^{\infty} \Delta K_{mi} \xi_i^2 . \tag{5.238}$$

After substituting expression (5.233) for strain and considering Eq. (5.217), it will be obtained that for a solid end-electroded bar at $l_e = l$

$$\Delta K_{3ii} = K_{3ii}^{E} \frac{k_{33}^2}{1-k_{33}^2}\left(1 - \frac{8}{i^2\pi^2}\right) \quad (i = 1,3,...), \tag{5.239}$$

$$\Delta K_{3ii} = K_{3ii}^{E} \frac{k_{33}^2}{1-k_{33}^2} \quad (i = 2,4,...), \tag{5.240}$$

$$\Delta K_{3mi} = -K_{3ii}^{E} \frac{k_{33}^2}{1-k_{33}^2} \frac{4}{\pi^2 mi} \quad (i = 1,3...;\ m = 1,3,...;\ m \neq i). \tag{5.241}$$

The fact that $\Delta K_{3mi} \neq 0$ at the odd values of i and m indicates that a coupling between these vibration modes exists due to nonuniformity of electromechanical conversion and, strictly speaking, the normal modes for a bar under the longitudinal piezoelectric effect and for a passive bar differ. Practically, this coupling is negligibly small.

As follows from formulas (5.237) and (5.240), at $i = 2, 4,...$

$$K_{3ii} = K_{3ii}^E + \Delta K_{3ii} = K_{3ii}^E \frac{1}{1-k_{33}^2} = K_{3ii}^D. \tag{5.242}$$

For a segmented bar at $l_e = l$ by means of formula (5.158) we obtain

$$\Delta K_{3ii} = K_{3ii}^E \frac{k_{33}^2}{1-k_{33}^2}\left[1-\beta\frac{\sin^2(i\pi/2N)}{(i\pi/2N)^2}\right], \tag{5.243}$$

where $\beta = 1$, if i/N is a fraction, and $\beta = 2$, if i/N is an odd number. As it was noted with regard to formula (5.95), $\Delta K_{3ii} \approx 0$ for $N/i \geq 6$ in the case that PZT ceramics is used.

If $l_e < l$, the values of ΔK_{3im} can be found by means of the formulas (5.142) and (5.238) in combination with Eq. (5.217), wherein integration should be performed over the segment confined between the electrodes. However, as l_e/l decreases, these values rapidly decrease and can be neglected in practical calculations at least when $l_e/l < 0.85$.

It is noteworthy that the segmented transducer designs are used for relatively low frequency applications. The end-electroded piezoelements for high frequency applications (starting from the height of a piezoelement that allows its convenient polarization) are used as solid, and all the "ΔK considerations" are applicable.

The electromechanical transformation coefficients $n_m(l_e)$ may be found by formula (5.211) after substituting expression (5.233) for strain and integrating over the segment limited by electrodes. These manipulations yield

$$n_m(l_e) = \frac{wtd_{3i}}{s_{ii}^E V} E_3\left[\theta_m\left(\frac{l+l_e}{2}\right) - \theta_m\left(\frac{l-l_e}{2}\right)\right] \quad (m = 1,2,3,...). \tag{5.244}$$

Here the mode of vibration is $\theta_m = \cos(m\pi x/l)$, and the electric field has the following values: $E_3 = V/t$ for the transverse piezoelectric effect; $E_3 = V/l_e$, $E_3 = VN/l_e$ for the longitudinal piezoelectric effect in the variants of the solid and segmented bar. From formula (5.244) we obtain

$$n_m(l_e) = \frac{wtd_{3i}}{s_{ii}^E}\sin\left(\frac{m\pi l_e}{2l}\right)\sin\left(\frac{m\pi}{2}\right) \tag{5.245}$$

for the variants shown in Figure 5.22 (a) and (b), and

$$n_m(l_e) = \frac{wtd_{33}N}{s_{33}^E l_e}\sin\left(\frac{m\pi l_e}{2l}\right)\sin\left(\frac{m\pi}{2}\right) \tag{5.246}$$

for the variant shown in Figure 5.22 (c).

After all the equivalent parameters are determined, vibrations of the bar can be calculated using the equivalent circuit of Figure 5.21, because with above made assumptions contours of the circuit (respective equations for the generalized velocities) are independent. With electrodes positioned symmetrically relative to the middle cross section of a bar the even (antisymmetric, at $m = 2k$) vibration modes are electromechanically passive as $n_m = 0$. Electromechanical activity of the odd (symmetric, at $m = 2k$-1) vibration modes depend on the electrode length l_e. Any mode of vibration at $m > 1$ can be suppressed (made electromechanically passive), if the length of electrodes meets condition

$$\sin\left(\frac{m\pi l_e}{2l}\right) = 0\text{, i.e., } \frac{l_e}{l} = \frac{2}{m}. \tag{5.247}$$

Thus, for example, if it is desirable to expend the frequency range, in which the first mode of vibration is dominant, the third mode of vibration can be suppressed by putting the length of the electrodes $l_e \approx 0.66l$. It is noteworthy that with this length of the electrodes the effective coupling coefficient of the first mode increases and almost reaches its optimal value, as it can be seen from Figure 5.23. The fact that even vibration modes are electromechanically passive at $l_{el} = l$ was physically explained in Section 2.4 with reference to Figure 2.7. Explanation of the same kind can be extended for the case of symmetrical electrodes of arbitrary size and can be illustrated with plots of the normal modes shown in Figure 5.23. So far as the electrodes are unipolar over the entire length (E_3 does not change its sign), signs of the charges formed in different parts of the electrodes coincide with the signs of strains and being averaged over the electrode surface they may be completely compensated, as it is in the variant of distribution of strains S_2 and S_4 in Figure 5.23 (a), and in the case that the third mode is intended to be suppressed, as it is in variant of S_3 at the length of the electrodes $l_e = 2l/3$. Electromechanical activity of the even modes can be restored, if to separate halves of electrodes electrically, and connect them in opposite phase, as it is qualitatively clear from the plots in Figure 5.23 (b). In general, the frequency response of a transducer can be changed by changing the electrodes configuration.

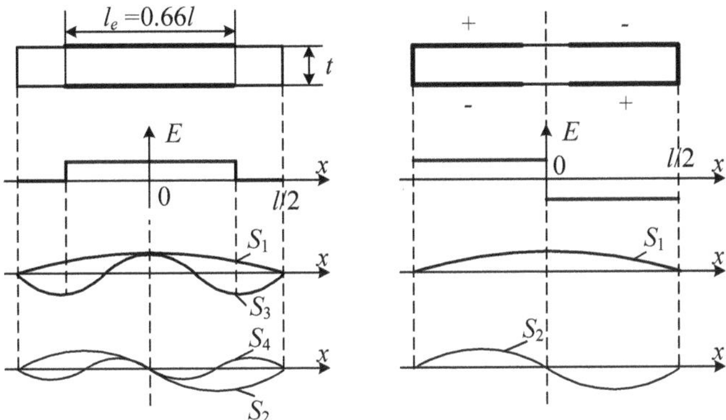

Figure 5.23: Natural modes of longitudinal vibration of bars: (a) unipolar electrodes, (b) halves of the electrodes are connected in opposite phase.

To qualitatively estimate effects of arbitrary electrodes geometry and electrical connection of their parts let us represent the electric field in the general case as

$$E_3 = \frac{V}{t}\Omega(x), \tag{5.248}$$

where the function $\Omega(x)$ depends on the electrode configuration and t is the separation between electrodes. We assume that $\Omega(x) > 0$ if $\boldsymbol{E}_3 \uparrow\uparrow \boldsymbol{P}$ and $\Omega(x) < 0$ if $E_3 \downarrow\uparrow P$. Thus, following the general expression (5.212), formulas for the transformation coefficient of a bar will be:

for the transverse piezoeffect

$$n_i = \frac{d_{31}S_{cs}}{s_{11}^{\mathrm{E}}t}\Omega_i, \tag{5.249}$$

for the longitudinal piezoelectric effect in the general case that the bar is segmented along its length

$$n_i = \frac{d_{33}S_{cs}N}{s_{33}^{\mathrm{E}}l_e}\Omega_i. \tag{5.250}$$

Factor Ω_i is

$$\Omega_i = \int_0^{l_e} \Omega(x)\theta_i'(x)dx\,. \tag{5.251}$$

Values of the transformation coefficients n_i depend significantly on what is the function $\Omega(x)$ that characterizes the electrodes configuration. Values of coefficients Ω_i for several variants of electrodes configuration are presented in Table 5.4. In the last row of the table the spectrum of the resonance frequencies (within the first three) is illustrated that corresponds to the active modes of vibration of a bar. Thus, by changing the function $\Omega(x)$ (by switching the parts of electrodes in particular), the frequency characteristics of the transducer can be governed.

Table 5.4: Spectrum of the natural frequencies of a bar vs. configuration of electrodes.

$\Omega(x)$	+ + - - Ω, x	+ - - + Ω, x	+ - Ω, x	+ - Ω, x_1, x_2, x
Ω_m, $k = 1,2,\ldots$	2 at $m = 2k-1$	4 at $m = 2(k-1)$	1 at $m = 2k-1$ 2 at $m = 2(k-1)$	$2\sin\dfrac{m\pi(x_1+x_2)}{2l}$ $\times\sin\dfrac{m\pi(x_2-x_1)}{2l}$
Spectrum	f_1 f_2 f_3	f_1 f_2 f_3	f_1 f_2 f_3	-

Consider now the case that the bar is loaded by impedances Z_0, Z_l, Z_x, and the external forces F_o, F_l, as shown in Figure 5.24. The equivalent external forces and load

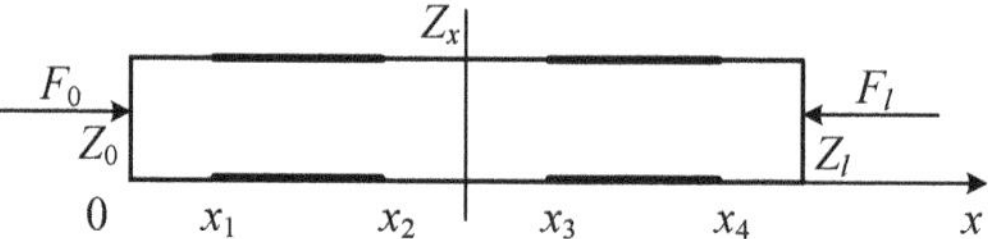

Figure 5.24: Bar transducer under actions of external loads and forces.

impedances that correspond to the real quantities can be determined by formulas (4.238) - (4.241). In this case, we will consider the forces and impedances being concentrated. If the real actions are distributed, then these forces and impedances must be considered as result of integrating the actions over surfaces, to which they are applied. In the particular cases $Z(l)$ may imitate the radiation impedance, $Z(x)$ - impedance of a supporting structure, and force $F(l)$ -

the equivalent force due to action of acoustic field. Considering that

$$\theta_i(x) = \cos(i\pi x / l), \quad \theta_i(0) = 1, \quad \theta_i(l) = (-1)^i , \tag{5.252}$$

we obtain

$$F_i = F_0 + (-1)^{i+1} F_l , \tag{5.253}$$

$$F_i = F_0 + (-1)^{i+1} F_l , \tag{5.254}$$

where the tensile forces should be regarded as positive. Besides, the mutual impedances

$$z_{im} = Z_0 + (-1)^{i+m} Z_l + Z_x \cos(m\pi x / l)\cos(i\pi x / l) \tag{5.255}$$

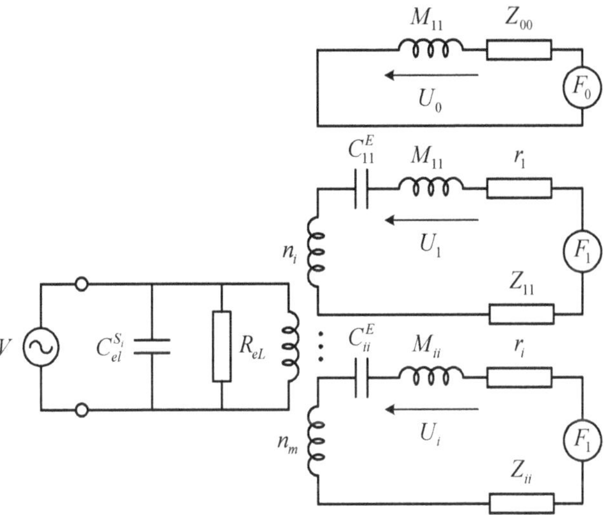

Figure 5.25: Multicontour equivalent circuit of a bar transducer.

exist, which characterize couplings between the vibration modes (contours of the equivalent circuit) that appear because the normal modes of the loaded bar, strictly speaking, differ from those for not loaded. Thus, using the equivalent circuit of Figure 5.20 in the general case of a loaded bar transducer becomes too complicated and virtually not appropriate. Vibration of the arbitrary loaded bar transducers will be considered in Section 5.8 based on the geometry coordinates approach. Under the assumption that loading impedances are small and interaction between the contours can be neglected, the general equivalent circuit of Figure 5.20 can be modified to the circuit shown in Figure 5.25 for the bar transducers. If the bar is fixed in the middle cross section ($Z(l/2)\to\infty$), then the only vibrations are possible, for which $\cos(i\pi/2)=0$, i.e., at $i=2m-1$. The even contours (including the one at $i=0$, which is responsible for movement of a bar without deformations) must be excluded from the circuit. In the frequency regions around resonance frequencies it is sufficient to consider only the contour that corresponds to the resonance vibration mode in the circuit of Figure 5.25. It is noteworthy that in this case the assumption of neglecting interaction between contours is especially true, because the actual effect of interaction depends on the introduced impedances $Z_{in\,mi}=z_{mi}(U_m/U_i)$, and in vicinity of the resonance frequency of ith mode $(U_m/U_i)\ll 1$.

When calculating transducer in the receive mode in the frequency range much below the first resonance, $(1/\omega C_{ii}^E)\gg\omega M_{ii}$, $1/\omega C_{ii}^E\gg|Z_{ii}|$. Suppose that the electrodes are of the full size (the modes are active at $i=2m-1$) and a force F_l is acting only on one end, while another end is free of an external action (that is $F_0=0$, $Z_0=0$). Retaining the elastic impedances only in the circuit of Figure 5.25 we obtain the output voltage in the following form

$$V_{out}=\frac{1}{\omega C_{el}^{S_1}}\frac{2wd_{31}}{s_{11}^E}F_l\sum_{m=1}^{\infty}\frac{2s_{11}^E l\omega}{\pi^2 S_{cs}}\frac{1}{(2m-1)^2}=\frac{1}{C_{el}^{S_1}}\frac{2wd_{31}}{s_{11}^E}\frac{s_{11}^E l}{4S_{cs}}F_l\,. \tag{5.256}$$

Here it is considered that

$$\sum_{m=1}^{\infty}\frac{1}{(2m-1)^2}=\frac{\pi^2}{8}\,. \tag{5.257}$$

The same result will be obtained using a different approach in Section 5.8.

So far, we have discussed calculating by means of the circuit shown in Figure 5.25 under the assumption that the loading impedances are small. If impedance at one of the ends is large ($Z_0\to\infty$), i.e., this end can be considered as fixed, then supporting system of normal modes must be used for calculating parameters of the equivalent circuit that matches this boundary condition, namely,

$$\theta_i(x)=\sin[(2i-1)\pi x/l]\,. \tag{5.258}$$

In this case using the above procedure for the side-electroded bar with full size electrodes we arrive at the following results:

$$\begin{gathered}M_{ii}=M/2,\quad K_{ii}^E=i^2\pi^2 S_{cs}/2ls_{11}^E,\\ n_i=(-1)^{m+1}d_{31}S_{cs}/s_{11}^E t\quad (i=2m-1;\ m=1,2,\dots).\end{gathered} \tag{5.259}$$

The equivalent forces and load impedances in this case are $F_i = (-1)^{m+1} F_l$, $Z_{ii} = Z_l$.

Everything that has been discussed with respect to the bar transducers vibrating longitudinally equally holds for the bar transducers performing the torsional vibrations. It is necessary only to make the following substitutions in all the formulas:

$$M_{ii} = \rho J_p l / 2, \quad K_{ii}^E = i^2 \pi^2 J_p / 2 s_{44}^E l, \quad n_i = \left(w^2 d_{24} / 4 s_{44}^E \right) \Omega_i . \tag{5.260}$$

5.7.4.2 Equations of Vibration in the plane of a Circular Disk Poled through its Thickness

Consider the radial vibration of a piezoceramic thickness polarized disk shown in Figure 5.26. We assume that the electrodes are axially symmetric and the edge of the disk is free. For the supporting functions will be taken the normalized normal modes of problem of vibration of a passive disk (see Section 4.4.2.2), namely,

$$\theta_i (r / a) = J_1(\lambda_i r / a) / J_1(\lambda_i) , \tag{5.261}$$

where λ_i are the eigenvalues, which by virtue of relation (4.166) for $\sigma = 0.3$ are equal to

$$\lambda_i = 2.05,\ 5.38,\ 8.57; \text{ and at } i > 3 \ \lambda_i \approx (i\pi - 0.9) . \tag{5.262}$$

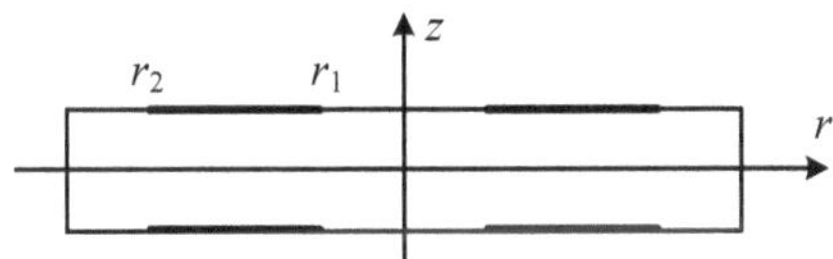

Figure 5.26: Circular disk with axially symmetric electrodes.

It is noteworthy that though for different piezoceramic materials $\sigma_1^E = s_{12}^E / s_{11}^E \neq 0.3$, it can be shown from the Eq. (4.167) for determining the eigenvalues that the values of λ_i change less than by 1% from their values at $\sigma = 0.3$ within limits of values of σ_1^E for the most usable PZT compositions (approximately between $\sigma_1^E = 0.27$ and 0.35). Thus, displacements $\xi(r)$ in the plane of a disk will be represented in the form of the series

$$\xi(r) = \sum_{i=1}^{\infty} \xi_{ai} \theta_i (r / a) = \sum_{i=1}^{\infty} \xi_{ai} J_i (k_i r) / J_1(k_i a) , \tag{5.263}$$

where $k_i = \lambda_i / a$ and ξ_{ai} are the radial displacements at $r = a$ for the different modes of vibration. By formula (5.205),

$$M_{mi} = 2\pi\rho t \int_0^a \theta_i \theta_m r dr \,, \tag{5.264}$$

where from follows that $M_{mi} = 0$ for $m \neq i$ due to orthogonality of the normal modes, and $M_{ii} = M(1 - 0.9/\lambda_i^2)$ with $M = \pi a^2 t\rho$. Thus,

$$M_{11} = 0.76M, \quad M_{22} = 0.97M, \quad M_{ii} \approx M \quad (i > 3) \,. \tag{5.265}$$

Other equivalent parameters of the disk can be found from expressions (5.206), (5.211) and (5.213). The quantities $W_e^{S_i}$, W_{em} and W_m^E that are involved in these expressions must be determined by formulas (5.139) through (5.141) after substituting the values of energy densities for the axially symmetrical deformation in the cylindrical coordinates from expressions (5.68) - (5.70). As the disk is poled in z direction and axis 3 of the crystallographic coordinate system goes in the same direction, the axes 1 and 2 are in the plane of a disk, $S_{rr} = S_1$ and $S_{\varphi\varphi} = S_2$. After substituting expressions (5.263) for $\xi(r)$ in formulas (4.145) we obtain (see properties of the Bessel functions in Appendix C.1)

$$S_1 = \frac{\partial \xi(r)}{\partial r} = \sum_1^{\infty} \xi_{ai} \frac{1}{J_1(k_i a)} \frac{\partial J_1(k_i r)}{\partial r} = \sum_1^{\infty} \xi_{ai} \frac{k_i}{J_1(k_i a)} \left[J_0(k_i r) - \frac{J_1(k_i r)}{k_i r} \right], \tag{5.266}$$

$$S_2 = \frac{\xi(r)}{r} = \sum_1^{\infty} \xi_{ai} \frac{k_i}{J_1(k_i a)} \frac{J_1(k_i r)}{k_i r} \,. \tag{5.267}$$

Thus,

$$S_1 + S_2 = \sum_{i=1}^{\infty} \xi_{ai} \frac{1}{r} \frac{d}{dr} [k_i r \theta(k_i r)] = \sum_{i=1}^{\infty} \xi_{ai} \frac{k_i J_0(k_i r)}{J_1(k_i a)} \,. \tag{5.268}$$

In the case that the electrodes completely cover the surfaces of the disk we obtain, using the expression for w_m^E from (5.69),

$$\begin{aligned} W_m^E = \frac{\pi t Y_1^E}{1 - \sigma_1^{E2}} \Bigg\{ & \int_0^a \left[\left(\frac{\partial \xi}{\partial r} \right)^2 + 2 \cdot 0.3 \frac{\xi}{r} \frac{\partial \xi}{\partial r} + \left(\frac{\xi}{r} \right)^2 \right] r dr - \\ & -2 \int_0^a (0.3 - \sigma_1^E) \frac{\xi}{r} \frac{\partial \xi}{\partial r} r dr \Bigg\} = \frac{1}{2} \sum_{i=1}^{\infty} K_{ii}^E \xi_i^2 + \sum_{m \neq i}^{\infty} K_{mi}^E \xi_i \xi_m \,. \end{aligned} \tag{5.269}$$

Since the supporting functions θ_i are defined as normal modes at $\sigma = 0.3$, the first integral in (5.269) contributes only to values of K_{ii}^E, defining the major part of these values. In the case that $\sigma_1^E \neq 0.3$ the second integral produces a certain change in K_{ii}^E and introduces the mutual rigidities K_{mi}^E, which characterize interaction between equations (5.225) and between contours of the equivalent circuit in Figure 5.20. Upon substituting in (5.269) displacement $\xi(r)$ in the form of series (5.261) and upon subsequent calculations, we obtain

$$K_{ii}^E = \frac{2\pi t Y_1^E\left(\beta_{ii} + \sigma_1^E - 0,3\right)}{1 - \sigma_1^{E2}}, \quad K_{mi}^E \approx K_{11}^E \frac{\sigma_1^E - 0,3}{1,6}, \tag{5.270}$$

where $\beta_{11} = 1.6$; $\beta_{22} = 14.1$; $\beta_{33} = 36$. For PZT piezoceramic compositions that are used in transducers $\left|\sigma_1^E - 0.3\right| < 0.05$ it can be assumed that

$$\begin{gathered} K_{11}^E = K_{eqv1}^E \approx \frac{10tY_1^E}{1-\sigma_1^{E2}}, \quad K_{22}^E = K_{eqv2}^E \approx 8.8K_{eqv1}^E, \\ K_{33}^E = K_{eqv3}^E \approx 22K_{eqv1}^E, \quad \left|K_{im}^E\right| < \frac{K_{11}^E}{30}. \end{gathered} \tag{5.271}$$

It can be seen that at $\sigma_1^E \neq 0.3$ some interaction between equations (5.225) exists due to a finite value of the mutual rigidities K_{mi}^E. Using the expressions for equivalent parameters of a disk and equations (5.231), the estmation of values of introduced impedances,

$$Z_{int\,i} = j\sum_{m \neq i}^{\infty}\left(U_m / U_i\right)\omega C_{mi}^E, \tag{5.272}$$

for the frequency ranges around the resonance frequencies can be obtain in the form

$$\left|Z_{int\,i}\right| < 0.02/\omega C_{ii}^E Q_m, \tag{5.273}$$

where Q_m is the quality factor of the mechanical system that has an order of magnitude higher than 10. Thus, the effect of $Z_{int\,i}$ can be neglected, and the equations (5.231)for the disk and corresponding contours of the equivalent circuit of Figure 5.20 can be regarded as independent. The values of natural frequencies are defined by the relation

$$\omega_i = \frac{\lambda_i}{a}\sqrt{\frac{Y_1^E}{\rho(1-\sigma_1^{E2})}} = \sqrt{\frac{K_{ii}^E}{M_{ii}}}. \tag{5.274}$$

For the first mode of vibration with M_{11} and K_{11}^E taken from expressions (5.265) and (5.271) we obtain

$$\omega_i = \frac{2.04}{a}\sqrt{\frac{Y_1^E}{\rho(1-\sigma_1^{E2})}}, \tag{5.275}$$

whereas the value obtained from differential equation for $\sigma_1^E = 0.3$ is $\lambda_1 = 2.05$ (see the note below expressions (5.262)).

When considering the quantities $C_e^{S_{12}}$ and n_i, we will assume that the electrodes may be partial. Strictly speaking, a change in the dimensions of the electrodes influences the energy W_m^E as well, however, we will not take this into account, as in the previous example of a bar. At least, W_m^E must remain unchanged, if we assume that electrodes on inactive parts of a disk exist, but they are electrically isolated and short circuited. Using formulas for densities of the electrical and electromechanical energies (5.68) and (5.70), we obtain from Eqs. (5.206) and (5.211)

$$C_e^{S_{1,2}} = \varepsilon_{33}^T(1-k_p^2)\pi\left(r_2^2 - r_1^2\right)/t, \tag{5.276}$$

$$W_{em} = \frac{1}{2}\frac{2\pi d_{31} t Y_1^E}{1-\sigma_1^E}\int_{r_1}^{r_2}(S_1 + S_2)E_3 r dr = \frac{1}{2}V\sum_{i=1}^{\infty} n_i \xi_{ai}. \tag{5.277}$$

After substituting expression for $(S_1 + S_2)$ by formula (5.268) the electromechanical transformation coefficients will be found as

$$n_i = \frac{2\pi d_{31} a Y_1^E}{1-\sigma_1^E}\left[\frac{r_2}{a}\theta_i\left(\frac{r_2}{a}\right) - \frac{r_1}{a}\theta_i\left(\frac{r_1}{a}\right)\right]. \tag{5.278}$$

In the case that $r_1 = 0$ and $r_2 = a$

$$n_i = \frac{2\pi d_{31} a Y_1^E}{1-\sigma_1^E}. \tag{5.279}$$

The values of the transformation coefficients at different correlations between radiuses r_2 and r_1 can be obtained using plots in Figure 5.27.

In particular, the electromechanical activity of the modes θ_2 and θ_3 (and whatever mode, in general) can be significantly increased (or suppressed) by appropriate choice of the size and location of parts of electrodes. Qualitatively this can be explained exactly in the same way, as it was done in the case of the longitudinal vibration of a bar, if to consider plots of functions $(S_1 + S_2)_i$ vs. ratio r/a, which are according to expression (5.268) proportional to $J_0(k_i r)$. Plots of these functions for the first three modes of vibration are shown in Figure 5.28 (note that the functions in the figure change signs, when passing the zero points).

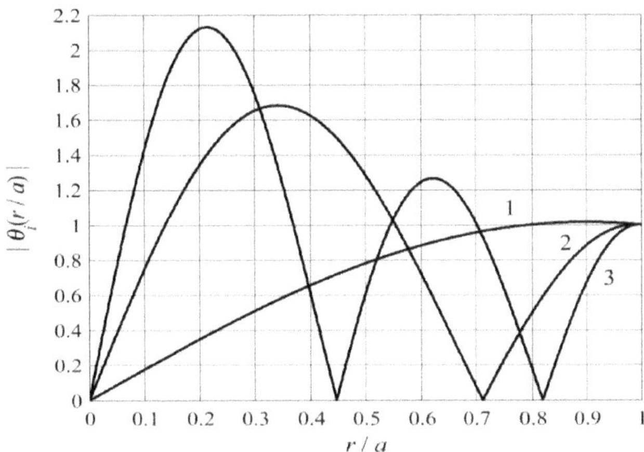

Figure 5.27: Natural modes of the radial vibration of a disk (i = 1, 2, 3).

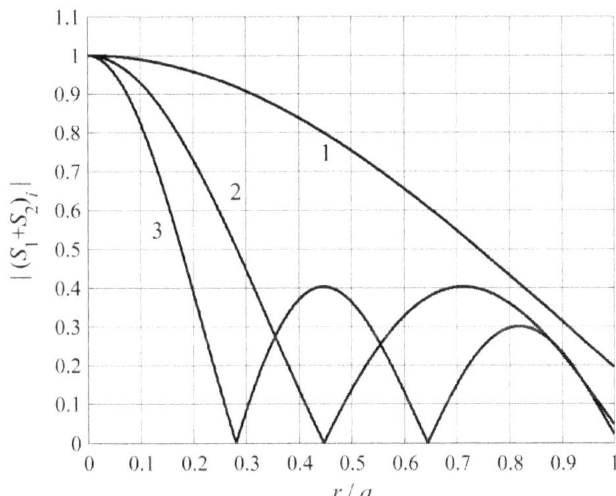

Figure 5.28: Plots of functions $(S_1+S_2)_i$ vs. r/a.

For quantitative estimation of quality of electromechanical conversion consider dependence of the effective coupling coefficients from location and size of the electrodes using expressions (5.178),

$$k_{eff}^2 = \frac{\alpha_c}{\alpha_c+1}, \text{ where } \alpha_c = \frac{n^2}{K_{eqv}^E C_e^{S_{1,2}}} = \frac{\alpha_c'}{K_{eqv}^E}. \tag{5.280}$$

The term

$$\alpha_c' = n^2 / C_e^{S_{1,2}} \tag{5.281}$$

depends on the location and size of the electrodes only. As it was noted previously, the equivalent rigidity, K_{eqv}^{E}, does not depend on geometry of the electrodes, unless difference between the elastic constants of polarized and not polarized ceramics on the parts of a piezoelement without the electrodes is taken into account. This difference is further neglected, moreover, usually the pats of a mechanical system are deprived of electrodes that contribute less to electromechanical conversion and therefore possess relatively small potential energy. The maximum value of k_{eff} is achieved at maximum value of coefficient α_c, and hence at maximum value of coefficient α'_c. At first, consider dependences of $\alpha'_{ci}(r)$ from radius of electrodes located in the center of a disk. Note that coefficients $\alpha'_i(a)$ don't depend on the number i of mode of vibration, and

$$\frac{\alpha'_{ci}(r)}{\alpha'_{c1}(a)} = \theta_i^2(r/a), \tag{5.282}$$

as it follows from expressions for $n_i(a)$ and $C_e^{S_{1,2}}$. Therefore, it can be concluded from plots in Figure 5.27 that the maximum values of $\alpha'_{ci}(r)$ are achieved for the first mode at $r = a$, for the second mode at $r = 0.35a$, and for the third mode at $r = 0.22a$. The corresponding values of the coefficient are

$$\alpha'_{c2}(0.35a) = \alpha'_{c1}(a)\theta_2^2(0.35) = 2.9\alpha'_{c1}(a), \tag{5.283}$$

$$\alpha'_{c3}(0.22a) = \alpha'_{c1}(a)\theta_2^2(0.22) = 4.6\alpha'_{c1}(a). \tag{5.284}$$

Consider now the dependences of this ratio from the relative width of concentric electrodes located around the nodal lines of the modes of vibration. Such electrodes may be used in some applications of disk transducers, e.g., for the piezoelectric transformers that will be considered in the next section. Denote the radiuses of the nodal lines r_{in}, where i is number of the mode and n is number of the nodal line. The radiuses of the nodal lines are: $r_{21} = 0.71a,\ r_{31} = 0.45a,\ r_{32} = 0.8a$. We will mark coefficients α_c and α'_c that correspond to the radiuses of the nodal lines with the same subscripts. Results of calculating the relative widths of electrodes $\Delta r/a$, at which the maximum value of ratios $\alpha'_{cin}/\alpha'_{ci0}$ is achieved are:

$$\begin{gathered}\alpha'_{c21\max} \approx 2.0\alpha'_{c20} \text{ at } \Delta r \approx 0.23a, \quad \alpha'_{c31\max} \approx 1.6\alpha'_{c30} \text{ at } \Delta r \approx 0.28a, \\ \alpha'_{c32\max} \approx 1.2\alpha'_{c30} \text{ at } \Delta r \approx 0.23a.\end{gathered} \tag{5.285}$$

Here $\alpha'_{c20} = \alpha'_{c2}(0.35a)$, $\alpha'_{30} = \alpha'_3(0.22a)$. It is convenient to express all the above coefficients through the coefficient $\alpha_{c1}(a)$ for fully electroded disk vibrating in the first mode. After substituting in the expression (5.280) values of K^E_{eqv1}, $C^{S_{1,2}}_{el1}$ and n_1 from (5.271), (5.276) and (5.279), this coefficient will be obtained in the form

$$\alpha_{c1}(a) = \frac{\alpha'_{c10}}{K^E_{eqv1}} = \frac{\pi}{5}(1+\sigma^E_1)\frac{k^2_p}{1-k^2_p}. \tag{5.286}$$

For PZT-4 ceramics $\alpha_{c1}(a) = 0.42$ (with $\sigma^E_1 = 0.33$, $k_p = 0.58$), and $k_{eff} = \sqrt{\alpha_{c1}/(1+\alpha_{c1})} = 0.54 = 0.94k_p$. Thus, the radial vibrating disk is a convenient configuration for experimental determining the planar coupling coefficient, k_p, by means of measuring k_{eff} by resonance- antiresonance method.

The coefficients α_{cin} that are necessary for calculating respective effective coupling coefficients will be obtained as follows. The coefficients α_{cin} are equal to α'_{cin}/K^E_{eqvi} under the assumption that the rigidities are independent on configuration of electrodes. The sequence of manipulations will be illustrated with example of calculating coefficients α_{c20} and α_{c21} for the second mode of vibration. The coefficient $\alpha_{c20} = \alpha'_{20}/K^E_{eqv2}$, and due to relation (5.283) $\alpha_{c20} = (2.9/K^E_{eqv2})\alpha'_1(a)$. The coefficient $\alpha_{c21} = \alpha'_{c21}/K^E_{eqv2}$, and due to relations (5.283) and (5.285) $\alpha_{c21} = (5.8/K^E_{eqv2})\alpha'_{c1}(a)$. Considering that $\alpha'_{c1}(a) = \alpha_{c1}(a)K^E_{eqv1}$ and $K^E_{eqv2}/K^E_{eqv1} = 8.8$ according to expressions (5.271), we finally obtain that $\alpha_{c20} = 0.33\alpha_{c1}(a)$ and $\alpha_{c21} = 0.66\alpha_{c1}(a)$. The expression for coefficient $\alpha_{c1}(a)$ is given by formula (5.286). In the case that PZT-4 ceramics is used $\alpha_{c1}(a) = 0.42$, $\alpha_{c20} \approx 0.13$, $\alpha_{c21} \approx 0.26$. The respective effective coupling coefficients are $k_{eff1} = 0.54$, $k_{eff20} = 0.34$, $k_{eff21} = 0.45$.

For the second mode of the fully electroded disk, $n_2(a) = n_1(a)$, $C^{S_{1,2}}_{e2}(a) = C^{S_{1,2}}_{e1}(a)$ and $\alpha_{c2}(a) = \alpha_{c1}(a)/8.8$. Thus, this mode is much less effective electromechanically. With PZT-4 used $\alpha_{c2}(a) \approx 0.05$ and $k_{eff2} = 0.21$ instead of 0.42 and 0.54, respectively, for the first mode. Thus, electromechanical conversion performed by the second mode transducer with the full-size electrodes can be greatly improved by using properly located partial electrodes. These results will be helpful in the following section.

5.7.4.3 Electro-Mechano-Electrical Transducers

6.1.3.1.5. Equivalent Circuit of the Transducer

So far, we assumed that the entire volume of a piezoceramic body is confined between one pair of electrodes. In some cases it proves necessary to use only a part of a piezoelement volume as active, or/and to divide electrodes into electrically insulated sections and to use them separately. In the last case we will call the transducers electro-mechano-electrical (for short EME). Electro-mechano-electrical transducers have numerous applications. They may be used in electromechanical transformers, as means for establishing electrical feedback in the devices that employ piezoceramic resonators, for performing control of parameters of a transducer in process of operating. Theoretical treatment of EME transducers requires considering internal energy of a piezoceramic body in case that it has separate operating electrodes, as shown in Figure 5.6. We will assume that in the figure part $\tilde{V}_1$ of the volume is the active part, in which electromechanical conversion takes place. Part of the volume $\tilde{V}_2$ has separate electrodes, electrical conditions of which can vary depending on the design task. Generally, they may be considered as open circuited, because thus obtained solution can be easily extrapolated to results of operation with any electrical load by using the Thevenin's theorem. Part of the volume $\tilde{V}_3$ is assumed to be without electrodes and imitates a passive portion of the piezoelement that is not polarized or previously polarized, but with removed electrodes. The value of the internal energy of the entire body that is consuming independent flow of the electric energy may be represented following expression (5.144), as $W_{int} = W_e^{S_i} + W_{em}$, where

$$W_{em} = W_{m1}^E + \Delta W_1 + \underline{W_{m2}} + \underline{W_{m3}} . \tag{5.287}$$

Here W_{m1}^E and ΔW_1 are determined by formulas (5.138) and (5.142); the energies $\underline{W_{m2}}$ and $\underline{W_{m3}}$ are underlined, because these parts of the volume consume independent flow of mechanical energy. In calculating $\underline{W_{m3}}$, two cases must be considered. If electrodes in this area were not applied in process of the piezoelement manufacturing, then $\underline{W_{m3}}$ must be determined with values of elastic moduli of non-polarized ceramics. If electrodes were removed after the piezoelement was polarized, then $D = 0$ due to absence of electrodes and therefore of free charges throughout this part of the volume. In this case the mechanical energy must be determined at the values of elastic constants corresponding to $D = 0$, i. e., $\underline{W_{m3}} = \underline{W_{m3}^D}$. The part $\tilde{V}_2$ of the volume is in the mode of mechanoelectrical conversion (is consuming the external mechanical energy and converts it into electrical form), therefore the internal energy of this part is $\delta w_{\text{int}} = \underline{\delta w_m}$. Using expression (5.52) for the mechanical energy of the volume element, we obtain according formula (5.227) that

$$\underline{W_{m2}} = \underline{W_{m2}^E} + \Delta W_2 + \underline{W_{me2}}, \quad \underline{W_{me2}} = \underline{W_{e2}^{S_i}}, \tag{5.288}$$

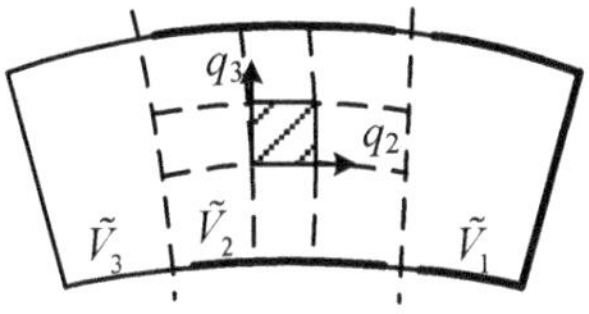

Figure 5.29: Configuration of a piezoceramic body of a general type with separate electrodes.

where $\underline{W_{m2}^E}$, ΔW_2 and $\underline{W_e^{S_i}}$ must be determined by formulas (5.139), (5.142) and (5.140), in which E_3 must be replaced by E_{3oc}. Under condition of short circuited electrodes $E_{3oc} = 0$, $\underline{W_{me2}} = \underline{W_{e2}^{S_i}} = 0$, and $\underline{W_{m2}} = \underline{W_{m2}^E} + \Delta W_2$. Thus, in the case under consideration expressions that characterize the internal energy and energy balance in course of vibration become

$$W_{em} = W_{m1}^E + \Delta W_1 + \underline{W_{m2}^E + \Delta W_2} + \underline{W_{m3}} + \underline{W_{me2}}, \quad \underline{W_{me2}} = \underline{W_{e2}^{S_i}}. \tag{5.289}$$

Here the result of electro-mechano-electrical conversion (“converted energy”) is $\underline{W_{me2}}$. Therefore, expression for the effective coupling coefficient (5.88) that corresponds to this conversion has to be modified to the form

$$k_{eff}^2 = \frac{\underline{W_{me2}}}{W_e^{S_i} + W_{em}} = \frac{\underline{W_{e2}^{S_i}}}{W_e^{S_i} + W_{m\Sigma oc}}. \tag{5.290}$$

For brevity the energy W_{em} in (5.289) is denoted as $W_{m\Sigma\, oc}$, the total mechanical energy of open circuited piezoelement.

It can be seen that in expression (5.289) the relations (5.144) for electromechanical and (5.227) for mechanoelectrical conversions are combined. After the equations and equivalent circuits that describe electromechanical and mechanoelectrical conversions are known, there is no need to repeat analogous derivation for EME transducers. The result can be obtained by combining already existing equivalent circuits for these two cases.

Usually, EME transducers operate near their resonance frequencies, or less often below the first resonance, and therefore they can be considered as having one mechanical degree of freedom. The equivalent circuit of the electro-mechano-electrical transducer that employs two pairs of electrically separated electrodes is shown in Figure 5.30. This circuit, one part of which operates in the mode of the electromechanical conversion (it will be conditionally denoted as input) and another operates in the mode of the mechanoelectrical conversion (output), allows calculating all the operating parameters of the transducers. The output and input parts can be used interchangeably. In this sense naming them "input" and "output" is conditional.

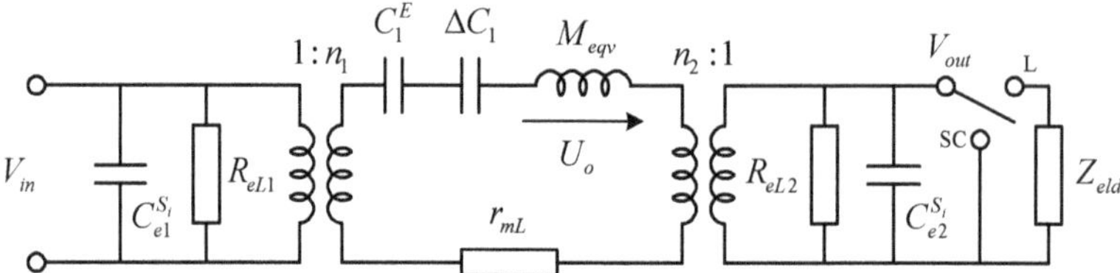

Figure 5.30: Equivalent circuit of electro-mechano-electrical transducer.

The following has to be noted regarding parameters of the circuit given that the mode of vibration of the mechanical system of the transducer is known. Values of the capacitances and transformation coefficients n_1, $C_{e1}^{S_i}$ and n_2, $C_{e2}^{S_i}$ must be determined by the general formulas for the respective segments of electrodes. The equivalent mass M_{eqv} does not depend on separation of electrodes and is determined by the mode of vibration only. The equivalent compliance $C_{\Sigma sc}^E = C_m^E = 1/(K_m^E + \Delta K)$ must be determined in accordance with relation

$$W_{m\Sigma sc} = (\xi_o^2/2)(K_{m1}^E + K_{m2}^E + K_{m3} + \Delta K_1 + \Delta K_2) \approx \xi_o^2/2)(K_m^E + \Delta K)\,. \tag{5.291}$$

Here subscript *sc* indicates that the output of the piezoelement is short circuited. Value of the term K_{m3}, strictly speaking, depends on the status of the passive part, as it was previously discussed, but the differences that may occur are not significant. We will assume for simplicity that electrodes on this part exist and are short circuited. Thus, $K_{m3} \approx K_{m3}^E$ and therefore $K_{m1}^E + K_{m2}^E + K_{m3} \approx K_m^E$, i.e., equal to the rigidity of the piezoelement with full size electrodes. In the term $\Delta K = \Delta K_1 + \Delta K_2$ the items ΔK_i exist only for the parts of piezoelement, where longitudinal piezoeffect is employed. Otherwise $\Delta K_i = 0$, as it was discussed previously in Section 5.5. As the result

$$C_{\Sigma sc}^E = C_m^E = 1/(K_m^E + \Delta K)\,. \tag{5.292}$$

In some applications the separated electrodes can be used for control of operating characteristics of a transducer, for feedback, or even for adjustment within certain limits of EME resonance frequency. The latter is due to effect that the electrical output of the transducer produces on parameters of the mechanical system. The effect is determined by value of impedance

$$Z_{int} = n_2^2 / \left(1 / R_{eLd} + j\omega C_{e2}^{S_i}\right) \tag{5.293}$$

introduced into the mechanical contour from the secondary electrical side, as shown in Figure 5.31. With short circuited electrodes $Z_{int} = 0$, with open electrodes $Z_{in} = n_2^2 / j\omega C_{e2}^{S_i}$. This leads to an increase of elastic rigidity, and hence, to an increase of the resonance frequency of the mechanical system.

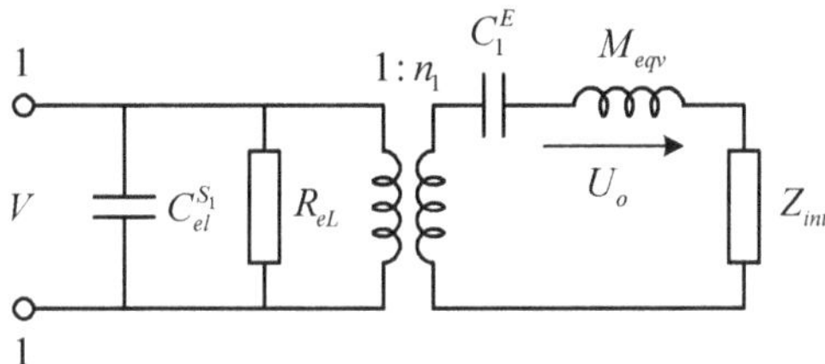

Figure 5.31: Equivalent electromechanical circuit of the electro-mechano-electrical transducer with impedance of the electrical output transformed into mechanical contour.

Piezoceramic transformers represent the most demanding application of EME, because they may perform transmission of significant amounts of energy, whereas other applications do not require large output signals. In the last case analogy can be drawn with electroacoustic receiver, if to replace the source of electrical energy by the acoustomechanical generator that is shown in Figure 1.8. The equivalent circuit of the transformer is analogous to those for electroacoustic projector with radiation impedance replaced by the impedance introduced from the electrical side. Therefore, all the considerations regarding properties of the projectors presented in Chapter 3 are applicable qualitatively to this case. The peculiarity of this case is that it is easier to change impedance of the electrical vs. acoustic load, and thus to achieve more favorable conditions of matching. In particular, the load can be made pure active by using inductance for compensating the reactive component of the electrical output. For example, with parallel inductance $L = 1/\omega_{op}^2 C_{e2}^S$, where ω_{op} is the operating frequency of the transformer, the introduced capacitance in the circuit of Figure 5.31 must be short circuited and introduced resistance will be $R_{int} = n_2^2 R_{Ld}$.

Although a detailed analysis of properties of the piezoelectric transformers is out of scope of this treatment, the piezoelements will be considered shown in Figure 5.32, as examples of those used in the typical transformer designs. The goal is to illustrate procedures of determining their effective coupling coefficients and resonance frequencies. The peculiarity is that the input parameters of the transformer depend on the status of its output. We will consider the cases that they are open (terminated by large impedance) or short circuited (terminated by small impedance). "Large" and "small" are meant being much larger and much smaller in comparison with $1/\omega C_{el}^{S_i}$.

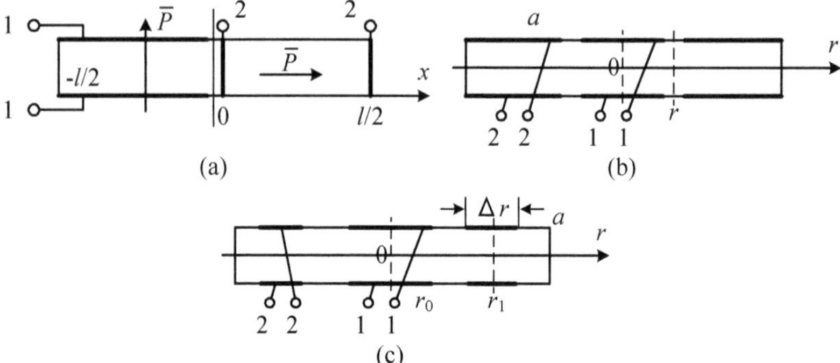

Figure 5.32: Piezoelements that are used in the electromechanical transformer designs: (a) longitudinal vibrating bar that employs both transverse and longitudinal piezoeffect (Rosen-type transformer), (b) radial vibrating in the first mode disk with split electrodes, (c) radial vibrating in the second mode disk with partial electrodes.

6.1.3.1.6. Longitudinally Vibrating Bar Piezoelement

With reference to expressions (5.237), (5.240) and (5.252) after integrating over the halves of the bar that operate in the transverse (input labeled *I*) and longitudinal (output labeled *II*) piezoeffect at the first mode of vibration, the following parameters of the equivalent circuit will be obtained.

The electromechanical transformation coefficients and capacitances are

$$n_I = \frac{d_{31} w}{s_{11}^E}, \quad n_{II} = \frac{2 d_{33} w t}{s_{33}^E l}; \tag{5.294}$$

$$C_{eI}^{S_1} = \varepsilon_{33}^T (1 - k_{31}^2) \frac{wl}{2t} \text{ and } C_{eII}^{S_3} = \varepsilon_{33}^T (1 - k_{33}^2) \frac{2wt}{l}. \tag{5.295}$$

The rigidities under the condition that output of the transducer is short circuited (loaded with a small impedance) are

$$K_{\Sigma sc} = K_m^E + \Delta K_2, \tag{5.296}$$

where

$$K_m^E = \frac{\pi^2 wt}{4l} \left(\frac{1}{s_{11}^E} + \frac{1}{s_{33}^E} \right), \tag{5.297}$$

and ΔK_2 is due to longitudinal piezoeffect in one half of the bar. Using formula (5.155) will be obtained that

$$\Delta K_2 = \frac{\pi^2 wt}{4ls_{33}^E} \frac{k_{33}^2}{1 - k_{33}^2} \left(1 - \frac{8}{\pi^2} \right). \tag{5.298}$$

Thus,

$$K_{\Sigma sc} = \frac{\pi^2 wt}{4l} \left[\frac{1}{s_{11}^E} + \frac{1}{s_{33}^E (1 - k_{33}^2)} \left(1 - \frac{8}{\pi^2} k_{33}^2 \right) \right] = \frac{\pi^2 wt}{4l} \left[\frac{1}{s_{11}^E} + \frac{1}{s_{33}^D} \left(1 - \frac{8}{\pi^2} k_{33}^2 \right) \right]. \tag{5.299}$$

The resonance frequency of the transducer under the condition that opposite sections are short circuited is

$$f_{r\,sc} = \frac{1}{2\pi} \sqrt{\frac{K_{\Sigma sc}}{M_{eqv}}} \tag{5.300}$$

where $M_{eqv} = 0.5wtl\rho$, i.e.,

$$f_{r\,sc} = \frac{1}{2\sqrt{2}\,l} \sqrt{\frac{1}{\rho} \left[\frac{1}{s_{11}^E} + \frac{1}{s_{33}^D} \left(1 - \frac{8}{\pi^2} k_{33}^2 \right) \right]}. \tag{5.301}$$

If output 2 -2 is open circuited (operates in the receive mode with large load)

$$K_{\Sigma ocI} = K_m^E + \Delta K_2 + \frac{n_{II}^2}{C_{eII}^{S_3}} = K_{\Sigma sc} + \frac{n_{II}^2}{C_{eII}^{S_3}}. \tag{5.302}$$

The last term, which is introduced from the secondary electrical circuit, after substituting expressions for n_{II} and $C_{eII}^{S_3}$ from (5.294) and (5.295) may be represented in the form

$$\frac{n_{II}^2}{C_{eII}^{S_3}} = \frac{\pi^2 wt}{4ls_{33}^D}\frac{8}{\pi^2}k_{33}^2. \tag{5.303}$$

Thus,

$$K_{\Sigma ocI} = \frac{\pi^2 wt}{4l}\left[\frac{1}{s_{11}^E}+\frac{1}{s_{11}^D}\right], \tag{5.304}$$

and formula for the resonance frequency becomes

$$f_{r\,sc} = \frac{1}{2\sqrt{2}l}\sqrt{\frac{1}{\rho}\left(\frac{1}{s_{11}^E}+\frac{1}{s_{33}^D}\right)}. \tag{5.305}$$

Determine the effective coupling coefficient of EME transducer in the case that its output is open circuited by formula (5.290). After substituting expressions for energies involved $W_{eI}^{S_1} = 0.5V^2C_{eI}^{S_1}$, $W_{eII}^{S_3} = 0.5(\xi_o n_{II})^2 / C_{eII}^{S_3}$, and $W_{m\Sigma\,oc} = 0.5\xi_o^2 K_{\Sigma ocI}$, taking into consideration that $(V/\xi_o) = (K_{\Sigma ocI}/n_I)$ and after some manipulations we will arrive at

$$k_{eff}^2 = \alpha_{cII}^{oc}\frac{\alpha_{cI}^{oc}}{1+\alpha_{cI}^{oc}}, \tag{5.306}$$

Where

$$\alpha_{cI}^{oc} = \frac{n_I^2}{C_{eI}^{S_1}K_{\Sigma ocI}}, \quad \alpha_{cII}^{oc} = \frac{n_{II}^2}{C_{eII}^{S_1}K_{\Sigma ocII}}. \tag{5.307}$$

The superscripts in coefficients α_c are introduced in order to distinguish their values determined under condition that electrodes on the remaining parts of a piezoelement are open circuited from commonly determined under condition that they are short circuited. Though the difference between these values can be neglected, when determining the coupling coefficients. All the parameters in expressions for coefficients α_{cI}^{oc} and α_{cII}^{oc} are known from relations (5.294), (5.295) and (5.302). The resulting values of these coefficients are

$$\alpha_{cI}^{oc} = \frac{8}{\pi^2}\frac{k_{31}^2}{1-k_{31}^2}\frac{1}{1+s_{11}^E/s_{33}^D}, \quad \alpha_{cII}^{oc} = \frac{8}{\pi^2}k_{33}^2\frac{1}{1+s_{33}^D/s_{11}^E}. \tag{5.308}$$

And the effective coupling coefficient is

$$k_{eff}^2 = \frac{8}{\pi^2}k_{33}^2\frac{1}{1+s_{33}^D/s_{11}^E}k_{effI}^2, \tag{5.309}$$

where k_{effI} is the effective coupling coefficient of electromechanical transducer with input I that has rigidity $K_{\Sigma ocI}$, namely,

$$k_{effI}^2 = \left[1 + \frac{\pi^2}{8}\frac{1-k_{31}^2}{k_{31}^2}(1 + s_{11}^E / s_{33}^D)\right]^{-1} . \quad (5.310)$$

If PZT-8 ceramics is used ($s_{11}^E = 11.5 \cdot 10^{-12}$ m²/N, $s_{33}^E = 8.5 \cdot 10^{-12}$ m²/N, $k_{31} = 0.3$, $k_{33} = 0.64$), then $k_{eff} = 0.08$. Thus, such the EME transducer is a poor energy converter due to double energy transformation.

Important characteristic of EME transducer is its turns ratio . As it follows from the equivalent circuit in Figure 5.31, the maximum value of the turns ratio is achieved with open circuited output, and expression for this value is

$$N_{tr} = \left|\frac{V_{IIoc}}{V_I}\right| = \frac{1}{V_I}\frac{|\dot{\xi}_o| n_{II}}{\omega C_{eII}^{S_3}}, \quad (5.311)$$

where

$$\alpha_{ci} . \quad (5.312)$$

(In course of the manipulations relation $(r_{mL} / K_{\Sigma oc}) = (1/Q_m \omega_{roc})$ was used).

Thus, expression for the turns ratio becomes

$$N_{tr} = \frac{n_I n_{II}}{C_{eII}^{S_3} K_{\Sigma ocI}} Q_m \frac{f_{roc} / f}{\sqrt{1 + Q_m^2[(f / f_{roc}) - (f_{roc} / f)]^2}} . \quad (5.313)$$

Using definitions (5.307) and considering that $K_{\Sigma ocI} \approx K_{\Sigma ocII}$, the first factor in this expression can be represented, as

$$\frac{n_I n_{II}}{C_{eII}^{S_3} K_{\Sigma ocI}} Q_m = \sqrt{\alpha_{cI}^{oc} \cdot \alpha_{cII}^{oc}} \cdot \sqrt{\frac{C_{eI}^{S_1}}{C_{eII}^{S_3}}} \cdot Q_m = N_{tr}(f_{res}) . \quad (5.314)$$

and finally

$$N_{tr} = N_{tr}(f_{roc}) \frac{f_{roc} / f}{\sqrt{1 + Q_m^2[1 - (f_{roc} / f)^2]^2}} . \quad (5.315)$$

Here $N_{tr}(f_{roc})$ is the maximum value of the turns ratio that can be achieved at resonance frequency with the open circuited output. For the particular case under consideration expressions for α_{ci}^{oc} are given by formulas (5.307) and for C_{ei}^S by formulas (5.295). After substituting these expressions into (5.315) will be obtained that

$$N_{tr}(f_{roc}) = \frac{2l}{\pi^2 t} Q_m \cdot k_{33} k_{31} \sqrt{\frac{1-k_{31}^2}{1-k_{33}^2}} \cdot \frac{1}{1+(s_{33}^E / s_{11}^E)(1-k_{33}^2)} . \tag{5.316}$$

(Remember that $s_{33}^D = s_{33}^E(1-k_{33}^2)$ according to (5.102)). The aspect ratio l/t depends on geometry of the piezoelement only, all the rest of the expression isë determined by properties of the ceramics used and by the quality factor of the transducer, in other words by the mechanical losses in the EME transducer design. In the above example with PZT-8 ceramics used

$$N_{tr}(f_{roc}) = 2.7 \cdot 10^{-2}(l/t)Q_m . \tag{5.317}$$

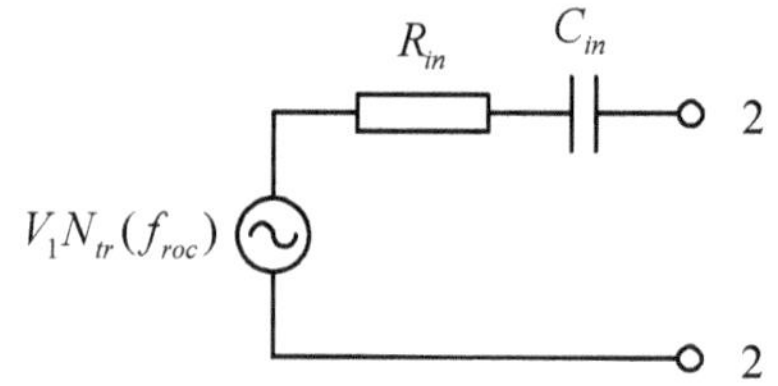

R_{in} C_{in} 2
$V_1 N_{tr}(f_{roc})$ 2

Figure 5.33: Representation of the EME transducer operating at resonance frequency as the equivalent generator.

The EME transducer can be represented as equivalent generator regarding the output 2 -2 that is shown in Figure 5.33 for the case that the transducer operates at the resonance frequency. According to the Thevenin's theorem the electromotive force of this generator is $V_{2oc} = V_1 N_{tr}(f_{roc})$, and the internal impedance must be determined as impedance between terminals 2 - 2 with terminals 1 - 1 short circuited. It can be shown after straightforward manipulations that at the resonance frequency $R_{in} = Q_m \alpha_{cII} / \omega_{roc} C_{eII}^{S_3}$, and $C_{in} \approx C_{eII}^{S_3}$. Thus, the rated power of EME transducer (maximum output power that can be delivered to the matched load $R_{Ld} = R_{in}$) is

$$\dot{W}_{\max} = \frac{[V_I N_{tr}(f_{roc})]^2}{4R_{in}} = \frac{1}{4} V_I^2 \frac{N_{tr}^2(f_{roc})}{Q_m \alpha_{cII}^{oc}} \omega_{roc} C_{eII}^{S_3} , \tag{5.318}$$

where parameter that depends on the EME converter properties may be denoted $\gamma_{\dot{W}}$ is

$$\gamma_{\dot{W}} = \frac{N_{tr}^2(f_{roc})}{Q_m \alpha_{cII}^{oc}} \omega_{roc} C_{eII}^{S_3} . \tag{5.319}$$

In the above considered example with coefficient $N_{tr}(f_{roc})$ by formula (5.317) this parameter is

$$\gamma_{\dot{W}} = 7.3 \cdot 10^{-4}(l/t)^2 Q_m \frac{\omega_{roc} C_{eII}^{S_3}}{\alpha_{cII}^{oc}}. \quad (5.320)$$

Values of $C_{eII}^{S_3}$, ω_{roc} and α_{cII}^{oc} are expressed through parameters of piezoelement by formulas (5.295), (5.105) and (5.308), respectively.

Seemingly EME transducers can transform significant electrical power, but the maximum (rated) power is limited by dynamic mechanical stress in the piezoelement and its heating, as the equal power is dissipated in the form of mechanical losses. The power limited by the mechanical strength can be estimated for the case under consideration as follows. Given that at the first mode of vibration $\xi(x) = \xi_o \sin(\pi x/l)$ and the strain is $S(x) = (\xi_o \pi/l)\cos(\pi x/l)$, at the resonance frequency

$$T_{\max} = Y \frac{\pi \dot{\xi}_o}{\omega_{roc} l}. \quad (5.321)$$

Here Y would be Young's modulus of material in case the elastic properties along the mechanical system were uniform. In our example the elastic modulus is $1/s_{11}^E$ within one half of the bar and $1/s_{33}^D$ within another. For estimations that involve value of maximum permissible dynamic stress, T_{pd}, which is known approximately, it may be acceptable to use for the Young's modulus the average value $Y_{av} = 0.5[(1/s_{11}^E) + (1/s_{33}^D)]$. Thus, the maximum power limited by the dynamic strength is

$$\dot{W}_{mT} = \dot{\xi}_o^2 r_{mL} = \frac{\omega_{roc}^2 l^2}{\pi^2 Y_{av}^2} r_{mL} T_{pd}^2. \quad (5.322)$$

Note that here $\dot{W}_{mT}$ is calculated on the mechanical side of the equivalent circuit, because the matched load transformed into mechanical contour is equal to r_{mL}, besides

$$r_{mL} = \frac{\omega_{roc} M_{eqv}}{Q_m} = \frac{\pi f_{roc} \tilde{V} \rho}{Q_m}, \quad (5.323)$$

where $\tilde{V}$ is volume of the piezoelement. Assuming also that $f_{roc} \approx (1/2l)\sqrt{Y_{av}/\rho}$, we finally obtain for the maximum power density

$$(\dot{W}_{mT}/\tilde{V}) = \frac{\pi}{Q_m Y_{av}} f_{roc} T_{pd}^2 \ \text{W/m}^3. \quad (5.324)$$

According to Ref. 2 (see also Ch. 11) it can be taken $T_{pd} \approx 25$ MPa. For PZT-4 and PZT-8 ceramics $Y_{av} \approx 10^{11}$ N/m^2. Thus, with these ceramics used

$$(\dot{W}_{\max} / \tilde{V}) \leq \frac{20}{Q_m} f_{r(kHz)} \text{ W/cm}^3. \tag{5.325}$$

It must be remembered that this power can be available at efficiency of electro-electrical energy conversion $\eta_{ee} = 0.5$ .This may result in significant overheating of the piezoelement. Increase of efficiency can be achieved for expense of the power converted. The trade-off between the efficiency and power available is the matter of a particular piezoelectric transformer design.

6.1.3.1.7. Radially Vibrating Circular Disk

We will consider only the electromechanical characteristics of the disk that are important from the point of view of analysis performed above for the longitudinally vibrating bar piezoelement under the assumption that the disk vibrates in the first mode and the electrodes are split at variable radius r (Figure 5.32 (b) in order to determine dependence of parameters of the piezoelement on the radius. Electromechanical characteristics of the radially vibrating circular disk with partial electrodes were considered in the preceding Section 5.7.4.2 under the condition that electrodes on the remaining part of the disk exist and are short circuited. Specifics of the piezoelement for EME conversion is that the output electrodes are assumed to be open circuited to be able to treat the case of arbitrary loaded output. In this section data are presented regarding dependencies of relative values of coefficients that characterize effective coupling coefficients associated with the partial electrodes from the relative widths $\Delta r / r_{in}$ of electrodes located around the nodal lines of higher (with number n) modes of vibration. These data allow one to estimate whether is it reasonable to use higher modes of vibration for designing EME with partial electrodes of the type shown in Figure 5.32 (c).

As both input and output parts of the circular disk piezoelements employ transverse piezoeffect, the equivalent circuit in Figure 5.30 simplifies to those shown in Figure 5.34 for open circuited output. Status of opposite electrical sides does not influence values of their electromechanical transformation coefficients and capacitances. According to expression (5.278) for the inner part confined between the central electrodes

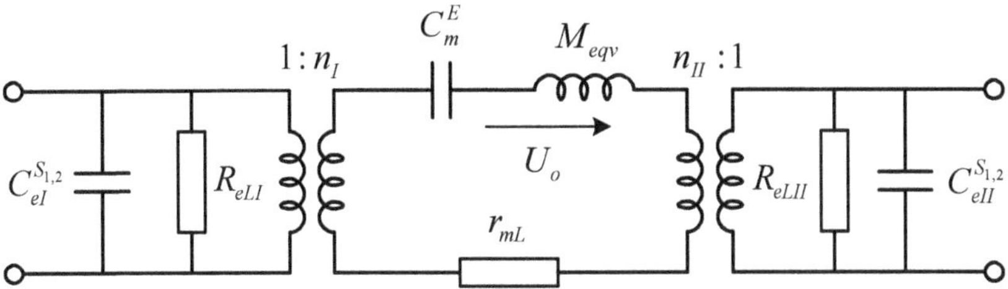

Figure 5.34: Equivalent circuit of EME radial disk converter with open circuited output.

$$n_I = \frac{2\pi d_{31} a Y_1^E}{1-\sigma_1^E} \frac{r}{a} \theta_1\left(\frac{r}{a}\right), \tag{5.326}$$

and for the outer part

$$n_{II} = \frac{2\pi d_{31} a Y_1^E}{1-\sigma_1^E}\left[1-\frac{r}{a}\theta_1\left(\frac{r}{a}\right)\right]. \tag{5.327}$$

The respective capacitances are

$$C_{eI}^{S_{1,2}} = \varepsilon_{33}^T(1-k_p^2)\pi r^2/t \text{ and } C_{eII}^{S_{1,2}} = \varepsilon_{33}^T(1-k_p^2)\pi(a^2-r^2)/t. \tag{5.328}$$

The equivalent rigidity for the case that the opposite electrical sections are short circuited according to (5.271) is

$$K_m^E = \frac{10tY_1^E}{1-\sigma_1^{E2}}. \tag{5.329}$$

With output electrodes open circuited

$$K_{m\,ocII} = K_m^E + \frac{n_{II}^2}{C_{eII}^{S_{1,2}}} = K_m^E\left(1+\frac{n_{II}^2}{C_{eII}^{S_{1,2}} K_m^E}\right). \tag{5.330}$$

After introducing the coefficients

$$\alpha_{cI} = \frac{n_I^2}{C_{eI}^{S_{1,2}} K_m^E} \text{ and } \alpha_{cII} = \frac{n_{II}^2}{C_{eII}^{S_{1,2}} K_m^E}, \tag{5.331}$$

these expressions become

$$\begin{aligned} K_{m\,ocI} &= K_m^E\left(1+\frac{n_{II}^2}{C_{cII}^{S_{1,2}} K_m^E}\right) = K_m^E\left(1+\alpha_{cII}\right), \\ K_{m\,ocII} &= K_m^E\left(1+\frac{n_I^2}{C_{eI}^{S_{1,2}} K_m^E}\right) = K_m^E\left(1+\alpha_{cI}\right). \end{aligned} \tag{5.332}$$

Here $K_{m\,oci}$ ($i = I, II$) are the rigidities determined under condition that electrodes on the opposite part are open circuited. Note that coefficients α_{ci}^{oc} in the previous case were determined by formula (5.307) under the condition that opposite terminals are open circuited. Following expressions (5.326)-(5.329),

$$\alpha_{cI} = A\theta_1^2\left(\frac{r}{a}\right), \quad \alpha_{cII} = A\frac{1}{1-(r/a)^2}\left[1-\frac{r}{a}\theta_1\left(\frac{r}{a}\right)\right]^2, \tag{5.333}$$

where

$$A = \frac{\pi}{5}(1+\sigma_1^E)\frac{k_p^2}{1-k_p^2}. \tag{5.334}$$

For PZT-4 and PZT-8 ceramics $A = 0.42$ and 0.29.

The resonance frequency of the piezoelement measured at the input terminals with output terminals open circuited is

$$\omega_{rI} = \sqrt{\frac{K_{mocI}}{M_{eqv}}} = \sqrt{\frac{K_m^E}{M_{eqv}}(1+\alpha_{cII})} = \omega_{rsc}\sqrt{1+\alpha_{cII}}, \tag{5.335}$$

where ω_{rsc} is the resonance frequency of the disk fully covered by electrodes, namely,

$$\omega_{rsc} = \frac{2.05}{a}\sqrt{\frac{Y_1^E}{\rho(1-\sigma_1^{E2})}}. \tag{5.336}$$

Note that the resonance frequency measured at the output terminals with input open circuited is

$$\omega_{rII} = \omega_{rsc}\sqrt{1+\alpha_{cI}}. \tag{5.337}$$

The coefficients α_{ci} are related to the effective coupling coefficients determined under condition that the opposite terminals are short circuited by formula (5.280), thus, α_{cI} corresponds to the effective coupling coefficient of the central part, and α_{cII} to the peripheral part.

Plots are presented in Figure 5.35 of dependences of coefficients α_{cI} and α_{cII} normalized to factor A from relative inner radius of the electrodes. It must be remembered that

$$\xi(r/a) = \xi_o\theta_1(r/a) = \xi_o\frac{J_1(2.05r/a)}{J_1(2.05)} = 1.74\xi_o J_1(2.05r/a). \tag{5.338}$$

Thus, relation between the effective coupling coefficients of the input and output sections of the disk can be changed in a broad range by changing inner radius of the electrodes. At $r/a \approx 0.46$ the effective coupling coefficients are equal. For PZT-4 they are $k_{eff} \approx 0.43$.

Expression for turns ratio of EME converter at resonance frequency that is obtained from the general formula (5.314) in this case looks like

$$N_{tr}(f_{roc}) = Q_m \frac{n_I n_{II}}{C_{eII}^{S_{1,2}} K_m^E (1+\alpha_{cII})} = Q_m \sqrt{\frac{C_{eI}^{S_{1,2}}}{C_{eII}^{S_{1,2}}}} \frac{\sqrt{\alpha_{cI} \cdot \alpha_{cII}}}{1+\alpha_{cII}}. \tag{5.339}$$

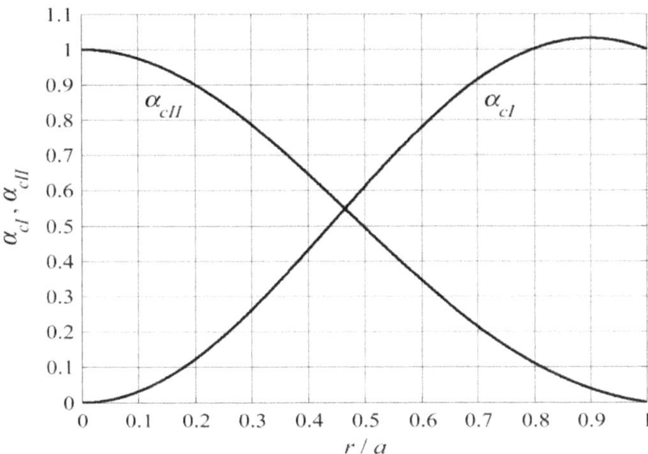

Figure 5.35: Dependences of coefficients α_{cI} and α_{cII} normalized to factor A from relative inner radius of the electrodes.

General formula for the turns ratio may be obtained after substituting expressions (5.328) for the capacitances and (5.333) for the coefficients α_{ci}. At $r = 0.46a$, $\alpha_{cI} = \alpha_{cII} = 0.53A$ and

$$N_{tr}(f_{roc}) = 0.52 Q_m \frac{0.53A}{1+0.53A}. \tag{5.340}$$

In case that PZT-4 or PZT-8 ceramics are used (with $A = 0.42$ and 0.29), $N_{tr}(f_{roc}) \approx 9 \cdot 10^{-2} Q_m$ and $N_{tr}(f_{roc}) \approx 7 \cdot 10^{-2} Q_m$, respectively.

In determining the maximum available output power, the same procedure can be used as was demonstrated for the bar piezoelement. In the factual calculations some changes in values of parameters involved must be made as follows. In formula (5.323) for r_{mL} the equivalent mass is $M_{eqv1} = 0.76M$ according to (5.265), and the resonance frequency ω_{roc} has to be determined by formula (5.335). Following the expression (5.338) the maximum stress in the center of the disk is

$$T_{\max} = Y_1^E \left.\frac{d\xi(r/a)}{dr}\right|_{r=0} = \frac{3.6}{a} Y_1^E \xi_o \,, \tag{5.341}$$

In practical designing an advantage may have EME converters that employ the second modes of vibration of the considered piezoelements. As the corresponding equivalent parameters were determined in the previous sections, calculations analogous to those performed here can be made in a straightforward way.

Considering issues related to principles of EME transducer designing is out of scope of this treatment. The only goal of analysis made in this section is to illustrate technique of estimations of properties of a piezoelement employed in such transducers that is based on their equivalent circuit representation. Information about different approaches to calculating various piezoelectric transformer designs, and detailed analyses of their parameters can be found in vast literature on these issues. See, for example, Ref. 9 and bibliography therein.

5.8 Equations of Vibration of Piezoceramic Bodies in Geometrical Coordinates

Solution to the problem of vibration of a piezoceramic body in the generalized coordinates is rather general and, in essence, comes down to a formal procedure of determining the equivalent parameters, if solution is known to the vibration problem for the identical body made of passive material. Yet there are cases, when it is preferable to derive equations of vibration for piezoceramic bodies directly in geometrical coordinates. For this purpose, we will use the variational principle in the same way, as it was done for bodies made from passive materials in Chapter 4. Initially, we will consider the entire volume of the body to be confined between solid electrodes so that there are no nonuniformities caused by the absence of electrodes on some part of the body, or by a difference in electric conditions on the parts of the electrodes that are electrically insulated from each other. Several peculiarities exist in applying the variational principle to deriving equations of motion for the piezoceramic bodies.

Firstly, we must consider density of the internal energy w_{int} instead of w_{pot}, therefore, expression for the Lagrangian will be (see (1.96))

$$L = w_{kin} - w_{int} + w_e \,. \tag{5.342}$$

When considering expression (5.34) for the internal energy, $w_{int} = w_e^{S_i} + w_{mech}^E$, it should be remembered that in calculating $w_e^{S_i} = \varepsilon_{33}^{S_i} E_3^2 / 2$ for a volume element located inside the piezoceramic body, E_3 must be replaced with E_3' in accordance with relation (5.133). (Note that we will further use subscript m instead of $mech$ for brevity, thus, the formula for the internal energy will be $w_{int} = w_e^{S_i} + w_m^E$).

Secondly, although the mechanical effects must be considered in the same way as for the passive bodies, differences arise in setting the boundary conditions, which must be formulated with regard to the piezoelectric effect. Besides, the electrical boundary conditions should also be considered.

For the bodies, motion of which is governed by equation (4.41) (bars in the longitudinal and torsional vibrations, plates in vibration in their plane), in the case that they are made from the active materials we will use Euler's equations in the form of (4.2)

$$\frac{d}{dt}\left(\frac{\partial w_{kin}}{\partial \dot{\xi}}\right) - \frac{\partial}{\partial x}\left(\frac{\partial w_{int}}{\partial \xi_x'}\right) = \frac{\partial w_e}{\partial \xi} \,, \tag{5.343}$$

where displacement ξ is a function of the geometry coordinates

5.8.1 Extensional and Torsional Vibrations of Piezoceramic Bars

5.8.1.1 Equations of Motion

We consider one-dimensional vibrations of the bars. Following formulas (5.132) and (5.133) in rectangular coordinates we obtain

$$D_3 = \frac{C_{e\Delta}^{S_i} V}{l_e} + \frac{n_\Delta}{l_e} \int_0^{l_e} S_i(x_i) dx_3 \,, \tag{5.344}$$

$$E_3' = E_3 + \frac{n_\Delta}{C_{e\Delta}^{S_i} l_e} \int_0^{l_e} S_i(x_i) dx_3 - \frac{n_\Delta}{C_{e\Delta}^{S_i}} S_i(x_i) \,, \tag{5.345}$$

where $S_i(x_i) = \xi'(x_i)$ is the working strain and l_e is separation between electrodes. For the transverse piezoelectric effect $S_i(x_i) - S_1(x_1)$, $l_e = t$, $E_3' = E_3$. For the longitudinal piezoelectric effect $S_i(x_i) = S_3(x_3)$, and at $l_e = l$, $E_3' \neq E_3$. Inside of a body that is segmented in the direction of the electrical lines of force, the expression for E_3' in rectangular coordinates within a segment of the length Δx_3, becomes at $\Delta x_3 \to 0$

$$E_3' = E_3 + \frac{n_\Delta}{C_{e\Delta}^{S_3} \Delta x_3} \int_{x_3}^{x_3+\Delta x_3} S_3(x_3) dx_3 - \frac{n_\Delta}{C_{e\Delta}^{S_3}} S_3(x_3) \approx E_3 + \frac{n_\Delta}{C_{e\Delta}^{S_3}} [S_3(x_3) - S_3(x_3)] = E_3 . \quad (5.346)$$

Actually, for fulfilment of this condition it is practically sufficient to have not less than 6 segments on half of wavelength of deformation, as it is shown in Section 5.5.2. The two variants differ in principle.

There is no distribution of strain $S = \xi'$ along the lines of force, and $E_3' = E_3$ (the transverse piezoelectric effect, the longitudinal piezoelectric effect in a body segmented along the lines of force as noted above). In this case

$$\frac{\partial w_{\text{int}}}{\partial \xi'} = \frac{\partial}{\partial \xi'} \left(w_e^{S_i} + w_m^E \right) = \frac{\partial w_m^E}{\partial \xi'} = K_\Delta^E \xi' , \quad (5.347)$$

and for all the bodies of this kind equation (5.343) coincides with Eq. (4.2) for analogous body made from a passive material, if w_{int} is substituted into the equation instead of w_{pot}. Therefore, all the equations of the kind of Eq. (4.40) also coincide, except that K_Δ should be replaced with the respective coefficients K_Δ^E taken from Table 5.3, so that

$$c^2 = K_\Delta^E / m_\Delta = (c^E)^2 . \quad (5.348)$$

There exists a strain distribution along the lines of force, E_3' and $w_{el}^{S_i}$ depend on ξ'. These are the cases of thickness vibrations of plates and vibrations of solid bars under the longitudinal piezoelectric effect. Substituting the value of E_3' from formula (5.346) into the expression for $w_e^{S_i}$ and using relations (5.95) and (5.100) in the process of manipulation, we

$$\frac{\partial}{\partial x} \left(\frac{\partial w_{\text{int}}}{\partial \xi'} \right) = \frac{\partial}{\partial x} \left[\frac{\partial \left(w_e^{S_i} + w_m^E \right)}{\partial \xi'} \right] = K_\Delta^E \xi'' + \frac{n_\Delta^2}{C_{e\Delta}^{\varepsilon_i}} \xi'' = K_\Delta^D \xi'' . \quad (5.349)$$

Thus, in this case the equations of motion also coincide with those for similar passive bodies under the action of the same system of external forces, if to replace K_Δ by K_Δ^D, which means that c^2 has to be replaced by

$$(c^D)^2 = K_\Delta^D / m_\Delta . \quad (5.350)$$

Since the forces of electromechanical origin do not exist in the equations of motion, they must appear in boundary conditions, so far as the electromechanical conversion does take place.

5.8.1.2 Boundary Conditions

6.1.3.1.8. Mechanical boundary conditions

At first, we will consider the ideal mechanical boundary conditions, under which no flux of mechanical energy flows through the boundaries, i.e., $W_e = 0$. In the variant of the longitudinal vibrations these conditions result from the relation $W_e = f\xi\big|_{x_i=0,l} = 0$, in the variant of the torsional vibration – from $W_e = M_f\varphi\big|_{x=0,l} = 0$. At the free end the displacements and rotations are possible, and the conditions $f = 0$, $M_f = 0$ must be met. Considering Eq. (5.84) in the variant of the longitudinal vibrations

$$f = S_{cs}T\big|_{x=0,l} = S_{cs}(K_\Delta^E\xi' - n_\Delta E_3)_{x=0,l}, \tag{5.351}$$

where from

$$\xi'\big|_{x=0,l} = E_3 n_\Delta / K_\Delta^E. \tag{5.352}$$

Values of n_Δ / K_Δ^E must be taken from Table 5.3 for a respective case. Thus, $(n_\Delta / K_\Delta^E) = d_{33}$, or d_{31}, or e_{33} / c_{33}^E for a bar under the longitudinal or transverse piezoelectric effect, and for a plate vibrating through the thickness, respectively.

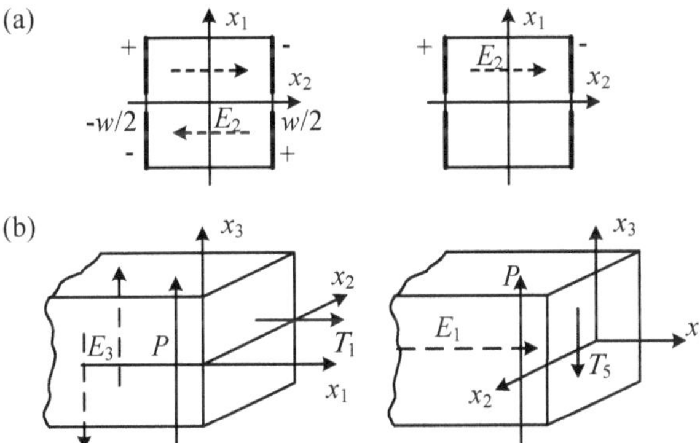

Figure 5.36: To the formulation of boundary conditions for a piezoceramic bar intended to vibrate: (a) in torsion (on the left-fully active bar, on the right-half passive); (b) in flexure (on the left - employing extensional deformations, on the right - employing shear deformations).

In the variant of the torsional vibration of a piezoceramic bar of square cross section poled along its axis x_3 perpendicular to the cross section, as shown in Figure 5.36 (a).

$$M_f = w\int_{-w/2}^{w/2} T_4 x_1 dx_1 + w\int_{-w/2}^{w/2} T_5 x_2 dx_2 \,. \tag{5.353}$$

(For the energy status of volume element of passive bar under the torsional deformation see Sec. 4.2.2, example 6). If there are no electrodes on the side surfaces of the bar, $T_4 = S_4 / s_{44}^D$ and $T_5 = S_5 / s_{44}^D$. By virtue of Eq. (4.24), $S_4 = -x_1\varphi'_{x_3}$ and $S_5 = x_2\varphi'_{x_3}$. Thus

$$M_f = \left(\varphi'_{x_3} / s_{44}^D\right) w \left(\int_{-w/2}^{w/2} x_1^2 dx_1 + \int_{-w/2}^{w/2} x_2^2 dx_2 \right) = \varphi'_{x_3} (J_p / s_{44}^D) \,. \tag{5.354}$$

It follows from the condition $M_f = 0$ that $\varphi'_{x_3} = 0$. If electrodes are applied to the side surfaces $x_2 = \pm w/2$ and $E_2 \neq 0$, then

$$T_4 = S_4 / s_{44}^E - (d_{15} / s_{44}^E) E_2, \quad T_5 = S_5 / s_{44}^D \,. \tag{5.355}$$

After substituting these expressions for stresses into formula (5.353), we obtain

$$M_f = \varphi'_{x_3} \frac{J_p}{2}\left(\frac{1}{s_{44}^E} + \frac{1}{s_{44}^D} \right) + M_{em} = G^E \varphi'_{x_3} + M_{em} \,, \tag{5.356}$$

where

$$M_{em} = -\frac{w d_{15}}{s_{44}^E} \int_{-w/2}^{w/2} E_2 x_1 dx_1 \tag{5.357}$$

is the moment of electromechanical origin, $G^E = J_p (s_{44}^E + s_{44}^D) / 2 s_{44}^E s_{44}^D$ is the torsional rigidity of a piezoceramic bar of the square cross section. (Sign minus shows that at the positive direction of vector $\boldsymbol{E}_2$ the generated moment is acting in the anticlockwise direction).

If the electrodes are applied and connected electrically in such a manner that $E_2(x_1)$ is not an even function, as this is shown in Figure 5.36 (a), then $M_{em} \neq 0$. In the case that two halves of the electrodes are connected in opposite, $M_{em} = -w^3 d_{15} E_2 / 4 s_{44}^E$. Given that for square cross section $J_p = w^4 / 6$, from condition $M_f = 0$ we find

$$\varphi'_{x_3} = 3 s_{44}^D d_{15} E_2 / w (s_{44}^E + s_{44}^D) \,. \tag{5.358}$$

Thus, the effect of electromechanical conversion is equivalent to those produced by the force $f_{em} = S_{cs} n_\Delta E_3$ in case of the longitudinal vibration or by the moment M_{em} in case of torsion, which are acting at the end of the piezoceramic bar. The electric energy that enters transducer through the electrodes appears as would be transformed due to electromechanical conversion into mechanical energy that enters through the ends. These energies are $W_{em} = S_{cs} n_{\Delta} E_3 \xi\big|_{x=0,l}$ and $W_{em} = M_{em}\varphi\big|_{x=0,l}$ for the longitudinal and torsional vibrations, respectively.

On the fixed end the conditions $\xi = 0$ and $\varphi = 0$ coincide with those for a passive body. No transformation of the electric energy into the mechanical energy that flows through the fixed end takes place, since at this end $f_{em}\xi = 0$ and $M_{em}\varphi = 0$. Thus, piezoceramic bar with fixed ends and with unipolar electrodes that fully cover the side faces is electromechanically passive. This fact was noted also, when considering examples in Section 5.8, where an explanation was made based on analysis of the quantity W_{em} by representing strains, as expansion in terms of natural vibration modes (see Figure 5.23). In order to make the electromechanical conversion efficient under the fixed boundary conditions, or in order to increase its efficiency under conditions (5.349) and (5.358), the electrodes configuration and electrical connections must be modified, as it was recommended in Section 5.8. The simplest way to achieve this is just to remove the portions of electrodes near the edges, or to separate the electrodes at the nodal lines of the strain distribution and connect their adjacent portions in reversed polarity. In both cases the equations of vibration must be written for each segment of the body with changed electrical boundary conditions, and mating conditions must be met at the boundaries between the segments. Thus, for example, for bars of the design shown in Figure 5.22 equations of vibration for segments 1 and 3 must be formulated as for the passive parts with propagation speed $c = \sqrt{Y / \rho}$, where Y and ρ are parameters of not polarized ceramics, strictly speaking. For segment 2 the propagation speed has to be determined, as for the active piezoelement: c^E for the designs in Figure 5.22 (a) and (c), and c^D for the design in Figure 5.22 (b). (Note that Figure 5.22 (a) is replicated as Figure 5.37 (a)). Analogous approach can be used regarding more general transducer design, in which case the passive parts of a bar have different cross sections, as shown in Figure 5.37 (b), under the assumption that these parts vibrate in the piston like mode (uniformly in plane of a cross section). Materials of the passive parts can also be different.

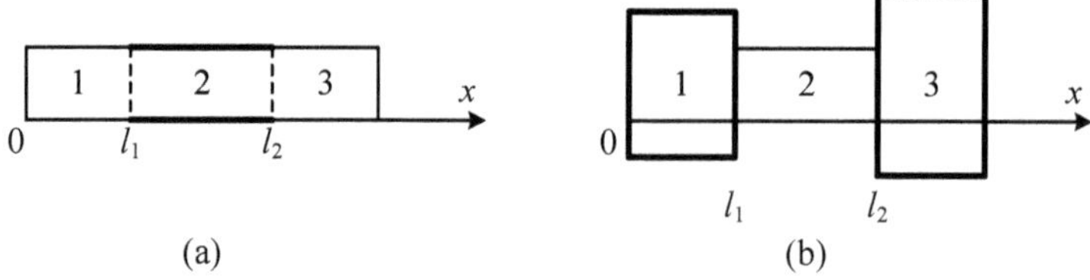

Figure 5.37: The transducer design with passive parts having the same (a) and different (b) cross sections as the piezoelement has.

For this design the mating conditions for the longitudinal vibrations are

$$\xi_1(l_1) = \xi_2(l_1), \quad \xi_2(l_2) = \xi_3(l_2), \tag{5.359}$$

$$S_{cs1}T_1(l_1) = S_{cs2}T_2(l_2), \quad S_{cs2}T_2(l_2) = S_{cs3}T_3(l_2). \tag{5.360}$$

In the example of Figure 5.37 (a) $S_{cs1} = S_{cs2} = S_{cs3}$. Due to the piezoelectric effect in the middle segment of the bar, T_2 should be used in the form that for a general case is represented by expression (5.84). If this is taken into account, then for the particular case of a bar the conditions (5.360) will be transformed to

$$S_{cs1}K_\Delta \xi_1'(x)\big|_{x=l_1} = S_{cs2}\left[K_\Delta^E \xi_2'(x)\big|_{x=l_1} - n_\Delta E_3'\right], \tag{5.361}$$

$$S_{cs3}K_\Delta \xi_3'(x)\big|_{x=l_2} = S_{cs2}\left[K_\Delta^E \xi_2'(x)\big|_{x=l_2} - n_\Delta E_3'\right]. \tag{5.362}$$

For the design shown in Figure 5.22 (b) (with longitudinal piezoeffect on the active part) in these expressions K_Δ^E should be replaced by K_Δ^D.

In the variant of the torsional vibrations displacement ξ in the mating conditions must be replaced by the angle φ, and the force $f = S_{cs}T$ must be replaced by the moment M_f. Finally, for the cross section with coordinate $x = l_1$ condition (5.362) becomes

$$(J_{p1} / s_{44})\varphi_1'(x)\big|_{x=l_1} = \left[J_{p2}(s_{44}^E + s_{44}^D) / 2s_{44}^E s_{44}^D\right]\varphi_2'(x)\big|_{x=l_1} + M_{em}, \tag{5.363}$$

where $(1/s_{44}) = \mu$ is the shear modulus of the passive material.

6.1.3.1.9. Electrical boundary conditions

Electrical boundary conditions may be ideal, if $\overline{W}_{el} = VI^* = 0$, and not ideal, if $W_{el} \neq 0$. In general, using relation (5.344) we obtain for one-dimensional longitudinal vibrations of a solid piezoelement

$$I = \int_{S_{el}} \dot{D}_3 dS_{el} = C_e^{S_i}\dot{V} + \frac{n_\Delta S_{el}}{l_e}\left[\dot{\xi}(l_2) - \dot{\xi}(l_1)\right] = j\omega C_e^{S_i} V + \frac{n_\Delta S_{el}}{l_e}\left[U(l_2) - U(l_1)\right], \quad (5.364)$$

where $l_e = l_2 - l_1$, and S_{el} is area of the electrode. If to substitute values of the velocities $U(l_1)$ and $U(l_2)$ into this equation, the input admittance of a transducer can be found, as $Z_{in} = V/I$.

Under the torsional vibration strains S_4 do not change in the direction of lines of force of the electric field E_2, therefore, $E_2' = E_2$, and expression for D_2 coincides with those in local piezoelectric equation, namely,

$$D_2 = (d_{24}/s_{44}^E)S_4 + \varepsilon_{11}^T E_2 = (d_{24}/s_{44}^E)\varphi'(x_3)x_1 + \varepsilon_{11}^T E_2 . \quad (5.365)$$

Considering that halves of the electrodes in Figure 5.36 (a) are connected in parallel, we obtain

$$I = \int_{S_{el}} \dot{D}_2 dS_{el} = \frac{\varepsilon_{11}^T S_{el}}{w}\dot{V} + \frac{w^2 d_{24}}{4 s_{44}^E}\left[\dot{\varphi}(l_2) - \dot{\varphi}(l_1)\right]. \quad (5.366)$$

Results for the ideal electric boundary conditions can be obtained from expressions (5.364) and (5.366). Thus, with open electrodes $I = 0$, and substituting $V = V_{oc}$ in relation (5.364), we arrive at

$$V_{oc} = -\frac{n_\Delta S_{el}}{l_e C_e^{S_i}}\left[\dot{\xi}(l_2) - \dot{\xi}(l_1)\right] = \frac{n_\Delta S_{el}}{l_e C_e^{S_i}})\left[U(l_1) - U(l_2)\right]\xi . \quad (5.367)$$

Note that condition $I = 0$ is not equivalent to D_3 = constant, since equalizing currents can flow in a piezoelectric element with open electrodes, as they do in the case of the transverse piezoelectric effect. With the electrodes short circuited it should be assumed that $V = 0$ in formula (5.364). In this case $E_3 = 0$ (recall that E_3 is the electric field generated in a clamped body by voltage applied between the electrodes). Under the transverse piezoelectric effect and in a sufficiently segmented piezoelement $E_3' = E_3 = 0$. Under the longitudinal piezoelectric effect in a solid (not segmented) piezoelement $E_3' \neq 0$, as it follows from the relation (5.133) by assuming that $E_3 = 0$.

5.8.2 Equations of Flexural Vibrations and of the Radial Vibrations of a Circular Disk

Equations for the flexural vibrations of piezoceramic beams and plates and for the radial vibrations of the piezoceramic disks can be derived directly from the variational principle in the form of (4.42), with Lagrangian taken in the form (5.342). Equations of motion can be obtained by using procedure analogous to that used for deriving Eqs. (4.48) and (4.188) for flexural vibrations of beams and circular plates and Eq. (4.160) for radial vibration of the circular disk. As a result it may be concluded that in these cases equations of vibration for the piezoceramic bodies fully coincide with the equations for the passive bodies of the same configuration, if to replace in the latter K_Δ by K_Δ^E (c by c^E) for the transverse piezoelectric effect and for the longitudinal piezoeffect in the beams in case that they are sufficiently segmented along the electrical lines of force. In all the cases forces of electromechanical origin are not presented in the equations of vibration. They are accounted for by the boundary conditions that must be formulated regarding the piezoelectric effect.

5.8.2.1 Mechanical Boundary Conditions

The ideal mechanical boundary conditions for Eq. (4.160) of radial vibration of the piezoceramic disk may be obtained assuming that there is no energy flux through the edge of the disk, i.e., $W_e = f\xi_r(a) = 0$, where from follows that either $\xi_r(a) = 0$, if the edge is fixed, or $f = S_b T_{rr} = 0$, if it is free. Here S_b is the area of the boundary surface of the disk. Considering Eqs. (5.11) and (5.12) we obtain condition for $T_{rr} = T_1$ in the form

$$T_1 = \frac{s_{11}^E}{s_{11}^{E2} - s_{12}^{E2}}\left(S_1 - \frac{s_{12}^E}{s_{11}^E} S_2\right) - \frac{d_{31}E_3}{s_{11}^E + s_{12}^E} = T_{em} - \frac{d_{31}E_3}{s_{11}^E + s_{12}^E} = 0\,. \tag{5.368}$$

Thus, the edge that is free from external action would be under the action of forces of the electromechanical origin $f_{em} = S_b T_{em}$, and electric energy supplied to the disk is transformed into mechanical energy

$$W_{em} = S_b d_{31} E_3 \xi_r(a) / (s_{11}^E + s_{12}^E) \tag{5.369}$$

that flows into the disk through its boundary surface. With the fixed boundary $W_{em} = 0$ due to $\xi_r(a) = 0$, and the disk is electromechanically passive.

Consider now the boundary conditions for a beam in flexure (Eq. (4.48)). The ideal mechanical boundary conditions at the ends of the beam follow from the condition that an external flux of mechanical energy is absent. Thus, from expression (4.52)

$$W_e = M_f \xi' + Q\xi = 0\,. \tag{5.370}$$

The values of moment and shearing force M_f and Q (see Figure 5.36 (b)) are related to mechanical stresses at the ends of a beam according to formulas

$$M_f = w\int_{-t/2}^{t/2} T_1 x_3 dx_3, \quad Q = w\int_{-t/2}^{t/2} T_5 dx_3\ , \tag{5.371}$$

where T_1 and T_5 are the normal and shear stresses. It follows from Eq. (5.84) that for the transverse piezoelectric effect $T_1 = S_1 / s_{11}^E - d_{31}E_3(x_3)/s_{11}^E$, where $S_1 = -x_3\xi_3''$ (see formula (4.28)). Thus, we obtain

$$M_f = (J/s_{11}^E)\xi_3'' - \frac{wd_{31}}{s_{11}^E}\int_{-t/2}^{t/2} E_3(x_3)x_3 dx_3\ . \tag{5.372}$$

The second term vanishes, and no electromechanical conversion takes place, if $E_3(x_3)$ is an even function. The function $E_3(x_3)$ must be non-symmetrical relative to the central plane of a beam. If the beam is mechanically uniform, this function must be odd. This can be achieved, if the bar consists of two bonded layers connected so that directions of vectors $\boldsymbol{E}$ and $\boldsymbol{P}$ coincide in one half and are opposite in the other half, e.g., as it is shown in Figure 5.36 (b). Assuming that $E_3(x_3) = E_3(-x_3)$, we obtain

$$M_f = (J/s_{11}^E)\xi_3'' - d_{31}E_3 t^2/4s_{11}^E = (J/s_{11}^E)\xi_3'' - M_{em}\ . \tag{5.373}$$

In the expression (5.371) for Q the stress T_5 must be substituted from Eq. (5.8), i.e.,

$$T_5 = c_{44}^E S_5 - e_{15}E_1\ , \tag{5.374}$$

where following Eq. (4.33)

$$S_5 = (t^2/4 - x_3^2)\xi_3'''/2\ . \tag{5.375}$$

Thus,

$$Q = Jc_{44}^E\xi_3''' - e_{15}wtE_1\ , \tag{5.376}$$

i.e., the electromechanical conversion generates a shearing force only with $E_1 \neq 0$, as shown in Figure 5.36 (b). For this purpose, upon poling the piezoelement in the direction x_3 the working electrodes must be applied to the surfaces perpendicular to axis x_1. In this case the electromechanical conversion is possible because of shear strain arising on the contour of the plate.

In the case of the simply supported end ($M_f = 0$, $\xi_3 = 0$) and of the free end ($M_f = 0$, $Q = 0$) it follows from formula (5.373) that at $E_3 \neq 0$ the boundary conditions are

$$\xi_3'' = 3d_{31}E_3 / wt \,. \tag{5.377}$$

(Remember that for the beam with rectangular cross section $J = wt^3/12$.) The same conclusion can be made, as in the case of the longitudinal vibration, that the electrical energy that is coming through the electrodes would be transformed into the mechanical energy, $W_{em} = M_{em}\xi_3'$, generated by the equivalent moment M_{em} acting on the boundary.

At the fixed end ($\xi_3 = 0$, $\xi_3' = 0$) $W_{em} = 0$, and no energy transformation takes place. The piezoceramic beam with two fixed ends is electromechanically passive. The boundary conditions for Eq. (4.187) of flexural vibrations of the circular plates may be formulated in the way analogous to those for the beams.

5.8.2.2 Electrical Boundary Conditions

The electrical boundary conditions may be formulated based on the expression for current through a transducer. For each particular case, the charge density D is determined by formula (5.132). Thus, in the case of axially symmetric vibrations in the plane of a circular disk (transverse piezoeffect),

$$D_3 = \varepsilon_{33}^{S_{1,2}} E_3 + \frac{d_{31}}{s_{11}^E + s_{12}^E}\left(\frac{\partial \xi_r}{\partial r} + \frac{\xi_r}{r}\right). \tag{5.378}$$

For the flexural vibrations of a rectangular bar per half of its thickness

$$D_3 = \varepsilon_{33}^T (1 - k_{31}^2) E_3 + \frac{d_{31}}{s_{11}^E}\frac{t^2}{4}\frac{\partial^2 \xi_3}{\partial x_3^2}. \tag{5.379}$$

For the flexural vibrations of a circular plate per one half-plate

$$D_3 = \varepsilon_{33}^{S_{1,2}} E_3 + \frac{t}{4}\frac{d_{31}}{s_{11}^E + s_{12}^E}\left(\frac{\partial^2 \xi_3}{\partial r^2} + \frac{1}{r}\frac{\partial \xi_3}{\partial r}\right). \tag{5.380}$$

Substituting solutions of equations of vibration for the displacements ξ_3 into expressions for charge density, we will obtain electrical boundary conditions in the form of expression for current

$$I = \int_{S_{el}} \dot{D}_3 dS_{el} \,. \tag{5.381}$$

The manipulations with this equation are straightforward, and the results obtained may be analyzed, as this is done in Section 5.8.1.2.2.

It is noteworthy that considering transducers vibration problems in geometrical coordinates it is difficult to determine, what measures can be taken for increasing efficiency of electromechanical conversion without recourse to the concept of solution in terms of the generalized coordinates, as it was previously illustrated. For example, it is hard to predict how the dimensions of electrodes and way of connecting their parts must be modified to optimize efficiency of electromechanical conversion for various transducer types. Even more challenging is that in these cases we would have to set up and to solve equations of types (4.47), (4.187) and (4.159) for segments of a mechanical system that belong to the partial electrodes, and to mate them at the boundaries of the segments. Due to these complications, it is much simpler to solve the corresponding problems in generalized coordinates from the very beginning.

An exception in this respect presents solution of equation (4.40) that is applicable to widely used designs of the type schematically shown in Figure 5.37 (b). The solution allows convenient interpretation in the form of the equivalent electromechanical circuit, which enables considering various loading of the transducers ends and changing the dimensions and ways of connecting the electrodes. Such the solution for the passive mechanical systems is illustrated in Section 4.3.3 with T-network mechanical equivalent circuit displayed in Figure 4.6 (a). Consider this approach for analogous mechanical systems that are fully or partially made of piezoelements.

5.8.3 Equivalent Three-Port Network of a Longitudinally Vibrating Piezoceramic Bar

Refer to Eq. (4.90) of the steady state harmonic longitudinal vibrations. The general solution of this equation for a passive bar has the form of Eq. (4.91). In order to obtain solution for a piezoceramic bar, we must replace therein the wave number k by k^E (c by c^E) for solid bars that employ the transverse piezoeffect and for the segmented bars (with sufficient number of segments according to Section 5.5.2) under the longitudinal piezoeffect, and by k^D (c^D) for the solid bars under longitudinal piezoeffect. In addition, the electromechanical boundary conditions (5.352) must be taken into account as well as mechanical boundary conditions. The latter arise in presence of the loads and forces acting on the ends of the bars in the form of Eqs. (4.96) and (4.97). At first, the solid piezoceramic bars performing longitudinal vibrations will be considered. Results of the analysis will be valid for all mechanical systems, for which Eq. (4.90) is applicable, if appropriate values of coefficients n_Δ, $C_{e\Delta}^{S_i}$ and K_Δ^E are used. For the piezoceramic bars under the transverse piezoeffect with full size electrodes ($l_e = l$) the boundary conditions obtained by combining Eqs. (4.96), (4.97) and (5.352) will be

$$K_\Delta^E S_{cs}\, \xi'(x)\big|_{x=0} = -Z_0\dot{\xi}_0 - F_0 + n_\Delta S_{cs}\, E_3'\big|_{x=0}, \tag{5.382}$$

$$K_\Delta^E S_{cs}\, \xi'(x)\big|_{x=l} = -Z_l\dot{\xi}_l - F_l + n_\Delta S_{cs}\, E_3'\big|_{x=l}. \tag{5.383}$$

Note that the sign of the strain at the left end of a bar (at $x = 0$) must be changed to opposite due to the accepted rule of signs (see Section 1.5.2 and 4.3.3), and the signs of the displacements and forces must be changed accordingly. In case of the longitudinal piezoeffect K_Δ^D must be used instead of K_Δ^E. Variants with the transverse and longitudinal piezoeffects will be considered separately in further discussion, since in these cases the values of $E_3'\big|_{x=0,l}$ are different. Namely, for the transverse piezoelectric effect $E_3' = E_3 = V/t$, and for the longitudinal piezoeffect

$$E_3' = \frac{V}{l} + \frac{n_\Delta}{C_{e\Delta}^{S_3} l}(\xi_l + \xi_0) = \frac{V}{l} + \frac{N_3}{j\omega C_e^{S_3}}(U_l + U_0), \tag{5.384}$$

as it follows from the general relation (5.346) with taking into account that $n_\Delta = d_{33}/s_{33}^E$ and $C_{e\Delta}^{S_3} = \varepsilon_{33}^T(1-k_{33}^2)$. (Note that the signs of displacement ξ_0 and velocity U_0 are already changed to opposite.) In Eq. (5.384)

$$N_3 = wtd_{33}/s_{33}^E l \text{ and } C_e^{S_3} = \varepsilon_{33}^T(1-k_{33}^2)wt/l. \tag{5.385}$$

Expression for the current flowing through transducer can be presented in the form analogous to Eq. (5.364)

$$I = j\omega C_e^{S_i} V + N_i(U_l + U_0), \tag{5.386}$$

where for the transverse piezoeffect ($i = 1$)

$$N_1 = wd_{31}/s_{11}^E, \quad C_e^{S_1} = \varepsilon_{33}^T(1-k_{31}^2)wl/t, \tag{5.387}$$

for the longitudinal piezoelectric effect ($i = 3$) N_3 and $C_e^{S_3}$ are given by formulas (5.385). Substituting expressions for $\xi(x)$ from Eq. (4.94) and for E_3' into the boundary conditions (5.382), we obtain relations similar to those presented by Eqs. (4.98) and (4.99). After introducing the designations for impedances according to formulas (4.100) we will obtain equations:

$$Z_1^E(U_l+U_0)+Z_2^E U_0+Z_0 U_o+F_0=VN_1, \quad (5.388)$$

$$Z_1^E(U_l+U_0)+Z_2^E U_l+Z_l U_l+F_l=VN_1, \quad (5.389)$$

for the transverse piezoelectric effect, and

$$\left(Z_1^D-\frac{N_3^2}{j\omega C_e^{S_3}}\right)(U_l+U_0)+Z_2^D U_0+Z_0 U_0+F_0=VN_3, \quad (5.390)$$

$$\left(Z_1^D-\frac{N_3^2}{j\omega C_e^{S_3}}\right)(U_l+U_0)+Z_2^D U_l+Z_l U_l+F_l=VN_3, \quad (5.391)$$

for the longitudinal piezoelectric effect. Designations for the impedances in above equations are analogous to those introduced by relations (4.100). Namely, they are

$$Z_1^E=-j\rho c^E S_{cs}/\sin(k^E l),\ Z_2^E=j\rho c^E S_{cs}\tan(k^E l/2), \quad (5.392)$$

$$Z_1^D=-j\rho c^D S_{cs}/\sin(k^D l),\quad Z_2^D=j\rho c^D S_{cs}\tan(k^D l/2). \quad (5.393)$$

Equations (5.388) – (5.391) for the transverse and longitudinal piezoeffect in combination with Eq. (5.386) for the electrical side solve the problem of the transducers calculating.

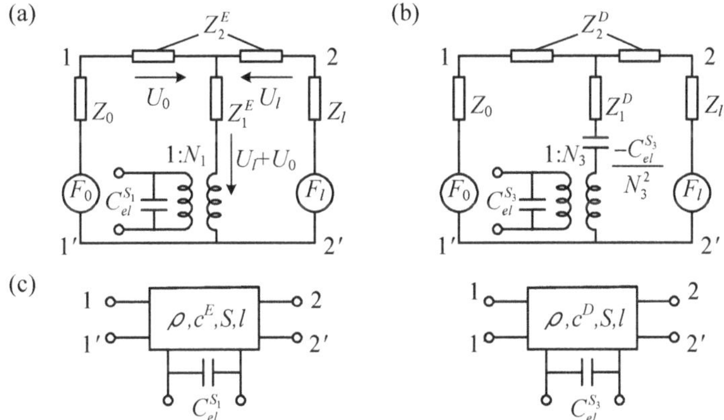

Figure 5.38: Equivalent electromechanical three-port networks: (a) for the transverse piezoeffect, (b) for the longitudinal piezoeffect, (c) schematic representation of the networks.

Comparing these equations with analogous equations for a passive bar, which were associated with the two-port network shown in Figure 4.6, we arrive at the conclusion that the equivalent electromechanical circuits describing vibrations in the piezoceramic bars can be represented in analogous way. The difference is that one more port must be introduced accounting for the effect of electromechanical conversion as it is shown in Figure 5.38 (a) and (b). The effect of electromechanical conversion is provided by introducing the ideal transformers with electromechanical transformation ratios N_i and by appearance of the "negative compliance" $C = -C_e^{S_3} / N_3^2$ in the case of longitudinal piezoeffect. Note that for a segmented bar under the longitudinal piezoeffect with number of segments more than 6 the circuit shown in Figure 5.38 (a) is valid, if to replace N_1 by N_3 and $C_e^{S_1}$ by $C_e^{S_3}$.

Instead of the three-port networks shown in Figure 5.38 (a) and (b) their schematic representations shown in Figure 5.38 (c) may be used for brevity. Correlation between the input electrical (V, I) and output mechanical (U_l, U_0 and hence also $\xi(x)$) quantities and *vice versa* can be found for various mechanical loads and electrical conditions using cascade connection of such circuits. To illustrate application of the cascade connection of the simplified networks consider several examples of piezoelements presented in Figures 5.39 and 5.40 that previously were treated in the generalized coordinates. In these examples the cascade connection of the

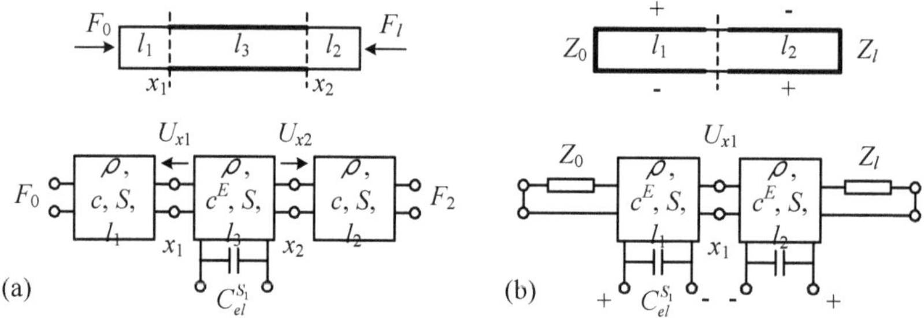

Figure 5.39: Variants of the piezoelements design and their corresponding cascade equivalent circuits representations: (a) piezoelement with partial electrodes, (b) piezoelement with split electrodes.

equivalent circuits can be used in the way, as it is illustrated in the Figures, instead of solving new equations of motion and considering mating conditions on the boundaries of the parts having different elastic and electromehcanical properties. The example presented in Figure 5.39 (a) corresponds to transducer in the form of a piezoceramic bar cemented with passive parts, and those presented in Figure 5.39 (b) corresponds to a piezoelement with electrically separated electrodes, to which different voltages can be applied. There is no need to consider analytically the mating conditions (5.361), (5.362) to be met over cross sections with the coordinates x_1 and x_2. In the circuits shown in the Figure they are met automatically. The vibration velocities of these cross sections correspond to the "currents" $\underline{U(x_1)}$ and $\underline{U(x_2)}$ in the respective branches of the equivalent circuit. Various voltages can be applied to the separated parts of the electrodes in Figure 5.39 (b). The case of electrical connection of the electrodes in the opposite polarities, which is shown in the Figure, illustrates how the effects of different electrodes arrangement on the frequency response and quality of the electromechanical conversion can be considered for a transducer in the shape of a bar.

Slightly changed circuits arrangement that is shown in Figure 5.40 represents the equivalent circuit of the Rosen-type EME converter that was considered in Section 5.7.4.3 using representation in the generalized coordinates. In this case the half-bar that employs the longitudinal effect is presented as mechanical load for the input half-bar vibrating in the transverse mode.

The advantage of the current approach is in its ability to perform calculation of transducer operating characteristics in a broad frequency range and universality in meeting mating and boundary conditions. But the universal nature of this technique comes for expense of physical clarity and simplicity of calculations in generalized coordinates in many practically important cases, in which this universality is not needed. In particular, these are the cases of operating in the frequency ranges around resonance frequencies and in a wide range below the first resonance.

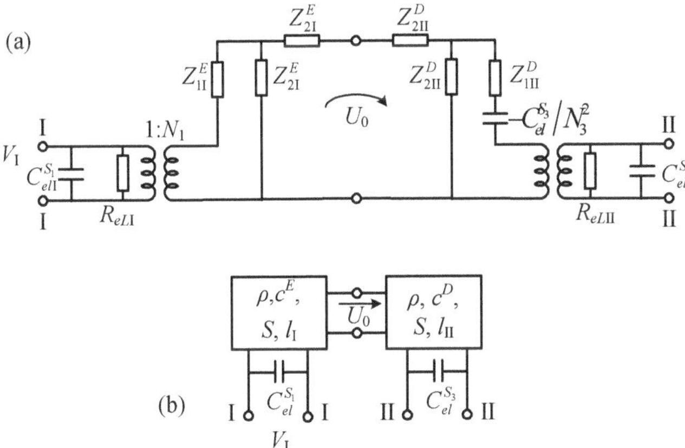

Figure 5.40: The equivalent circuit of the Rosen-type EME converter: (a) full size equivalent circuit, (b) cascade representation.

In this chapter the general questions regarding electromechanical conversion in the piezoelectric ceramic bodies were considered. The transducer examples were involved for purposes of illustrating different aspects of the problem. The electromechanical transducers of different type under specific loading and acting forces that are typical for their applications to underwater acoustics will be analysed in detail in Part III of the treatment. Prior to this the issues related to interaction of electromechanical transducers with acoustic field will be considered in the next chapter.

5.9 References

1. G. A. Smolensky, V. A. Bokov, V. A. Isupov, N. N. Krainik, R. E. Pasynkov, A. I. Sokolov, and N. K. Yushin, *Ferroelectrics and Related Materials* (Gordon and Breach, New York, 1984).
2. D. A. Berlincourt, D. R. Curran, and H. Jaffe, *Piezoelectric and Piezomagnetic Materials and their Function in Transducers*, in Physical Acoustics, Vol. I, Part A, edited by W. P. Mason (Academic, New York, 1964).
3. B. S. Aronov, "Energy analysis of a piezoelectric body under nonuniform deformation," J. Acoust. Soc. Am. **113**(5), 2638-2646 (2003).
4. W. G. Cady, *Piezoelectricity* (McGraw-Hill, New York, 1946).
5. R. Holland, E. P. EeR Nisse, *Design of Resonant Piezoelectric Devices*, (MIT, Cambridge, 1969).
6. B. S. Aronov, "On the optimization of the effective electromechanical coupling coefficients of a piezoelectric body," J. Acoust. Soc. Am. **114**(2), 792-800 (2003).
7. A. F. Ulitko, "On the theory of the electromechanical conversion of energy in the nonuniformly deformed piezoceramic bodies," Prikladnaya Mekhanika **13**(10), 115-123 (1977) (in Russian), available in English as International Applied Mechanics.
8. J. A. Lewis, "The Effect of Driving Electrode Shape on the Electrical Properties of Piezoelectric Crystals," Bell System Tech. J. **40**(5), 1259-1280 (1961).
9. E. M. Syed, F. P. Dawson, and E. S. Rogers, "Analysis and modeling of a Rozen type piezoelectric transformer," Proc. IEEE PESC, 4, 1761-1766 (2001).

CHAPTER 6

ACOUSTIC RADIATION

6.1 Introduction

6.1.1 Scope of the Chapter

In this Chapter the acoustic field related parameters of electromechanical transducers that are necessary for their calculation as electroacoustic transducers are considered. Typical geometries and wave sizes of radiating surfaces are listed, and expressions for the acoustic field related parameters are summarized. Unfortunately, there is no such a literature source on the radiation problems, with reference to which all the needed parameters of transducers can be readily obtained. Though a great number of references exists, in which various radiation problems are solved for different purposes. To make it easier using the results available, a brief information on the general theory of radiation and on the methods of solving radiation problems is presented. In terms of statement of the radiation problems we will define transducer as a part of transmit or receive system that operates with a single power amplifier or preamplifier. It can be made of a single piezoelement, or of several mechanically isolated parts (elementary transducers). The latter variant can be used out of technological considerations, or in order to avoid harmful effects of coupled vibrations in the mechanical system of a transducer. We will assume that all the elementary transducers are supplied with the same voltage, if they are connected in parallel, or with the same current, if connected in series. When considering radiation of a transducer we will assume that all the comprising elementary transducers vibrate with no amplitude and phase distributions imposed intentionally. Though an unintended not uniform distribution of velocities over the transducer surface may occur in the case that the transducer is comprised of several mechanically isolated parts as result of their acoustic interaction. Therefore, the acoustic interaction between elementary transducers is also considered among the radiation problems.

6.1.2 Geometries and Wave Sizes of Radiating Surfaces

Typical configurations of radiating surfaces of the transducers can be classified in several groups, as shown in Figure 6.1. The following types of mechanical systems of transducers and configurations of their radiating surfaces will be considered.

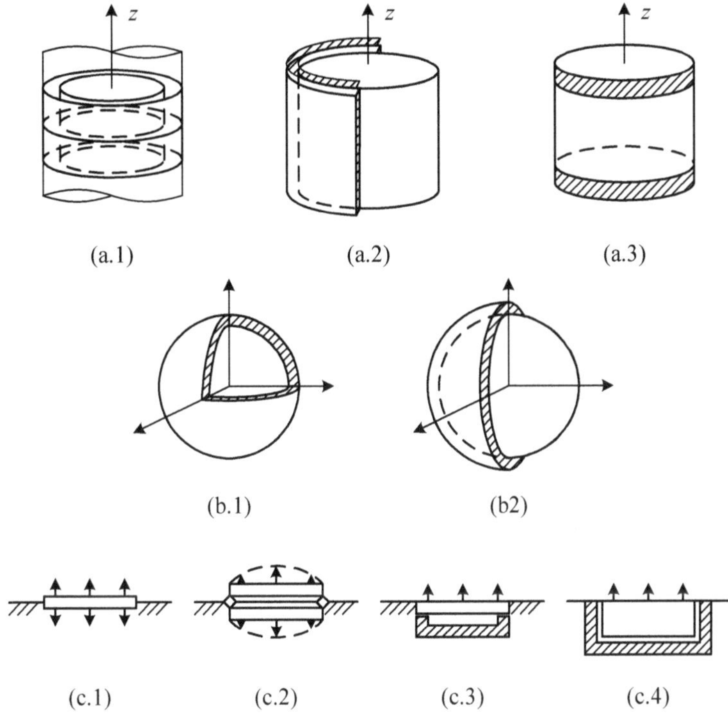

Figure 6.1: Configurations of the transducers radiating surfaces: (a) cylindrical surfaces, (b) spherical surfaces, (c) flat surfaces.

Cylindrical transducers made of solid piezoelectric ceramic cylinders or composed of the elementary ring transducers (Figure 6.1(a.1)). The transducers may employ different circumferential modes of vibration, among which the "pulsating" (zeroth mode) and "oscillating" (first mode) are the most widely used. Transducers comprised of the rings performing flexural vibrations (solid and slotted) and of the incomplete rings also fall in this category. A part of radiating surface of the transducers of this kind may be covered with a baffle for achieving unidirectional radiation in the horizontal plane (Figure 6.1(a.2)). The models of infinitely long cylindrical radiator and of the transducer of a finite height embedded into infinitely long rigid cylindrical baffle of the same diameter are useful for approximate estimation of acoustic field related parameters of cylindrical transducers, though the real transducers have a finite height, as shown in Figure 6.1(a.3).

Spherical transducers made of piezoelectric ceramic spheres (Figure 6.1(b.1)). They may be used in the pulsating and oscillating modes of vibration (employment of the higher modes can be also imagined). The spherical transducers can be partially baffled for achieving the unidirectional radiation (Figure 6.1(b.2)).

Transducers with flat radiating surface. Piston like pulsating transducers (Figure 6.1(c.1)) and the flexural type transducers made of circular or rectangular plates vibrating with nonuniform distribution of velocity over radiating surface (Figure 6.1(c.2)). Being used in the double sided design they can be considered as embedded flash into the infinite absolutely rigid baffle due to symmetry. The dimensions of a single elementary transducer of the flexural type usually are small compared with acoustic wavelength. Transducers of this type having one sided design can be used with the finite size baffles flash with their radiating surface (Figure 6.1(c.3)). The widest used transducers with piston like vibrating surfaces are the transducers of the Tonpilz design. Their radiating surface may have circular or rectangular (square) configuration with dimensions typically less than $\lambda/2$. When used as a single projector the transducer can be supplied with a baffle, as shown in Figure 6.1(c.4).

Transducers with flat radiating surface vibrating in the "piston like" mode with their back side baffled (Figure 6.1(c.4)). They can be made of the rectangular piezoceramic bars or circular plates vibrating through their thickness. In this case the transducers have relatively high operating frequencies, and dimensions of the radiating surfaces may reach many wavelengths, as for example, in transducer for side scan sonar (though such size is more typical for the arrays, but this case falls into category of a transducer by our classification).

A brief outline of parameters of transducers that are needed for their calculating as electroacoustic and their general definitions will be considered in the next section. Partially these issues were considered also in Part I.

6.1.3 Acoustic Field Related Parameters of Transducers

6.1.3.1 Transducers Having a Single Mechanical Degree of Freedom

Transducers of this kind have a fixed velocity distribution on the radiating surface in operating frequency range that can be represented as

$$U(\boldsymbol{r}_\Sigma,\omega)=U_o(\omega)\cdot\theta(\boldsymbol{r}_\Sigma,\omega), \tag{6.1}$$

where $\boldsymbol{r}_\Sigma$ is the radius vector defining the points on the transducer surface Σ shown in Figure 6.3 and U_o is the velocity of the reference point on the surface that has radius vector $\boldsymbol{r}_o$.

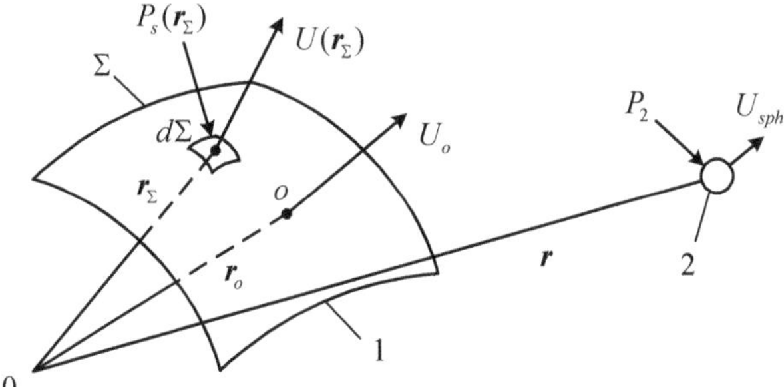

Figure 6.2: Illustration of the mechano-acoustic system consisting of the surface of radiating transducer #1 and pulsating sphere of a small radius # 2.

Note that all the quantities in this chapter will be used in the complex form. The sound pressure in the radiated acoustic field, $P(\boldsymbol{r})$, and on the transducer surface in particular, $P(\boldsymbol{r}_\Sigma)$, has to be determined by means of the acoustic radiation theory so far as distribution of velocity on the transducer surface is known as one of the boundary conditions.

6.1.3.1.10. Sound Pressure and Diffraction Coefficient

The sound pressure generated by a vibrating surface may be represented in general form as

$$P(\boldsymbol{r},\omega)=\frac{\rho c}{r}U_o e^{-j(kr-\pi/2)}\chi(\boldsymbol{r},\omega), \tag{6.2}$$

where function $\chi(\boldsymbol{r},\omega)$ depends on the radiating surface configuration and mode of vibration. We will refer to this function as the diffraction function of a transducer in the transmit mode. In the far field this function becomes independent of distance r and represents the directional factor of the transducer, $H(\boldsymbol{r},\omega)$,

$$H(\boldsymbol{r},\omega)=\left.\frac{\chi(\boldsymbol{r},\omega)}{\chi(\boldsymbol{r}_n,\omega)}\right|_{|r|\to\infty}. \tag{6.3}$$

Here $\boldsymbol{r}_n$ is the radius vector pointed in direction of acoustic axis of the transducer.

For uniformly vibrating (pulsating) spherical radiator of radius a (see Section 2.2)

$$P(\boldsymbol{r},\omega)=\frac{\rho c}{r}e^{-j(kr-\pi/2)}U_o\frac{ka^2}{1+jka}e^{-jka}, \tag{6.4}$$

and the diffraction function is

$$\chi=\frac{ka^2}{1+jka}e^{-jka}. \tag{6.5}$$

Further we introduce concept of the "referred volume velocity," $U_{\tilde{V}r}$, for an arbitrary vibrating surface defined as

$$U_{\tilde{V}r}=U_o S_\Sigma, \tag{6.6}$$

where S_Σ is the total area of the radiating surface. This quantity does not depend on the mode of vibration in contrast to the "real" volume velocity, or to the source strength, which is defined as

$$U_{\tilde{V}}=U_o\int_\Sigma \theta(\boldsymbol{r})\,d\Sigma=U_o S_{av}. \tag{6.7}$$

Here

$$S_{av}=\int_\Sigma \theta(\boldsymbol{r}_\Sigma)\,d\Sigma \tag{6.8}$$

is the average radiating surface area. Thus, for the oscillating sphere $U_{\tilde{V}r}=U_o 4\pi a^2$, whereas $U_{\tilde{V}}=0$. In the case that the wave size of the sphere is small ($ka\to 0$), it follows from expressions (6.4) and (6.6) that sound pressure generated by the sphere is

$$P_0=\frac{\rho c}{r}e^{-j(kr-\pi/2)}U_o ka^2=\frac{\rho c}{r}e^{-j(kr-\pi/2)}U_{\tilde{V}r}\frac{1}{2\lambda}. \tag{6.9}$$

The ratio of the sound pressure generated by an arbitrary transducer to the sound pressure generated by a small pulsating sphere that has the same referred volume velocity we define as the diffraction coefficient of the transducer in the transmit mode, $k_{dif.t}$. Using expressions (6.2) and (6.9), we obtain

$$\frac{P(\boldsymbol{r},\omega)}{P_0(U_{\tilde{V}_r})}=\frac{2\lambda\,\chi(\boldsymbol{r},\omega)}{S_\Sigma}=k_{dift}(\boldsymbol{r},\omega) \tag{6.10}$$

In the case of the spherical transducer

$$k_{dift} = \frac{1}{1+jka} e^{jka} . \tag{6.11}$$

6.1.3.1.11. Radiation Impedance

The acoustic power radiated by a transducer can be found by integrating the acoustic power density $P(\overline{r}_\Sigma)U^*(\overline{r}_\Sigma)\,d\Sigma$ over the transducer's radiating surface Σ. In the complex form the acoustic radiating power is

$$\bar{\dot{W}}_{ac} = \int_\Sigma P(\boldsymbol{r}_\Sigma)U^*(\boldsymbol{r}_\Sigma)\,d\Sigma = U_o^* \int_\Sigma P(\boldsymbol{r}_\Sigma)\theta(\boldsymbol{r}_\Sigma)\,d\Sigma , \tag{6.12}$$

where the sound pressure $P(\boldsymbol{r}_\Sigma)$ on the transducer surface has to be determined by expression (6.2). The acoustic power radiated can be represented in the form

$$\bar{\dot{W}}_{ac} = Z_{ac}\left|U_o\right|^2 , \tag{6.13}$$

which can be considered as the definition for the transducer radiation impedance, Z_{ac} . Equating expressions (6.12) and (6.13) we obtain

$$Z_{ac} = \frac{1}{U_o}\int_\Sigma P(\boldsymbol{r}_\Sigma)\cdot\theta(\boldsymbol{r}_\Sigma)d\Sigma . \tag{6.14}$$

In the case that projector is composed of several elementary transducers the radiation impedance of the elementary transducers can be different under the assumption that their velocities are the same since under this assumption the sound pressure on the surface of the transducer may have nonuniform distribution. If the transducer has a solid mechanical system, then the sound pressure is averaging on its surface, which results in some value of the radiation impedance determined by formula (6.14). The elementary transducers, for which the averaging of pressure takes place on their surfaces, may have different radiation impedances. This may cause a difference in the velocities of vibration of the elementary transducers under the same voltage applied, which contradicts the initial assumption regarding uniformity of their vibration. To make possible calculating a real distribution of velocities between the elementary transducers in this case the mutual radiation impedances between the transducers are introduced.

The mutual impedance between two elementary transducers in assembly vibrating with the same velocity distribution $U = U_0\,\theta(\boldsymbol{r})$ can be determined by the expression (Refs. 25, 26)

$$z_{12} = \frac{1}{U_0} \int_{\Sigma_2} P_1(\boldsymbol{r}_{\Sigma 2}) \theta(\boldsymbol{r}_{\Sigma}) d\Sigma, \tag{6.15}$$

where $P_1(\boldsymbol{r}_{\Sigma 2})$ is the sound pressure generated by transducer #1 on the blocked surface of transducer #2 at the condition that all the other members of the assembly also are blocked.

6.1.3.1.12. Directivity of a Transducer (D).

The property of a real transducer to generate in the far field larger intensity in direction of acoustic axis in comparison with intensity that omnidirectional transducer radiating the same acoustic power would generate in the same point is called directivity (*D*) (or the coefficient of acoustic energy concentration). According to this definition

$$D = \frac{|P_{tr}(\boldsymbol{r}_n, \omega)|^2}{|P_{omni}(\boldsymbol{r}_n, \omega)|^2}, \tag{6.16}$$

where $P_{tr}(\boldsymbol{r}_n, \omega)$ is given by expression (6.2), $\boldsymbol{r}_n$ is the radius vector pointed in direction of acoustic axes of the transducer (direction of maximum of the directional factor) and P_{omni} is the sound pressure generated in the same point by an omnidirectional transducer. Values of the sound pressures should meet the condition that the total active acoustic powers radiated are the same in both cases. For the directional transducer

$$\dot{W}_{ac\,tr} = \frac{|P_{tr}(\boldsymbol{r}_n, \omega)|^2 r^2}{\rho c} \int_{4\pi} |H(\boldsymbol{r}, \omega)|^2 \, d\Omega, \tag{6.17}$$

where Ω is the solid angle and integration is assumed over the sphere of radius *r*. For the omnidirectional transducer after integrating over the sphere of the same radius we obtain

$$\dot{W}_{ac\,omni} = \frac{4\pi r^2}{\rho c} |P_{omni}(\boldsymbol{r}_n, \omega)|^2. \tag{6.18}$$

Equating relations (6.17), (6.18) and using the definition (6.16) we arrive at expression for the directivity

$$D = \frac{4\pi}{\int_{4\pi} |H(\boldsymbol{r}, \omega)|^2 d\Omega}. \tag{6.19}$$

Useful correlation can be obtained between the radiation resistance, r_{ac}, directivity and value of the diffraction function in direction of acoustic axis, $\chi(\boldsymbol{r}_n, \omega)$. Following expression

(6.2),

$$|P_{tr}(\boldsymbol{r}_n,\omega)|^2 = \left(\frac{\rho c}{r}\right)^2 |U_o|^2 \, |\chi(\boldsymbol{r}_n,\omega)|^2 \,. \tag{6.20}$$

Combining this expression with (6.17), (6.19) and the alternative expression for the acoustic power

$$\dot{W}_{ac\,tr} = r_{ac}|U_o|^2 \,, \tag{6.21}$$

we arrive at the relation

$$r_{ac} = 4\pi\rho c \frac{1}{D} \cdot |\chi(\boldsymbol{r}_n,\omega)|^2 \,. \tag{6.22}$$

This formula can be useful for determining the directivity in the case that the radiation resistance and the value of the diffraction function on the acoustic axis of a transducer are known.

6.1.3.1.13. Equivalent Force and Diffraction Coefficient in Receive Mode

In the receive mode the acoustic field constitutes the source of energy, W_{am}, supplied to the transducer. The mode of vibration of the transducer mechanical system under action of acoustic field is expressed by the same formula (6.1) as for the transmit mode, due to condition that the system has a single degree of freedom. Therefore, the acoustomechanical power in the complex form can be determined as

$$\bar{\dot{W}}_{am} = \int_\Sigma P(\boldsymbol{r}_\Sigma) U^*(\boldsymbol{r}_\Sigma)\, d\Sigma = U_o^* \int_\Sigma P(\boldsymbol{r}_\Sigma)\, \theta(\boldsymbol{r}_\Sigma)\, d\Sigma \,, \tag{6.23}$$

where $P(\boldsymbol{r}_\Sigma)$ is the sound pressure on the surface of the transducer. The sound pressure may be represented as

$$P(\boldsymbol{r}_\Sigma) = P^U(\boldsymbol{r}_\Sigma) - P_{br}(\boldsymbol{r}_\Sigma) \,, \tag{6.24}$$

where $P^U(\boldsymbol{r}_\Sigma)$ is the sound pressure on the blocked transducer surface (at $U = 0$) and $P_{br}(\boldsymbol{r}_\Sigma)$ is the sound pressure due to the back radiation generated by vibration of the transducer surface. Upon substituting expression (6.24) for $P(\boldsymbol{r}_\Sigma)$ into Eq. (6.23) we obtain

$$\bar{\dot{W}}_{am} = U_o^* \int_\Sigma P^U(\boldsymbol{r}_\Sigma)\theta(\boldsymbol{r}_\Sigma) d\Sigma - U_o^* \int_\Sigma P_{br}(\boldsymbol{r}_\Sigma)\,\theta(\boldsymbol{r}_\Sigma) d\Sigma \,, \tag{6.25}$$

or
$$\bar{\dot{W}}_{am} = \bar{\dot{W}}^{U}_{am} - \bar{\dot{W}}_{ac\,br}\,. \tag{6.26}$$

The second term on the right side of the relation (6.25) is the acoustic energy of the back radiation, and it can be represented as

$$\bar{\dot{W}}_{ac\,br} = Z_{ac}\left|U_o\right|^2\,, \tag{6.27}$$

where Z_{ac} is defined by formula (6.14). We denote the integral in the first term as the equivalent force, F_{eqv}. Thus,

$$F_{eqv} = \int_{\Sigma} P^{U}(\boldsymbol{r}_{\Sigma})\theta(\boldsymbol{r}_{\Sigma})\,d\Sigma\,, \tag{6.28}$$

and from relation (6.26) follows that

$$\bar{\dot{W}}^{U}_{am} = U_o^{*}F_{eqv}\,. \tag{6.29}$$

Now the expression (6.26) for the acoustomechanical power can be rewritten in the form

$$\bar{\dot{W}}_{am} = (F_{eqv} - Z_{ac}U_o)U_o^{*}\,. \tag{6.30}$$

On the other hand, if to denote the input impedance of the mechanical system as Z_m, the power supplied to the mechanical system is

$$\bar{\dot{W}}_{am} = Z_m U_o U_o^{*}\,. \tag{6.31}$$

Comparing relations (6.29) and (6.30) we arrive at

$$(Z_m + Z_{ac})U_o = F_{eqv}\,. \tag{6.32}$$

This relation may be interpreted by the circuit of "acoustomechanical generator" with mechanomotive force F_{eqv} and internal impedance Z_{ac}, which is equivalent to action of the acoustic field. Calculating the equivalent force, F_{eqv}, is the subject of radiation theory. In the case that dimensions of a transducer are small compared with the wavelength of sound, we have $P^{U}(\boldsymbol{r}_{\Sigma}) \approx P_0$, where P_0 is the sound pressure in the propagating wave, and

$$F_{eqv} = P_0\int_{\Sigma}\theta(\bar{r}_{\Sigma})\,d\Sigma = P_0 S_{av}\,. \tag{6.33}$$

If the dimensions of the transducer are comparable with the wavelength, the equivalent force may be represented as

$$F_{eqv} = P_0 k_{dif\,r} S_\Sigma , \tag{6.34}$$

where $k_{dif\,r}$ is the diffraction coefficient in the receive mode and S_Σ is the total radiation surface area. The diffraction coefficient $k_{dif\,r}$ may be calculated by equating the formulas (6.34) and (6.28) after the sound pressure distribution $P^U(\boldsymbol{r}_\Sigma)$ is found by solving the diffraction problem for the blocked transducer surface. But in fact, if the radiation problem is already solved, the diffraction coefficient $k_{dif\,r}$ can be determined by applying the reciprocity principle to the mechanoacoustic system consisting of two transducers: transducer # 1 with surface Σ, on which the distribution of velocity is specified as $U(\boldsymbol{r}_\Sigma) = U_o\,\theta(\boldsymbol{r}_\Sigma)$, and pulsating sphere # 2 of small radius a located at a large distance from the transducer (as shown in Figure 6.3). This is done in Section 2.2 and results in the relation (2.37),

$$F_{eqv} = 2\lambda\chi(\boldsymbol{r},\omega)P_0 . \tag{6.35}$$

The function $\chi(\boldsymbol{r},\omega)$ is determined by Eq. (6.2) from solution to the radiation problem. By comparing expressions (6.34) and (6.35) we obtain the diffraction coefficient for the transducer in the receive mode in the form

$$k_{dif\,r} = \frac{2\lambda\,\chi(\boldsymbol{r},\omega)}{S_\Sigma} . \tag{6.36}$$

that coincides with expression (6.10) for the difraction coefficient, $k_{dif\,t}$, introduced for the transducer in the transmit mode. Therefore the distunguishing subscripts t and r will be further omitted.

6.1.3.2 Transducers with Mechanical Systems Having Multiple Degrees of Freedom

In general, the mechanical systems of electroacoustic transducers must be considered as having multiple degrees of freedom. In other words, the velocity distribution on the surface of mechanical system may be represented as function of a number of independent variables (generalized velocities). If the actual velocity distribution on the transducer surface is expressed as

$$U(\boldsymbol{r}_\Sigma,\omega) = \sum_{i=1}^{N} U_i(\omega)\cdot\theta_i(\boldsymbol{r}_\Sigma) , \tag{6.37}$$

where $\theta_i(\boldsymbol{r}_\Sigma)$ is a set of the linearly independent functions satisfying the boundary conditions for the mechanical system, then the quantities $U_i(\omega)$ can be considered as the generalized velocities for the system. In this case the acoustic radiation related parameters of a transducer must be expressed using the generalized velocities.

The sound pressure radiated by a transducer having distribution of velocity on its surface in the form of the series (6.37) can be represented as

$$P(\boldsymbol{r},\omega)=\sum_{i=1}^{N}P_i(\boldsymbol{r},\omega), \tag{6.38}$$

where $P_i(\boldsymbol{r},\omega)$ is the modal sound pressure corresponding to the generalized velocity U_i.

According to Eq. (6.2)

$$P_i(\boldsymbol{r},\omega)=\frac{\rho c}{r}U_i\,e^{-j(kr-\pi/2)}\,\chi_i(\boldsymbol{r},\omega), \tag{6.39}$$

and $\chi_i(\boldsymbol{r},\omega)$ is the diffraction function that corresponds to the modal distribution of velocity $\theta_i(\boldsymbol{r})$. Expression for the modal diffraction coefficients will be obtained in the form analogous to Eq. (6.10)

$$k_{dif\,i}=\frac{P_i(\boldsymbol{r},\omega)}{P_0(U_{\tilde{V}r})}=\frac{2\lambda\,\chi_i(\boldsymbol{r},\omega)}{S_\Sigma}. \tag{6.40}$$

After substituting the distribution of velocity (6.37) and sound pressure expressed by formula (6.38) under the integral in Eq. (6.12), the acoustic power radiated may be represented as

$$\bar{\dot{W}}_{ac}=\int_\Sigma\left[\sum_{l=1}^{N}P_l^*(\boldsymbol{r}_\Sigma,\omega)\right]\left[\sum_{i=1}^{N}U_i\,\theta_i(\boldsymbol{r}_\Sigma)\right]d\Sigma=\sum_{i=0}^{N}Z_{acii}U_i\cdot U_i^*+\sum_{i\neq l}^{N}z_{acil}\cdot U_i\cdot U_l^*. \tag{6.41}$$

In this expression Z_{acii} is the self (modal) radiation impedance of the mode of vibration θ_i and z_{acil} is the mutual (inter-modal) impedance between modes θ_i and θ_l. In the case that the supporting functions θ_i are orthogonal on the surface of a transducer (this is a preferable choice of the system of supporting functions), then the mutual impedances disappear.

The modal equivalent force, F_{eqvi}, will be determined from expression for the acoustomechanical energy, $\bar{\dot{W}}_{am}^U$, by formula

$$F_{eqvi}=\frac{\partial\bar{\dot{W}}_{am}^U}{\partial U_i}. \tag{6.42}$$

The acoustomechanical energy becomes represented as

$$\bar{\dot{W}}_{am}^{U} = \int_{\Sigma} P^{U}(\boldsymbol{r}_{\Sigma}) \left[\sum_{i=0}^{N} U_i \theta_i (\boldsymbol{r}_{\Sigma}) \right]^{*} d\Sigma \tag{6.43}$$

after substituting $U(\boldsymbol{r}_{\Sigma})$ in the form of the series (6.37) into formula (6.23). Thus, the modal equivalent force is

$$F_{eqvi} = \int_{\Sigma} P^{U}(\boldsymbol{r}_{\Sigma}) \theta_i (\boldsymbol{r}_{\Sigma}) d\Sigma \,. \tag{6.44}$$

6.2 Formulation of the Radiation Problem

6.2.1 Acoustic Wave Equation

Acoustic radiation theory considers acoustical processes as linear and occurring in ideal (inviscid) fluid. Although real fluids possess some viscosity, it is small enough for not to be considered in linear acoustic equation of motion. Some rotational effects and related shear deformations in the real fluid are confined to a thin layer near boundaries (boundary layer), where they may produce an energy loss. Under the assumption of linearity the acoustic fields can be described by a single scalar function, the velocity potential $\Phi(\boldsymbol{r},t)$, to which the particle velocity, $\boldsymbol{v}(\boldsymbol{r},t)$, and the sound pressure, $p(\boldsymbol{r},t)$, in the acoustic field are related as

$$\boldsymbol{v} = -\boldsymbol{grad}\,\Phi = -\nabla\Phi \,, \tag{6.45}$$

$$p = \rho_0 \frac{\partial \Phi}{\partial t} \,, \tag{6.46}$$

where ρ_0 is the fluid density in the state of equilibrium.

It is noteworthy that in some references (1, 2) the relations are accepted

$$\boldsymbol{v} = \boldsymbol{grad}\,\Phi = \nabla\Phi \,, \tag{6.47}$$

$$p = -\rho_0 \frac{\partial \Phi}{\partial t} \,, \tag{6.48}$$

In the vector analysis the convention is that direction of vector coincides with direction of the steepest decrease of the potential (i.e., should be $\boldsymbol{v} = -\nabla\Phi$), and descriptions of the electric fields in terms of electric potential comply with this definition. For acoustic fields the relations (6.45), (6.46) and (6.47), (6.48) can be used interchangeably, if consistent throughout a treatment. Indicative of this is that direction of the vector of acoustic energy flux

$$\mathbf{\Omega} = p\mathbf{v} = -\rho\nabla\Phi\frac{\partial\Phi}{\partial t} \tag{6.49}$$

is invariant to this choice. In fact, the choice of the system of relations has to do with sign convention for independent variables used.

Deriving the acoustic wave equation (Refs. 1, 2) involves knowing the relations between the basic quantities that characterize the acoustic field in the elementary volume of a fluid, namely, the particle velocity, $\mathbf{v}$, the sound pressure $p = \tilde{P} - \tilde{P}_o$, where $\tilde{P}$ is the instantaneous pressure, and $\tilde{P}_o$ is the equilibrium pressure; and the condensation (compression or dilatation), $s = (\rho_0 - \rho)/\rho_o$, where ρ is the instantaneous density of the fluid, and ρ_o is the equilibrium density. These relations are:

The equation of state

$$p = Bs\,, \tag{6.50}$$

where $B = \rho_0(\partial\tilde{P}/\partial\rho)_{\rho_o}$ is the adiabatic bulk modulus (all the acoustic processes in fluid can be considered adiabatic in the practical range of frequencies). The assumption is that $|s| \ll 1$. Note that in the phase of compression (at $\rho/\rho_0 > 1$) both condensation and sound pressure are negative, and in the phase of dilatation ($\rho/\rho_0 < 1$) they are positive. This is analogous to sign convention for the strain and stress in the mechanical systems.

The linearized (at $|s| \ll 1$) equation of continuity (the mass conservation law)

$$\frac{\partial\rho}{\partial t} = -\mathbf{div}(\rho_0\mathbf{v})\,, \tag{6.51}$$

or

$$\frac{\partial s}{\partial t} + \mathbf{div}\,\mathbf{v} = 0\,. \tag{6.52}$$

It follows from the mass conservation law that

$$d(\rho\tilde{V}) = \rho_0 d\tilde{V} + \tilde{V}d\rho = 0\,, \tag{6.53}$$

and given that $ds = -d\rho/\rho_0$ we obtain from Eq. (6.53) that

$$p = B\frac{d\tilde{V}}{\tilde{V}}\,. \tag{6.54}$$

This equation is the acoustic equivalent of the Hook's law.

After applying the Newton's second law to an element of ideal fluid we obtain the force equation (linearized Euler's equation)

$$\rho_o \frac{\partial \boldsymbol{v}}{\partial t} = -\boldsymbol{\nabla} p . \tag{6.55}$$

Combining Eqs. (6.45), (6.46), and (6.55), we arrive at wave equation for the velocity potential

$$\nabla^2 \Phi - \frac{1}{c^2}\frac{\partial^2 \Phi}{\partial t^2} = 0 , \tag{6.56}$$

where ∇^2 is the differential operator (Laplace operator) that in the rectangular coordinates is presented as

$$\nabla^2 = \frac{\partial^2}{\partial x^2} + \frac{\partial^2}{\partial y^2} + \frac{\partial^2}{\partial z^2} , \tag{6.57}$$

and c is the sound speed in the fluid equal to

$$c = \sqrt{B / \rho_0} . \tag{6.58}$$

Note that this expression is analogous to expression for the sound speed of the longitudinal wave in a bar, $c = \sqrt{Y / \rho}$. Thus, the bulk modulus B can be considered as analogous to the Young's modulus. Taking into consideration the expressions (6.50) and (6.58) we obtain relation between the sound pressure and condensation

$$p = c^2 \rho_0 s . \tag{6.59}$$

An alternative way of deriving the acoustic wave equation can be used that employs the variational Least Action principle instead of the Newtonian approach[3]. One of the reasons in favor of this derivation is that it is in line with general application of the energy method to treatment of the electroacoustic transducers including related radiation problems.

After the Lagrangian for small vibrations of element of inviscid fluid is represented in the form

$$L = w_{kin} - w_{pot} , \tag{6.60}$$

where w_{kin} and w_{pot} are the kinetic and potential energies of the element of volume (energy densities). The kinetic energy density of the fluid after using expression (6.45) will be

$$w_{kin} = \frac{1}{2}\rho_0 v^2 = \frac{1}{2}\rho_0 [(\Phi'_x)^2 + (\Phi'_y)^2 + (\Phi'_z)^2]. \tag{6.61}$$

The potential energy of a unit volume is the work required for changing its state to the state characterized by condensation s from the state of equilibrium ($s = 0$), that is

$$w_{pot} = \int_0^s p ds. \tag{6.62}$$

After substituting expression (6.59) for p we obtain

$$w_{pot} = \int_0^s \rho_0 c^2 s ds = \frac{\rho_0 c^2 s^2}{2} = \frac{p^2}{2\rho_0 c^2}, \tag{6.63}$$

or with reference to expression (6.46)

$$w_{pot} = \frac{\rho_0}{2c^2}\left(\frac{\partial \Phi}{\partial t}\right)^2. \tag{6.64}$$

Thus, we obtain expression for the Lagrangian in the form of

$$L = \frac{\rho_0}{2}\left\{[(\Phi'_x)^2 + (\Phi'_y)^2 + (\Phi'_z)^2] - \frac{1}{c^2}\left(\frac{\partial \Phi}{\partial t}\right)^2\right\}. \tag{6.65}$$

The procedure of deriving the Euler's equation of the problem in the geometry coordinates is the same, as for the longitudinally vibrating mechanical systems (bars, for example).

The Euler's equation for this case is

$$\frac{\partial}{\partial t}\left(\frac{\partial L}{\partial \Phi_t}\right) + \frac{\partial}{\partial x}\left(\frac{\partial L}{\partial \Phi'_x}\right) + \frac{\partial}{\partial y}\left(\frac{\partial L}{\partial \Phi'_y}\right) + \frac{\partial}{\partial z}\left(\frac{\partial L}{\partial \Phi'_z}\right) = 0. \tag{6.66}$$

The wave equation (6.56) follows from this equation after fulfilling the prescribed differentiations.

For the steady state conditions, we can use the complex representation of the velocity potential $\Phi(\boldsymbol{r},t) = \Phi(\boldsymbol{r})e^{j\omega t}$ and the wave equation becomes the Helmholtz equation

$$(\nabla^2 + k^2)\Phi = 0, \tag{6.67}$$

or due to relation (6.46)

$$(\nabla^2 + k^2)P = 0, \tag{6.68}$$

where $k = \omega / c$ is the wave number. The time dependent factor $e^{j\omega t}$, which is used in this form in considering the electromechanical transduction, is further omitted. Note that in many references on the radiation problems per se the time dependent factor is used in the form $e^{-j\omega t}$. As the physical meaning has only the real part of a solution, this difference does not influence the final results of a treatment. To bring all the manipulations made in these references to the form used in our case, just (-*j*) must be changed to (+*j*).

The Laplacian ∇^2 in the axisymmetric cylindrical coordinates is

$$\nabla^2 = \frac{1}{r}\frac{\partial}{\partial r}\left(r\frac{\partial}{\partial r}\right) + \frac{1}{r^2}\frac{\partial^2}{\partial \varphi^2} + \frac{\partial^2}{\partial z^2}, \tag{6.69}$$

and in the axisymmetric spherical coordinates

$$\nabla^2 = \frac{1}{r^2}\frac{\partial}{\partial r}\left(r^2\frac{\partial}{\partial r}\right) + \frac{1}{r^2 \sin\varphi}\frac{\partial}{\partial \varphi}\left(\sin\varphi\frac{\partial}{\partial \varphi}\right). \tag{6.70}$$

6.2.2 Sources of Acoustic Field, Boundary Conditions

The homogeneous wave equation (6.56) does not account for the acoustic field generation. In the acoustic applications of electromechanical transduction the acoustic fields are generated by the vibrating surfaces. This can be an entire transducer surface, or a vibrating part of the transducer structure that can also include passive elements of transducer designs, such as baffles and caps. In the ideal fluid only normal component of the surface velocity produces acoustic radiation. Effects of a small (although finite) viscosity in a real fluid are confined to a very thin layer at the boundary, within which a loss of energy may occur. This issue does not influence process of radiation and will be considered separately in a due place.

In the most typical case the radiation problem is in determining the acoustic field (the velocity potential $\Phi(\boldsymbol{r})$ that satisfies Eq. (6.67)) under the condition that the normal component of the surface velocity, U_n, is a known function $f_1(\boldsymbol{r}_\Sigma)$, i.e.,

$$U_n(\boldsymbol{r}_\Sigma) = -\left.\frac{\partial \Phi}{\partial n}\right|_\Sigma = f_1(\boldsymbol{r}_\Sigma). \tag{6.71}$$

($\boldsymbol{n}$ is the outward normal to the surface). This problem is called the Newman's problem, or the boundary-value problem of the second kind. Such is the problem of radiation by the vibrating tall cylindrical transducers (Figure 6.1 (a)) and spheres (Figure 6.1 (b)) without baffles and with perfectly rigid baffles on a part of their surfaces, and the transducers with flat surfaces (Figure 6.1 (c)) embedded in the rigid baffle of a large size. On the rigid baffles velocity is zero $[U(\boldsymbol{r}_{\Sigma b})=0]$.

Determining the acoustic field in the case that the potential (sound pressure P related to the potential by formula (6.46) $P=j\omega\rho_0\Phi$) is given on the surface, i.e.,

$$\Phi|_{\Sigma}=f_2(\boldsymbol{r}_{\Sigma}), \tag{6.72}$$

constitutes the Dirichlet's problem, or the boundary-value problem of the first kind. In the particular case of the absolutely compliant or "pressure-released" surface $f_2(\boldsymbol{r}_{\Sigma})=0$. This problem is not typical for transducers radiation per se. It can arise as the problem of diffraction of sound wave on a "pressure-released" body located close to a transducer surface.

In the most general case, the normal velocity is specified over a part of the transducer surface and the sound pressure or local impedance over another. This condition can be formulated as

$$\left(\frac{\partial\Phi}{\partial n}+\alpha\Phi\right)_{\Sigma}=f_3(\boldsymbol{r}_{\Sigma}), \tag{6.73}$$

where α is constant over a part of the surface. This is the mixed boundary-value problem, or the boundary-value problem of the third kind. Such situation takes place, for example, when the baffles in the case of the transducers shown in Figure 6.1 are made from a "pressure-release" material. From the formulation (6.73) all the boundary-value problems follow at different values of the coefficient α. In particular, the problems of Newman and Dirichlet follow at $\alpha\to 0$ and $\alpha\to\infty$. On the part of a surface (on the baffles or other structural elements of a transducer), over which $\alpha(r_{\Sigma})=const.$, the local input impedance $Z_b(r_{\Sigma})$ is specified (subscript b stands for "baffle"), where

$$Z_b(\boldsymbol{r}_{\Sigma})=\frac{P(\boldsymbol{r}_{\Sigma})}{U_n(\boldsymbol{r}_{\Sigma})}=-j\omega\rho\Phi/(\partial\Phi/\partial n)\Big|_{\Sigma}. \tag{6.74}$$

The concept of the local input impedance is applicable in the case that the parts of a baffle can move under the action of applied sound pressure independently in normal direction to the baffle surface. For example, the baffle design with heavy inserts (pieces of lead) encapsulated in polyurethane shown in Figure 6.3 can be approximated, as locally reacting. In general, the structural elements of a transducer design (such as the caps) are the elastic bodies and formulating the mixed boundary conditions on their surfaces complicates.

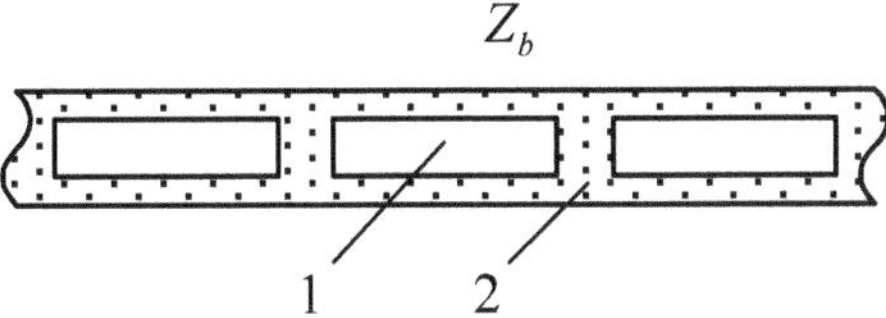

Figure 6.3: Example of a baffle design: fragments of lead (1), encapsulating material (2).

6.2.3 Sommerfeld Radiation Condition

For solution of the Helmholtz's equation to be unique, it should satisfy the condition of radiation that has to be fulfilled at great distances from the radiating body. This condition was formulated by Sommerfeld as the statement "...The energy, which is radiated from the sources must scatter to infinity, no energy may be radiated from infinity to the field." The analytical expression of this statement for 3D field (in the spherical coordinates) is

$$\lim r\left(\frac{\partial\Phi}{\partial r}+jk\Phi\right)_{r\to\infty}=0\,. \tag{6.75}$$

This condition can be formulated in the equivalent form of

$$\lim\Phi\big|_{r\to\infty}=A(\varphi,\theta)\frac{e^{-jkr}}{r}, \tag{6.76}$$

which means that the sound field at large distances from any radiating body is the outgoing spherical wave. The first factor in Eq. (6.76) does not depend on distance, and it represents the directional factor of a projector.

For the 2D acoustic field (in the cylindrical coordinates) the radiation condition takes form

$$\lim\Phi=A(\varphi)\frac{e^{-jkr}}{\sqrt{r}}, \tag{6.77}$$

and the field represents outgoing cylindrical wave with magnitude changing as $1/\sqrt{r}$ in the far field (though this case is not realistic for the transducers of a finite size).

6.2.4 Solving the Radiation Problems by Separation of Variables

The method of separation of variables in the Helmholtz equation is well suited for solving the radiation problems for a transducer in the case that its surface matches coordinate surface of a coordinate system, in which the Helmholtz equation is separable. The wave dimensions of the single transducers are usually relatively small, and the series of functions that represent solutions for a boundary-value problem quickly converge. It appears very often that retaining several terms of the series is sufficient for obtaining accurate enough results (this will be shown with below considered examples).

At first consider the Helmholtz equation in the cylindrical coordinates (the Laplacian is given by expression (6.69)),

$$\frac{1}{r}\frac{\partial}{\partial r}\left(r\frac{\partial P}{\partial r}\right)+\frac{1}{r^2}\frac{\partial^2 P}{\partial \varphi^2}+\frac{\partial^2 P}{\partial z^2}+k^2 P=0\,. \tag{6.78}$$

We represent the assumed solution for the equation as a product of functions depending on a single coordinate each, $P=F_1(r)\cdot F_2(\varphi)\cdot F_3(z)$. Due to 2π periodicity of the coordinate system over the coordinate φ it can be assumed that

$$F_2(\varphi)=A_2 e^{jn\varphi}\,, \tag{6.79}$$

where n is an integer number. After substituting $P=F_1(r)F_3(z)A_2e^{jn\varphi}$ into Eq. (6.78) and dividing all the terms by P we obtain

$$\left[\frac{1}{F_1}\frac{1}{r}\frac{\partial}{\partial r}\left(r\frac{\partial F_1}{\partial r}\right)-\frac{n^2}{r^2}\right]+\frac{1}{F_3}\frac{\partial^2 F_3}{\partial z^2}+k^2=0\,. \tag{6.80}$$

The term in brackets depends on coordinate r only and the second term depends only on coordinate z. The Eq. (6.80) can hold in all the range of the coordinates only in the case that each of these terms is constant. Denote these constants for the first and second terms as $-k_r^2$ and $-k_z^2$, respectively. Then we obtain the equations

$$\frac{1}{r}\frac{\partial}{\partial r}\left(r\frac{\partial F_1}{\partial r}\right)+\left(k_r^2-\frac{m^2}{r^2}\right)F_1=0\,, \tag{6.81}$$

and

$$\frac{d^2F_3}{dz^2}+k_z^2F_3=0\,, \tag{6.82}$$

where $k_r^2+k_z^2=k^2$.

The general solution for Eq. (6.82) is

$$F_3(z)=A_3e^{-jk_z z}+B_3e^{jk_z z}\,, \tag{6.83}$$

where A_3 and B_3 are the arbitrary constants. The solution represents two plane waves propagating in different directions of axis z. In the case that the cylindrical shell is infinitely long, and magnitude of vibration does not depend on z, $k_z = 0$.

The Eq. (6.81), which can be transformed to the more common form of

$$\frac{1}{y}\frac{d}{dy}\left(y\frac{dF_1}{dy}\right)+\left(1-\frac{n^2}{y^2}\right)F_1=0, \tag{6.84}$$

where $y = k_r r$, is the Bessel equation. The solutions of this equation are combinations of the Bessel functions, $J_n(k_r r)$, and Neumann functions, $N_n(k_r r)$. Subscript n indicates the order of the functions. The solution for the waves spreading out of a cylinder can be presented in terms of the Hankel functions of the second kind (due to $e^{j\omega t}$ time dependence)

$$H_n^{(2)}(k_r r) = J_n(k_r r) - jN_n(k_r r). \tag{6.85}$$

In the case that the cylindrical shell is infinitely long and magnitude of its vibration does not depend on z (two-dimensional acoustic wave) $k_r = k$ and the partial solutions of the Eq. (6.78) for outgoing waves are the functions

$$P_n(r,\varphi) = (A_n \cos n\varphi + B_n \sin n\varphi)H_n^{(2)}(kr). \tag{6.86}$$

In the general case that distribution of velocity on the infinitely long shell in the axial direction exists,

$$P_n(r,\varphi) = (A_n \cos m\varphi + B_n \sin n\varphi)H_n^{(2)}\left(\sqrt{k^2-k_z^2}\cdot r\right)e^{\pm jk_z z}. \tag{6.87}$$

Consider now the Helmholtz equation in the axial symmetrical spherical coordinates (the Laplacian is given by expression (6.70)),

$$\frac{1}{r^2}\frac{\partial}{\partial r}\left(r^2\frac{\partial P}{\partial r}\right)+\frac{1}{r^2\sin\varphi}\frac{\partial}{\partial\varphi}\left(\sin\varphi\frac{\partial P}{\partial\varphi}\right)+k^2P=0. \tag{6.88}$$

If to represent the sound pressure as the product of two functions $P(r,\varphi) = R_1(r)R_2(\varphi)$ and substitute this expression into the Helmholtz equation in the spherical coordinates, then the equations for R_1 and R_2 can be separated as follows

$$\frac{1}{R_1}\frac{1}{r^2}\frac{\partial}{\partial r}\left(r^2\frac{\partial R_1}{\partial r}\right)+k^2R_1=-\frac{1}{R_2}\frac{1}{\sin\varphi}\frac{\partial}{\partial\varphi}\left(\sin\varphi\frac{\partial R_2}{\partial\varphi}\right). \tag{6.89}$$

Here the left part does not depend on φ and the right side does not depend on r, and they should be equal to the same constant. This constant should be equal to $m(m+1)$ (this is shown in Ref.5), and one of equations becomes

$$\frac{d^2 R_1}{dz^2}+\frac{2}{z}\frac{dR_1}{dz}+\left[1+\frac{m(m+1)}{z^2}\right]=0\,, \tag{6.90}$$

where z=kr and $m=0, 1, 2, \ldots$.

Thus, two equations will be obtained for determining functions R_1 and R_2:

$$\frac{1}{r^2}\frac{d}{dr}\left(r^2\frac{dR_1}{dr}\right)+\left[k^2-\frac{m(m+1)}{r^2}\right]R_1=0\,, \tag{6.91}$$

$$\frac{1}{\sin\varphi}\frac{\partial}{\partial\varphi}\left(\sin\varphi\frac{\partial R_2}{\partial\varphi}\right)+m(m+1)\,R_2=0\,. \tag{6.92}$$

After substituting $\cos\varphi=x$ the Eq. (6.92) can be represented in the form of the Legendre equation

$$\frac{d}{dx}\left[(1-x^2)\frac{dR_2}{dx}\right]+m(m+1)R_2=0\,. \tag{6.93}$$

The partial solutions of Eq. (6.90), where it is denoted $z=kr$, are the spherical Bessel functions of order m:

of the first kinde,

$$j_m(z)=\sqrt{\pi/2z}\,J_{m+1/2}(z)\,; \tag{6.94}$$

of the second kind,

$$n_m(z)=\sqrt{\pi/2z}\,N_{m+1/2}(z)\,; \tag{6.95}$$

and the spherical Hankel functions that for outgoing wave are

$$h_m^{(2)}(z)=j_m(z)-jn_m(z)=\sqrt{\pi/2z}\,H_{m+1/2}^{(2)}(z)\,. \tag{6.96}$$

The partial solutions for Eq. (6.93) are the Legendre polynomials of order m,

$$P_m(x)=P_m(\cos\varphi)\,. \tag{6.97}$$

Thus, the partial solutions for the Eq. (6.87) for outgoing waves are

$$P_m(r,\varphi)=C_m P_m(\cos\varphi)\cdot h_m^{(2)}(kr)\,. \tag{6.98}$$

Properties of the special functions: cylindrical Bessel functions, spherical Bessel functions, and Legendre polynomials can be found in Refs. 5 and 6. Some of the properties of these functions that are used throughout this treatment are presented in Appendix C.

To get a unique solution for a particular radiation problem, a combination of the partial solutions given by expressions (6.86) and (6.98) must be matched to the boundary conditions on the surfaces of corresponding transducers. Examples of solving the radiation problems and calculating the radiation related parameters for the cylindrical and spherical transducers will be considered in the following sections.

6.3 Radiation of the Cylindrical Transducers.

Radiation of the cylindrical transducers will be analyzed under different boundary conditions.

At first, we consider the acoustic field radiated by an idealized model of infinitely long transducer vibrating with velocity independent on z coordinate and having axial symmetric distribution over circumference with respect to axis $\varphi = 0$ that is shown in Figure 6.4. The distribution of velocity can be represented as

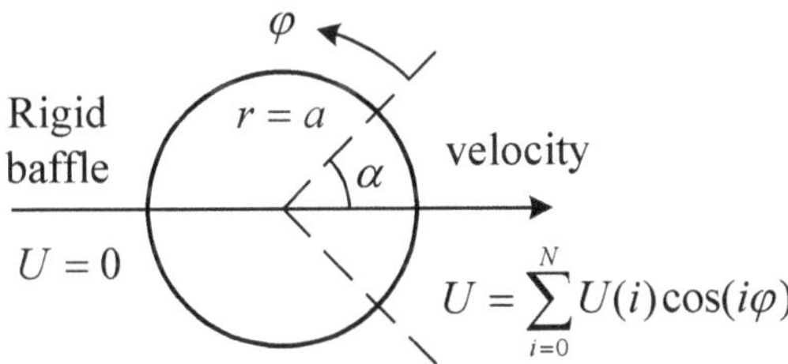

Figure 6.4: Illustration of the cylindrical shell (two-dimensional case) vibrating with arbitrary axial symmetric with respect to axis $\varphi = 0$ distribution of velocity over the circumference.

$$U(\varphi) = \sum_{i=0}^{N} U_i \cos i\varphi \text{ at } |\varphi| \le \alpha \text{ and } U(\varphi) = 0 \text{ at } |\varphi| > \alpha , \tag{6.99}$$

where $U(\varphi)$ is the complex amplitude of velocity (remember that time depending on factor $e^{j\omega t}$ is omitted for brevity), α is the "opening angle." The radiation problem is two-dimensional,

and its general solution for the sound pressure according to expression (6.86) is

$$P(r,\varphi)=\sum_{n=0}^{\infty}A_n H_n^{(2)}(kr)\cos n\varphi\,. \tag{6.100}$$

Here A_n are the arbitrary constants, which have to be determined from the condition that the velocity in the sound field at $r=a$ should be equal to the velocity $U(\varphi)$ of the radiating surface. Thus, the condition on the boundary is

$$U(r,\varphi)\big|_{r=a}=-\frac{1}{j\omega\rho}\frac{\partial P}{\partial r}\bigg|_{r=a}=-\frac{k}{j\omega\rho}\sum_{n=0}^{\infty}A_n H_n^{(2)\prime}(ka)\cos n\varphi\,. \tag{6.101}$$

After representing $U(\varphi)$ as an expansion into a Fourier series

$$U(\varphi)=\sum_{n=0}^{\infty}[\sum_{i=1}^{N}\frac{\varepsilon_n}{2\pi}U_i\, a_{ni}(\alpha)]cosn\varphi\,. \tag{6.102}$$

where $\varepsilon_n=1$ at $n=0$, $\varepsilon_n=2$ at $n\geq 1$ and

$$a_{00}(\alpha)=2\alpha,\quad a_{nn}(\alpha)=\alpha+\frac{\sin 2n\alpha}{2n},$$
$$a_{ni}(\alpha)=\int_{-\alpha}^{\alpha}\cos i\varphi\,\cos n\varphi\, d\varphi=\alpha\left[\frac{\sin(n+i)\alpha}{(n+i)\alpha}+\frac{\sin(n-i)\alpha}{(n-i)\alpha}\right], \tag{6.103}$$

and equating relations (6.102) and (6.101) we arrive at expression for values of constants A_n

$$A_n=-\,j\rho c\sum_{i=0}^{N}\frac{\varepsilon_n U_i\, a_{ni}(\alpha)}{2\pi\, H_n^{(2)\prime}(ka)}\,. \tag{6.104}$$

The important note must be made regarding notations for the transducer radiation related quantities. All of them depend on the wave size *ka* of a transducer. Therefore, the wave size will be omitted for brevity from arguments, of which these quantities depend. Thus, the sound pressure generated by a transducer will be denoted below as $P_i(r,\varphi,\alpha)$ instead of $P_i(ka,r,\varphi,\alpha)$. The analogous abbreviations will be used regarding the radiation impedances, diffraction coefficients and directional factors.

Upon substituting expression (6.104) into Eq. (6.100) we obtain

$$P(r,\varphi,\alpha)=-j\rho c\sum_{n=0}^{\infty}\left[\sum_{i=0}^{N}\frac{\varepsilon_n U_i\, a_{ni}(\alpha)}{2\pi}\right]\frac{H_n^{(2)}(kr)}{H_n^{(2)\prime}(ka)}\cos n\varphi=\sum_{i=0}^{N}P_i(r,\varphi,\alpha)\,, \tag{6.105}$$

where

$$P_i(r,\varphi,\alpha)=-j\rho c\frac{U_i}{2\pi}\sum_{n=0}^{\infty}a_{ni}(\alpha)\frac{\varepsilon_n H_n^{(2)}(kr)}{H_n^{(2)\prime}(ka)}\cos n\varphi \tag{6.106}$$

is the modal sound pressure generated by a single mode of vibration defined in the interval of values of φ $(-\alpha \le \varphi \le \alpha)$. At large distances from the cylinder (at $r \to \infty$)

$$H_n^{(2)}(kr) \to \sqrt{\frac{2}{\pi kr}}\, e^{-j(kr-\frac{\pi n}{2}-\frac{\pi}{4})}, \tag{6.107}$$

and we arrive at the following expression for the modal sound pressure

$$P_i(r,\varphi,\alpha) = -j\rho c\sqrt{\frac{2}{\pi kr}}\, e^{-j(kr-\pi/4)}\, U_i \sum_{n=0}^{\infty} \frac{a_{ni}(\alpha)\,\varepsilon_n}{2\pi} \frac{e^{j\pi n/2}}{H_n^{(2)\prime}(ka)} \cos n\varphi\,. \tag{6.108}$$

The sound field at large distances from the cylinder (in the “far zone”) according to Sommerfeld radiation condition expressed in the form of relation (6.77) can be described as product of two functions, one of which is nondimensional and depends only on the coordinate φ (the directional factor of the transducer, $H(\varphi)$) and another is a function of the distance r, the sound pressure on the acoustical axis of the transducer, $P(r,0)$. Following expressions (6.105) and (6.108) we obtain

$$P(r,0,\alpha) = -j\rho c\sqrt{\frac{2}{\pi kr}}\, e^{-j(kr-\pi/4)} \sum_{i=0}^{N} U_i \sum_{n=0}^{\infty} \frac{a_{ni}(\alpha)\,\varepsilon_n}{2\pi} \frac{e^{j\pi n/2}}{H_n^{(2)\prime}(ka)}\,. \tag{6.109}$$

Thus, the directional factor is

$$H(\varphi,\alpha) = \frac{\displaystyle\sum_{i=0}^{N} U_i \sum_{n=0}^{\infty} a_{ni}(\alpha)\,\varepsilon_n \frac{e^{-j\pi n/2}}{H_n^{(2)\prime}(ka)} \cos n\varphi}{\displaystyle\sum_{i=0}^{N} U_i \sum_{n=0}^{\infty} a_{ni}(\alpha)\,\varepsilon_n \frac{e^{j\pi n/2}}{H_n^{(2)\prime}(ka)}}\,. \tag{6.110}$$

The total power radiated per unit length of the cylinder may be found as

$$\begin{aligned} \bar{\dot{W}}_{ac} &= \int_{-\alpha}^{\alpha} P(a,\varphi,\alpha)\, U^*(\varphi)\, a\, d\varphi = \\ &= -j\frac{\rho c a}{2\pi} \sum_{n=0}^{\infty} \frac{\varepsilon_n H_n^{(2)}(ka)}{H_n^{(2)\prime}(ka)} \int_{-\alpha}^{\alpha} \left[\sum_{i=o}^{N} U_i a_{ni}(\alpha)\right]\left[\sum U_l^* \cos l\varphi\right] \cos n\varphi\, d\varphi = \\ &= -j\frac{\rho c a}{2\pi}\left[\sum_{i=0}^{N}\sum_{n=0}^{\infty} a_{ni}^2(\alpha) \frac{\varepsilon_n H_n^{(2)}(ka)}{H_n^{(2)\prime}(ka)} |U_i|^2 + \sum_{l\ne i}^{N} U_l U_i^* \sum_{n=0}^{\infty} a_{nl}(\alpha)\, a_{ni}(\alpha) \frac{\varepsilon_n H_n^{(2)}(ka)}{H_n^{(2)\prime}(ka)}\right]. \end{aligned} \tag{6.111}$$

This expression can be rewritten in the form

$$\bar{W}_{ac} = \sum_{i=0}^{N} \left[Z_{acii}(\alpha) + \sum_{l \neq i}^{N} z_{acil}(\alpha) \frac{U_l}{U_i} \right] |U_i|^2 = \sum_{i=0}^{N} Z_{aci}(\alpha) |U_i|^2 . \tag{6.112}$$

Here it is denoted

$$Z_{acii}(\alpha) = -j \frac{\rho c a}{2\pi} \sum_{n=0}^{\infty} a_{ni}^2(\alpha) \frac{\varepsilon_n H_n^{(2)}(ka)}{H_n^{(2)\prime}(ka)} , \tag{6.113}$$

as the modal radiation impedance for i-th mode of vibration defined in the interval $[-\alpha \le \varphi \le \alpha]$, and

$$z_{acil}(\alpha) = -j \frac{\rho c a}{2\pi} \sum_{n=0}^{\infty} a_{nl}(\alpha)\, a_{ni}(\alpha) \frac{\varepsilon_n H_n^{(2)}(ka)}{H_n^{(2)\prime}(ka)} , \tag{6.114}$$

as the intermodal impedance between modes i and l ($i \neq l$). Thus, the radiation impedance associated with the generalized velocity U_i is

$$Z_{aci}(\alpha) = Z_{acii}(\alpha) + \sum_{l \neq i}^{N} z_{acil}(\alpha) \frac{U_l}{U_i} . \tag{6.115}$$

In application to practical cylindrical transducer designs equating to zero velocity of vibration on a part of the transducer surface can be imagined as result of covering this part of the surface by infinitely thin absolutely rigid baffle. In this case angle α specifies sector of the transducer surface free of baffle. We consider the most important variants of velocity distributions that correspond to transducer without a baffle and with 180° baffle coverage, at $\alpha = \pi$ and $\alpha = \pi/2$, respectively.

6.3.1 Radiation of a Cylindrical Shell without Baffles

According to formula (6.103) at $\alpha = \pi$ all the coefficients $a_{ni}(\pi) = 0$ at $n \neq i$, $a_{nn}(\pi) = \pi$ at $n \neq 0$, and $a_{00}(\pi) = 2\pi$. We consider the single mode radiation. The modal sound pressure for the pulsating mode of vibration $U(\varphi) = U_0$ is

$$P_0(r, \pi) = -j\rho c\, U_0 \frac{H_0^{(2)}(kr)}{H_0^{(2)\prime}(ka)} . \tag{6.116}$$

For the modes of vibration $U(\varphi) = U_n \cos n\varphi$ at $n \neq 0$

$$P_n(r, \pi) = -j\rho c\, U_i \frac{H_n^{(2)}(kr)}{H_n^{(2)\prime}(ka)} \cos n\varphi . \tag{6.117}$$

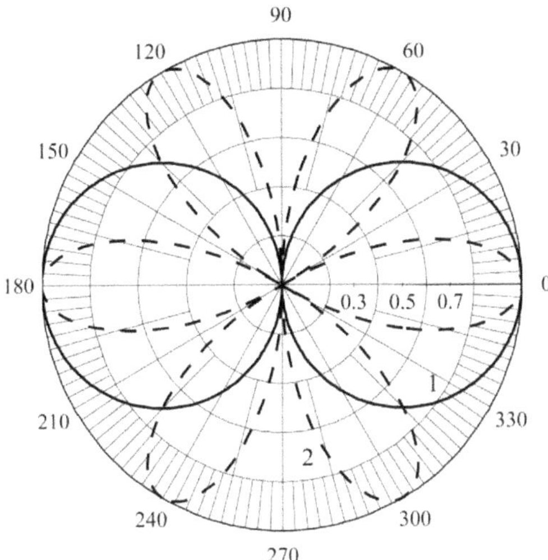

Figure 6.5: The modal directional factors of cylindrical transducers without baffles at the 1st (solid line) and 3rd mode (dashed line 2),

The directional factor of a cylinder vibrating in a single mode is $H(\varphi,\pi)=\cos n\varphi$. The directional factors corresponding to several modes of vibration are shown in Figure 6.5.

Consider sound pressure generated by the pulsating cylinder of small radius at $ka \ll 1$. Considering that $H_0^{(2)\prime}(ka)=-H_1^{(2)}(ka)$, and $H_1^{(2)}(ka)\approx j2/\pi ka$ (see Appendix C.1), from expression (6.116) will be obtained that

$$P_0(r,\pi)=\rho c\frac{k}{4}U_{\tilde{V}r}H_0^{(2)}(kr)\,, \tag{6.118}$$

where $U_{\tilde{V}r}=2\pi a U_0$ is the reference volume velocity (see Eq. (6.6)) per unit length for the pulsating cylinder. This is the sound pressure generated by the cylindrical (two-dimensional) simple source.

The diffraction coefficient for a single mode transducer is by the definition (6.10)

$$k_{dif\,n}(\pi)=\frac{P_n(r,0,\pi)}{P_0(r,ka\ll 1)}=-j^{n+1}\frac{2}{\pi ka}\frac{1}{H_n^{(2)\prime}(ka)}\,. \tag{6.119}$$

The diffraction coefficients for the cylindrical transducers vibrating in the single modes vs. ka are illustrated in Figure 6.6.

Thus, the modal sound pressure on the acoustic axis is

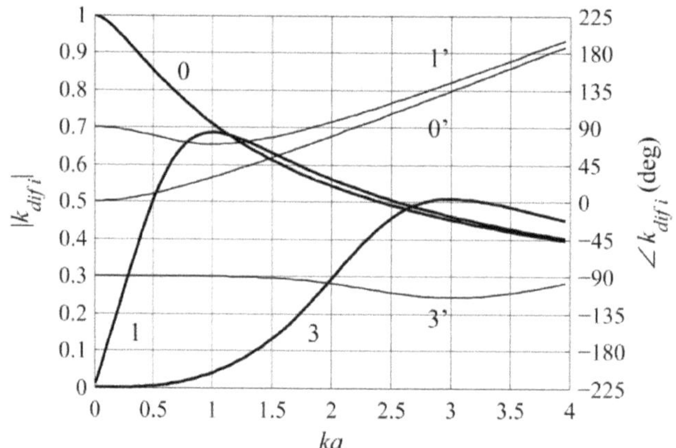

Figure 6.6: The modal diffraction coefficients of cylindrical transducers without baffles for $i = 0$, 1, 3 (phase is labeled with ').

$$P_n(r,0,\pi) = \frac{k\rho c}{4}\sqrt{\frac{2}{\pi kr}}\,2\pi a U_n\, k_{dif.n}(ka)e^{-j(kr-\pi/4)}\,. \tag{6.120}$$

Consider the radiation impedances of the cylindrical transducer per unit height associated with the generalized coordinate U_i. So far as the coefficients $a_{ni}(\pi) = 0$ at $n \neq i$, all the intermodal impedances $z_{acil}(\pi)$ in expression (6.115) vanish and $Z_{aci}(\pi) = Z_{acii}(\pi)$. According to formula (6.113) in this case the modal impedance for i^{th} mode of vibration is

$$Z_{acii}(\pi) = -j\rho ca\,\frac{a_{ii}^2(\pi)\,\varepsilon_i}{2\pi}\,\frac{H_i^{(2)}(ka)}{H_i^{(2)\prime}(ka)}\,. \tag{6.121}$$

It can be concluded from this relation that

$$-j\frac{\rho c\, a}{2\pi}\,\frac{H_n^{(2)}(ka)}{H_n^{(2)\prime}(ka)} = \frac{1}{a_{nn}^2(\pi)\,\varepsilon_n}\,Z_{acnn}(\pi)\,, \tag{6.122}$$

and after substituting into expression (6.113) of the term corresponding to the left side of this equality by its right side we arrive at

$$Z_{acii}(\alpha) = \sum_{n=0}^{\infty}\frac{a_{ni}^2(\alpha)}{a_{nn}^2(\pi)}\,Z_{acnn}(\pi)\,. \tag{6.123}$$

Thus, if the modal impedances for the unbaffled cylindrical shell (at $\alpha = \pi$) are determined, calculating the modal radiation impedances for an arbitrary baffle coverage (angle α) may be reduced to a mere calculating the coefficients $a_{ni}(\alpha)$.

It is worthwhile discussing in more detailed way the radiation impedances for the widest used 0 and 1 modes of vibration.

For the zero ("pulsating") mode of vibration $U(\varphi)=U_0$, and

$$Z_{ac00}(\pi)=-j\rho ca\frac{a_{00}^2(\pi)\,\varepsilon_0}{2\pi}\frac{H_0^{(2)}(ka)}{H_0^{(2)\prime}(ka)}=-j\rho c\,2\pi a\frac{H_0^{(2)}(ka)}{H_0^{(2)\prime}(ka)}. \tag{6.124}$$

(Note that $a_{00}(\pi)=2\pi$ and $\varepsilon_0=1$). Expression (6.124) may be modified as follows

$$\begin{aligned}Z_{ac00}(\pi)&=j\rho c\,2\pi a\frac{J_0(ka)-jN_0(ka)}{J_1(ka)-jN_1(ka)}=r_{ac00}(\pi)+jx_{ac00}(\pi)=\\&=\rho c2\pi a\,[\alpha_{00}(\pi)+j\beta_{00}(\pi)]\end{aligned}. \tag{6.125}$$

Here r_{ac00} and x_{ac00} are the active and reactive components of the radiation impedance, α_{00} and β_{00} are the nondimensional coefficients. If to take into account that

$$J_{n=1}(x)N_n(x)-J_n(x)N_{n+1}(x)=\frac{2}{\pi x}, \tag{6.126}$$

(see Appendix C.1), then we obtain from (6.125)

$$\alpha_{00}(\pi)=\frac{2}{\pi ka}\frac{1}{J_1^2(ka)+N_1^2(ka)}, \tag{6.127}$$

$$\beta_{00}(\pi)=\frac{J_0(ka)J_1(ka)+N_0(ka)N_1(ka)}{J_1^2(ka)+N_1^2(ka)}. \tag{6.128}$$

In the "long-wave" approximation (at $ka<<1$) $J_0(ka)\approx 1$, $N_0(ka)\approx(2/\pi)\ln ka$, $J_1(ka)\approx ka/2$, $N_1(ka)\approx -2/\pi ka$, and from equations (6.127) and (6.128) will be obtained $\alpha_{00}(\pi)\doteq\pi ka/2$, $\beta_{00}(\pi)\doteq ka\ln(1/ka)$, or

$$r_{ac00}(\pi)\approx\frac{\pi}{2}\rho c\frac{(2\pi a)^2}{\lambda}, \tag{6.129}$$

$$x_{ac00}(\pi)\approx[2\rho\pi a^2\ln(1/ka)]\,\omega=m_{ac}\,\omega, \tag{6.130}$$

where $m_{ac}=2M_w\ln(1/ka)$ and $M_w=\rho\pi a^2$ is the mass of water in the volume of the cylinder per unit length.

In the "short-wave" limit (at $ka\to\infty$) the asymptotic approximation (6.107) for the Hankel functions must be used (with substitution of *kr* by *ka*.) It will be obtained as the result that $\alpha_{00}\to 1$ and $\beta_{00}\to 0$. Thus, at $ka\to\infty$

$$r_{ac00}(\pi)\approx\rho c\,2\pi a,\quad x_{ac00}(\pi)\to 0. \tag{6.131}$$

For the first ("oscillating") mode of vibration $U(\varphi)=U_1\cos\varphi$, and after substituting $a_{11}(\pi)=\pi$ and $\varepsilon_1=2$ we obtain from formula (6.121) that

$$Z_{ac11}(\pi)=-j\rho c\pi a\frac{H_1^{(2)}(ka)}{H_1^{(2)\prime}(ka)}, \tag{6.132}$$

where
$$H_1^{(2)\prime}(x)=H_0^{(2)}(x)-\frac{1}{x}H_1^{(2)}(x). \tag{6.133}$$

In the long-wave approximation, at $ka\ll 1$,

$$H_1^{(2)}(ka)\approx\frac{ka}{2}+j\frac{2}{\pi ka},\quad H_1^{(2)\prime}(ka)\approx-j\frac{2}{\pi(ka)^2} \tag{6.134}$$

And

$$Z_{ac11}(\pi)=\rho c\pi a[\frac{\pi}{4}(ka)^3+jka]. \tag{6.135}$$

Thus,

$$r_{ac11}(\pi)\approx\rho c\pi^2 a(ka)^3/4, \tag{6.136}$$

$$x_{ac11}(\pi)\approx\rho c\,\pi a\,(ka)=\rho\pi a^2\,\omega=M_w\,\omega\,. \tag{6.137}$$

The ratio of the active components of the radiation impedances for the first and zero mode transducers of small size is

$$r_{ac11}(\pi)/r_{ac00}(\pi)=(ka)^2/4\,. \tag{6.138}$$

In the short-wave limit, at $ka\to\infty$, considering the asymptotic approximations for $H_1^{(2)}(ka)$ and $H_1^{(2)\prime}(ka)$ we arrive at

$$r_{ac11}(\pi)\approx\rho c\pi a=\rho cS_{eff1},\quad x_{ac11}(\pi)\approx 0\,. \tag{6.139}$$

In the expression for the radiation resistance the effective area of radiating surface per unit length, S_{eff}, is introduced, which is defined in the general case as

$$S_{eff}=\int_{\Sigma}\theta^2(r_\Sigma)d\Sigma, \tag{6.140}$$

where Σ is the radiating surface and $\theta(r_\Sigma)$ is the mode of vibration of the surface. For the cylindrical transducers with modes of vibration $\theta_n(\varphi)=\cos n\varphi$

$$S_{eff\,n}=\int_{-\alpha}^{\alpha}\cos^2 n\varphi\, ad\varphi, \tag{6.141}$$

and at $\alpha = \pi$ $$S_{eff\,0} = 2\pi a, \quad S_{eff\,n} = \pi a. \tag{6.142}$$

The nondimensional coefficients of the radiation impedances per unit length of the cylindrical shells vibrating in the single modes vs. *ka* are presented in Figure 6.7

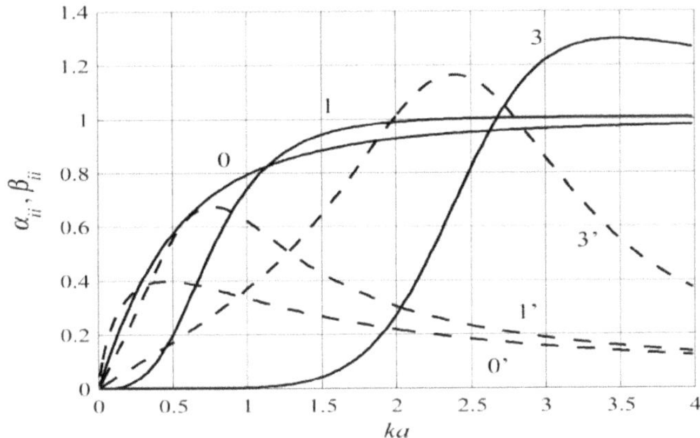

Figure 6.7: The modal nondimensional coefficients of the radiation impedances of cylindrical transducers without baffles for $i = 0, 1, 3$ (β_{ii} are labeled with ')

6.3.2 Radiation of Cylindrical Shell with Rigid Baffle at $\alpha = \pi/2$.

If to suppose that velocity distribution within interval $[-\pi/2 \le \varphi \le \pi/2]$ is $\theta(\varphi) = \cos i\varphi$, then the coefficients a_{ni} defined by formula (6.103) will be

$$a_{00}(\pi/2) = \pi, \; a_{ii}(\pi/2) = \pi/2, \; a_{ni}(\pi/2) = \frac{1}{n+i}\sin(n+i)\frac{\pi}{2} + \frac{1}{n-i}\sin(n-i)\frac{\pi}{2}. \tag{6.143}$$

All the coefficients at $n+i=2l$ $(n \ne i, \; l = 1,2,...)$ will be zero. For the zero and first modes the coefficients are:

$$a_{00}(\pi/2) = \pi, \; a_{01}(\pi/2) = 2, \; a_{02}(\pi/2) = 0, \; a_{03}(\pi/2) = -2/3, \; ... \tag{6.144}$$

$$a_{10}(\pi/2) = 2, \; a_{11}(\pi/2) = \pi/2, \; a_{12}(\pi/2) = -2/3, \; a_{13}(\pi/2) = 0, ... \tag{6.145}$$

Sound pressure $P_i(r,\varphi,\pi/2)$ for a single mode transducer will be defined by formula (6.108) with coefficients $a_{ni}(\pi/2)$ calculated according to formulas (6.144). The corresponding diffraction coefficients will be determined as the ratio (see the note related to expression (6.119))

$$k_{dif.i}(\pi/2)=\frac{P_i(r,0,\pi/2))}{P_0(r,ka\ll 1)}=\frac{2}{\pi^2 ka}\sum_{n=0}^{\infty} j^{n-1}\frac{a_{in}(\pi/2)\,\varepsilon_n}{H_n^{(2)\prime}(ka)}. \quad (6.146)$$

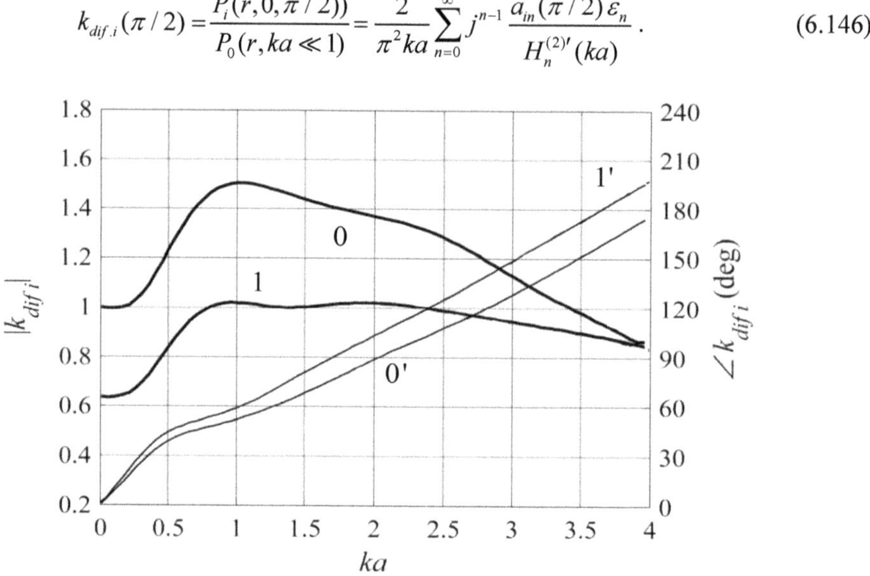

Figure 6.8: The modal diffraction coefficients of cylindrical transducers with baffles at $\alpha=\pi/2$ for the 0 and 1 modes (0' and 1' for the phase).

When calculating sound pressure $P_0(r,ka\ll 1)$ generated by the pulsating cylinder of a small radius the reference volume velocity per unit length is $U_{\tilde{V}r}=U_i\pi a$. In general, if diffraction coefficients are determined for a baffled cylinder at arbitrary angle of the baffle opening, the modal sound pressures on the acoustic axis can be calculated as

$$P_i(r,0,\alpha)=k_{dif.i}(\alpha)\,P_0(r,ka\ll 1), \quad (6.147)$$

and the reference volume velocity must be used in the form $U_{\tilde{V}r}=2a\cdot\alpha$, when calculating P_0 by formula (6.118).

The diffraction coefficients for the baffled cylinders vibrating in the single modes vs. *ka* are presented in Figure 6.8.

Directional factors of the cylinders at $\alpha=\pi/2$ may be calculated by formula

$$H_i(\varphi,\pi/2)=\frac{\sum_{n=0}^{\infty}(j)^n a_{in}(\pi/2)\,\varepsilon_n\cos n\varphi\,/\,H_n^{(2)\prime}(ka)}{\sum_{n=0}^{\infty}(j)^n a_{in}(\pi/2)\,\varepsilon_n\,/\,H_n^{(2)\prime}(ka)}. \quad (6.148)$$

The directional factors of the cylinders vibrating in the zero and first modes for several values of ka are illustrated in Figure 6.9.

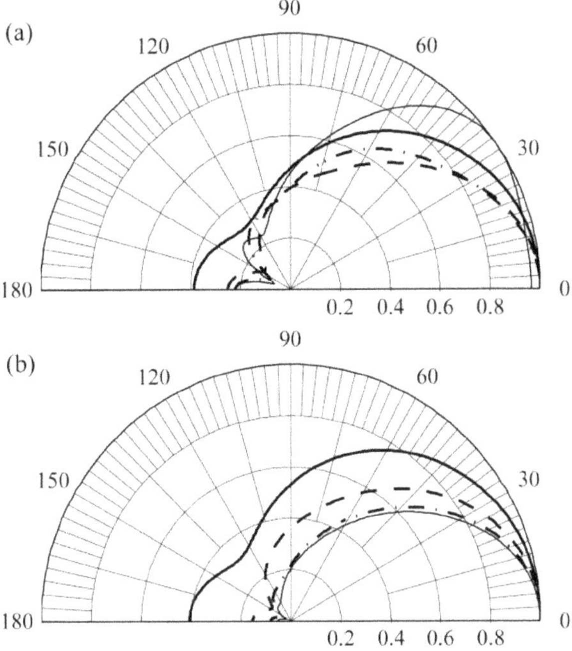

Figure 6.9: The modal directional factors of transducers at $\alpha = \pi/2$ baffle coverage for $ka = 1$ (solid line), $ka = 2$ (dashed line), $ka = 3$ (dash-dotted line), $ka = 4$ (thin solid line): (a) zero mode, (b) the first mode.

The radiation impedance for the cylindrical shell vibrating in a single mode will be determined, if to substitute the coefficients $a_{ni}(\pi/2)$ calculated by formula (6.143) into expression (6.113) or (6.123). Thus, for the zero and first modes we obtain

$$Z_{ac00}(\pi/2) = \frac{1}{2} Z_{ac00}(\pi) + \frac{2}{\pi} Z_{ac11}(\pi) - \frac{2}{3} Z_{ac33}(\pi) + \ldots \tag{6.149}$$

$$Z_{ac11}(\pi/2) = \frac{1}{\pi} Z_{ac00}(\pi) + \frac{1}{2} Z_{ac11}(\pi) + \frac{2}{3\pi} Z_{ac22}(\pi) + \ldots \tag{6.150}$$

Nondimensional coefficients of the radiation impedances are represented in Figure 6.10, given that $S_{eff0}(\pi/2) = \pi a$ and $S_{eff1}(\pi/2) = \pi a/2$.

Consider now simultaneous radiation of several modes of vibration. We assume that the velocity distribution is

$$U(\varphi)=\sum_{i=0}^{N}U_i\cos i\varphi \text{ at } |\varphi|\le\pi/2 \text{ and } U(\varphi)=0 \text{ when } |\varphi|>\pi/2. \quad (6.151)$$

The magnitudes of vibration U_i can be considered as the generalized velocities in the problem of calculating the cylindrical transducers. They must be determined by solving the Lagrange equations that describe the transducers operation.

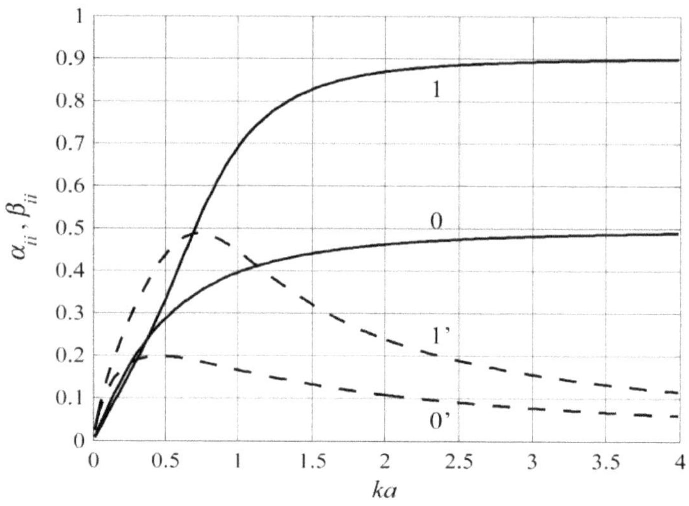

Figure 6.10: Nondimensional coefficients of the modal radiation impedances for zero (0) and first (1) modes at $\alpha=\pi/2$ (β_{ii} labeled with ').

According to Eq. (6.105) the sound pressure generated by the cylindrical shell may be found as superposition of the modal sound pressures $P_i(r,\varphi,\pi/2)$ represented by formula (6.108). Considering expressions (6.146) and (6.147) for the diffraction coefficient and sound pressure, respectively, finally we obtain

$$P_i(r,\varphi,\pi/2)=\rho c\frac{\pi ka}{4}\sqrt{\frac{2}{\pi kr}}e^{-j(kr-\pi/4)}U_i\,k_{dif.i}(\pi/2)\cos i\varphi, \quad (6.152)$$

and the directional factor may be represented as follows

$$H(\varphi,\pi/2)=\frac{\sum_{i=0}^{N}U_i\,k_{dif.i}(\pi/2)\cos i\varphi}{\sum_{i=0}^{N}U_i\,k_{dif.i}(\pi/2)}. \quad (6.153)$$

The radiation impedance associated with velocity U_i in this case will be represented following general formula (6.115) as

$$Z_{aci}(\pi/2) = Z_{acii}(\pi/2) + \sum_{l\neq i}^{N} z_{acil}(\pi/2)\frac{U_l}{U_i}, \tag{6.154}$$

where the self-modal impedances $Z_{acii}(\pi/2)$ are illustrated with the examples by formulas (6.149), (6.150) and Figure 6.10. The intermodal impedances z_{acil} given by general formula (6.114), after using relation (6.122) can be modified as follows

$$z_{acil}(\pi/2) = \sum_{n=0}^{\infty} \frac{a_{ni}(\pi/2)a_{nl}(\pi/2)}{a_{nn}^2(\pi)} Z_{acnn}(\pi). \tag{6.155}$$

Thus, for example,

$$\begin{aligned} z_{ac01}(\pi/2) = {} & \frac{a_{00}(\pi/2)a_{01}(\pi/2)}{a_{00}^2(\pi)} Z_{ac00}(\pi) + \frac{a_{10}(\pi/2)a_{11}(\pi/2)}{a_{11}^2(\pi)} Z_{ac11}(\pi) + \\ & + \frac{a_{20}(\pi/2)a_{21}(\pi/2)}{a_{22}^2(\pi)} Z_{ac22}(\pi) + \ldots, \end{aligned} \tag{6.156}$$

or taking into account expressions (6.143)-(6.145) and the fact that all coefficients $a_{ni}(\pi/2)$ are zero, for which $n \neq i$ and $n+i$ is even,, we obtain

$$z_{ac01}(\pi/2) = \frac{1}{2\pi} Z_{ac00}(\pi) + \frac{1}{\pi} Z_{ac11}(\pi). \tag{6.157}$$

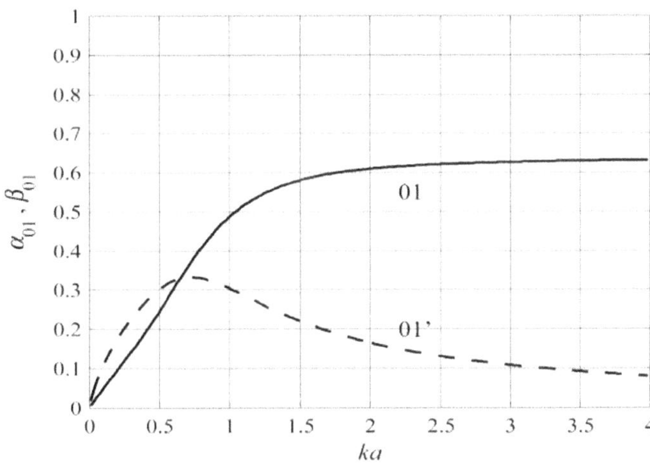

Figure 6.11: Nondimensional coefficients of intermodal radiation impedances at $\alpha = \pi/2$ (β_{01} labeled with ')

After substituting expressions for the modal impedances

$$\begin{aligned} Z_{ac00}(\pi) &= (\rho c)_w 2\pi a[\alpha_{00}(\pi) + j\beta_{00}(\pi)] \text{ and} \\ Z_{ac11}(\pi) &= (\rho c)_w \pi a[\alpha_{11}(\pi) + j\beta_{11}(\pi)] \end{aligned} \tag{6.158}$$

it may be concluded that

$$z_{ac01}(\pi/2) = (\rho c)_w \pi a\{[\alpha_{00}(\pi) + \alpha_{11}(\pi)]/\pi + j[\beta_{00}(\pi/2) + \beta_{11}(\pi/2)]/\pi\}. \tag{6.159}$$

Thus, the nondimensional coefficients of intermodal impedance $z_{ac01}(\pi/2)$ are

$$\alpha_{01}(\pi/2) = \frac{1}{\pi}[\alpha_{00}(\pi) + \alpha_{11}(\pi)] \text{ and } \beta_{01}(\pi/2) = \frac{1}{\pi}[\beta_{00}(\pi) + \beta_{11}(\pi)]. \tag{6.160}$$

These coefficients vs. *ka* are presented in Figure 6.11.

6.3.3 Radiation of an Infinite Cylindrical Shell with Compliant Baffle

In the case that the baffles applied to a cylindrical transducer surface are compliant, the better approximation to a real situation gives the assumption that a part of the surface is covered by absolutely compliant material. This assumption brings us to the radiation problem with mixed boundary conditions, in which case on one part of the surface the radial velocity is specified and on the rest of the surface the sound pressure is supposed to be zero, as shown in Figure 6.12.

Thus, the boundary conditions for the velocity potential are

$$\left.\frac{\partial \Phi}{\partial r}\right|_{r=a} = -U(\varphi), \quad \alpha \le |\varphi| < \pi, \tag{6.161}$$

$$\Phi(a,\varphi) = 0, \quad |\varphi| < \alpha. \tag{6.162}$$

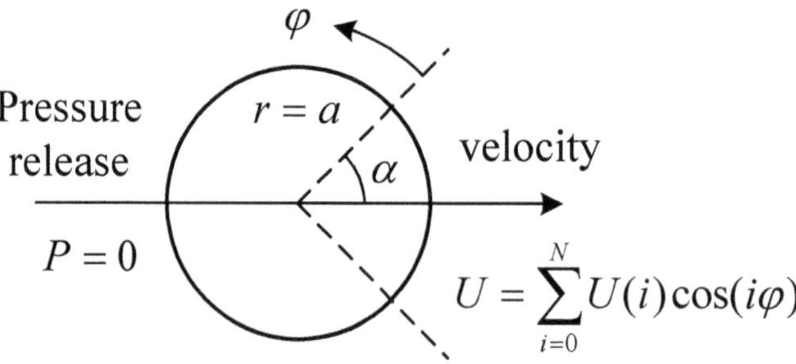

Figure 6.12: Illustration of the mixed boundary conditions.

The problem of radiation under these boundary conditions was considered in the previous works[7, 8]. A brief description of procedure used for getting the solution in Ref. 7 is as follows. Assuming that the distribution of velocity is symmetrical in respect to φ axis, a general solution to wave equation $\Delta\Phi + k^2\Phi = 0$ can be represented by expression (6.100) as

$$\Phi(r,\varphi) = \sum_{n=0}^{\infty} A_n\, H_n^{(2)}(kr)\cos n\varphi\,, \tag{6.163}$$

where the coefficients A_n must be found using the boundary conditions (6.161) and (6.162).

For getting an approximate solution to the problem the polynomials are introduced

$$\Phi_N(r,\varphi) = \sum_{n=0}^{N} A_n\, H_n^{(2)}(kr)\cos n\varphi\,. \tag{6.164}$$

These functions satisfy the wave equation, but don't satisfy the boundary conditions. However, the coefficients A_n can be determined in such a way as to best approximate the boundary conditions. The best approximation of the functions Φ_N to the solution of the problem will be defined as those minimizing the functional

$$F_N(A_0, A_1, ..., A_N) = \int_0^{\alpha} \left|\Phi(a,\varphi) - \Phi_N(a,\varphi)\right|^2 d\varphi + \int_{\alpha}^{\pi} \left| U(\varphi) - \left(\frac{\partial \Phi_N}{\partial r}\right)_{r=a} \right|^2 d\varphi \tag{6.165}$$

by proper chois of coefficients A_n. In general, these coefficients are complex quantities and may be represented as $A_n = a_n + jb_n$. The conditions of minimum for the functional F_N is formulated as

$$\frac{\partial F_N}{\partial a_i} = 0, \quad \frac{\partial F_N}{\partial b_i} = 0, \quad (i = 0, 1, ..., N)\,. \tag{6.166}$$

The following set of equations for determining the coefficients A_n follows from Eqs. (6.166),

$$\sum_{n=0}^{N} c_{ni}(\alpha, ka)\, A_n = \frac{k\pi}{2} d_i(\alpha)\, H_i'^{*}(ka), \quad (i = 0, 1, ..., N)\,. \tag{6.167}$$

In these equations (*) is the sign of complex conjugate, superscript $^{(2)}$ for Hankel function is omitted for brevity, and coefficients c_{ni} and d_i are:

$$d_0 = \frac{2}{\pi}\int_{\alpha}^{\pi} U(\varphi)\,d\varphi, \quad d_i = \frac{2}{\pi}\int_{\alpha}^{\pi} U(\varphi)\cos i\varphi\,d\varphi$$

$$c_{00}(\alpha, ka) = \alpha\left|H_0(ka)\right|^2 + k^2(\pi-\alpha)\left|H_0'(ka)\right|^2;$$

$$c_{ii}(\alpha, ka) = \frac{1}{2}(\alpha + \frac{\sin 2i\alpha}{2i})\left|H_i(ka)\right|^2 + \frac{k^2}{2}(\pi-\alpha-\frac{\sin 2i\alpha}{2i})\left|H_i'(ka)\right|, \tag{6.168}$$

$$(i = 1, 2, ..N);$$

$$c_{ni}(\alpha, ka) = \frac{1}{2}[\frac{\sin(n+i)\alpha}{n+i} + \frac{\sin(n-i)\alpha}{n-i}]\times$$

$$\times[H_n(ka)\,H_i^*(ka) - k^2 H_n'(ka)\,H_i'^* ka)], \quad (n \neq i).$$

Consider, for example, the case that $\alpha = \pi/2$, and $N=3$ as the first approximation. In accordance with formulas (6.168)

$$d_0 = 1, \quad d_1 = -\frac{2}{\pi}, \quad d_3 = \frac{2}{3\pi}, \quad d_2 = d_4 = 0,$$

$$c_{00}(\pi/2, ka) = \frac{\pi}{2}\left|H_0(ka)\right|^2 + k^2\frac{\pi}{2}\left|H_0'(ka)\right|^2,$$

$$c_{ii}(\pi/2,\, ka) = \frac{\pi}{4}\left|H_i(ka)\right|^2 + k^2\frac{\pi}{4}\left|H_i'(ka)\right|^2,$$

$$c_{ni} = 0 \text{ at } n+i = 2p, \quad (p = 1, 2, ...),$$

$$c_{01} = H_0(ka)\,H_1^*(ka) - k^2\,H_0'(ka)\,H_1'^*(ka), \tag{6.169}$$

$$c_{03} = -\frac{1}{3}[H_0(ka)\,H_3^*(ka) - k^2 H_0'(ka)\,H_0'^*(ka)],$$

$$c_{12} = \frac{1}{3}[H_1(ka)\,H_2^*(ka) - k^2 H_1'(ka)\,H_2'^*(ka)],$$

$$c_{23} = \frac{3}{5}[H_2(ka)\,H_3^*(ka) - k^2 H_2'(ka)\,H_3'^*(ka)],$$

$$c_{32} = c_{23}, \quad c_{30} = c_{03}.$$

The first approximation is

$$\Phi_3(r, \varphi) = \sum_{n=0}^{3} A_n\,H_n^{(2)}(kr)\cos n\varphi\,, \tag{6.170}$$

where coefficients A_n must be determined from the set of equations

$$\begin{array}{ccccccccc} c_{00}A_0 & + & c_{10}A_1 & + & 0 & + & c_{30}A_3 & = & \frac{k\pi}{2}H_0'^*(ka), \\ c_{01}A_0 & + & c_{11}A_1 & + & c_{21}A_2 & + & 0 & = & -k\,H_1'^*(ka), \\ 0 & + & c_{12}A_1 & + & c_{22}A_2 & + & c_{32}A_3 & = & 0, \\ c_{03}A_0 & + & 0 & + & c_{23}A_2 & + & c_{33}A_3 & = & \frac{k}{3}H_3'^*(ka). \end{array} \tag{6.171}$$

Results of calculations made for the case that the compliant baffles are used in comparison with those obtained with the rigid baffles are presented in Figure 6.13-Figure 6.15.

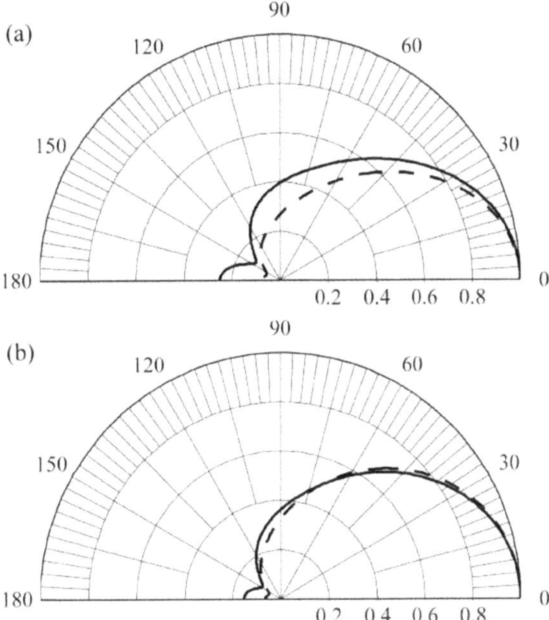

Figure 6.13: Directional factors of 180° baffled cylindrical transducers with the rigid (solid line) and compliant (dashed line) baffles at $ka = 2.0$: (a) zero mode of vibration, (b) the first mode of vibration.

The results of calculations show that significant difference between the two cases in terms of active radiation impedance and diffraction coefficient exists at small ka ($ka < 0.3 - 0.5$). Otherwise, the data are close. In terms of the directional factors, the main difference is in the back radiation, which is much smaller in the case of the compliant baffle.

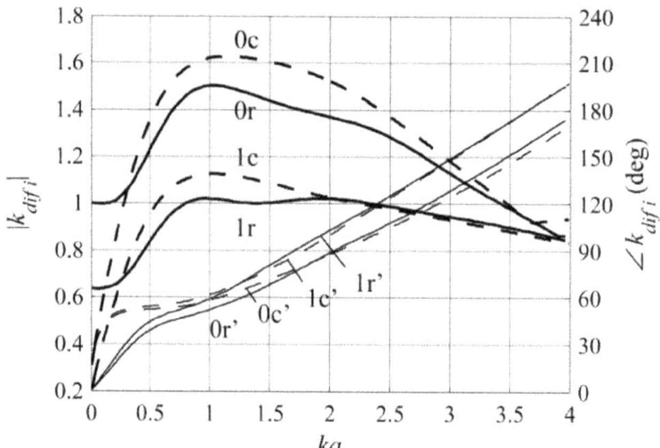

Figure 6.14: The diffraction coefficients of 180° baffled cylindrical transducers with the rigid (r – solid line) and compliant (c – dashed line) baffles at $ka = 2.0$: (a) zero (0) mode of vibration, (b) the first (1) mode of vibration (phase labeled with ').

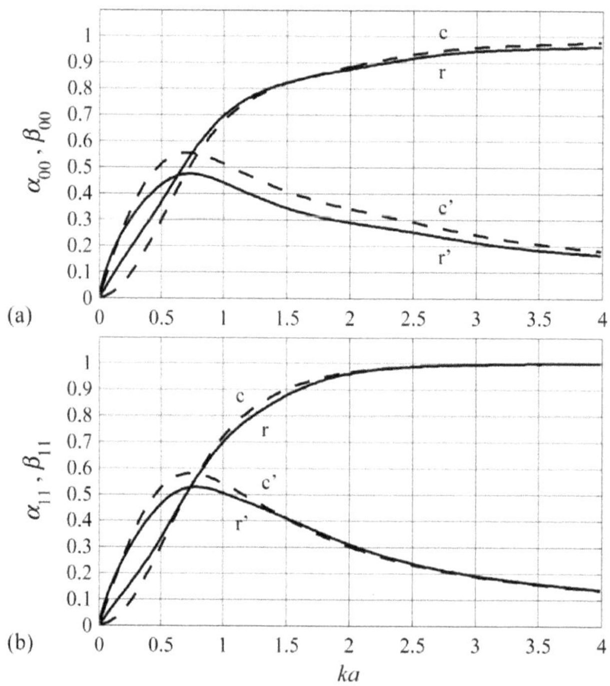

Figure 6.15: The nondimensional coefficients of radiation impedances of 180° baffled cylindrical transducers with the rigid and compliant baffles: (a) zero mode of vibration, (b) the first mode of vibration.

This conclusion is important for transducers modeling because the real designs of compliant baffles can be built (for example, out of corprene encapsulated in polyurethane) unlike the rigid baffles that can be imagined only theoretically. As the operating range of the piezoceramic cylindrical projectors is practically at $ka > 1$, all the calculations of acoustic field related parameters of the baffled transducers can be made using relatively simple expressions that are valid for the rigid baffles, rather than by employing the above technique specified for the mixed boundary conditions. It is noteworthy that the conclusion regarding comparison of effects of the rigid and compliant baffles does not mean that results of calculations made for a not ideal baffle (having finite impedance) can be expected to lie in between. Both perfectly rigid and perfectly compliant baffles are ideal and don't transfer acoustic energy, whereas the real baffles transfer a portion of acoustic energy and this may change the situation.

6.3.4 Radiation of a Finite Size Cylinder in an Infinitely Long Rigid Cylindrical Baffle

The model of infinitely long cylindrical surface with arbitrary velocity distribution along the circumference is very useful. It appears that the results obtained regarding the radiation impedances per unit length are applicable sufficiently accurately to the finite height transducers in case that their height is comparable with wavelength of sound. More accurate and applicable to transducers of smaller height is the model of finite height cylinder with infinitely long rigid cylindrical extensions on the ends that is shown in Figure 6.16).

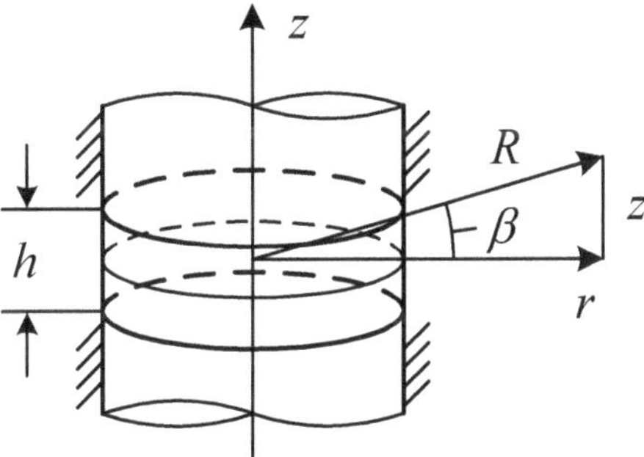

Figure 6.16: The finite height cylinder with rigid cylindrical baffle extensions.

Related problems are considered in several works [9-12]. A brief outline of the solution to the radiation problem is as follows.

A partial solution for the three dimensional Helmholtz equation in the cylindrical coordinates can be represented following the expressions (6.83) - (6.87), as

$$P_m(r,\varphi,z) = (A_1 e^{-jm\varphi} + B_1 e^{jm\varphi}) H_m^{(2)}(k_r r)(A_2 e^{-jk_z z} + B_2 e^{jk_z z}), \tag{6.172}$$

where $k_r^2 + k_z^2 = k^2$. Thus, the general solution for an arbitrary distribution of velocity along z axis can be presented in the form of

$$P(r,\varphi,z) = \sum_{m=-\infty}^{\infty} a_m e^{jm\varphi} \int_{-\infty}^{\infty} A(k_z) H_m^{(2)}(\sqrt{k^2 - k_z^2}\, r)\, e^{jk_z z} dk_z\,. \tag{6.173}$$

Here a_m and $A(k_z)$ must be determined from the boundary conditions for a particular velocity distribution. The term

$$H_m^{(2)}(\sqrt{k^2 - k_z^2}\, r)\, e^{\pm jk_z z} \tag{6.174}$$

in the expression (6.173) represents two cylindrical waves with wave numbers k_z that propagate in the positive and negative directions of the z axis. At k_z real (i.e., at $k > k_r$) they represent traveling waves propagating out of the source. At k_z imaginary ($k < k_r$) they form inhomogeneous waves that are dying down in radial direction. These waves contribute to the sound pressure in the near field and to the reactive (inertia) component of the radiation impedance. Suppose that arbitrary distribution of velocity on the surface at $r = a$ is a separable function of φ and z

$$U(\varphi,z) = U_0 \theta_1(\varphi)\theta_2(z), \tag{6.175}$$

where U_0 is velocity of the reference point at $\varphi = 0$, θ_1 and θ_2 are the normalized velocity distributions along the circumference and z axis. Thus, the boundary condition is

$$U\big|_{r=a} = U_0 \theta_1(\varphi)\theta_2(z) = -\frac{1}{j\omega\rho}\frac{\partial P}{\partial r}\bigg|_{r=a}. \tag{6.176}$$

Suppose that

$$b_m = \frac{1}{2\pi}\int_0^{2\pi} \theta_1(\varphi) e^{-jm\varphi} d\varphi \text{ and } B(k_z) = \int_{-\infty}^{\infty} \theta_2(z) e^{-jk_z z} dz \tag{6.177}$$

are the Fourier coefficients and Fourier transform of the velocity distributions in the circumference and *z* directions, respectively. Let us suggest for simplicity that the transducer is a cylinder of height *h* performing axisymmetric vibration with uniform distribution over the height, i.e.,

$$\theta_1(\varphi)=1, \quad \theta_2(z)=\begin{cases}1 & at\ |z|<h/2\\ 0 & at\ |z|>h/2\end{cases}. \tag{6.178}$$

Thus, by formulas (6.177)

$$b_0=1, \quad b_m=0 \text{ at } m\neq 0 \text{ and } B(k_z)=h\cdot\frac{\sin(k_z h/2)}{k_z h/2}=h\cdot H_h(k_z). \tag{6.179}$$

The designation $[\sin(k_z h/2)/k_z h/2]=H_h(k_z)$ is introduced for brevity.

After substituting expressions (6.179) into the boundary conditions and determining the unknown coefficients a_m and $A(k_z)$, we arrive at the following expression for the sound pressure generated by the cylindrical transducer embedded into the rigid cylindrical baffle,

$$P(r,z)=\frac{-j\omega\rho h U_0}{2\pi}\int_{-\infty}^{\infty} H_h(k_z)\frac{H_0^{(2)}(\sqrt{k^2-k_z^2}\,r)\,e^{-jk_z z}}{\sqrt{k^2-k_z^2}\,H_0^{(2)\prime}(\sqrt{k^2-k_z^2}\,a)}dk_z=\frac{-j\omega\rho h U_0}{2\pi}\cdot I. \tag{6.180}$$

Employing this function allows in principle calculating all the acoustic field related parameters listed in Section 6.1.3 by formulas presented therein for the cylindrical transducers of finite height embedded into the infinite rigid cylindrical baffle. But calculation of the integral in formula (6.180) is not straightforward. The manipulations of the integral that result in formulas, which can readily be used for calculating the acoustic field in the near zone (that is needed for calculating the radiation impedances) and in the far zone (that is needed for calculating the directional factors and diffraction coefficients) can be fulfilled as follows.

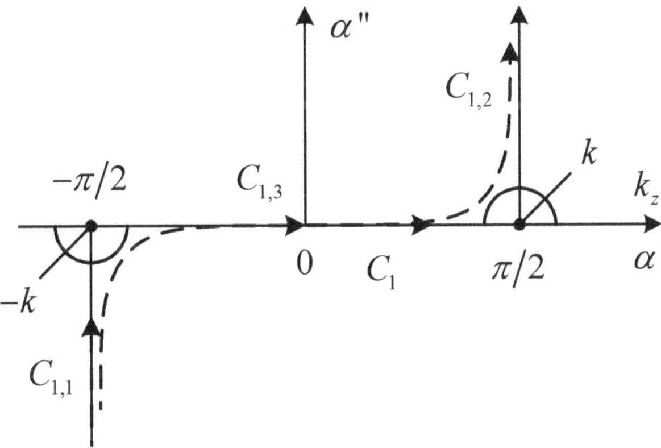

Figure 6.17: The contour of integration in the complex plane.

Consider the integral I from expression (6.180),

$$I = \int_{-\infty}^{\infty} H_h(k_z) \frac{H_0^{(2)}(\sqrt{k^2 - k_z^2}\, r)\, e^{-jk_z z}}{\sqrt{k^2 - k_z^2}\, H_0^{(2)\prime}(\sqrt{k^2 - k_z^2}\, a)} dk_z\,, \tag{6.181}$$

following Ref. 9. The expression under the integral has singularities at the points $k_z = \pm k$. Let us change variables by introducing $k_z = k\sin\alpha$. As integration goes over the interval $-\infty < k_z < \infty$ (Figure 6.17), it is supposed to be $|\sin\alpha| = |(e^{j\alpha} - e^{-j\alpha})/2j| > 1$, and the variable α should be complex. Let it be $\alpha = \alpha' + j\alpha''$. Then

$$k_z = k(\sin\alpha'\cosh\alpha'' + j\cos\alpha'\sinh\alpha''). \tag{6.182}$$

As k_z should be a real number, it follows that $\cos\alpha'\sinh\alpha'' = 0$ and

$$k_z = k\sin\alpha'\cosh\alpha''. \tag{6.183}$$

At $|k_z| < k$ from this condition follows that $\alpha'' = 0$ and α' changes in the interval $(-\pi/2) < \alpha' < (\pi/2)$. At the large values of $|k_z| > k$ it is obviously $\alpha'' \neq 0$ (otherwise it would be $|\sin\alpha'| > 1$) and it follows from the condition (6.183) that $\alpha' = \pm(\pi/2)$, and $k_z = \pm k\cosh\alpha''$. The function $\cosh\alpha''$ is even, and question arises about the sign of α'' (about the direction, in which the branches $\alpha' = \pm(\pi/2)$ go on the complex plane).

It can be shown from consideration of convergence of the Hankel function at $kr \gg 1$ that these branches have to go as it is shown in Figure 6.17. The singularity points $\pm k$ ($\alpha' = \pm(\pi/2)$) must be passed over the arcs of infinitely small radius ε. Thus, the integral (6.181) becomes

$$I = \int_{-\infty}^{\infty} H_h(k\sin\alpha) \frac{H_0^{(2)}(k\, r\cos\alpha)\, e^{-jkz\sin\alpha}}{H_0^{(2)\prime}(ka\cos\alpha)} d\alpha\,, \tag{6.184}$$

where integration goes over the contour C_1 in the complex plane shown in Figure 6.17, which includes the following intervals:

Interval $C_{1,1}$, over which $-\infty < \alpha'' < 0$, $\alpha' = -\pi/2$, $\sin\alpha = -\cosh\alpha''$, $\cos\alpha = \sinh\alpha''$;

Interval $C_{1,2}$, over which $0 < \alpha'' < \infty$, $\alpha' = \pi/2$, $\sin\alpha = \cosh\alpha''$, $\cos\alpha = -\sinh\alpha''$;

Interval $C_{1,3}$, over which $\alpha'' = 0$, $-\pi/2 < \alpha' < \pi/2$.

It can be shown that integral (6.148) over the arcs with radius $\varepsilon \to 0$ vanishes. Thus, the integral (6.148) may be presented, as

$$I = \int_{-\infty}^{\infty} H_h(k\cosh\alpha'') \frac{H_0^{(2)}(kr\sinh\alpha'')\,e^{-jkz\cosh\alpha''}}{H_0^{(2)\prime}(ka\sinh\alpha'')}d\alpha'' + \int_{-\pi/2}^{\pi/2} H_h(k\sin\alpha) \frac{H_0^{(2)}(kr\cos\alpha)\,e^{-jkz\sin\alpha}}{H_0^{(2)\prime}(ka\cos\alpha)}d\alpha. \tag{6.185}$$

In course of further manipulations with the comprising integrals the following relations for the Bessel functions will be used (Appendix C.3):

$$H_m^{(2)}(-jx) = -\frac{2}{j\pi}e^{j\frac{\pi m}{2}}K_m(x), \tag{6.186}$$

where $K_m(z)$ are the modified Bessel functions of the second kind,

$$H_0^{(2)\prime}(x) = -H_1^{(2)}(x), \quad K_0'(x) = -K_1(x). \tag{6.187}$$

After manipulations involving above relations, we finally arrive at the expression for the integral (6.185)

$$I = -\int_{-\infty}^{\infty} H_h(ka\cosh\alpha'') \frac{K_0(kr\sinh\alpha'')}{K_1(ka\sinh\alpha'')}e^{-jkz\cosh\alpha''}d\alpha'' - \int_{-\pi/2}^{\pi/2} H_h(ka\sin\alpha') \frac{H_0^{(2)}(kr\cos\alpha')}{H_1^{(2)}(ka\cos\alpha')}e^{-jkz\sin\alpha'}d\alpha'. \tag{6.188}$$

This expression can be used for calculating the acoustic field regardless of a distance from the source, but it is most suitable for determining the near field parameters. For calculating the sound pressure in the far field, the integral (6.184) can be further transformed using the asymptotic representation of the Hankel function.

6.3.4.1 Near Field of the Finite-Height Cylinder and Radiation Impedance

The sound pressure (6.180) generated by the cylinder becomes after substituting expression (6.188) for the integral I

$$P(r,z) = \frac{j\omega\rho hU_0}{2\pi}\left[\int_{-\infty}^{\infty} H_h(ka\cosh\alpha'') \frac{K_0(kr\sinh\alpha'')}{K_1(ka\sinh\alpha'')}e^{-jkz\cosh\alpha''}d\alpha'' + \int_{-\pi/2}^{\pi/2} H_h(ka\sin\alpha') \frac{H_0^{(2)}(kr\cos\alpha')}{H_1^{(2)}(ka\cos\alpha')}e^{-jkz\sin\alpha'}d\alpha'\right]. \tag{6.189}$$

The radiation impedance can be found by the relation

$$Z_{ac} = \frac{2\pi a}{U_0} \int_{-h/2}^{h/2} P(a,z)dz = r_{ac} + jx_{ac} \tag{6.190}$$

After performing the integration under the integrals in expression (6.189), and some straightforward manipulations that include: replacement the variables $\alpha' \to \theta$ and $\alpha'' \to \psi$, substituting H_h from expression (6.179), and presenting the Hankel functions through the Bessel and Neumann functions, the following expressions will be obtained for the radiation resistances and reactances:

$$r_{ac} = \frac{16\omega\rho}{\pi k^3} \int_0^{\pi/2} A d\theta, \tag{6.191}$$

where
$$A = \frac{\sin^2[(kh/2)\sin\theta]}{\sin^2\theta \cdot \cos\theta[J_1^2(ka\cos\theta) + N_1^2(ka\cos\theta)]}; \tag{6.192}$$

$$x_{ac} = \frac{8a\omega\rho}{k^2}\left[\int_0^{\pi/2} B d\theta + \int_0^{\infty} C d\psi\right], \tag{6.193}$$

where

$$B = \frac{\sin^2[(kh/2)\sin\theta][J_0(ka\cos\theta)J_1(ka\cos\theta) + N_0(ka\cos\theta)N_1(ka\cos\theta)}{\sin^2\theta \cdot \cos\theta[J_1^2(ka\cos\theta) + N_1^2(ka\cos\theta)]}, \tag{6.194}$$

$$C = \frac{\sin^2[(kh/2)\cosh\psi]K_0(ka\sinh\psi)}{\cosh^2\psi K_1(ka\sinh\psi)}. \tag{6.195}$$

Expressions (6.191) - (6.195) coincide with expressions for the radiation impedance derived in a different way in Ref. 10. Note that function C rapidly goes to zero with increase of the argument ψ due to existence of $\cosh^2\psi$ in the denominator.

Results of calculation of nondimensional (normalized to the surface area of the cylinder) coefficients of the radiation resistance and reactance vs. ka are presented in Figure 6.18).

At small ka

$$J_1(ka\cos\theta) \to 0, \quad N_1(ka\cos\theta) \approx 2/\pi ka\cos\theta, \tag{6.196}$$

and after substituting $\sin\theta = x$ expression (6.191) can be transformed to

$$r_{ac} = \frac{\rho c}{\pi h^2} S_{cyl}^2 \int_0^1 \frac{\sin^2(kh/2)x}{x^2} dx. \tag{6.197}$$

After integrating by parts (putting $u = \sin^2(kh/2)x$ and $dv = z^{-2}dz$), finally can be obtained the following expression for the radiation resistance

$$r_{ac} = \rho c S_{cyl} \frac{2a}{h}[kh \cdot Si(kh) - 2\sin^2(kh/2)], \tag{6.198}$$

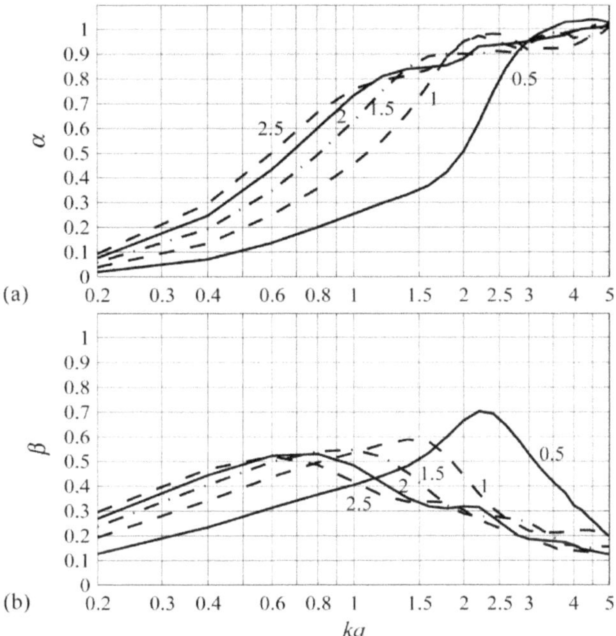

Figure 6.18: Dependences of the nondimensional coefficients of the radiation resistance (a) and reactance (b) on ka for aspect ratios $h/2a = 0.5, 1.0, 1.5, 2.0, 2.5$.

where the function

$$Si(kh) = \int_0^{kh} \frac{\sin y}{y} dy \tag{6.199}$$

is the sine integral [5, 6]. Dependence of nondimensional coefficient of the radiation resistance on kh for different aspect ratios of cylinders at small ka are shown in Figure 6.19.

When analyzing the data on the radiation impedances it should be kept in mind that especially important to have these data accurate for the ranges of ka around the resonance wave sizes of the transducers. Thus, for transducers made of PZT compositions that employ the extensional vibrations of rings $ka > 2$, for the transducers of the flexural type (including slotted rings) $ka \approx (0.2 - 0.4)$.

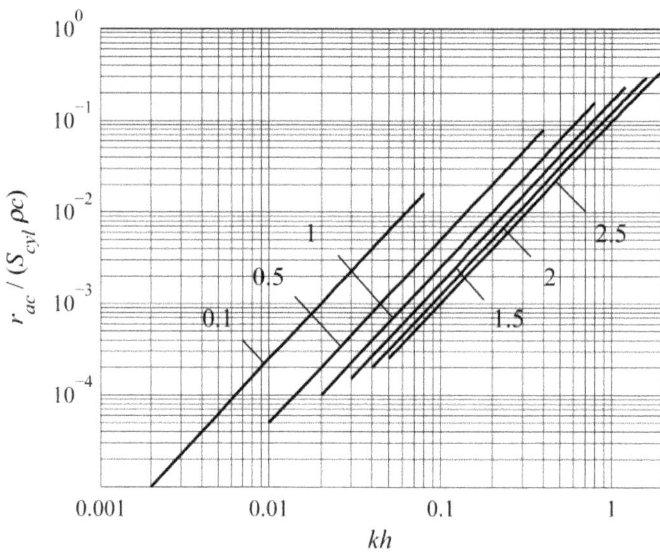

Figure 6.19: Curves showing dependence of radiation resistance from *kh* for small *ka* at different cylinder aspect ratios, $h/2a = 0.1, 0.5, 1.0, 1.5, 2.0, 2.5$.

6.3.4.2 Far Field of the Finite-Height Transducer and the Directional Factor

For calculating the sound pressure in the far field integral (6.184) can be transformed using the asymptotic representation of the Hankel function

$$H_m^{(2)}(kr\cos\alpha) \approx \sqrt{\frac{2}{\pi kr\cos\alpha}} e^{-jkr\cos\alpha} e^{jm\pi/2 + j\pi/4}. \tag{6.200}$$

After substituting this expression into the integral (6.184) we obtain at $m = 0$

$$I \approx e^{j\pi/4} \int_{C_1} H_h(k\sin\alpha) \sqrt{\frac{2}{\pi kr\cos\alpha}} \frac{e^{-jk(r\cos\alpha + z\sin\alpha)}}{H_0^{(2)\prime}(ka\cos\alpha)} d\alpha. \tag{6.201}$$

The exponential term in the numerator represents a plane wave propagating under angle α in respect to the perpendicular to the cylinder axis. Thus, the integral summarizes a set of the waves propagating under different angles. They include the regular traveling waves propagating away from the source (they correspond to real values of α) and inhomogeneous waves, which are dying out in direction perpendicular to the axis (they correspond to the imaginary values of α). Indeed, when integrating along the branches $C_{1.1}$ and $C_{1.2}$ ($\alpha = -(\pi/2) - j\alpha''$ and $\alpha = (\pi/2) + j\alpha''$, respectively) $\cos\alpha = -j\sinh\alpha''$ and the factor $e^{-jkr\cos\alpha} = e^{-kr\sinh\alpha''}$

vanishes with increase of r. The corresponding waves propagate along the cylinder axis and decay in the radial direction. By the way, this shows that direction of vertical branches in Figure 6.17 is chosen correctly. Otherwise, this factor would be $e^{kr\sinh\alpha''}$, and the wave would infinitely increase with increase of r in violation of the radiation principle.

For the far field approximation, we assume that the distance to an observation point is very large ($R \gg h$). Under this assumption the stationary phase method can be used for approximate calculating integral (6.201). This method is applicable to calculating integrals of the kind

$$I = \int_C F(y) e^{jpf(y)} dy, \tag{6.202}$$

where parameter p is large, $f(y)$ is an analytical function. The stationary phase point y_0 is a root of equation $f'(y) = 0$. If $F(y)$ changes slowly in vicinity of the point $y = y_0$, then the asymptotic formula can be obtained for integral (6.202)

$$I \approx \frac{\sqrt{2\pi}\, F(y_0) e^{jpf(y_0)}}{\sqrt{pe^{j\pi} f''(y_0)}}. \tag{6.203}$$

At $kr \gg 1$ integral (6.201) is of the (6.202) kind with $p = kr$ and $f(\alpha) = \cos a + (z/r)\sin\alpha$. The exponential term rapidly oscillates, while the remainder of the integrant function changes slowly in vicinity of the stationary point, which being found from $f'(\alpha_0) = 0$ happens to be at

$$\alpha_0 = \arctan\frac{z}{r} = \beta \tag{6.204}$$

Thus, the stationary point indicates direction to an observation point (Figure 6.16). Using formula (6.203), we arrive at the approximate value of integral (6.201) (note that $r = R\cos\beta$, $z = R\sin\beta$)

$$I \approx \frac{2e^{j\pi/4}}{k\cos\beta H_0^{(2)\prime}(ka\cos\beta)} H_h(k\sin\beta)\frac{e^{-jkR}}{R}. \tag{6.205}$$

The farther an observation point is from the source, the more accurate is this approximation.

Consider now the expression (6.180) for the sound pressure in the far field. After substituting the expression for integral I from Eq. (6.205), representing $H_h(k\sin\beta)$ by formula

(6.179), and remembering that $H_0^{(2)\prime} = -H_1^{(2)}$, it will be obtained

$$P(R,\beta) = \frac{\rho ch}{\pi} U_0 e^{j\pi/4} \left[\frac{1}{\cos\beta H_1^{(2)}(ka\cos\beta)} \cdot \frac{\sin[(kh/2)\sin\beta]}{(kh/2)\sin\beta} \right] \frac{e^{-j(kR-\pi/2)}}{R}. \tag{6.206}$$

The factor in the brackets does not depend on distance to an observation point and characterizes the directivity factor of a finite height cylinder imbedded in the rigid cylindrical baffle. The last factor characterizes the outgoing spherical wave. This result is in accordance with the radiation principle.

The directional factor of the cylinder in the vertical plane is

$$H_{cb}(\beta) = \frac{P(R,\beta)}{P(R,0)} = \frac{H_1^{(2)}(ka)}{\cos\beta H_1^{(2)}(ka\cos\beta)} \cdot \frac{\sin[(kh/2)\sin\beta]}{(kh/2)\sin\beta}. \tag{6.207}$$

At $ka \ll 1$

$$H_1^{(2)}(ka\cos\beta) \approx \frac{2j}{\pi ka\cos\beta}, \tag{6.208}$$

and the directional factor of a thin cylinder in a rigid baffle is the same as for a line segment of length h in the free space,

$$H_h(\beta) = \frac{\sin[(kh/2)\sin\beta]}{(kh/2)\sin\beta}. \tag{6.209}$$

At $kh \ll 1$ from formula (6.207) follows that the factor

$$H_{rb}(\beta) = \frac{H_1^{(2)}(ka)}{\cos\beta H_1^{(2)}(ka\cos\beta)} \tag{6.210}$$

determines directionality in the vertical plane of a short ring embedded in the rigid cylindrical baffle. For comparison, the directional factor of a short ring in the free space is

$$H_{rf}(\beta) = J_0(ka\cos\beta). \tag{6.211}$$

Plots of the directional factors of a short ring in the vertical plane are shown in Figure 6.20 for different values of ka in comparison with those for the ring in the free space.

Thus, the directional factor of a cylindrical transducer in the vertical plane, $H_{cb}(\beta)$, is the product of the directional factors of a low ring embedded in a rigid baffle and of the line segment in the free space that has the height of the transducer,

$$H_{cb}(\beta) = H_{rb}(\beta) H_h(\beta)\,. \tag{6.212}$$

Being illustrated for simplicity with example of the uniformly vibrating cylinder, this product theorem is valid for an arbitrary distribution of velocity on a cylinder surface so far as the distribution of the velocity is separable, i.e., can be represented as $U(\varphi, z) = U_0\, \theta_1(\varphi)\theta_2(z)$ .This conclusion follows from the procedure of deriving the result, though it requires more cumbersome manipulations.

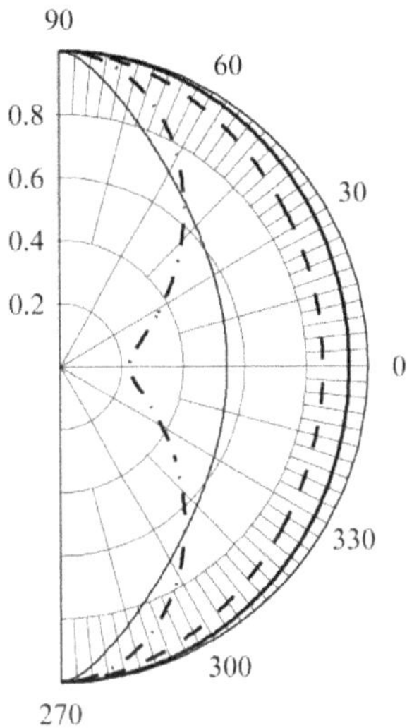

Figure 6.20: Plots of the directional factors of a short ring in the vertical plane at $ka = 0.5$ in the free space (solid line), at $ka = 0.5$ embedded in a rigid baffle (dashed line), at $ka = 2$ in free space (dash-dotted line), and at $ka = 2$ embedded in a rigid baffle (thin solid line).

Directionality of a finite cylinder vibrating in a rigid baffle with arbitrary distribution of velocity over circumference in the horizontal plane (at $\beta = 0$) is the same, as for the infinitely long cylinder having the same distribution of velocity. As to the directional factor of a baffled ring in the vertical plane, it can be qualitatively estimated that direction of maximum radiation deviates from the axis $\beta = \pi / 2$ in the plane of symmetry of the baffled ring, and its relative magnitude reduces.

The comparisons made show that at wave sizes of a cylinder approximately $(h/\lambda) > 0.8$ and $(h/2a) > 1$ the results obtained for the model of finite height cylinder vibrating in the rigid cylindrical baffle are close enough (for radiation impedances per unit height) to those obtained for much simpler model of infinite cylinder. At smaller wave heights and aspect ratios the model of finite size cylinder is more appropriate (though more complicated). But it is not clear to what extent this model can be accurate enough for short cylinders. In order to make this estimation and to provide means for modeling acoustic parameters of the short rings, the radiation problem has to be considered for cylinders vibrating in the free space (without the rigid baffle extensions). This problem is much more complicated for analytical solution. It does not allow for separation of variables in the Helmholtz equation and, strictly speaking, requires involvement of numerical methods for its solution. Nevertheless, in order to complete analysis of various variants of radiating by the cylindrical transducers the method of solving this problem and results obtained for representative aspect ratios of the cylindrical transducers is considered in the next section.

6.3.5 Radiation of a Finite-Size Cylinder in the Free Space

A number of works were devoted to solving the radiation problem for the finite size cylinders. Review of related bibliography on this issue can be found in Refs. 13, 14, which present probably the most comprehensive, easy to implement and physical clear interpretation of the problem. The technique described in Refs. 13, 14 and in more detailed way in Ref. 15, provides sufficient accuracy over a wide range of ka and aspect ratios $h/2a$ under arbitrary boundary conditions. The results obtained converge to the limiting cases of a long ($h/2a \gg 1$) and short ($h/2a \ll 1$) cylinders, and were experimentally verified for several cylindrical transducer designs having intermediate aspect ratios in Ref. 16. The brief outline of the technique used therein for solving the finite cylinder radiation problem is as follows.

The geometry of the cylinder in the cylindrical coordinate system is shown in Figure 6.21 (a). The solution is found by considering the integral Helmholtz equation

$$\varepsilon\Phi(x) - \iint_S K(x,x_0)\Phi(x_0)dS_0 = f(x)\,, \tag{6.213}$$

where

$$K(x,x_0) = \frac{1}{4\pi}\frac{\partial g(x,x_0)}{\partial n_0}\,, \tag{6.214}$$

$$f(x) = \frac{1}{4\pi}\iint_S U(x_0) g(x, x_0) dS_0 \,, \tag{6.215}$$

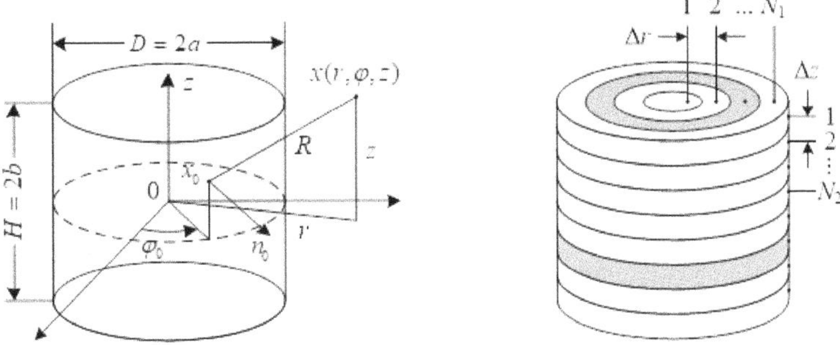

Figure 6.21: (a) Geometry of the cylinder with dimensions and reference coordinates, (b) partitioning of the cylindrical surface into rings and bands.

$$g(x, x_0) = \frac{e^{-jkR}}{R}, \quad R = |x - x_0| \,, \tag{6.216}$$

$$\varepsilon = \begin{cases} 1, & x \in D_{out}, \\ 1/2, & x \in S, \\ 0, & x \in D_{in}. \end{cases} \tag{6.217}$$

In these equations: $\Phi(x)$ is the velocity potential, $g(x, x_0)$ is the three-dimensional Green's function, U is the surface velocity, S is the surface of the cylinder, x and x_0 are the coordinates of observation point and of a point on the surface, R is the distance between observation and surface points; ε is coefficient that has a value based on where the observation point is taken (either outside the surface, on the surface, or inside the surface), n_0 is the outer normal to the surface.

Substituting expressions (6.214)- (6.217) into Eq. (6.213) yields

$$4\pi\varepsilon\Phi(x) - \iint_S \frac{\partial}{\partial n_0}\left[\frac{e^{-jkR}}{R}\right]\Phi(x) dS_0 = \iint_S U(x_0)\left[\frac{e^{-jkR}}{R}\right] dS_0 \,, \tag{6.218}$$

where the distance R given in cylindrical coordinates is known to be

$$R = \sqrt{(r^2 - r_0^2) - 2rr_0 \cos(\varphi - \varphi_0) + (z - z_0)^2} \,. \tag{6.219}$$

For most transducers intended for omnidirectional radiation in the horizontal plane, vibrations are axisymmetric, and this significantly reduces numerical computations. We will consider this case in order to avoid complications unnecessary for illustrating approach to solving the problem.

The first step in solving Eq. (6.218) is to divide the cylindrical surface into top, bottom and side surfaces ($S = S_t + S_b + S_s$) in order to evaluate the left-hand side and right-hand side surface integrals. After letting $\Phi(r,\varphi,x) = \Phi(r,x)$ and $U(r,\varphi,x) = U(r,x)$, and some manipulation, Eq. (6.218) can be represented as

$$\begin{aligned}
&4\pi\varepsilon\Phi(r,z) - \int_0^a \Phi(r_0,b)(z-b)\int_0^{2\pi} \frac{e^{-jkR_t}}{R_t^3}(1+jkR_t)r_0 d\varphi_0 dr_0 + \\
&+\int_0^a \Phi(r_0,-b)(z+b)\int_0^{2\pi} \frac{e^{-jkR_b}}{R_b^3}(1+jkR_b)r_0 d\varphi_0 dr_0 + \\
&+\int_{-b}^b \Phi(a,z_0)a\int_0^{2\pi} \frac{e^{-jkR_s}}{R_s^3}(1+jkR_s)(a-r\cos\varphi_0)d\varphi_0 dz_0 = \\
&=\int_0^a U(r_0,b)\int_0^{2\pi} \frac{e^{-jkR_t}}{R_t} r_0 d\varphi_0 dr_0 + \int_0^a U(r_0,-b)\int_0^{2\pi} \frac{e^{-jkR_b}}{R_b} r_0 d\varphi_0 dr_0 + \\
&+\int_{-b}^b U(a,z_0)a\int_0^{2\pi} \frac{e^{-jkR_s}}{R_s} d\varphi_0 dz_0.
\end{aligned} \tag{6.220}$$

where

$$\begin{aligned}
R_t &= \sqrt{r^2 + r_0^2 - 2rr_0\cos\varphi_0 + (z-b)^2}, \\
R_b &= \sqrt{r^2 + r_0^2 - 2rr_0\cos\varphi_0 + (z+b)^2}, \\
R_s &= \sqrt{r^2 + a^2 - 2ra\cos\varphi_0 + (z-z_0)^2}.
\end{aligned} \tag{6.221}$$

The next step is to partition the cylinder into discrete rings (on the top and bottom surfaces) and bands (on the lateral surface), as shown in Figure 6.21 (b), for numerical integration. Note that if in the general case there exists a velocity (or potential) distribution along the circumference of the cylinder (i.e., vibrations are not axisymmetric), then the rings and bands need to be further partitioned along the azimuth. Extending this analysis to the general case is straightforward.

With an axisymmetric vibration the velocity potential and velocity are constant over each ring and band (shaded in grey). There are N_1 rings on each of the top and bottom surfaces (over [0, *a*]) and $2N_2$ bands on the side surface (over [-*b*, *b*]) for a total of $N = 2(N_1 + N_2)$ discrete partitions. Solution to the integral Helmholtz equation will be found by solving *N* linear equations in the form

$$(4\pi\varepsilon - Y_{ij})\Phi_i = V_{ij}U_j \text{ where } i = 1\ldots N,\ j = 1\ldots N. \tag{6.222}$$

The complex $N \times N$ matrices Y_{ij} and V_{ij} are determined by summing all the non-planar contributions from all the partitions for each observation point. Note that Eqs. (6.222) can be used to determine the potential on the surface if distribution of velocities is given, or to determine the velocity on the surface if distribution of potentials is given. This can be done simply by interchanging columns of the two matrices.

In order to compute the radiation impedance of the cylinder, the observation points are chosen to be located on the surface at the center of each ring or band. The combined effect on the observation points from all other partitions vibrating forms each complex matrix element. For more accurate results, the contributing ring or band is further divided into M_1 smaller rings or M_2 smaller bands and the contributing point is swept over the azimuth in M_3 discrete angles.

The discrete implementation of Eq. (6.220) can be used to calculate coefficients of radiation impedance for finite cylinders by expression

$$Z_{ac} = \frac{1}{U_0 U_0^*} \iint_S P(x_0) U^*(x_0) dS_0 =$$
$$= \frac{2\pi}{|U_0|^2} \left[\int_0^a P(r_0, b) U^*(r_0, b) r_0 dr_0 + \int_0^a P(r_0, -b) U^*(r_0, -b) r_0 dr_0 \right. \tag{6.223}$$
$$\left. + \int_{-b}^{b} P(a, z_0) U^*(a, z_0) a dz_0 \right],$$

where U_0 is the reference velocity at the center of the lateral surface and $P = jk(\rho c)_w \Phi$ is the sound pressure. Plots of the nondimensional radiation coefficients of a uniformly vibrating ($U / U_0 = 1$ on the lateral surface) finite cylinder with rigid end caps ($U(r_o, b) = 0$ on the top and bottom surfaces) for height to diameter aspect ratios of $h/2a$ = 0.2, 0.3, 0.5, 0.7, 1.0, 1.5, 2.0 compared to the infinitely long cylinder case ($h / 2a \to \infty$) are shown in Figure 6.22.

Comparison of the plots for the nondimensional coefficients of radiation impedances of the finite size cylinders vibrating in the rigid cylindrical baffle (Figure 6.18) and in the free space at the same aspect ratios is presented in Figure 6.23

In addition to the plots in Figure 6.23 it has to be noted that at $h / \lambda > 0.8$ the nondimensional coefficient $\alpha(ka, h / \lambda)$ can be taken as $\alpha(ka)$ for the infinite cylinder of the same diameter within 10% accuracy. And with increasing ka, the value of h / λ, for which this approximation holds, decreases and $h/\lambda \approx 0.3$ with $ka \to \infty$.

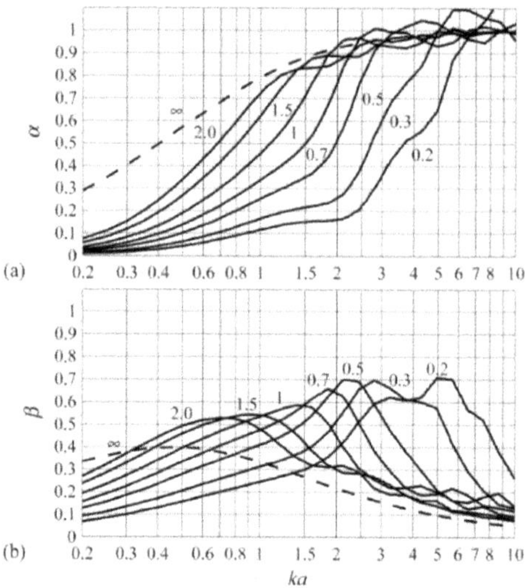

Figure 6.22: Nondimensional radiation coefficients of finite cylinders with rigid end caps for various aspect ratios: $h/2a = 0.2$, 0.3, 0.5, 0.7, 1.0, 1.5, 2.0, ∞ (labeled on the plots).

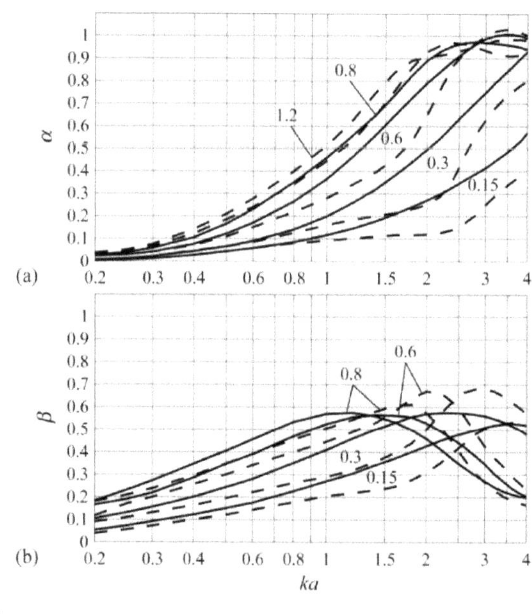

h

Figure 6.23: Comparison of the values of nondimensional coefficients of radiation impedances for finite pulsating cylinders in a rigid cylindrical baffle (solid lines) and without a baffle with fixed ends (dotted line). $h/2a = 0.15$ (1); 0.3 (2); 0.6 (3); 0.8 (4); 1.2 (5).

It is of interest for practical transducers designing to compare the values of radiation impedances for the transducers with different conditions on the ends, such as with rigid and compliant caps. For the latter case the boundary conditins on the ends are $P(r_0,|b|)=0$ instead of $U(r_0,|b|)=0$. Results of calculating the radiation resistance for the compliant ends are given in Figure 6.24. Comparison of the radiation impedances for the tranducers with rigid and comploant end caps at aspect ratio *h/2a* = 0.5 is presented in Figure 6.25.

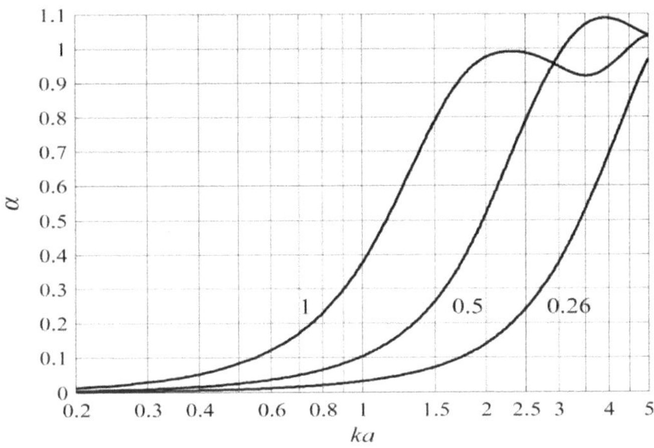

Figure 6.24: Nondimensional coefficients of radiation resistance for a transducer with compliant caps for *h/2a* = 0.26, 0.5, 1.0.

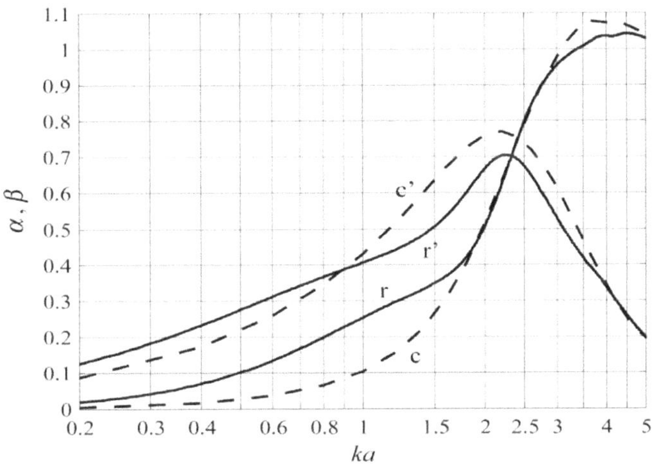

Figure 6.25: The nondimensional coefficients of radiation resistance and reactance for *h/2a* = 0.5. Rigid ends (r, r' – solid lines), compliant ends (c, c' – dashed lines), reactance is labeled with '.

6.4 Radiation of the Spherical Transducers

6.4.1 General Case

Consider the acoustic field radiated by a spherical shell (Figure 6.26) vibrating under arbitrary axial symmetric velocity distribution over its surface (for a thin-walled sphere it can be considered that $r_{out} = a + t/2 \approx a$)

$$U(a,\varphi) = \dot{\xi}_r(\varphi) = \sum_{i=0}^{N} U_i P_i(\cos\varphi), \tag{6.224}$$

where $\xi_r(\varphi)$ is the radial displacement of a spherical shell, $U_i = \dot{\xi}_i$ is the generalized velocity (the modal velocity) at $\varphi = 0$. In the general case a part of the sphere can be covered with a baffle. We will assume that the baffle is absolutely rigid and covers a segment of the sphere at $\varphi \geq \varphi_b$, as it is shown in Figure 6.26. Therefore, the condition on the surface is

$$U(a,\varphi) = \sum_{i=0}^{N} U_i P_i(\cos\varphi) \text{ at } \varphi \leq \varphi_b \text{ and } U(\varphi) = 0 \text{ at } \varphi > \varphi_b. \tag{6.225}$$

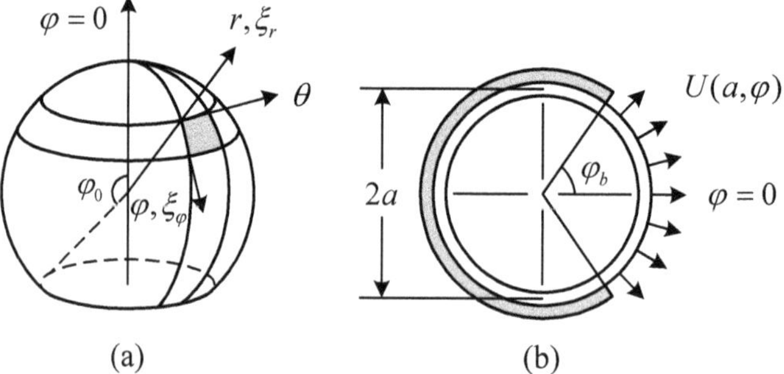

Figure 6.26: (a) Geometry of the spherical shell and coordinate system, (b) illustration of the spherical shell with a conformal baffle.

The general solution for the radiation problem is

$$P(r,\varphi) = \sum_{l=0}^{\infty} A_l h_l^{(2)}(kr) P_l(\cos\varphi). \tag{6.226}$$

Here $h_l^{(2)}(kr)$ are the spherical Hankel functions of the second kind, $P_l(\cos\varphi)$ are the Legendre polynomials of order l. The arbitrary constants A_l must be determined from condition of matching to the boundary conditions (6.225). i.e.,

$$U(a,\varphi) = -\frac{1}{j\omega\rho}\frac{\partial P}{\partial r}|_{r=a} = -\frac{k}{j\omega\rho}\sum_{l=0}^{\infty} A_l h_l^{(2)\prime}(ka)\, P_l(\cos\varphi), \tag{6.227}$$

where the prime (') in the superscript denotes a derivative with respect to ka. After representing $U(a,\varphi)$ at $0 \le \varphi \le \pi$ as a series in terms of the Legendre polynomials

$$U(a,\varphi) = \sum_{l=0}^{\infty}\left[\sum_{i=0}^{N} U_i a_{il}\right] P_l(\cos\varphi), \tag{6.228}$$

where

$$a_{il} = (l + 1/2)\int_0^{\varphi_b} P_i(a,\varphi)\ P_l(\cos\varphi)\sin\varphi d\varphi \tag{6.229}$$

and after equating this series to the series in the right side of relation (6.227), we arrive at

$$A_l = -j\rho c\sum_{i=0}^{N}\frac{U_i\ a_{il}(\varphi_b)}{h_l^{(2)\prime} ka)}\ . \tag{6.230}$$

Upon substituting A_l from Eq. (6.230) into expression (6.226) we obtain

$$P(r,\varphi,\varphi_b) = -j\rho c\sum_{i=0}^{N} U_i\left[\sum_{l=0}^{\infty} a_{il}(\varphi_b)\frac{h_l^{(2)}(kr)}{h_l^{(2)\prime}(ka)} P_l(\cos\varphi)\right] = \sum_{i=0}^{N} P_i(r,\varphi,\varphi_b), \tag{6.231}$$

where $P_i(r,\varphi,\varphi_b)$ is the modal sound pressure generated by the single mode of vibration defined at $\varphi \le \varphi_b$. At sufficiently large distances from a sphere (at $k\,r \to \infty$)

$$h_l^{(2)}(kr) \to \frac{j}{kr} e^{-j(kr - l\pi/2)}, \tag{6.232}$$

and we have the following expression for the modal sound pressure,

$$P_i(r,\varphi,\varphi_b) = \rho c\ \frac{1}{kr} e^{-jkr}\ U_i\sum_{l=0}^{\infty} a_{il}(\varphi_b)\frac{e^{j\pi l/2}}{h_l^{(2)\prime}(ka)} P_l(\cos\varphi)\,. \tag{6.233}$$

Using expressions (6.231) and (6.233) we obtain for the sound pressure on the acoustical axis of the transducer, $P(r,0,\varphi_b)$, and for the directional factor of the transducer, $H(\varphi,\varphi_b)$,

$$P(r,0,\varphi_b) = \rho c\frac{1}{kr} e^{-jkr}\sum_{i=0}^{N} U_i\ \sum_{l=0}^{\infty} a_{il}(\varphi_b)\ \frac{e^{j\pi l/2}}{h_l^{(2)\prime}(ka)}, \tag{6.234}$$

$$H(\varphi,\varphi_b) = \frac{\displaystyle\sum_{i=0}^{N} U_i\ \sum_{l=0}^{\infty} a_{il}(\varphi_b)\ \frac{e^{j\pi l/2}}{h_l^{(2)\prime}(ka)}\ P_l(\cos\varphi)}{\displaystyle\sum_{i=0}^{N} U_i\sum_{l=0}^{\infty} a_{il}(\varphi_b)\frac{e^{j\pi l/2}}{h_l^{(2)\prime}(ka)}}\,. \tag{6.235}$$

The modal diffraction coefficients of a spherical transducer relative to a plane wave propagating in the direction of axis $\varphi = 0$ was previously defined (see Eq. (6.10)) as the ratio of modal sound pressure generated by a transducer (Eq. (6.233)) to the sound pressure generated by a small pulsating sphere having the same referred volume velocity, (Eq. (6.9)). It will be found in the form

$$k_{dif\,i} = \frac{P_i(r,0,\varphi_b)}{P_0(r)} = -j\frac{1}{(ka)^2}\sum_{l=0}^{\infty} a_{il}(\varphi_b)\frac{e^{j\pi l/2}}{h_l^{(2)\prime}(ka)}. \tag{6.236}$$

Note, that the referred volume velocity for the ith mode is $U_{\tilde{V}i} = 4\pi a^2 U_i$, and the sound pressure generated by a small pulsating sphere having this volume velocity is

$$P_0(r) = j(\rho c)kU_{\tilde{V}i}e^{-jkr}/4\pi r. \tag{6.237}$$

The diffraction coefficient changes in accordance with the directional factor.

The total power radiated by a vibrating sphere may be found as

$$\begin{aligned}\bar{\tilde{W}}_{ac} &= 2\pi a^2\int_0^{\pi} P(a,\varphi,\varphi_b)\,U^*(a,\varphi)\sin\varphi d\varphi = \\ &= -j\rho c\,2\pi a^2\int_0^{\pi}\left[\sum_i^N U_i\sum_{l=0}^{\infty} a_{il}(\varphi_b)\frac{h_l^{(2)}(ka)}{h_l^{(2)\prime}(ka)}P_l(\cos\varphi)\right]\times \\ &\quad\times\left[\sum_{p=0}^{N} U_p^*\sum_{l=0}^{\infty} a_{pl}(\varphi_b)P_l(\cos\varphi)\right]\sin\varphi d\varphi = \\ &= -j\rho c 2\pi a^2\sum_{i=0}^{N}\left\{\left[\sum_{l=0}^{\infty}\frac{2}{2l+1}a_{il}^2(\varphi_b)\frac{h_l^{(2)}(ka)}{h_l^{(2)\prime}(ka)}\right]|U_i|^2 + \right. \\ &\quad\left. +\sum_{p\neq i}^{N}\left[\sum_{l=0}^{\infty}\frac{2}{2l+1}a_{pl}(\varphi_b)\,a_{il}(\varphi_b)\frac{h_l^{(2)}(ka)}{h_l^{(2)\prime}(ka)}\right]U_p^*U_i\right\}.\end{aligned} \tag{6.238}$$

This expression can be rewritten in the form

$$\bar{\tilde{W}}_{ac} = \sum_{i=0}^{N}\left[Z_{acii}(\varphi_b) + \sum_{p\neq i}^{N} z_{acip}(\varphi_b)\frac{U_p}{U_i}\right]|U_i|^2 = \sum_{i=0}^{N} Z_{aci}(\varphi_b)|U_i|^2, \tag{6.239}$$

where

$$Z_{acii}(\varphi_b) = -j\rho c\,4\pi a^2\sum_{l=0}^{\infty}\frac{1}{2l+1}a_{il}^2(\varphi_b)\frac{h_l^{(2)}(ka)}{h_l^{(2)\prime}(ka)}, \tag{6.240}$$

is the modal (self) radiation impedance for i^{th} mode of vibration defined for $\varphi \le \varphi_b$, and

$$z_{acip}(\varphi_b) = -j\rho c 4\pi a^2 \sum_{l=0}^{\infty} \frac{1}{2l+1} a_{pl}(\varphi_b)\, a_{il}(\varphi_b) \frac{h_l^{(2)}(ka)}{h_l^{(2)\prime}(ka)}, \tag{6.241}$$

is the intermodal (mutual) impedance between modes i and m ($m \ne i$). Thus, the radiation impedance associated with the generalized velocity U_i is

$$Z_{aci}(\varphi_b) = Z_{acii}(\varphi_b) + \sum_{p \ne i}^{N} z_{acip}(\varphi_b) \frac{U_p}{U_i}. \tag{6.242}$$

It is convenient to represent the self and mutual radiation impedances in the form

$$Z_{acii} = \rho c S_{effi}(\alpha_{ii} + j\beta_{ii}), \quad z_{acip} = \rho c S_{effi}(\alpha_{ip} + j\beta_{ip}), \tag{6.243}$$

where α and β are the nondimensional resistance and reactance coefficients and

$$S_{effi} = 2\pi a^2 \int_0^{\varphi_b} P_i^2(\cos\varphi)\sin\varphi d\varphi \tag{6.244}$$

will be defined as the effective radiating surface area.

6.4.2 Radiation of the Spherical Shell without Baffles

For the spherical transducers without baffles, i.e., at $\varphi_b = \pi$, all the coefficients a_{ip} at $i \ne p$ vanish due to orthogonality of the Legendre polynomials, and $a_{ii} = 1$. The modes of vibration that will be generated depend on the geometry of electrodes. We consider the two most common electrode configurations: unipolar electrodes on the whole surface of the spherical shell and the electrodes split in halves that are connected in opposite phase (bipolar electrodes).

6.4.2.1 Transducers with Unipolar Electrodes on the Whole Surface

In this case only the isolated zero mode of vibration is generated, $U(a,\varphi) = U_0$, $a_{00} = 1$, $S_{eff\,0} = 4\pi a^2$. This is a classic example of a transducer with a single mechanical degree of freedom that was considered in Chapter 2. The diffraction coefficient, sound pressure generated in the far field and radiation impedance are

$$k_{dif\,0} = e^{jka} \frac{1}{1 + jka} = \frac{1}{\sqrt{1+(ka)^2}} e^{j(ka - \arctan ka)}, \tag{6.245}$$

$$P_0(r,0,\pi) = \frac{\rho c}{2\lambda r} 4\pi a^2 U_0 e^{j(kr-\pi/2)} k_{dif\,0}, \tag{6.246}$$

$$Z_{ac0} = \rho c 4\pi a^2 (\alpha_0 + j\beta_0), \tag{6.247}$$

where the nondimensional coefficients of the radiation impedance are

$$\alpha_0 = \frac{(ka)^2}{1+(ka)^2} \text{ and } \beta_0 = \frac{ka}{1+(ka)^2}. \tag{6.248}$$

The plots of nondimensional coefficients and diffraction coefficient are depicted in Figure 6.27 and Figure 6.28 at $i = 0$. At low frequencies (at $ka < 0.2$)

$$\alpha_0 \approx (ka)^2 \text{ and } \beta_0 \approx ka. \tag{6.249}$$

The acoustic mass for the pulsating sphere of a small wave size is $m_{ac0} = \tilde{V}_{sph}\rho$, i.e., equal to mass of water in volume of the sphere.

6.4.2.2 Transducers with Bipolar Electrodes

In this case the modes of vibration at i = 1, 3, 5... are generated. From expression (6.244) it follows that $S_{eff\,i} = 4\pi a^2/(2i+1)$. For the first mode of vibration the diffraction coefficient and radiation impedance are

$$k_{dif1} = 1/(ka)^2 h_1^{(2)\prime}(ka), \tag{6.250}$$

$$Z_{ac1} = (\rho c)\frac{4\pi a^2}{3}(-j)\frac{h_1^{(2)}(ka)}{h_1^{(2)\prime}(ka)} = (\rho c)S_{eff1}(\alpha_{11} + j\beta_{11}). \tag{6.251}$$

Detailed information on the properties of the spherical Bessel functions and tabulated data for the functions can be found in Ref. 5. Summary of the properties is presented in Appendix C.2. In particular,

$$h_1^{(2)\prime}(ka) = \frac{1}{3}[h_0^{(2)}(ka) - 2h_2^{(2)}(ka)]. \tag{6.252}$$

where $h_i^{(2)}(ka) = j_i(ka) - jy_i(ka)$. The dependences of the modal diffraction coefficients $k_{dif\,i}$ and nondimensional coefficients α_{ii} and β_{ii} on ka for the transducers without baffles are plotted in Figure 6.27 and 6.28.

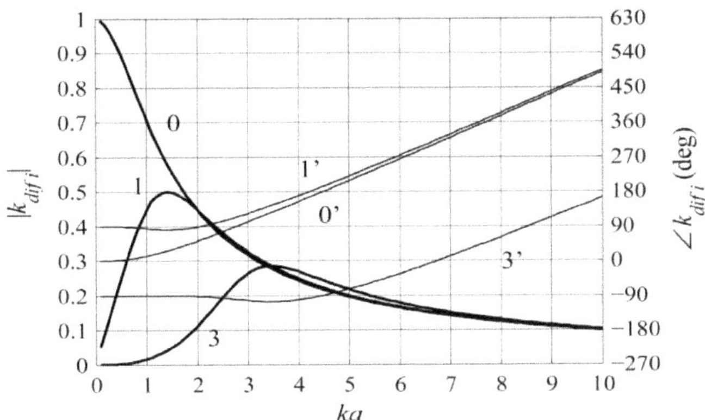

Figure 6.27: The modal diffraction coefficients $k_{dif\,i}$ for spheres without baffles for i = 0, 1, 3 (phase labeled with ').

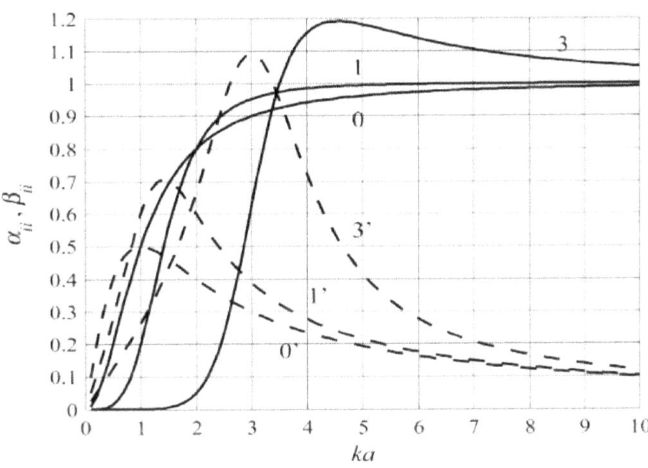

Figure 6.28: Nondimensional modal radiation impedance coefficients for spheres without baffles for i = 0, 1, 3 (β_{ii} labeled with ').

At $ka < 0.3$

$$j_m(ka) \approx \frac{(ka)^m}{1 \cdot 3 \cdot 5 \cdots (2m+1)}, \tag{6.253}$$

$$y_m(ka) \approx -\frac{1 \cdot 3 \cdot 5 \cdots (2m-1)}{(ka)^{m+1}}, \tag{6.254}$$

$$h_1^{(2)} \approx \frac{1}{3}\left[ka + j\frac{3}{(ka)^2}\right], \quad h_1^{(2)\prime}(ka) \approx \frac{1}{3}\left[1 - j\frac{6}{(ka)^3}\right], \tag{6.255}$$

and with aid of Eq. (6.251) we obtain

$$\alpha_1 \approx \frac{(ka)^4}{4} \text{ and } \beta_1 \approx \frac{ka}{2}. \tag{6.256}$$

The acoustic mass of a small oscillating sphere (that is called a simple dipole source) is

$$m_{ac} = \frac{1}{\omega}\rho c \cdot \frac{4\pi a^2}{3} \cdot \frac{ka}{2} = \frac{1}{2}\tilde{V}_{sph}\rho, \tag{6.257}$$

i.e., half of mass of water in the volume of the sphere. Comparison of the radiation resistances of the oscillating and pulsating simple sources, which is

$$\frac{\alpha_1}{\alpha_0} = \frac{(ka)^2}{4}, \tag{6.258}$$

shows that the oscillating source is much worse projector.

6.4.3 Radiation of a Spherical Shell with Baffles

The baffling of the spherical shells is intended for achieving a unidirectional radiation of the spherical transducers. These issues will be considered in Chapter 8. The most interesting for the practical applications are the variants of baffling a hemisphere (at $\varphi_b = \pi / 2$) and of a segment that corresponds to $\varphi_b = \pi / 3$.

First, we will assume that the baffles are ideally rigid, i.e., that the velocity is zero on the surface of the baffle. In the general expressions (6.234), (6.236) and (6.240) for the sound pressure, diffraction coefficient, and radiation impedance the angle φ_b must be taken according to the baffle coverage. Thus, for the case that the baffle covers a hemisphere, which we will consider for illustration with the numerical examples, $U(\varphi) = 0$ at $\pi / 2 \le \varphi \le \pi$. To apply the general expressions to calculating parameters of radiation for a particular baffle coverage and mode of the surface vibration, the coefficients a_{il} have to be found using formula (6.229). Thus, in the variant that electrodes are unipolar (zero mode is excited) at $\varphi_b = \pi / 2$ we find: $a_{11} = 1/2$, $a_{00} = 1/2$, $a_{01} = 3/4$, $a_{03} = -7/16$, $a_{05} = 11/32$,... and $a_{0l} = 0$ for l odd. In the variant of bipolar electrodes connection (1, 3, 5... modes are excited) we find $a_{12} = 5/16$, $a_{10} = 1/4$, ...,and $a_{1l} = 0$ for l even.

It is noteworthy that in both variants the non-zero "intermodal" coefficients a_{il} exist. This means that acoustic interaction takes place between electromechanically active zero mode and passive odd modes in the case of unipolar sphere excitation, as it follows from Eq. (6.241). And in the case of the bipolar excitation the active first mode generates electromechanically passive zero mode through the acoustic interaction.

The analogous calculations and conclusions can be made for the variant of baffling at $\varphi_b = \pi/3$. The results of calculations for both variants of baffling are presented here to be referred to in Chapter 8.

Dependences of the modal diffraction coefficients, nondimensional coefficients of self and mutual radiation impedances and directional factors from ka are presented for the baffled spherical transducers at $\varphi_b = \pi/2$ in Figure 6.29 - Figure 6.34 and at $\varphi_b = \pi/3$ in Figure 6.35 and Figure 6.36.

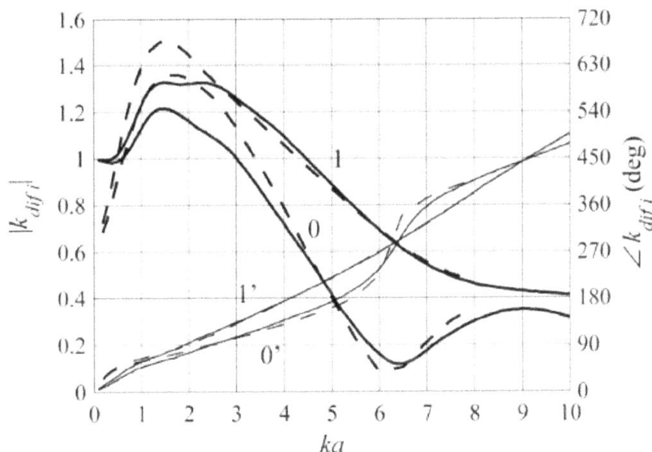

Figure 6.29: Modal diffraction coefficients of a baffled ($\varphi_b = \pi/2$) sphere: $|k_{dif\,i}|$ (thick lines) and phase $\angle k_{dif\,i}$ (thin lines, labeled with ') for $i=0$, 1 with rigid baffle (solid lines) and compliant baffle (dashed lines).

Calculations show that the magnitudes of the mutual impedances between modes drop very quickly, as separation between modes increases, and especially so the higher the orders of the modes are. As it follows from plots presented in Figure 6.31, only the mutual impedance z_{ac01} between the zero and first modes has significant value, and z_{ac13} can be already practically neglected.

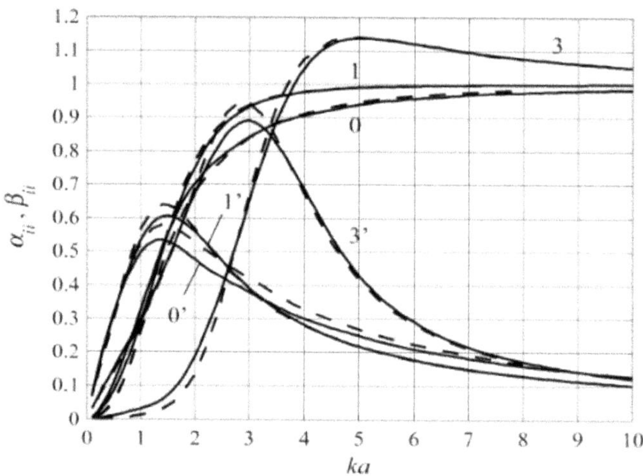

Figure 6.30: Nondimensional coefficients of the modal self-radiation impedances of a baffled ($\varphi_b = \pi / 2$) sphere: α_{ii} and β_{ii} (labeled with ') for i = 0, 1, 3 with rigid baffle (solid lines) and compliant baffle (dashed lines).

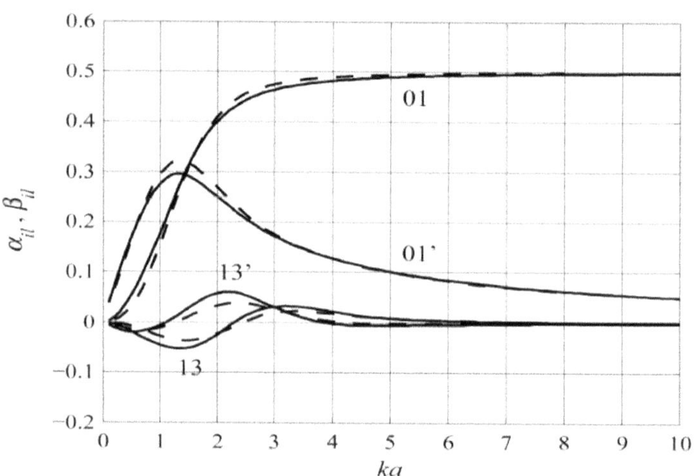

Figure 6.31: Nondimensional coefficients of the modal mutual radiation impedances of a baffled ($\varphi_b = \pi / 2$) sphere: α_{il} and β_{il} (labeled with ') for i = 0, 1, 3 with rigid baffle (solid lines) and compliant baffle (dashed lines).

The modal directional factors corresponding to zero and first modes of vibration at different *ka* are shown in Figure 6.32 and Figure 6.33, and for the second and third modes they are presented in Figure 6.34

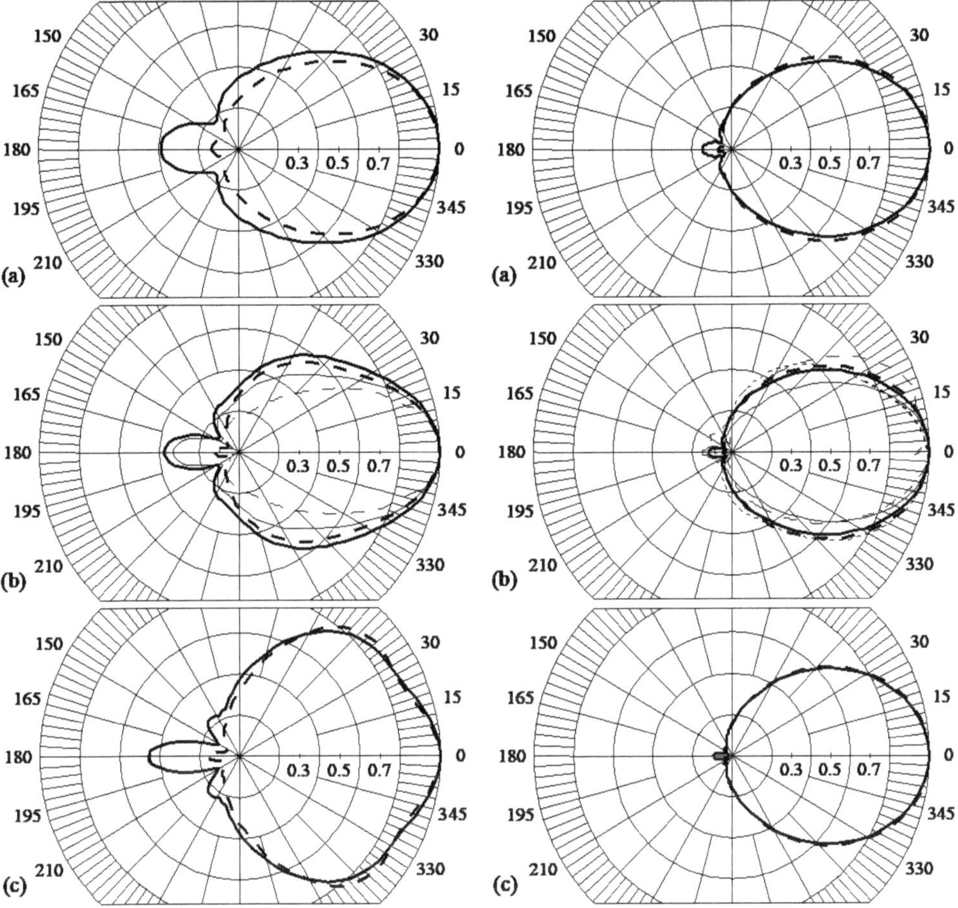

Figure 6.32 (on the left): Directional factors for the zero mode of vibration of a baffled ($\varphi_b = \pi / 2$) spherical transducer at different ka: (a) ka = 2.5, (b) ka = 3.5, and (c) ka = 4.5. Shown are the modal directional factors with rigid baffle (thick solid lines) and compliant baffle (thick dashed lines). Plot (b) shows the directional factors calculated for spherical transducer #1 with rigid baffle (thin solid line) and measured with compliant baffle (thin dashed line).

Figure 6.33 (on the right): Directional factors for the first mode of vibration of a baffled ($\varphi_b = \pi / 2$) spherical transducer at different ka: (a) ka = 3.5, (b) ka = 4.5, and (c) ka = 5.5. Shown are the modal directional factors with rigid baffle (thick solid lines) and compliant baffle (thick dashed lines). Plot (b) shows the directional factors calculated for spherical transducer #1 with rigid baffle (thin solid line) and measured for hemispherical transducer (thin dashed line)

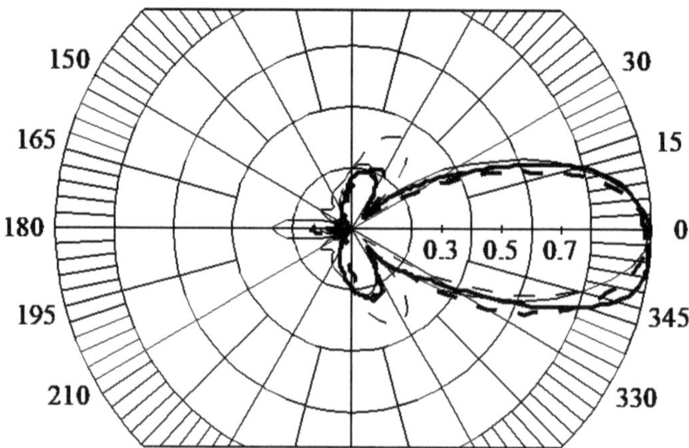

Figure 6.34: Modal directional factors for a baffled ($\varphi_b = \pi / 2$) sphere corresponding to the second and third modes of vibration: second mode at $ka = 6.5$ (thin solid line), third mode at $ka = 8.5$ (thin dashed line). For comparison the measured directional factor of spherical transducer operating in the third mode (thick solid line) and the measured directional factor of the hemispherical transducer operating in the third mode (thick dashed line) are also shown.

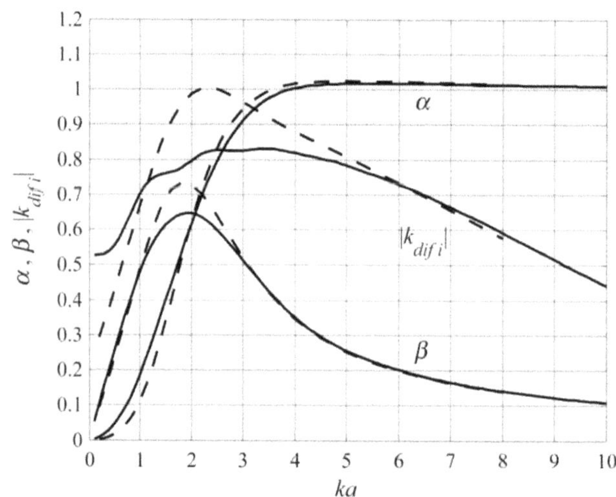

Figure 6.35: Diffraction coefficients and nondimensional radiation impedance coefficients for a baffled ($\varphi_b = \pi / 3$) sphere

The variant of baffling at $\varphi_b = \pi / 3$ may be of a practical interest in case that the mode of vibration is $U(\varphi) = U_r\, P_{1.6}(\cos\varphi)$, as it will be shown in Chapter 8. The plots in Figure 6.35 and Figure 6.36 are calculated for this mode of vibration. The effective surface area for the open

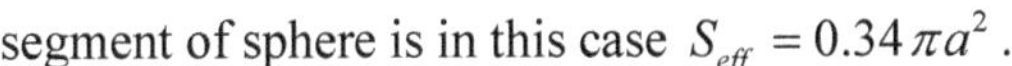
segment of sphere is in this case $S_{eff} = 0.34\pi a^2$.

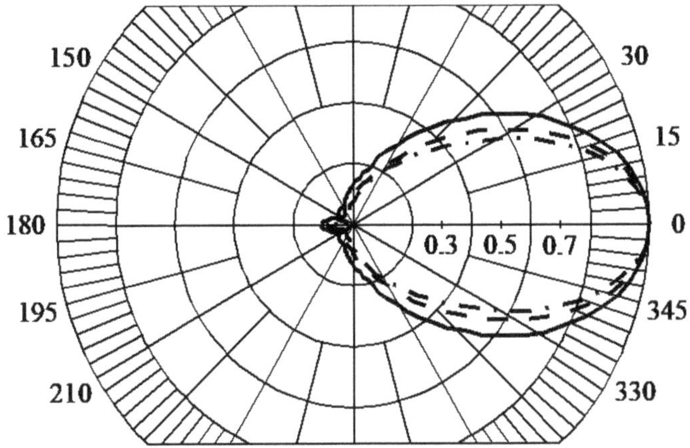

Figure 6.36: Directional factors of baffled ($\varphi_b = \pi / 3$) sphere at different *ka*: 4.5 (solid line), 5.5 (dashed line), and 6.5 (dot-dashed line)

A remarkable property of the directional factors of the baffled spherical transducers is that they remain almost unchanged in a broad frequency range around the operating resonance frequencies of the transducers, as this can be seen from the above figures.

To this point the assumption was made that the velocity on the baffled part of a sphere was zero due to employing the ideally rigid baffle. While a useful approximation, it is not clear how to practically realize such a baffle. It is more practical to assume that the baffle is ideally compliant, and that the sound pressure is zero on the baffled part. This brings us to the case of a radiation problem with mixed boundary conditions. For solving this problem for the sphere, the method is used that previously was employed in Section 6.3.3 regarding the cylindrical shell. Omitting the mathematical manipulations, the results of calculations obtained for the cases that $\varphi_b = \pi / 2$ and $\varphi_b = \pi / 3$ are presented in Figure 6.29 through Figure 6.36. The results show very close agreement with those obtained under assumption of absolutely rigid baffle (except for the drop of the level of back radiation, which is especially prominent for zero mode velocity distribution). The results also show good agreement with experimental data, which were obtained with baffles made of corprene (a rubber-cork composition) that functions closely to an ideally compliant material.

6.5 Radiation of Transducers having Flat Surfaces

6.5.1 General Considerations

Solutions to majority of radiation problems for the planar surfaces are obtained under the assumption that the surfaces are flash with infinite rigid plane baffle (Figure 6.1 (c)). They are based on application of Huygens's principle, one of formulations of which is the integral formula (*Rayleigh's integral*)

$$P(\boldsymbol{r}) = \frac{j\omega\rho}{2\pi}\int_{\Sigma} U_n(\boldsymbol{r}_\Sigma)\frac{e^{-jkr}}{r}d\Sigma. \tag{6.259}$$

Here Σ is the surface of the infinite plane, $U_n(\boldsymbol{r}_\Sigma)$ is the normal velocity of the points of the surface with coordinates $\boldsymbol{r}_\Sigma$, and $\boldsymbol{r}(x_1, y_1, z_1)$ is the radius vector of an observation point with $r = \sqrt{(x-x_1)^2 + (y-y_1)^2 + (z-z_1)^2}$. The velocity in this formula must be known on the entire surface of the infinite plane. As the velocity is known on the surface of a transducer only, the existence of the infinite rigid baffle on which the velocity is zero is crucial for using the formula.

It is noteworthy that another formulation of the Huygens's principle is

$$P(\boldsymbol{r}) = \frac{j\cos\theta}{\lambda}\int_{\Sigma} P(\boldsymbol{r}_\Sigma)\frac{e^{-jkr}}{r}d\Sigma, \tag{6.260}$$

where θ is the angle between normal to the plane and direction on the observation point.

Using this formula requires knowing the sound pressure on the entire infinite plane surface, therefore it can be applied to the case that transducer is embedded into the absolutely compliant flat baffle. Factor $\cos\theta$ shows that whatever small wave size of a transducer placed in the compliant baffle is it does not radiate (receive) acoustic energy along the baffle.

For the two-dimensional radiation problem, such as determining the field of vibrating strip infinite in direction of axis y embedded into the rigid plane, formulation of the Huygens's principle for sound pressure per unit length along the y axis is

$$P(\boldsymbol{r}) = \frac{\rho\omega}{2}\int_{-\infty}^{\infty} U_n(x)H_0^{(2)}(kr)\,dx. \tag{6.261}$$

Here $r = \sqrt{(x-x_0)+z_0^2}$, where x_0 and z_0 are the coordinates of an observation point. In particular, the rigid baffle flash with the surface of a transducer can be simulated by its plane of symmetry in the case that the transducer is double-sided pulsating piston of infinitesimal wave thickness, as it is shown in Figure 6.1 (a). In the cases that thickness of the symmetrical pulsating transducers is significant, or dimensions of real baffles are small compared with wavelength, or real transducers are used without the baffles, the radiation problem for the planar transducers complicates, and these cases must be considered on the separate issues. This will be done in Sect. 6.5.8.4.

6.5.2 Radiation of a Circular Pulsating Piston

The radiation impedance of the thin pulsating piston was first calculated by Rayleigh[17]. Because of special importance of this case, and of originality of the derivation performed we will briefly reproduce this derivation. Consider the circular disk that vibrates uniformly with the normal velocity U_0 embedded in the absolutely rigid plane baffle as shown in Figure 6.37. The sound pressure on the surface of the disk can be calculated according to formula (6.259) as

$$P(\boldsymbol{r}_\Sigma) = \frac{j\omega\rho}{2\pi} U_0 \int_\Sigma \frac{e^{-jkr_\Sigma}}{r_\Sigma} d\Sigma . \tag{6.262}$$

At first, we calculate the sound pressure on the surface of a thin ring having radius r_x produced by uniformly vibrating elements of the disk of this radius (denoted $P(r_x)$). If to place the center of polar coordinate system at some point of the ring (point o in Figure 6.37), then the pressure at this point, $P_o(r_x)$, can be found as (note that the radius of the arc l is $r = 2r_x \cos\theta$)

$$P_o(r_x) = \frac{j\omega\rho}{2\pi} U_0 \int_{-\pi/2}^{\pi/2} \int_0^{2r_x\cos\theta} e^{-jkr} dr d\theta = \frac{\rho c}{\pi} U_0 \int_0^{\pi/2} [1 - e^{-j2kr_x\cos\theta}] d\theta . \tag{6.263}$$

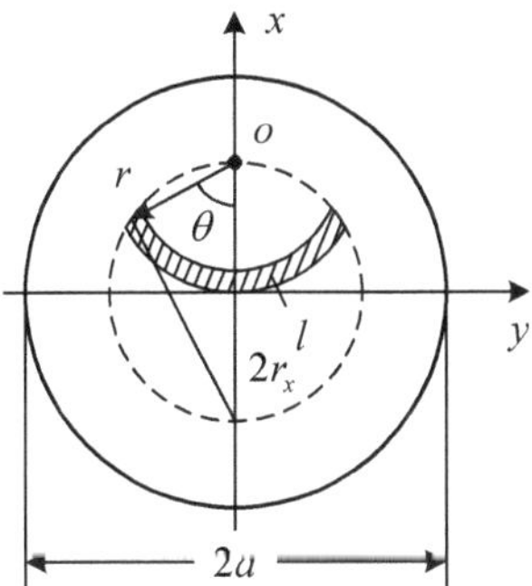

Figure 6.37: Illustration of geometry considerations used for calculating the radiation impedance. The "ring" is marked by the dashed line, l is the arc of radius r.

Taking into consideration expressions for the Bessel and Struve functions (Ref. 5, 6, see also Appendix C.1),

$$J_0(z) = \frac{2}{\pi}\int_0^{\pi/2} \cos(z\cos\theta)d\theta, \tag{6.264}$$

$$S_0(z) = \frac{2}{\pi}\int_0^{\pi/2} \sin(z\cos\theta)d\theta, \tag{6.265}$$

the sound pressure at any point of the ring can be represented as

$$P_o(r_x) = \frac{\rho c}{2}U_0\{[1 - J_0(2kr_x)] + jS_0(2kr_x)\}. \tag{6.266}$$

Thus, the force that is acting on the entire ring is

$$F(r_x) = P_o(r_x)\cdot 2\pi r_x dr_x, \tag{6.267}$$

and the total force acting on the surface of a disk of the radius a will be found as

$$F_{disk} = 2\pi\int_0^a P_o(r_x)r_x dr_x = \pi\rho c U_0\int_0^a \{[1 - J_0(2kr_x)] + jS_0(2kr_x)\}r_x dr. \tag{6.268}$$

After performing integration keeping in mind that

$$\int zJ_0(z)dz = zJ_1(z) \text{ and } \int zS_0(z)dz = zS_1(z), \tag{6.269}$$

for the radiation impedance will be obtained

$$Z_{ac} = \frac{F_{disk}}{U_0} = \rho c\pi a^2\left\{\left[1 - \frac{2J_1(2ka)}{2ka}\right] + j\frac{2S_1(2ka)}{2ka}\right\} = \rho c\pi a^2(\alpha + j\beta). \tag{6.270}$$

The nondimensional coefficients of radiation impedance α and β can be represented in the form of the series (for brevity we denote $2ka = z$)

$$\alpha = 1 - \frac{2J_1(z)}{z} = \frac{z^2}{2^2 1!2!} - \frac{z^4}{2^4 2!3!} + \dots, \tag{6.271}$$

$$\beta = \frac{2S_1(z)}{z} = \frac{4}{\pi}\left(\frac{z}{3} - \frac{z^3}{3^2 5} + \frac{z^5}{3^2 5^2 7} - \dots\right). \tag{6.272}$$

Their plots are shown in Figure 6.38.

At $ka < 0.5$ (with accuracy not less than 5%)

$$\alpha \approx \frac{1}{2}(ka)^2 \text{ and } \beta \approx \frac{8}{3\pi}(ka) . \tag{6.273}$$

Thus, the radiation resistance is

$$r_{ac} \approx \frac{1}{2}\pi a^2 \rho c (ka)^2 = 2\pi\rho c \frac{S^2}{\lambda^2} . \tag{6.274}$$

This is the general expression for radiation resistance of a simple source given that in this case $S_{av} = S$.

The radiation reactance and the acoustical mass are,

$$x_{ac} \approx \pi a^2 \rho c \frac{8}{3\pi} ka \text{ and } m_{ac} \approx \pi a^2 \rho \frac{8}{3\pi} a = \frac{2}{\pi} \tilde{V}_{sph} \rho , \tag{6.275}$$

where $\tilde{V}_{sph}$ is the volume of sphere having the same radius.

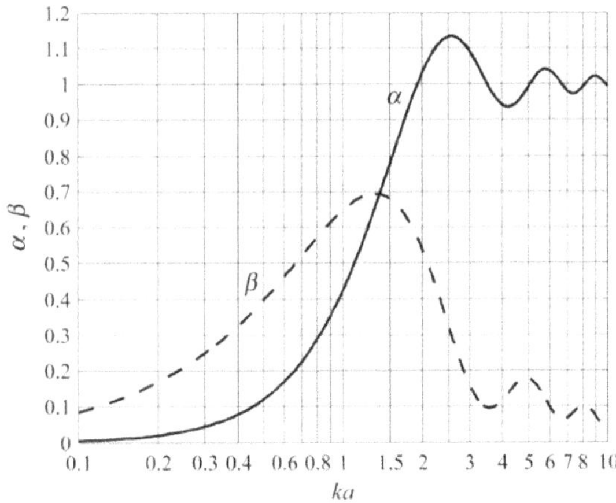

Figure 6.38: The nondimensional coefficients of radiation impedance of pulsating disk imbedded in the infinite rigid baffle.

When determining sound pressure generated by the disk in the far field, modulus of the radius vector of an observation point in formula (6.259), which will be denoted here as $\boldsymbol{R}$, is much greater than radius of the disk, $|\boldsymbol{R}| \gg a$. Due to the axial symmetry of radiation, it is sufficient considering that the observation point is in plane perpendicular to the plane of the disk, as shown in Figure 6.39. First, consider the sound pressure that is generated by an elementary ring of radius r. The sound pressure of an element of the ring having coordinates x and φ is proportional to its volume velocity, $U(x,\varphi) = U_0 r d\varphi dr$. Distance r_x between the observation point and the point with coordinate $x = r\cos\varphi$ on the surface of the disk can be expressed as

$$r_x = \sqrt{(R\cos\theta)^2 + (R\sin\theta - x)^2} \approx R\sqrt{1 - 2(x/R)\sin\theta} \tag{6.276}$$

or, given that $(x/R) \ll 1$,

$$r_x \approx R[1 - (x/R)\sin\theta]\,. \tag{6.277}$$

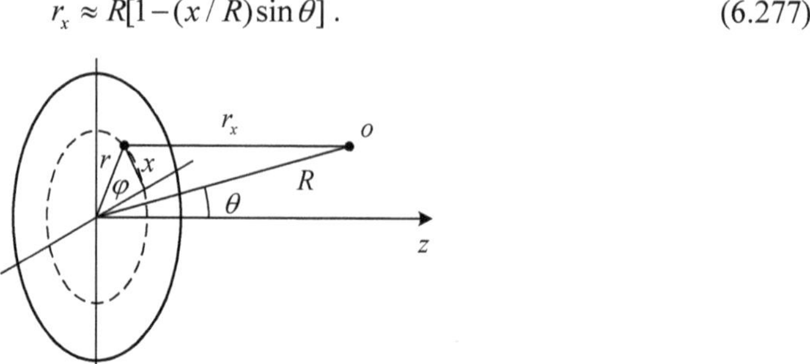

Figure 6.39: Geometry for determining the sound pressure of a disk in the far field.

The sound pressure generated by the element of the ring in the far field, being obtained by using formula (6.259), is

$$P(r,\varphi) = \frac{j\omega\rho}{2\pi} U_0 r dr \frac{e^{-jkR}}{R} e^{jkr\sin\theta\cos\varphi} d\varphi. \tag{6.278}$$

And the total sound pressure at the point of observation will be found as result of integrating this expression over the ring circumference,

$$P(r,\varphi) = j\omega\rho U_0 r dr \frac{e^{-jkR}}{R} \left[\frac{1}{2\pi} \int_0^{2\pi} e^{jkr\sin\theta\cos\varphi} d\varphi \right]. \tag{6.279}$$

Integral in the brackets is the Bessel function $J_0(kr\sin\theta)$. Thus, expression for the sound pressure generated by the ring in the far field becomes

$$P(r,\theta) = j\rho ckrdrU_0 \frac{e^{-jkR}}{R} J_0(kr\sin\theta)\,, \tag{6.280}$$

and the directional factor of the ring is

$$H_{ring}(\theta) = J_o(kr\sin\theta)\,. \tag{6.281}$$

In order to obtain the sound pressure radiated by the entire disk having radius *a*, expression (6.280) must be integrated by radius. Given that

$$\int zJ_0(z) = zJ_1(z), \tag{6.282}$$

the result is

$$P_{disk}(\theta) = j\rho cka^2 U_0 \frac{e^{-jkR}}{R} \frac{J_1(ka\sin\theta)}{ka\sin\theta}. \tag{6.283}$$

The directional factor of the disk is

$$H_{disk}(\theta) = \frac{2J_1(ka\sin\theta)}{ka\sin\theta}. \tag{6.284}$$

The diffraction coefficient at $\theta = 0$ for all the transducers imbedded in the infinite rigid plane baffle is $k_{dif} = 2$, as the sound pressure in incoming plane wave doubles on the surface of the baffle. Using formula (6.22), expression (6.270) for r_{ac}, and Eq. (6.283) for the sound pressure on the axis, the following relation can be obtained for the directivity of the disk

$$D = \frac{4\pi \cdot \pi a^2}{\lambda^2}\left[1 - \frac{2J_1(ka)}{ka}\right]^{-1}. \tag{6.285}$$

At $a > \lambda/2$ the directivity differs less than by 6% from its value by formula

$$D = \frac{4\pi \cdot \pi a^2}{\lambda^2} = \frac{4\pi \cdot S}{\lambda^2}. \tag{6.286}$$

6.5.3 Radiation of a Thin Ring Pulsating in the Axial Direction

Radiation of a thin ring pulsating in axial direction may be considered as the extreme case of radiation of a radially vibrating cylinder of small height having rigid caps, as shown in Figure 6.40, using symmetry considerations.

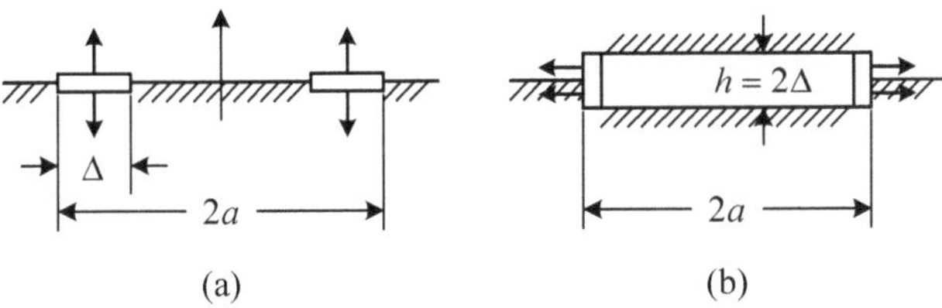

Figure 6.40: (a) Thin ring pulsating in axial direction, (b) radially vibrating cylinder of a small height. They can be considered as vibrating in the infinite rigid plane baffle by the symmetry considerations.

The geometry relations in the figure are $\Delta \ll a,\ \Delta \ll \lambda,\ h = 2\Delta$. The procedures analogous

to those used for calculating radiation of a disk can be used in this case. The sound pressure generated by a ring in the far field and directional factor accordingly were already obtained in process of determining these characteristics for a disk and expressed by formulas (6.280), (6.281).

Determining the radiation impedance requires a little different calculation. The geometry considerations for solving this problem are illustrated with Figure 6.41. The sound pressure on the element o of the ring exerted by elements 1 and 2 remote from point o by distance $r_x = 2a\cos\varphi$ that have the volume velocity $U_{\tilde{v}} = U_0 r_x dr_x d\varphi$ is

$$P_o(r_x,\varphi) = \frac{j\omega\rho}{\pi} U_0 e^{-j2ka\cos\varphi} dr_x d\varphi . \tag{6.287}$$

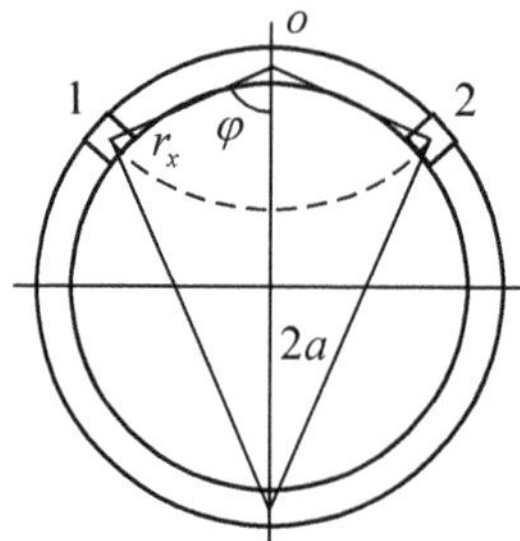

Figure 6.41: Illustration of the geometry considerations used for calculating radiation impedance of a ring.

Thus, the total sound pressure at the arbitrary point of the ring will be obtained as

$$P_o(r_x,\varphi) = \frac{j\omega\rho}{\pi} U_0 \int_0^{2a} \int_0^{\pi/2} e^{-j2ka\cos\varphi} dr_x d\varphi , \tag{6.288}$$

or taking into consideration the definitions (6.264) and (6.265)

$$P_o = \rho c U_0 \cdot ka\left[S_0(2ka) + jJ_0(2ka)\right] . \tag{6.289}$$

The force acting on the entire surface of the ring and the radiation impedance are

$$F_{ring} = 2\pi a\Delta \cdot P_o , \tag{6.290}$$

$$Z_{ac} = \frac{F_{ring}}{U_0} = \rho c S_{ring} \cdot ka\left[S_0(2ka) + jJ_0(2ka)\right] . \tag{6.291}$$

Thus, the nondimensional coefficients of radiation impedance are

$$\alpha = ka \cdot S_0(2ka)\,, \tag{6.292}$$

$$\beta = ka \cdot J_0(2ka)\,. \tag{6.293}$$

For a cylinder of the height $h = 2\Delta$ with rigid caps formula (6.291) for the radiation impedance remains the same with replacement of S_{ring} by $S_{cyl} = 2\pi ah$ due to the fact that $h \ll \lambda$ and the sound pressure on its surface is the same as on the surface of the ring. Therefore, formulas (6.292) and (6.293) for the nondimensional coefficients of radiation impedance are valid for this case also.

6.5.4 Radiation of an Infinitely Long Pulsating Strip

The radiation problem for the vibrating strip embedded into the rigid plane that is infinite in direction of axis y is two-dimensional. The geometry of the problem is illustrated in Figure 6.42.

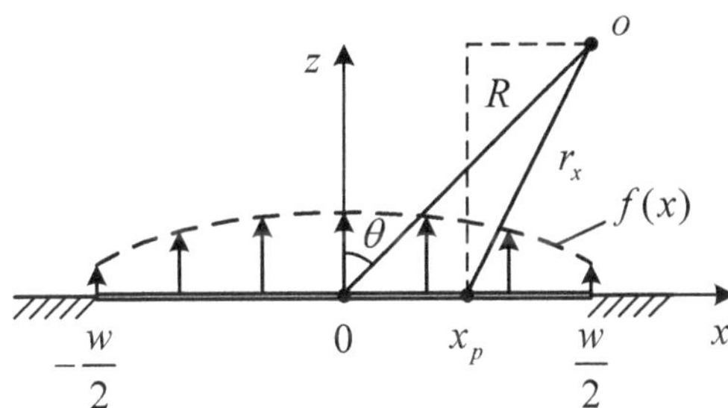

Figure 6.42: To the radiation of the infinite strip embedded into the rigid baffle.

Therefore formulation (6.261) of the Huygens's principle for determining the sound pressure radiated by element Δy of length along y axis of the strip is

$$P(\mathbf{r}) = \frac{\rho\omega}{2} \cdot \Delta y \int_{-W/2}^{W/2} U_n(x) H_0^{(2)}(k\mathbf{r})\, dx\,. \tag{6.294}$$

Here $U_n(x) = U_0 f(x)$ is distribution of the normal velocity over the width of the strip. In the case of uniform vibration $f(x) = f_0(x) = 1$. Also of interest is the distribution of velocity $f(x) = f_1(x) = \cos(\pi x / W)$, if to consider radiation of the column like double-sided rectangular plate bender transducer that is composed of simply supported beams.

By the definition of the radiation impedance

$$Z_{ac} = \frac{\bar{\dot{W}}_{ac}}{|U|^2} = \frac{1}{U_0} \int_{-W/2}^{W/2} P(x_p) f(x_p)\, dx\,, \tag{6.295}$$

where $P(x_p)$ is the sound pressure generated by vibration of the strip in the point x_p on its surface. For the sound pressure in the point x_p we obtain from formula (6.294)

$$P(x_p) = \frac{\omega\rho}{2} U_0 \int_{-W/2}^{W/2} f(x) H_0^{(2)}[k|x - x_p|]\, dx\,. \tag{6.296}$$

The resultant expression, from which the radiation impedance can be calculated, is

$$Z_{acil} = \frac{\omega\rho}{2} \int_{-W/2}^{W/2} \int_{-W/2}^{W/2} H_0^{(2)}(k|x - x_p|) f_i(x) f_l(x_p) dx dx_p, \quad i, l = 0, 1\,. \tag{6.297}$$

Note that at $x \to x_p$ $N_0[k|x - x_p|] \to \infty$ as $\ln k|x - x_p|$, therefore a small interval around this point must be excluded from calculation. The subscripts 0 and 1 correspond to the uniform and cosine distributions, respectively. Thus, Z_{00} and Z_{11} are the radiation impedances for the case of uniform and cosine velocity distributions. Z_{01} is the mutual radiation impedance between the modes in the case that the distribution is superposition of uniform and cosine modes of vibrations having equal magnitudes. The radiation impedance per unit length may be represented as

$$Z_{acil}(kW) = \rho c S_{effi} [\alpha_{il}(kW) + j\beta_{il}(kW)]\,, \tag{6.298}$$

where $\alpha_{il}(kW)$ and $\beta_{il}(kW)$ are nondimensional coefficients of the radiation resistance and reactance, S_{effi} is the effective radiating surface area per unit length,

$$S_{effi} = \int_{-W/2}^{W/2} f_i(x)^2\, dx\,. \tag{6.299}$$

Dependences of the nondimensional coefficients of the self-radiation impedances from kW at the uniform distribution ($S_{eff0} = W$ per unit length), at the cosine distribution ($S_{eff1} = W/2$ per unit length), and of the coefficients for the mutual impedances between these modes of vibration (in which case it is taken that $S_{eff01} = W$) are shown in Figure 6.43. The numerical values of the coefficients for $kW < 2$ are presented in Table 6.1. This range of values of kW, at which $W/\lambda < 1/3$, is typical for the rectangular bender transducers.

Table 6.1: Impedances and nondimensional coefficients for the cases of uniform and sinusoidal distributions of velocities, and the mutual impedances between these modes of vibration.

kW	Z_{ac00}		Z_{ac11}		Z_{ac01}	
	α_{00}	β_{00}	α_{11}	β_{11}	α_{01}	β_{01}
0.1	0.050	0.123	0.040	0.105	0.031	0.079
0.2	0.100	0.203	0.080	0.178	0.063	0.132
0.3	0.149	0.266	0.121	0.236	0.095	0.175
0.4	0.199	0.318	0.161	0.286	0.126	0.210
0.5	0.247	0.361	0.201	0.329	0.158	0.239
0.6	0.296	0.398	0.241	0.366	0.188	0.264
0.7	0.343	0.428	0.280	0.399	0.219	0.286
0.8	0.389	0.453	0.319	0.428	0.249	0.304
0.9	0.435	0.474	0.357	0.452	0.279	0.319
1.0	0.480	0.490	0.395	0.474	0.308	0.332
1.1	0.523	0.503	0.433	0.493	0.336	0.342
1.2	0.565	0.512	0.470	0.509	0.364	0.350
1.3	0.606	0.518	0.506	0.522	0.391	0.356
1.4	0.645	0.521	0.541	0.534	0.418	0.361
1.5	0.683	0.521	0.576	0.543	0.444	0.363
1.6	0.720	0.519	0.610	0.550	0.468	0.364
1.7	0.755	0.515	0.643	0.555	0.493	0.364
1.8	0.788	0.509	0.676	0.558	0.516	0.362
1.9	0.819	0.501	0.707	0.559	0.538	0.358
2.0	0.849	0.491	0.738	0.559	0.559	0.354

The sound pressure in the far field can be found by using formula (6.294) and the asymptotic expression for the Hankel function

$$H_0^{(2)}(kr)_{kr\to\infty} \to \sqrt{\frac{2}{\pi kr}}e^{-j\left(kr-\frac{\pi}{4}\right)}. \tag{6.300}$$

Thus, the sound pressure generated by an element Δy of the strip in the normal direction (at $\theta = 0$) in the far field (at $R \gg W$) is

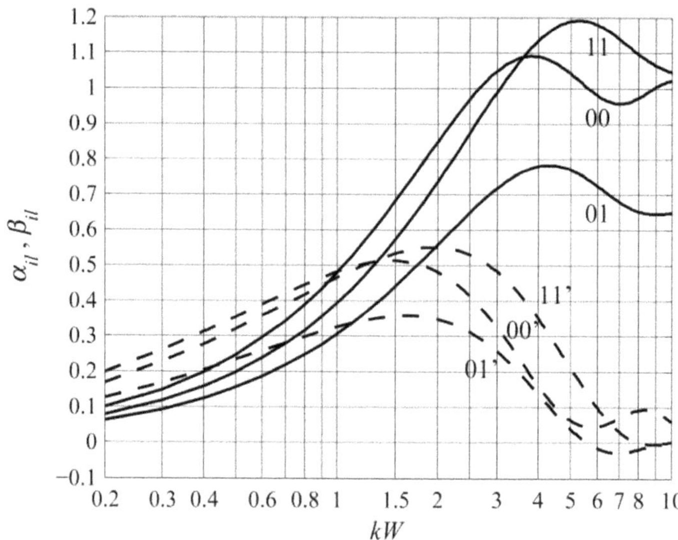

Figure 6.43: Dependences of the nondimensional coefficients of the active (solid lines) and reactive (dashed lines) components of radiation impedances for an infinitely long strip vibrating in a rigid plane.

$$P(R,0) = \frac{\rho c}{\sqrt{\lambda}} U_{\tilde{V}} \cdot \frac{e^{-j(kR - \frac{\pi}{4})}}{\sqrt{R}} . \tag{6.301}$$

Here $U_{\tilde{V}}$ is the volume velocity, $U_{\tilde{V}} = U_0 S_{av}$, and

$$S_{av} = \Delta y \cdot \int_{-W/2}^{W/2} f(x) dx \tag{6.302}$$

is the average surface area of an element of the strip: $S_{av} = W \cdot \Delta y$ for uniform vibration and $S_{av} = (2W / \pi) \cdot \Delta y$ for the cosine distribution.

The directional factor of a strip in the plane perpendicular to the strip is the same as for the segment of length W due to symmetry. Using the geometry considerations that follow from Figure 6.42, the sound pressure at the observation point can be represented as

$$P(R,\theta) = P(R,0) \left[\int_{-W/2}^{W/2} f(x) dx \right]^{-1} \cdot \int_{-W/2}^{W/2} f(x) e^{-jk(r_x - R)} dx , \tag{6.303}$$

where $r_x = R - x \sin\theta$ analogous to (6.277). Thus, the directional factor is in general

$$H(\theta) = \left[\int_{-W/2}^{W/2} f(x)dx \right]^{-1} \cdot \int_{-W/2}^{W/2} f(x)e^{-jkx\sin\theta}dx \,. \tag{6.304}$$

In the case of uniform velocity distribution ($f(x) = 1$) we obtain the well-known expression

$$H(\theta) = \frac{\sin[(kW/2)\sin\theta]}{(kW/2)\sin\theta}, \tag{6.305}$$

and for the cosine distribution $f(x) = \cos(\pi x / w)$

$$H(\theta) = \frac{\cos[(kW/2)\sin\theta]}{1-[(kW/\pi)\sin\theta]^2} \,. \tag{6.306}$$

6.5.5 Radiation of the Rectangular Pulsating Pistons

The radiation impedance of a rectangular piston vibrating uniformly in the infinite rigid plane baffle was considered in several works by numerical calculations based on using Rayleigh's integral (6.259). The most complete results of the calculations produced for pistons in the wide range of their aspect ratios wave sizes are presented in Ref. 18, 19. This resalts will be considered in this section from the point of view of application to transducer designs.

Plots of the nondimensional coefficients of radiation impedances of the rectangular pistons for their different widths to lengths aspect ratios, $R = W / L$, are shown in Figure 6.44. Several conclusions can be drawn from analysis of these data.

Comparison with results presented in Figure 6.38 shows that the nondimensional coefficients of radiation impedance of the square pistons (at $R = 1$) to a great accuracy are the same as for the circular pistons having equal surface area, i.e., at $ka = kW / \sqrt{\pi}$. The nondimensional coefficients of radiation impedance of rectangular pistons having the wave width $kW > 1$ at $R = W / L > 4$ behave like those for the infinite strips (at $R \to \infty$). Therefore, the data presented in Figure 6.43 and in Table 6.1 can be used for calculating radiation impedances of the rectangular pistons with such dimensions.

The sound pressure generated by the rectangular piston in the far field can be determined by direct applying the integral formulation of Huygens's principle (6.259). In the case that non-uniform distribution of velocity exists over surface of the piston its volume velocity is $U_{\tilde{V}} = S_{av}U_o$, where S_{av} is the average surface area. For a uniform distribution $U_{\tilde{V}} = WLU_o$.

Thus, in this case

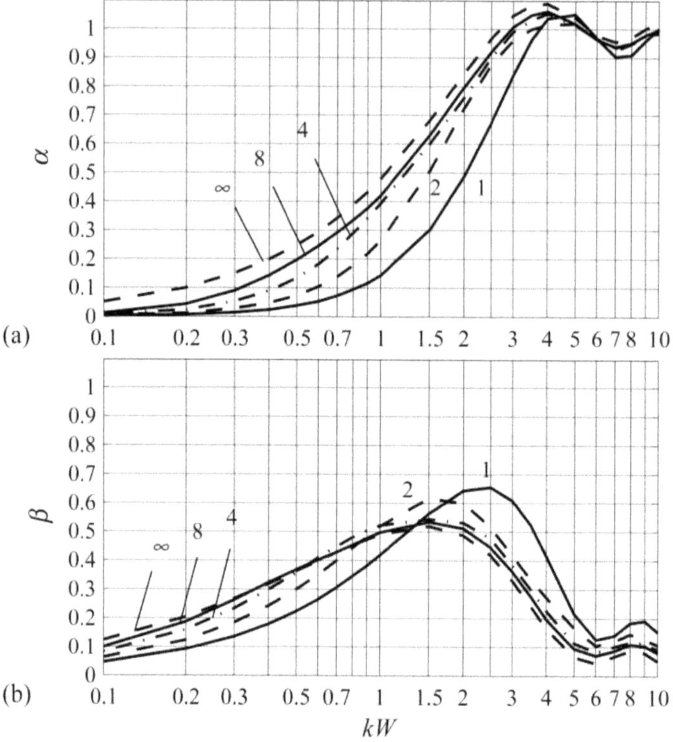

Figure 6.44: Nondimensional coefficients of the radiation impedances of the rectangular radiating surfaces vs. wave width kW at different aspect ratios $R = W/L$.

$$P(0) = j\frac{k\rho c}{2\pi}WLU_o \cdot \frac{e^{-jkR}}{R}. \tag{6.307}$$

Expressions for the directional factor of the rectangular piston that has geometry shown in Figure 6.45 will be obtained by using formulas (6.305) and (6.306) for directional factors of the linear segments without and with distributions of velocity over length. Following the product theorem, in the case that vibration is uniform.

$$H(\theta,\varphi) = \frac{\sin[(kW/2)\sin\theta]}{(kW/2)\sin\theta} \cdot \frac{\sin[(kL/2)\sin\varphi]}{(kL/2)\sin\varphi}. \tag{6.308}$$

In the case of velocity distribution over the width $U(x) = U_o \cos(\pi x / W)$, which is typical of the rectangular bender transducers,

$$H(\theta,\varphi)=\frac{\cos[(kW/2)\sin\theta]}{1-[(kW/\pi)\sin\theta]^2}\cdot\frac{\sin[(kL/2)\sin\varphi]}{(kL/2)\sin\varphi}. \tag{6.309}$$

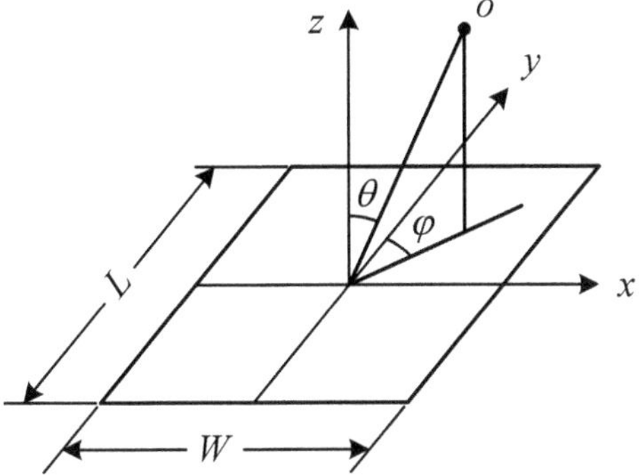

Figure 6.45: Rectangular piston and spherical coordinate system for representing its directional factor (point o is the observation point).

6.5.6 Radiation of the Oscillating Disk

Surfaces of the oscillating disk vibrate in phase, as shown in Figure 6.46 (b), where the main types of the flat piston-like projectors are schematically depicted. It is noteworthy that in all the cases the thickness of the projectors is assumed to be infinitesimal, and in this sense it would be more appropriate to call them membranes. Real projectors have a finite thickness. Effects of the finite thickness will be considered in Section 6.5.8.4.

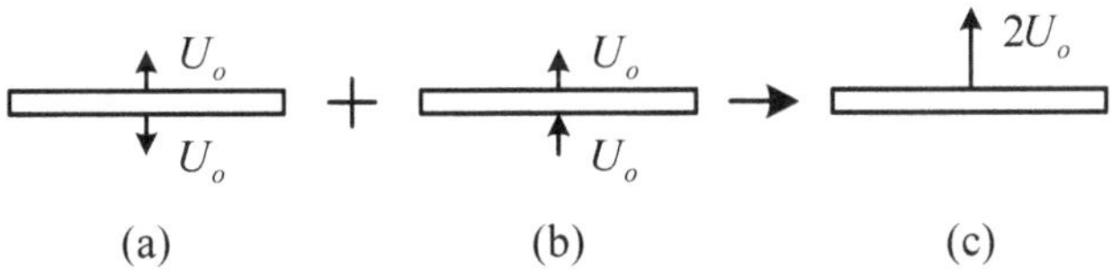

U_o, U_o + U_o, U_o → $2U_o$

(a) (b) (c)

Figure 6.46: Types of the circular disks vibration: (a) pulsating disk, (b) oscillating disk, (c) one-sided (vibrating on one side) disk as the superposition of (a) and (b).

The sound pressure on the surfaces of the oscillating disk (Figure 6.46 (b)) is in opposite phase (when compression occurs on one side, expansion takes place on another). Due to symmetry the sound pressure on the plane that is continuation of the disk surface is zero. This means that the oscillating disk can be imagined as uniformly vibrating membrane embedded in the infinite absolutely compliant plane baffle. The radiation problem for the oscillating disk was considered by L. Y. Gutin in Ref. 20. The solution was presented in the form of expansion into a series in terms of the oblate spheroidal functions. The coefficients of the series are presented in Ref. 20 for a number of terms that is sufficient for calculating radiation parameters for the range of *ka* having practical interest for transducer designing. The results of calculating the radiation parameters are summarized as follows.

The nondimensional coefficients of radiation impedance are plotted in Figure 6.47 and given in Table 6.2.

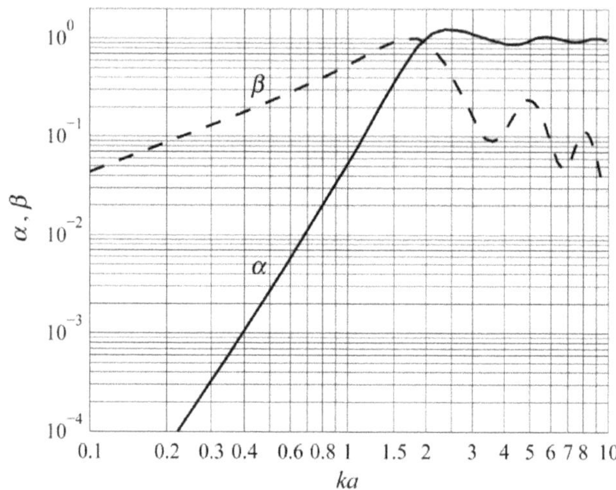

Figure 6.47: Nondimensional coefficients α and β of radiation impedance of the oscillating circular disk.

Table 6.2: Nondimensional coefficients of radiation impedance of the oscillating circular disk.

ka	α	β	*ka*	α	β
0.1	4.4e-6	0.044	3	1.12	0.17
0.5	3.7e-3	0.23	5	0.96	0.23
1	0.06	0.54	8	0.95	0.11
2	1	0.9	10	0.97	0.03

The main application of the results obtained for the oscillating disk is to design the pressure gradient transducers, which predominantly have a small wave size. At $ka \ll 1$ (practically up to $ka \approx 0.3$)

$$\alpha \approx \frac{8}{27\pi^2}(ka)^4, \quad \beta \approx \frac{4}{3\pi}(ka). \tag{6.310}$$

From $x_{ac} = \rho c \pi a^2 \beta = \omega m_{ax}$ follows that the acoustic (radiation) mass is

$$m_{ac} = \frac{4}{3} a^3 \rho = \frac{1}{\pi} \tilde{V}_{sph} \rho \,, \tag{6.311}$$

where $\tilde{V}_{sph}$ is the volume of sphere having the same radius. Thus, the acoustic mass is twice smaller than for the pulsating disk (see (6.275)).

Expression for the radiation resistance

$$r_{ac} = \rho c \pi a^2 \frac{8}{27\pi^2} (ka)^4 \tag{6.312}$$

seemingly does not comply with the general expression for radiation resistances of the simple sources

$$r_{ac} = 2\pi \rho c \frac{S_{av}^2}{\lambda^2} \,, \tag{6.313}$$

but for the oscillating disk $S_{av} = 0$, which means that $r_{ac} = 0$ to the first approximation, and the oscillating disk is very poor projector at low frequencies. Expression (6.312) is for the radiation resistance to the second approximation.

The sound pressure in the far field can be presented in the form[20]

$$P(ka,\theta) = \frac{\rho c}{R} e^{-j(kR - \frac{\pi}{2})} U_o \frac{1}{k} F(ka) \cdot E(ka,\theta) \,, \tag{6.314}$$

where at $ka < 4$

$$E(ka,\theta) = P_1(\cos\theta) + \gamma(ka) P_3(\cos\theta) \,, \tag{6.315}$$

and functions $F(ka)$ and $\gamma(ka)$ have values that are presented in Table 6.3.

Table 6.3: Values of functions $F(ka)$ and $\gamma(ka)$.

ka	0.5	1.0	1.5	2.0	2.5	3.0
$F(ka)$	$j0.027$	$0.02(1+j12.6)$	$0.36(1+j3.2)$	$2.2(1+j0.9)$	$3.7(1+j0.45)$	$5.0(1+j0.18)$
$\gamma(ka)$	0.01	0.04	0.09	0.15	0.23	0.32

At $ka \ll 1$ (practically, at $ka < 0.5$)

$$P(ka,\theta) \approx \frac{\rho c}{R} e^{-j(kR - \frac{\pi}{2})} U_o \frac{j}{k} \cdot \frac{2}{3\pi} (ka)^3 \cos\theta \,. \tag{6.316}$$

The directional factor is

$$H(ka,\theta)=\frac{E(ka,\theta)}{E(ka,0)}=\frac{P_1(\cos\theta)+\gamma(ka)P_3(\cos\theta)}{1+\gamma(ka)}. \tag{6.317}$$

At $ka \ll 1$

$$H(ka,\theta)=H(\theta)=\cos\theta, \tag{6.318}$$

i.e., directional factor of small oscillating disk is that of the dipole. The directional factor practically does not change at least up to $ka=1$, as it follows from Figure 6.48, where the directional factors of an oscillating disks are shown at different values of *ka*.

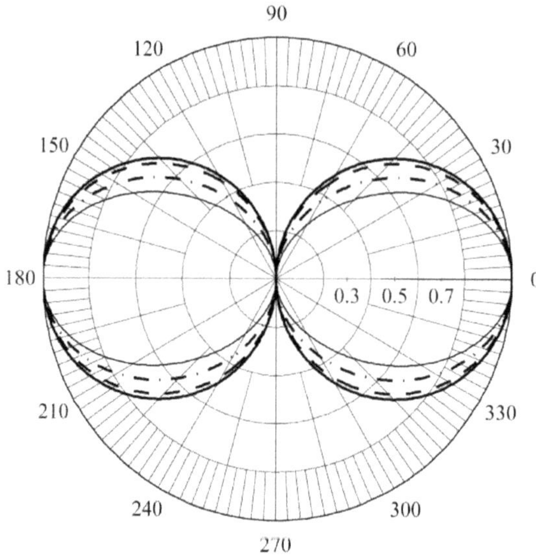

Figure 6.48: Directional factors of the oscillating disk: *ka* = 0.25 (solid line, $\cos\theta$), *ka* = 1 (dashed line), *ka* = 2 (dash-dotted line), *ka* = 3 (thin solid line).

Comparing expressions (6.314) and (6.2) for the sound pressure in the far field we can conclude that the diffraction function for the oscillating disk in direction of acoustic axis (at $\theta=0$) is

$$\chi(ka)\big|_{\theta=0}=\frac{1}{k}F(ka)\cdot E(ka,0)=\frac{1}{k}F(ka)\cdot[1+\gamma(ka)]. \tag{6.319}$$

Using the general formula for the diffraction coefficient (6.10) and taking into consideration that the total surface area of the disk is $S_\Sigma=2\pi a^2$, we arrive at the expression

$$k_{dif\,o}=\frac{2\lambda}{S_\Sigma}\cdot\chi(ka)_{\theta=0}=\frac{2}{(ka)^2}F(ka)\cdot[1+\gamma(ka)]. \tag{6.320}$$

From Eq. (6.316) follows that for $ka < 0.5$

$$\chi(ka,\theta) = \frac{j}{k} \cdot \frac{2}{3\pi}(ka)^3 \cos\theta \,, \tag{6.321}$$

and after substituting in the general formula for the diffraction coefficient we obtain

$$k_{dif}(\theta) = j\frac{4}{3\pi}(ka)\cos(\theta) \,. \tag{6.322}$$

Moduli of the diffraction coefficients of disks at $\theta = 0$ as function of ka are presented in Figure 6.49.

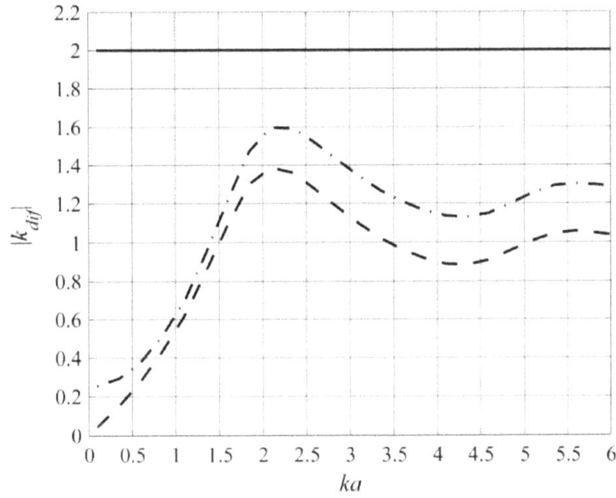

Figure 6.49: Modulus of diffraction coefficients of circular disks as functions of *ka*: pulsating disk in infinite planar baffle (solid line), oscillating disk (dashed line), disk radiating from one side (dash-dotted line).

6.5.7 Radiation of the Disk Vibrating on One Side (Gutin's Superposition Concept)

The radiation problem for the disk vibrating on one side without baffles was solved by superposing results of already available solutions for the pulsating and oscillating disks in Ref. 20. The idea is self-explanatory from Figure 6.46. This technique became obvious after it was first suggested by L. Y. Gutin in Ref. 20. Sometimes it is referred to as "the Gutin's concept" (Ref. 21), but mostly is used without any reference (for example, in Ref. 22).

Superposing the potentials of acoustic fields generated by the pulsating and oscillating disks vibrating with equal velocities leads to the following results for the disk vibrating on one side with the same velocity.

The nondimensional coefficients of the radiation impedance (denoted α_1 and β_1) are

$$\alpha_1 = \frac{1}{2}(\alpha_p + \alpha_o) \text{ and } \beta_1 = \frac{1}{2}(\beta_p + \beta_o), \tag{6.323}$$

where subscripts p and o correspond to pulsating and oscillating disks. Coefficients α_p and β_p are given by expressions (6.271) and (6.272), and plotted in Figure 6.50.

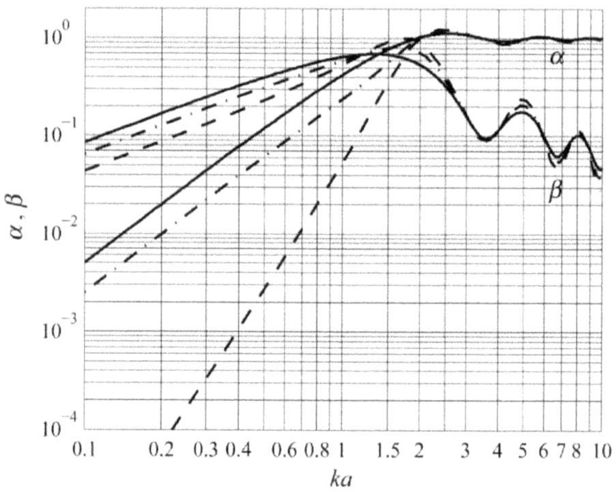

Figure 6.50: Nondimensional coefficients of radiation impedances of the circular disks: pulsating (solid line), oscillating (dashed line), one-sided (dash-dotted line).

Coefficients α_o, β_o are presented in Table 6.2 and shown in Figure 6.47. The values of coefficients α_1 and β_1 are summarized in Table 6.4.

Table 6.4: Nondimensional coefficients of radiation impedance of the disk radiating from one side.

ka	α_1	β_1	ka	α_1	β_1
0.1	2.5e-3	0.065	2	1.02	0.71
0.5	0.065	0.31	5	0.98	0.21
1	0.24	0.59	10	0.98	0.037

At $ka \ll 1$

$$\alpha_1 = \frac{1}{2}\left[\frac{1}{2}(ka)^2 + \frac{8}{27\pi}(ka)^4\right] \approx \frac{1}{4}(ka)^2, \tag{6.324}$$

$$\beta_1 = \frac{1}{2}\left[\frac{8}{3\pi}(ka) + \frac{4}{3\pi}(ka)\right] = \frac{2}{\pi}(ka). \tag{6.325}$$

The acoustic mass is

$$m_{ac} = \frac{3}{2\pi}\tilde{V}_{sph}. \tag{6.326}$$

Expression for the sound pressure in the far field being found as superposition of expressions (6.283) and (6.314) is

$$P_1(ka,\theta) = \frac{\rho c}{R} e^{-j(kR-\frac{\pi}{2})} U_0 \frac{ka^2}{4}\left[\frac{2J_1(ka\sin\theta)}{ka\sin\theta} + \frac{2}{(ka)^2}F(ka)\cdot E(ka,\theta)\right]. \tag{6.327}$$

The sound pressure on the axis is

$$P_1(ka,0) = \frac{\rho c}{R} e^{-j(kR-\frac{\pi}{2})} U_0 \frac{ka^2}{4}\left\{1 + \frac{2}{(ka)^2}F(ka)\cdot[1+\gamma(ka)]\right\}. \tag{6.328}$$

The directional factor is

$$H(ka,\theta) = \frac{P_1(ka,\theta)}{P_1(ka,0)}. \tag{6.329}$$

Comparison of the directional factors of the circular disks at $ka = 3$ is made in Figure 6.51. The diffraction coefficient on the axis will be found by formula (6.10) as

$$k_{dif} = \frac{2\lambda}{S_\Sigma}\cdot \chi(ka)_{\theta=0}, \tag{6.330}$$

where $S_\Sigma = \pi a^2$ (the disk radiates from one side) and

$$\chi(ka)_1 = \frac{(ka)^2}{4}\left\{1 + \frac{2}{(ka)^2}F(ka)\cdot[1+\gamma(ka)]\right\}. \tag{6.331}$$

Thus,

$$k_{dif1} = 1 + \frac{2}{(ka)^2}F(ka)\cdot[1+\gamma(ka)]. \tag{6.332}$$

Taking into consideration that diffraction coefficient for the pulsating disk is $k_{dif\,p} = 2$, and for the oscillating disk ($k_{dif\,o}$) is given by expression (6.320), we can conclude that

$$k_{dif1} = \frac{k_{dif\,p} + k_{dif\,o}}{2} . \tag{6.333}$$

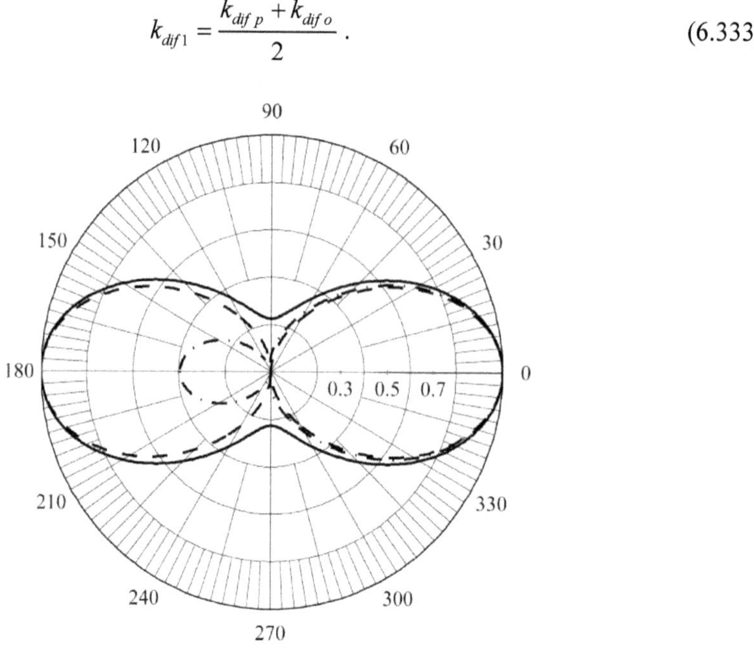

Figure 6.51: Directional factors of the circular disks at $ka = 3$: pulsating disks (solid line), oscillating disks (dashed line), disk vibrating on one side (dash-dotted line).

Table 6.5: Diffraction coefficient of the one-sided disk as function of *ka*.

ka	$\lvert k_{dif}\rvert$	ka	$\lvert k_{dif}\rvert$	ka	$\lvert k_{dif}\rvert$
0.1	0.25	1	0.62	10	1.22
0.2	0.27	2	1.55		
0.5	0.34	5	1.23		

At $ka < 0.5$

$$k_{dif1} = 1 + j\frac{2}{3\pi}(ka) . \tag{6.334}$$

Values of the diffraction coefficient of the one-sided disk as function of *ka* are presented in Table 6.5.

6.5.8 Radiation of Disks Embedded in the Baffles of Finite Size

6.5.8.1 Introduction

The preceding analysis shows significant difference between characteristics of acoustic energy radiation in free space of the circular disks of small wave size having different velocity distribution (pulsating, oscillating and one-sided). This is illustrated with data regarding the nondimensional coefficients of radiation resistance and ratio of the reactive to active power radiated that are presented in Table 6.6. The comparison shows that the characteristics of pulsating disk are comparable with those of the pulsating sphere, which can be considered as an ideal source. And the oscillating disk of small wave size is an extremely poor source of radiation. At large wave sizes (practically at $ka \geq 3$ or at $(2a/\lambda) > 1$) the difference in radiating characteristics of all the disks vanishes.

Table 6.6: Ratios of the reactive to active power radiated by different sources.

Projector Type	ka	α	β/α
Pulsating Disks	< 0.5	$(ka)^2/2$	$16/3\pi ka$
	0.5	0.125	3.4
	3	1.09	0.15
Oscillating Disk	< 0.5	$8(ka)^4/27\pi^2$	$9\pi/2(ka)^3$
	0.5	$1.9 \cdot 10^{-3}$	110
	3	1.17	0.18
One-Sided Disk	< 0.5	$(ka)^2/4$	$8/\pi ka$
	0.5	0.062	5.1
	3	1.14	0.16
Sphere	< 0.5	$(ka)^2/[1+(ka)^2]$	$1/ka$
	0.5	0.2	2
	3	0.9	0.3

Transducers with flat radiating surfaces shown in Figure 6.1(c) usually have dimensions of their surfaces relatively small compared with wavelength at operating frequencies. Thus, the simply supported circular plates of a double plate bender transducer have $D/\lambda < 0.2$. (This follows from formula for its resonance frequency $f_r = 0.45(t/a^2)c_{ceramic}$ at usually adopted

$t/a < 1/5$). The overall thickness of a single double plate bender transducer, t_{tr}, is usually about two thicknesses of the comprising plates, i.e., $t_{tr}/\lambda < 0.4$. As to the rectangular plate benders that are made of beams, we will assume that a single transducer unit has approximately the same wave size as a circular plate having the same resonance frequency ($l = w = D$). Though this assumption is not quite rigorous, it gives sufficiently accurate estimation of the order of quantities. Diameters of the piston like vibrating surfaces of transducer configurations shown in Figure 6.1 (c.2) and (c.3), which are typical for the Tonpilz transducer design enclosed in a housing, usually have size $D/\lambda < 0.5$ ($ka < 1.5$). (This requirement is out of consideration of steering directivity pattern of an array populated by the transducers of this kind).

Thus, all the listed transducers fall into the category of transducers with radiating surfaces small compared with wavelength, when operating as single transducer units. Their radiation characteristics in this case can be improved by using rigid baffles flash with their radiating surfaces. The ideal results obviously would be obtained with baffles of infinite size, but sufficiently good improvements can be achieved with real baffles of properly chosen finite size.

The quantitative estimations of effects that rigid baffles of finite size exert on the characteristics of acoustic radiation of the circular disks were produced in several works[15, 21, 23]. The problem was solved in Ref. 23 in the oblate spheroidal functions in the way similar to those used in Ref. 20 for calculating radiation characteristics of the oscillating disk without a baffle. The most comprehensive analysis of the related problems based on employing integral equations to calculating acoustic fields was performed in Ref. 15. The numerical results regarding the radiation characteristics of the disks embedded in the rigid circular baffles that are presented in this section are due to this work.

6.5.8.2 Radiation of the Oscillating Disks Embedded in the Rigid Baffles of Finite Size

A qualitative explanation of difference in radiation characteristics of oscillating disks of small and significant wave size can be done with help of Figure 6.52.

At small wave size the wave radiated from the front surface propagates round the edge of the disk and comes to the back surface practically in the same phase, in which this surface vibrates. Thus, the main fluid flow goes from one side of the disk to another in the process of vibration,

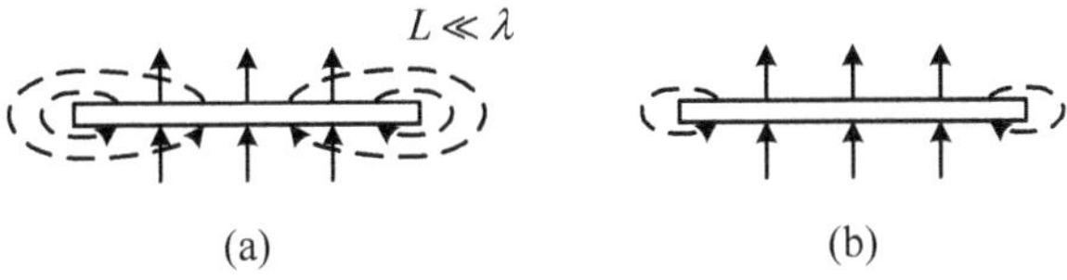

Figure 6.52: Qualitative illustration of diffraction on the oscillating disk: (a) at small wave size (low frequencies), (b) at high frequencies.

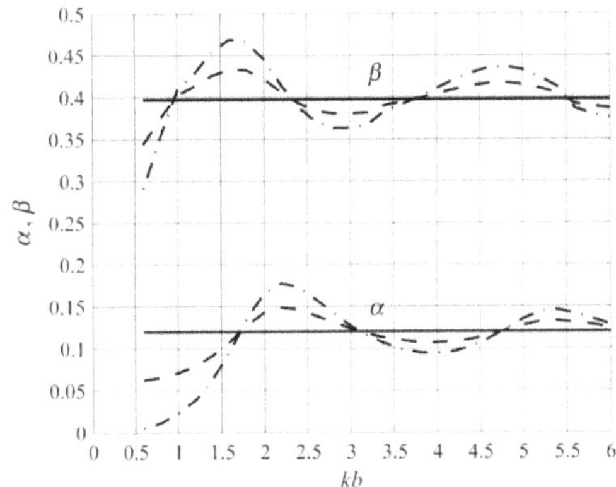

Figure 6.53: Dependences of the nondimensional coefficients of the radiation impedances of disk with $ka = 0.5$ on the wave radius of the rigid baffle for the oscillating disks (dash-dotted lines) and one-sided disks (dashed lines). Solid lines correspond to the pulsating disk, for which characteristics don't depend on the size of the baffles.

and only a small portion of the radiation goes to infinity. Acoustic short-circuiting takes place. With increase of the disk wave size the portion of radiation that comes over the edge to the back surface in phase with its vibration reduces and eventually becomes negligible. Radiations from the disk surfaces become independent and take place in respective half spaces. For the one-sided disk radiation in rear half space due to diffraction over the edge exists though it is much less pronounced than in case of the oscillating disk. It can be expected that the harmful effect of diffraction can be reduced by placing the disks in a circular rigid baffle and thus moving the edge of the entire structure away from the source of radiation. Thus, for the pulsating disk there is no diffraction over the edge, because in terms of radiation in half space it is equivalent to radiation of disk embedded in the infinite rigid baffle due to symmetry. As it follows from the Table 6.6, the acoustic energy radiation characteristics of the disks of all the configurations become quantitatively practically the same at $ka \geq 3$. This allows to assume that the wave diameter of the baffle sufficient for achieving the goal of effective suppressing the effect of diffraction on the acoustic energy radiation may be about $d / \lambda = 1$.

Dependences of the nondimensional coefficients of radiation impedance for the disks having different wave radiuses *ka* from the wave radiuses of the rigid baffles, *kb*, are presented in Figure 6.53 and Figure 6.54. In Figure 6.53 they are shown for small disk with $ka = 0.5$ in the more detailed way.

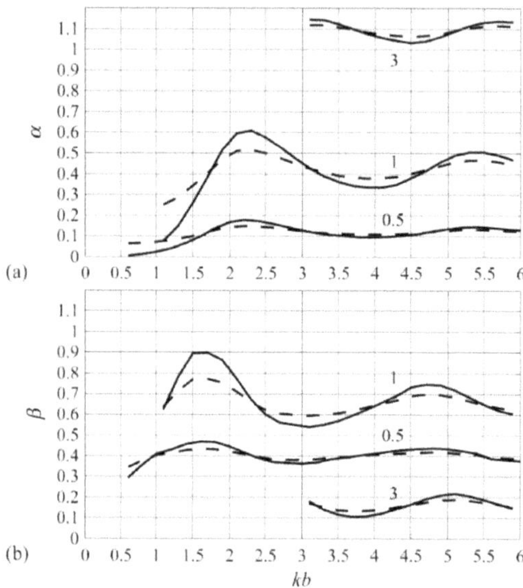

Figure 6.54: Dependences of the nondimensional coefficients of the radiation impedances of disks of different wave size on the radius of the rigid baffle: (a) α , (b) β for the oscillating disks (solid lines) and one-sided disks (dashed lines). Numbers on the curves correspond to the wave radiuses (*ka*) of the disks.

As can be seen from Figure 6.53, at $kb \approx 2$ the normalized components of radiation impedance reach the values that correspond to those for the pulsating disk, around which they oscillate with increase of the size of a baffle. Analogous dependences are shown in Figure 6.54 for a wider range of the disks wave dimensions. They confirm that the wave radius of the rigid baffle that is sufficient for achieving values of the coefficients, around which they oscillate with further increase of the baffle size, is about $kb = 2 \div 3$. Dependence of the nondimensional coefficients of radiation impedances of differently vibrating disks on the wave size in case that they are embedded in the rigid baffle having wave size $kb = 3.0$ are presented in Figure 6.55.

Both components of the radiation impedance up to $ka \approx 0.5$ and the active component up to $ka \approx 1.0$ behave like in the infinite baffle. At $ka > 3.0$ active component of radiation of the disks practically does not depend on the size of the baffle (slightly oscillates around its value at the infinite baffle, as shown in Figure 6.54). The reactive component at $ka > 3.0$ oscillates more significantly, but its average value becomes small.

For the disks with baffles of small wave sizes (at $kb < 0.5$) the following approximate expressions for the nondimensional coefficients of radiation impedance may be used as it is shown in Ref. 21,

$$\alpha = \frac{8}{27\pi^2}\left\{1 + 2.26\left[\frac{b^2}{a^2} - 1\right]\right\}(ka)^4, \tag{6.335}$$

$$\beta = \frac{8}{3\pi} \cdot \frac{3(b/a) - 2}{3(b/a) - 1}(ka). \tag{6.336}$$

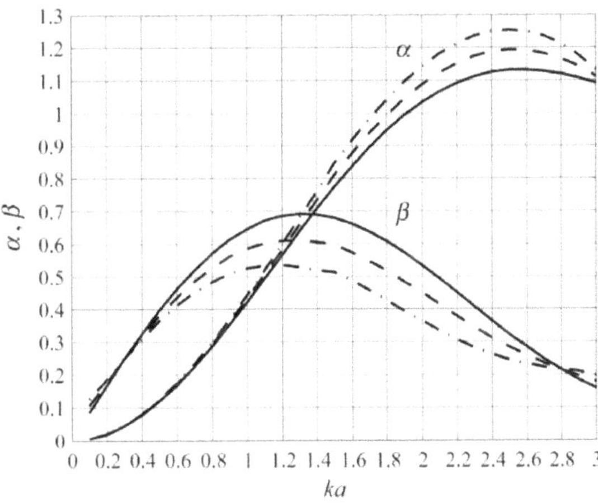

Figure 6.55: Nondimensional coefficients of the radiation impedance for the disks embedded in the rigid baffle at $kb = 3.0$: pulsating disk (the same as in the infinite baffle) (solid line), one-sided disk (dashed line), oscillating disk (dash-dotted line).

The algorithms developed in Ref. 15, 23 allow a detailed analysis of peculiarities of the far field characteristics of the disks vibrating in the rigid baffles, namely, of the sound pressure generated on the acoustical axis and directional factors.

As it is shown with results of calculations made therein at small wave dimensions of the disks these characteristics depend significantly on the wave size of a baffle. At $ka \geq 3$ they are determined mainly by dimensions of the projector itself. As well as the radiation impedance, the sound pressure on the axis sharply increases with increase of the wave size of a baffle up to $kb \approx 3$ ($b \approx \lambda/2$), and then its value starts to oscillate. The oscillations are more pronounced (up to about 6 dB) for the oscillating disks of small wave size (with $ka < 1.5$) and less pronounced for the one-sided disks of the same size. At value $kb \approx 6.3$ ($b \approx \lambda$) the sound pressure achieves minimum. The qualitative explanation of this behavior that takes place due to diffraction of sound on the edge of the disks can be done with help of the Figure 6.56.

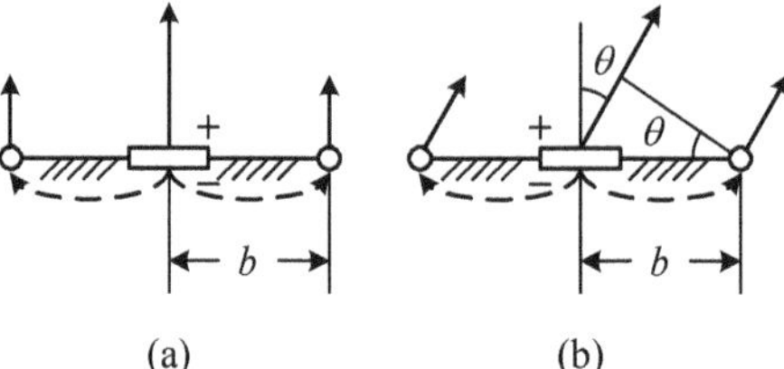

Figure 6.56. Illustration of the effects of diffraction on the sound pressure on the axis (a) and on the directional factor (b) of an oscillating disk of small wave size embedded in the finite size rigid baffle.

The sound pressure on the axis is a superposition of the direct radiation from the front surface of the disk and of radiation of its back surface that comes over the edge. This back radiation can be imagined as produced by an additional projector located on the edge. If the width of the baffle is $b \approx \lambda/2$, then given that the surfaces of the disk vibrate in antiphase the additional projector radiates in phase with the front surface, and this results in maximum of sound pressure on the axis, which is greater than it would be produced with the infinite baffle. In the case that $b \approx \lambda$, the phase of the additional projector changes by 180°, and sound pressure on the axis drops to its minimum.

The effect of the baffle on the directionality depends mainly on the dimensions of the projector. At small wave size ($ka < 1.5$) the influence is very strong. Thus, at $kb \approx 6.3$ very sharp drop appears in the directivity pattern in a narrow range of angles close to the axis in addition to greatly reduced value of sound pressure on the axis. These angles (around $\theta \approx 7^o$ in Figure 6.56 (b)) correspond approximately to radius *b*, on which the phase shift brings radiation of the additional projector on the edge to antiphase with the direct radiation of the disk. At larger dimensions of the disks their directional factors practically are not influenced by the baffles.

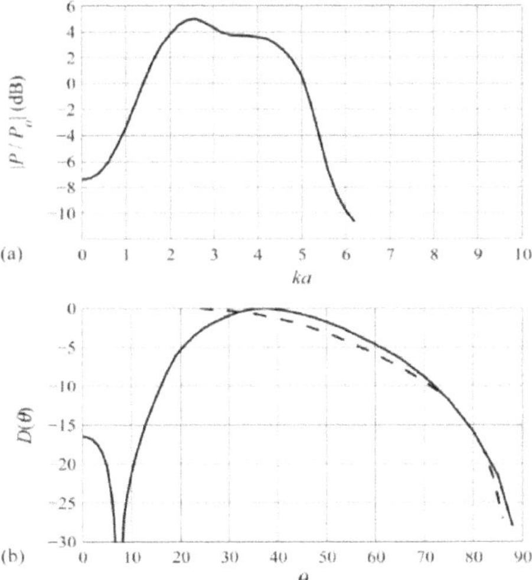

Figure 6.57: (a) Relative values of sound pressure on the axis of small size ($ka < 1.5$) oscillating disk embedded in the finite size baffle. P_o is sound pressure generated by the disk in the infinite baffle. (b) Directional pattern of small ($ka < 1.5$) oscillating disk vibrating in the rigid baffle. Size of the baffle is $kb \approx 6.3$.

The peculiarities of behavior of the sound pressure on the axis and of the directional factors of oscillating disks of small size ($ka < 1.5$) embedded in the finite size baffles are qualitatively illustrated by plots in Figure 6.57. The values of parameters are approximate. More accurate results of calculations can be found in Ref. 15. The qualitative understanding of impact of baffles of finite size based on considering the ideal baffles may be even more important than knowing the exact values thus calculated, because properties of the real baffles usually are not known to a sufficient accuracy.

It is noteworthy that conclusions regarding large oscillations of sound pressure and deep drops in the directional characteristics that are made for the ideally rigid circular baffles occur since all the points of the additional projector on the edge vibrate in phase because of symmetry. For baffles of different configuration (for square baffles for example) these effects may be significantly reduced. Besides, these effects may be reduced due to a finite compliance and active losses that the real baffles may possess.

It is noteworthy that conclusions regarding large oscillations of sound pressure and deep drops in the directional characteristics that are made for the ideally rigid circular baffles occur since all the points of the additional projector on the edge vibrate in phase because of symmetry. For baffles of different configuration (for square baffles for example) these effects may be significantly reduced. Besides, these effects may be reduced due to a finite compliance and active losses that the real baffles may possess.

It is noteworthy that conclusions regarding large oscillations of sound pressure and deep drops in the directional characteristics that are made for the ideally rigid circular baffles occur since all the points of the additional projector on the edge vibrate in phase because of symmetry. For baffles of different configuration (for square baffles for example) these effects may be significantly reduced. Besides, these effects may be reduced due to a finite compliance and active losses that the real baffles may possess.

6.5.8.3 One-Sided Disk in the Rigid Baffle of Finite Size

After dependence of radiation characteristics of oscillating disk from dimensions of the rigid baffles is determined, the analogous characteristics of the one-sided disk can be obtained by employing Gutin's concept, as result of superposing the acoustic fields generated by pulsating disk and oscillating disk embedded in the rigid baffle of corresponding size, as illustrated in Figure 6.58.

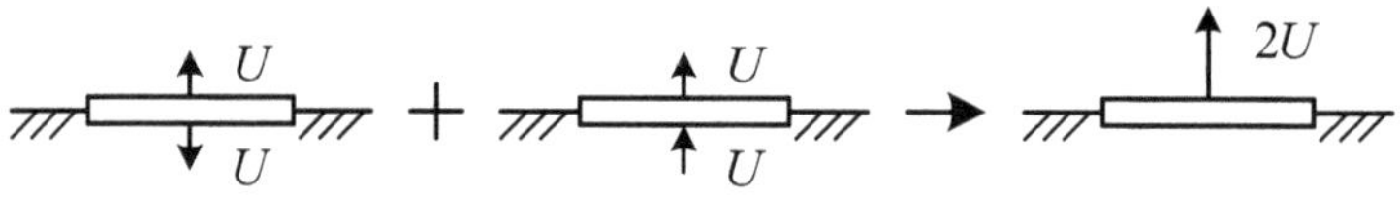

Figure 6.58: Illustration of superposing the acoustic fields generating by the disks with the finite rigid baffles.

Analogous to situation for the disks without baffles in this case

$$P_{o-s,b} = \frac{1}{2}(P_{puls} + P_{osc,b}), \tag{6.337}$$

$$Z_{o-s,b} = \frac{1}{2}(Z_{puls} + Z_{osc,b}), \tag{6.338}$$

where subscripts "*o-s,b*" and "*osc,b*" denote parameters related to one-sided and oscillating baffled disks, respectively. Dependences of the nondimensional coefficients of radiation impedances of one-sided dusks from wave size of the baffles are shown in Figure 6.53 and Figure 6.54 by dashed lines. For the case that $kb = 3.0$ values of these coefficients are shown in Figure

6.55 as functions of the wave sizes of the disks. As follows from the Figures, the values of nondimensional coefficients of one-sided disks are intermediate of those for pulsating and baffled oscillated disks up to $kb \approx 2.0$. At $kb > 2.0$ they oscillate around common value for all the disks. The magnitude of oscillations for one-sided disk is smaller than for the oscillating. Influence of the baffles on the directional characteristics of the one-sided disk is the most pronounced at $ka < 1.5$. They significantly reduce the back radiation at angles $\theta° > 90°$ compared with those without baffles. At $ka > 3.0$ characteristics are practically the same, as without baffles. Peculiarity is that in all the cases maximum of radiation exists at angle $\theta° = 180°$. This is because all the points of the additional projector that is formed as result of diffraction on the edge of a baffle radiate in phase in this direction. Note that the same effect takes place in the case of one-sided disk without the baffles (see Figure 6.51).

6.5.8.4 Radiation of Disks Having a Finite Thickness

The projector can be imagined in this case as a finite size cylinder with vibrating ends: pulsating (with ends vibrating in phase) and oscillating (with ends vibrating in antiphase). The same algorithm was employed for solving this radiation problem, as was used for calculating radiation impedances of the finite size cylinder radiating by its side surface in Ref. 15 (see Section 6.3.2.5).

Radiation by the ends was considered under the conditions that the side surface of the cylinder is rigid ($U_r = 0$), or compliant ($P_r = 0$). The results obtained for the nondimensional coefficients of the radiation impedances are presented in Figure 6.59 and Figure 6.60.

It is interesting to note that up to values $ka = 1$ the active component of the radiation impedance of the pulsating disk with rigid surface reduces with increase of separation between vibrating ends up to $h / a = 2$, while in the case of oscillating disk it increases. The clear physical explanation to this difference can be done, if to imagine that the radiation impedance of each end consists of its self-radiation impedance and mutual impedance between the ends. (For the acoustic interaction between transducers see the next section.) The sign of active component of the mutual impedance at these separations is positive for sources vibrating in phase and negative for those vibrating in antiphase (that have the dipole nature), and the magnitude of the active component reduces with increase of the separation. At values $ka > 2$ the interaction between ends becomes negligible, and they vibrate independently. The nondimensional coefficient of the active component becomes close to unity. At smaller *ka* ($ka < 1$) an increase of the thickness from $h/a = 0$ (that corresponds to the thin disk vibrating in infinite rigid baffle) to $h/a = 2$ (that is close to one-sided disk vibrating in the rigid cylindrical baffle) results in reducing the radiation resistance in two times approximately.

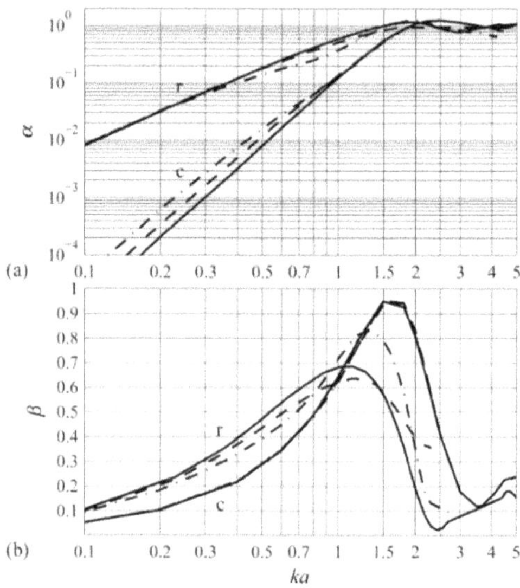

Figure 6.59: Nondimensional coefficients of the radiation impedance for the pulsating disks having finite thickness at $h/a = 0.5$ (solid line), $h/a = 1$ (dashed line), $h/a = 2$ (dash-dotted line) with rigid (r) and compliant (c) side surfaces.

In the case that the side surface is compliant, radiation resistance is significantly smaller (up to order of magnitude at small *ka*). At $ka > 2$ it becomes the same as with the rigid side surface. As to the nondimensional coefficients of the radiation reactance, their magnitudes remain close for both the rigid and compliant side surfaces.

With data for the radiation impedances of the pulsating and oscillating disks of finite thickness having rigid side surface known, the result can be readily obtained for the disk vibrating in one end of a rigid cylindrical housing by employing the Gutin's superposition concept. This situation is typical for the case that the housing encloses transducer of Tonpilz type, as is shown schematically in Figure 6.1(c.4).

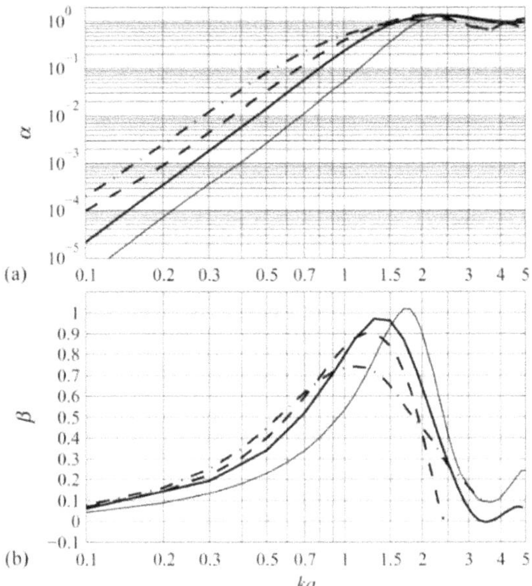

Figure 6.60: Nondimensional coefficients of the active (a) and reactive (b) components of the oscillating disks having finite thickness with rigid side surface for $h/a = 0.5$ (solid line), $h/a = 1$ (dashed line), $h/a = 2$ (dash-dotted line), $h/a = 0$ (thin solid line).

6.6 Acoustic Interaction Between Transducers

6.6.1 Introduction

The mechanical systems of the transducers may be often composed of several parts mechanically isolated and connected electrically. This can be done due to technological reasons, when it is hard to manufacture an entire transducer as a solid structure, or in order to avoid harmful effects of coupled vibration that take place because of unfavorable aspect ratio of mechanical system of the transducer. Typical examples of mechanical systems composed of several parts. are shown in Figure 6.61. In all these cases effect of the sound pressure that develops on the surface of a transducer by its uniform vibration does not average, as it would be on the surface of the solid mechanical system.

The situation arises that at equal velocities of vibration of the transducer parts the averaged sound pressures on the parts may be different and they appear to be loaded by different radiation impedances. The effects of interaction between transducer parts can be often estimated analytically, as it will be shown below.

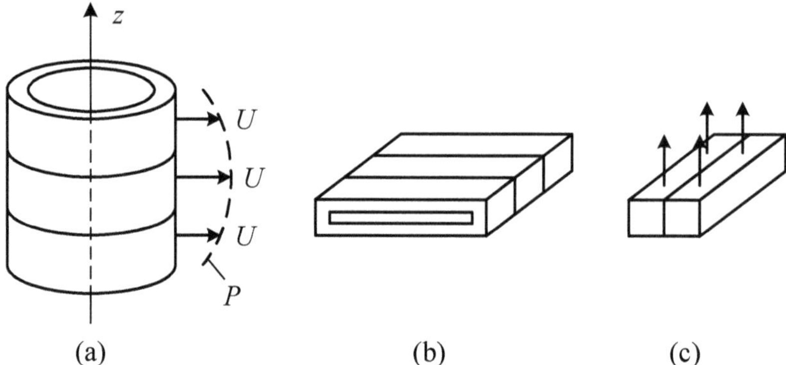

Figure 6.61: Examples of mechanical systems of transducers that are composed of elementary parts: (a) cylindrical transducer, (b) rectangular flexural plate bender, (c) side-scan sonar transducer.

The analogous effects of acoustic interaction exist between the transducers in arrays. An analytical determining the interactions in arrays can be significantly complicated. This is especially true for arrays populated by transducers having a non-planar configuration and those furnished with special baffles, which may in particular be intended for reducing an effect of interaction. The most practical way to determine mutual impedances between transducers in these cases is through experimentation. Certain peculiarities are inherent in procedures of measuring the mutual impedances that are related to properties of transducers involved in this process. Therefore, one of objectives of this Section is to consider the methods for measuring the mutual radiation impedances.

6.6.2 General Considerations and Definitions

In order to show, how the effects of the acoustic interaction can be accounted for in calculating the transducers, we consider a system consisting of two identical transducers T_1 and T_2 (Figure 6.62 (a)) that vibrate with velocities $U_1(\boldsymbol{r}_{\Sigma 1})$ and $U_2(\boldsymbol{r}_{\Sigma 2})$. Considering that the mutual influence on parameters of transducers has the most significant effect in the close to their resonance frequency region, we will assume that interacting transducers have one degree of freedom.

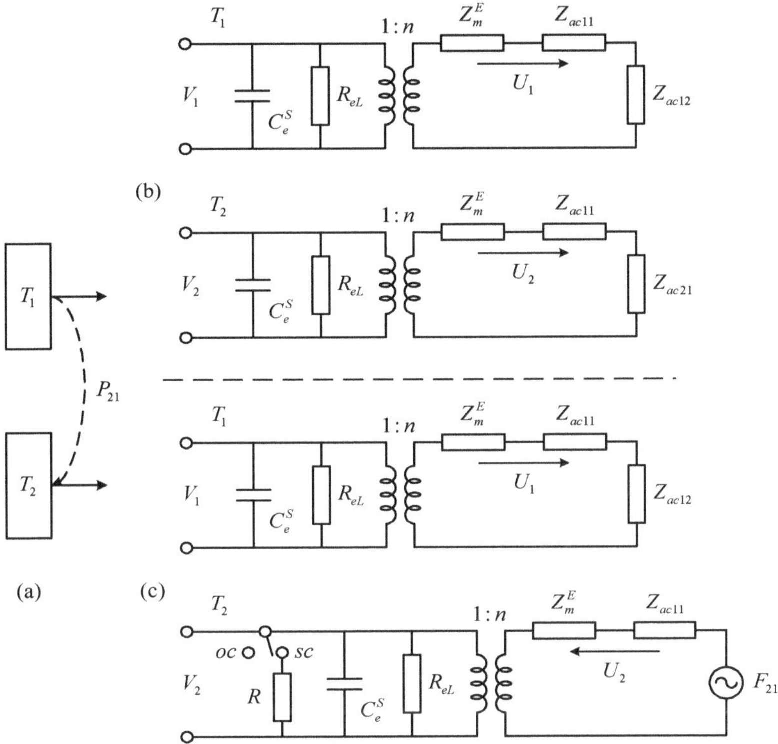

Figure 6.62: To the interaction of transducers: (a) interacting transducer; (b) and (c) are variants of the equivalent circuits of interacting transducers with introduced impedances, Z_{ac12}, and equivalent force, $F_{21} = z_{21}U_1$. (Note that variants (b) and (c) are separated by the dashed line.)

The equivalent circuits of the transducers in this case can be presented as one contour circuits, and the vibration velocities as having the form $U_1(\boldsymbol{r}_{\Sigma1}) = U_1\theta(\boldsymbol{r}_{\Sigma1})$ and $U_2(\boldsymbol{r}_{\Sigma2}) = U_2\theta(\boldsymbol{r}_{\Sigma2})$.

The mode of the transducers operating can be considered as superposition of the two modes:

a. Transducer T_2 is clamped ($U_2 = 0$), transducer T_1 vibrates with velocity $U_1(\boldsymbol{r}_{\Sigma1})$ and generates the acoustic pressure $P_{11}(\boldsymbol{r}_{\Sigma1})$ on its own surface and $P_{21}(\boldsymbol{r}_{\Sigma2})$ at the clamped surface of transducer T_2.
b. Transducer T_2 vibrates with velocity $U_2(\boldsymbol{r}_{\Sigma2})$ and generates the acoustic pressures $P_{22}(\boldsymbol{r}_{\Sigma2})$ on its own surface and $P_{21}(\boldsymbol{r}_{\Sigma1})$ at the clamped surface of transducer T_1.

Superposition of these two modes results in the acoustic pressure on surfaces of the simultaneously operating transducers T_1 and T_2. The power of their acoustic radiation can be presented as follows:

$$P_1(\boldsymbol{r}_{\Sigma 1}) = P_{11}(\boldsymbol{r}_{\Sigma 1}) + P_{12}(\boldsymbol{r}_{\Sigma 1}), \tag{6.339}$$

$$\bar{\dot{W}}_{ac1} = \int_{\Sigma_1} P_{11}(\boldsymbol{r}_{\Sigma_1}) U_1^* \theta(\boldsymbol{r}_{\Sigma_1})\, d\Sigma_1 + \int_{\Sigma_1} P_{12}(\boldsymbol{r}_{\Sigma_1}) U_1^* \theta(\boldsymbol{r}_{\Sigma_1})\, d\Sigma_1 ; \tag{6.340}$$

$$P_2(\boldsymbol{r}_{\Sigma_2}) = P_{22}(\boldsymbol{r}_{\Sigma_2}) + P_{21}(\boldsymbol{r}_{\Sigma_2}), \tag{6.341}$$

$$\bar{\dot{W}}_{ac2} = \int_{\Sigma_2} P_{22}(\boldsymbol{r}_{\Sigma_2}) U_2^* \theta(\boldsymbol{r}_{\Sigma_2})\, d\Sigma_2 + \int_{\Sigma_2} P_{21}(\boldsymbol{r}_{\Sigma_2}) U_2^* \theta(\boldsymbol{r}_{\Sigma_2})\, d\Sigma_2 . \tag{6.342}$$

The first integrals in expressions (6.340) and (6.342) represent the quantities $\bar{\dot{W}}_{acii} = Z_{acii} |U_i|^2$, where Z_{acii} are the self-radiation impedances of one of the transducers in presence of another transducer clamped. Taking into consideration that pressure P_{12} is proportional to velocity U_2, the second integral in expression (6.340) can be represented in the form of $\bar{\dot{W}}_{ac12} = z_{ac12} U_2 U_1^*$, where

$$z_{ac12} = \frac{1}{U_2} \int_{\Sigma_1} P_{12}(\boldsymbol{r}_{\Sigma_1}) \theta(\boldsymbol{r}_{\Sigma_1})\, d\Sigma_1 . \tag{6.343}$$

By the reciprocity $z_{ac12} = z_{ac21}$. Thus, the expressions (6.340) and (6.342) can be represented as

$$\bar{\dot{W}}_{aci} = \left(Z_{acii} + z_{acil} \frac{U_l}{U_i} \right) |U_i|^2, \quad i,l = 1,2 . \tag{6.344}$$

The quantity

$$z_{acil} = r_{acil} + jx_{acil} \tag{6.345}$$

is called the mutual radiation impedance, where r_{acil} and x_{acil} are the mutual radiation resistance and reactance, respectively. The impedance

$$Z_{acil} = z_{acil} \frac{U_l}{U_i} = R_{acil} + jX_{acil} \tag{6.346}$$

may be called the introduced impedance, where R_{acil} and X_{acil} are the introduced radiation resistance and reactance.

Relations (6.344) can be alternatively expressed as

$$\bar{\dot{W}}_{aci} = Z_{acii} |U_i|^2 + F_{acil} U_i^* , \tag{6.347}$$

where $F_{acil} = z_{\text{àñil}} U_l$ is the equivalent "acoustomotive" force, with which transducer T_l acts on the transducer T_i during their simultaneous operation. By comparing with expression (6.343)

$$F_{acil} = \int_{\Sigma_i} P_{il}(\boldsymbol{r}_{\Sigma_i})\theta(\boldsymbol{r}_{\Sigma_i})d\Sigma_i \,. \tag{6.348}$$

In accordance with expressions (6.344) and (6.347) the interaction between transducers T_1 and T_2 can be accounted for by introducing the impedance Z_{ac21} (as is shown in Figure 6.62 (b)) or the equivalent force F_{ac21} (as it is shown in Figure 6.62 (c)) in the equivalent circuit of the transducer T_2. Both representations of interaction are equivalent, but the first is more convenient in evaluating the changes in the transducer parameters operating in the transmit mode in the close to resonance range, while the second is preferable in the case of operating in the receive mode in the frequency range below resonance. The transducer T_2 either generates additional energy or consumes energy of the acoustic field in the process of interaction depending on the signs of introduced impedance Z_{ac21} and force F_{ac21}.

The mutual impedance $z_{àñil}$ is inherent in the transducers geometry, mode of vibration and the relative location of the transducers. And the introduced impedance depends substantially on relative velocities of the transducers vibration. Magnitude of the introduced impedance can be changed by changing the ratio of magnitudes of the transducers vibration, as it follows from relation (6.346). In the case that the phase between the velocities changes, a correlation between the introduced resistance and reactance changes. According to expression (6.346)

$$R_{acil} = r_{acil} \cdot \mathrm{Re}(U_l / U_i) - x_{acil} \cdot Im(U_l / U_i)\,, \tag{6.349}$$

$$X_{acil} = x_{acil} \cdot \mathrm{Re}(U_l / U_i) + r_{acil} \cdot \mathrm{Im}(U_l / U_i)\,. \tag{6.350}$$

Usually, the active component of the total radiation impedance and therefore the active component of the introduced impedance are of a particular interest in the process of transducer designing due to their effect on the radiation of the acoustic energy. But in the case that velocities of vibration of transducers in array differ in phase (and the differences may change because of steering) both the active and reactive components of the mutual radiation impedance must be known to determine the total radiation resistance. Thus, for example, if the phase shift between velocities is 90°, i.e., $(U_2 / U_1) = -j|U_2 / U_1|$, then

$$R_{acil} = x_{acil} |U_l / U_i|, \tag{6.351}$$

$$X_{acil} = r_{acil} |U_l / U_i|. \tag{6.352}$$

In accordance with the equivalent electromechanical circuits in Figure 6.62 experimental methods of determining the mutual impedances can be based on measuring the input impedance or the output voltage of one of interacting transducers under some special conditions of operating the neighboring transducer. These issues will be discussed in Section 6.4.10. The possibility to increase magnitude of introduced impedance and to convert its reactive component into active (that is easier to measure) proves to be helpful in terms of increasing accuracy of the measurements.

General properties of the mutual radiation impedances may be illustrated with examples of the acoustic interaction between two transducers, for which the theoretical solutions are available. The easiest for considering and yet very informative are examples of the transducers having dimensions small in respect to wavelength (simple sources).

6.6.3 Interaction between Simple Sources

Suppose that two small identical transducers are in free space at a distance d from each other and their surfaces, having an arbitrary shape, vibrate with the volume velocities $U_{\tilde{v}} = U_0 S_{av}$. The pressure generated by one of transducers (#2) at the surface of another (#1) is (see formula (6.246) at $k_{dif\,0} = 1$)

$$P_{12} = \frac{\rho c}{2\lambda r} U_0 S_{av} e^{-j(kr - \pi/2)}, \tag{6.353}$$

where $r = d + \Delta r$ is the distance between the surface elements of the transducers. Assuming that $|\Delta r| \ll d$ and $k \cdot \Delta r \ll 1$, Δr can be neglected and we obtain using expression (6.343) that

$$z_{àñ12} = j\pi\rho c \frac{S_{av}^2}{\lambda^2} \frac{e^{-jkd}}{kd}. \tag{6.354}$$

Considering that

$$\pi\rho c \frac{S_{av}^2}{\lambda^2} = r_{ac11} \tag{6.355}$$

is the radiation resistance of a transducer with small wave dimensions, we have

$$(z_{12} / r_{11}) = \frac{\sin kd}{kd} + j\frac{\cos kd}{kd}. \tag{6.356}$$

(Note that the part *ac* in the subscripts will be further omitted for brevity, as the acoustical quantities only will be considered.)

It is noteworthy that modulus of the mutual impedance drops with separation between the sources as $|z_{12}/r_{11}| = 1/kd$. It is physically clear that reduction of modulus of mutual impedance between projectors of a finite size should be more rapid.

In the case that one of transducer dimensions significantly exceeds another, more appropriate is to use for estimating the interaction between transducers expression for the normalized mutual radiation impedance per unit length of the infinitely long cylinders having diameter small compared to the wavelength (cylindrical simple sources). Starting from expression (6.118) for the sound pressure generated by the cylinder of small wave size

$$P_0(r, \pi) = \rho c \frac{k}{4} U_{\tilde{V}r} H_0^{(2)}(kr) , \tag{6.357}$$

and considering that its self-radiation resistance (6.129) is

$$r_{ac11} = \frac{\pi}{2} \rho c \frac{S_{av}^2}{\lambda} \tag{6.358}$$

we will obtain in the same way, as for the three-dimensional simple source, that

$$(z_{12}/r_{11}) = J_0(kd) - jN_0(kd) . \tag{6.359}$$

Plots of the active and reactive components of the normalized mutual impedances by formulas (6.356) and (6.359) are presented in Figure 6.63 and Figure 6.64.

Dependencies of the mutual impedances from separation between simple sources located on the surface of perfectly rigid plane are the same as those for sources in the free space. This can be shown by analogous derivation. But the normalizing radiation resistances of the sources (denote them r_{11}^*) are different. In terms of acoustic field generated by a source the perfectly rigid plane boundary can be replaced by the image of the source vibrating with the same velocity, as shown in Figure 6.65 (a). Thus, we may obtain r_{11}^* as result of interaction between the source and its image, namely,

$$r_{11}^* = r_{11}\left(1 + \frac{\sin 2kd_1}{2kd_1}\right) \tag{6.360}$$

for the 3D source, and

$$r_{11}^* = r_{11}[1 + J_0(2kd_1)] \tag{6.361}$$

for cylindrical (2D) source.

At $kd_1 \rightarrow 0$, $r_{11}^* \rightarrow 2r_{11}$. Thus, for 3D source

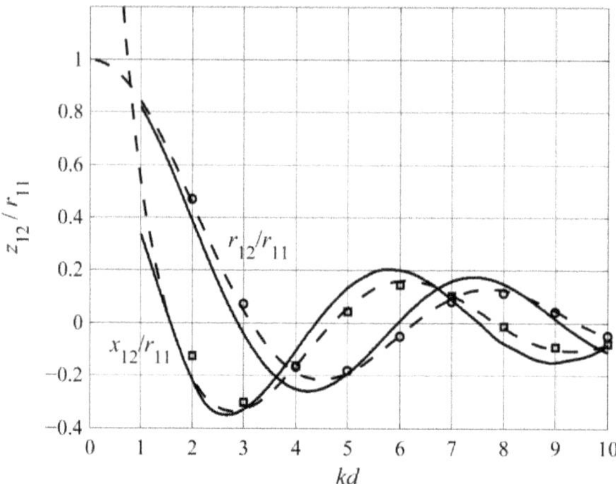

Figure 6.63: A typical behavior of the normalized mutual impedances for different transducers: simple sources (formula (6.356)) (dashed lines); circular pistons with $2a/\lambda = 0.3$ in the rigid plane (circles and squares); finite cylinders with $h/\lambda = 0.36$ and $2a/\lambda = 0.73$ in the rigid cylindrical baffle (solid line).

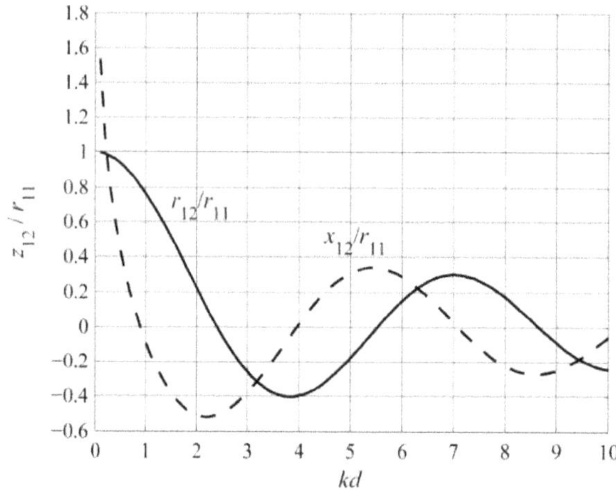

Figure 6.64: Normalized mutual impedances for the two-dimensional simple sources.

$$r_{11}^* = 2\pi\rho c \frac{S_{av}^2}{\lambda^2} \tag{6.362}$$

and for the cylindrical source per unit length

$$r_{11}^* = \pi\rho c \frac{S_{av}^2}{\lambda}. \tag{6.363}$$

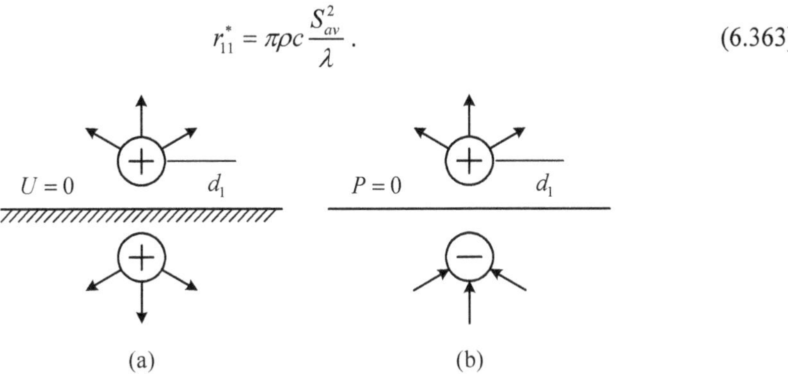

Figure 6.65: Simple sources located close to the perfectly rigid (a) and compliant (b) plane boundaries and their images.

Expressions (6.360) and (6.361) show, how the radiation resistance of a source changes with distance from the rigid baffle. They will be valid for the variant of compliant baffle, if to change signs plus to minus, as in this case velocity of the image has the opposite sign (Figure 6.65 (b)). The expression

$$r_{11}^* = r_{11}\left(1 - \frac{\sin 2kd_1}{2kd_1}\right) \tag{6.364}$$

can be useful for determining preferable distance from surface of water for measuring radiation resistance. At $2kd_1 = 0$, i.e., at $d_1 = \lambda / 4$ the radiation resistance is the same as in the free space ($r_{11}^* = r_{11}$). Thus, the radiation resistances of the bender transducers that have small resonance dimensions with respect to the wavelength can be measured at relatively small depth.

6.6.4 Interaction between Transducers of Finite Size

Universal importance of the expressions (6.356) and (6.359) for the normalized mutual impedances is due to the fact that these expressions are reasonably valid even in the case that dimensions of the transducers are not very small provided that values for r_{11} correspond to the particular transducer type. This can be illustrated with examples of transducer configurations, for which results of analytical calculations for the mutual impedances are available.

6.6.4.1 Interaction between the Circular Disks

One of the methods of calculating the mutual impedances between plane radiators vibrating in the infinite rigid plane and particular results for the circular piston-like vibrating disks are presented in Ref. 24. Values of the normalized mutual radiation impedances calculated for $ka = 1$ ($2a/\lambda \approx 0.32$) are shown in Figure 6.63 by the circles. They are in almost complete agreement with those for the simple sources. The same approximation remains practically valid for even larger values of *ka*. It is noteworthy that in application to underwater transducers uniformly vibrating circular disks imitate radiating surfaces of the heads of Tonpilz transducers. Their diameters usually should not exceed half of wavelength from consideration of steering directivity patterns of an array. To this extent formula (6.356) can be used for estimating effects of transducers interaction with sufficient accuracy.

For the flexural disk (bender) transducers that vibrate with velocity distribution over surface the calculations of the normalized mutual impedances are applicable, as this was shown in Ref. 25. The results obtained for the case of the simply supported edge up to $ka = 1$ are in complete agreement with formula (6.356), and value of the normalizing self-radiation resistance of the disk is determined by formula (6.362) for a simple source, where $S_{av} = 0.46\pi a^2$ for simply supported edge. It should be noted that dimensions of the bender transducers are usually much smaller than the wavelength. Maximum size of the flexural circular disk is typically $(2a/\lambda) < 0.2$ (see the comment under Table 6.6). Thus, the simple source approximation for determining the mutual impedances is also good enough for practical applications in this case.

6.6.4.2 Interaction between the Infinitely Long Strips

The problem is two-dimensional due to symmetry and interaction between strips per unit length can be found as for the segments shown in Figure 6.66.

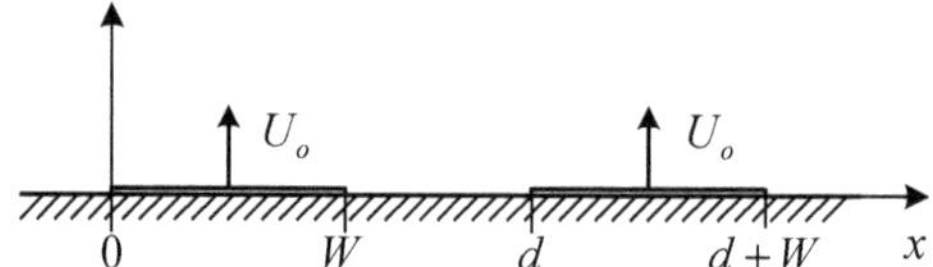

Figure 6.66: To the acoustic interaction between infinitely long strips.

The procedure used for determining self-radiation impedance of the strip in Section 6.3.4.4 will result in the expression for the mutual radiation impedance,

$$z_{12} = \frac{\omega\rho}{2}\int_0^W\int_d^{d+W} H_0^{(2)}(k|x_1 - x_2|)\,dx_1 dx_2\,, \tag{6.365}$$

where $d > W$. Formula (6.297) for the self-radiation impedance in the case that vibration over the strip is uniform ($f_i(x) = f_l(x) = 1$),

$$Z_{acil} = \frac{\omega\rho}{2} \int_{-W/2}^{W/2} \int_{-W/2}^{W/2} H_0^{(2)}(k|x_1 - x_2|) dx_1 dx_2 \ . \qquad (6.366)$$

Thus, the normalized mutual impedance will be

$$\frac{z_{12}}{r_{11}} = \frac{\int_0^W \int_d^{d+W} H_0^{(2)}(k|x_1 - x_2|)\, dx_1 dx_2}{\text{Re}\left\{ \int_0^W \int_0^W H_0^{(2)}(k|x_1 - x_2|)\, dx_1 dx_2 \right\}} = \frac{\alpha_{12}}{\alpha_{11}} + j\frac{\beta_{12}}{\alpha_{11}} \ . \qquad (6.367)$$

Expressions for $(\alpha_{12} / \alpha_{11})$ and $(\beta_{12} / \alpha_{11})$ are plotted as functions of kd at values of kW =1, 2, 4 and 8 in Figure 6.67.

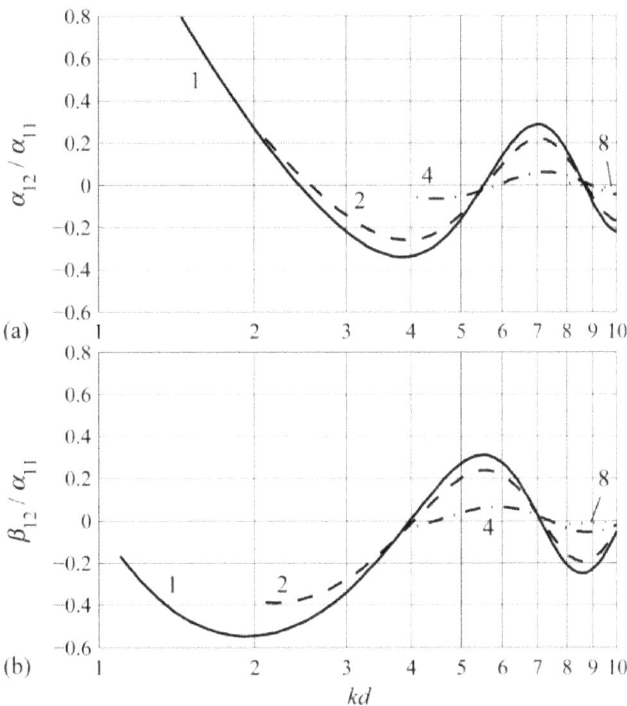

Figure 6.67: Normalized active (a) and reactive (b) components of the mutual radiation impedance of the strips as functions of separation kd for different values of kW (kW = 1, 2, 4, 8).

6.6.4.3 Interaction between the Square Pistons

The normalized mutual radiation impedances between the square pistons were calculated in Ref. 26. The results for square pistons practically coincide with those for the simple sources shown in Figure 6.63 at least up to $kW \approx 3$. Especially important for transducers designing is the case that the pistons are closely spaced. Obviously, the effect of acoustic interaction in this case is the strongest and should be reduced with separation between them. Consider the mutual radiation impedances of two adjacent identical square pistons shown in Figure 6.68. They can be calculated in a straightforward way after the self-radiation impedances of rectangular pistons at different aspect ratios are determined (see Sect. 6.5.5). Radiation impedance of the rectangular piston Z_{1+2} can be expressed as

$$Z_{1+2} = 2Z_{11} + 2Z_{12}. \tag{6.368}$$

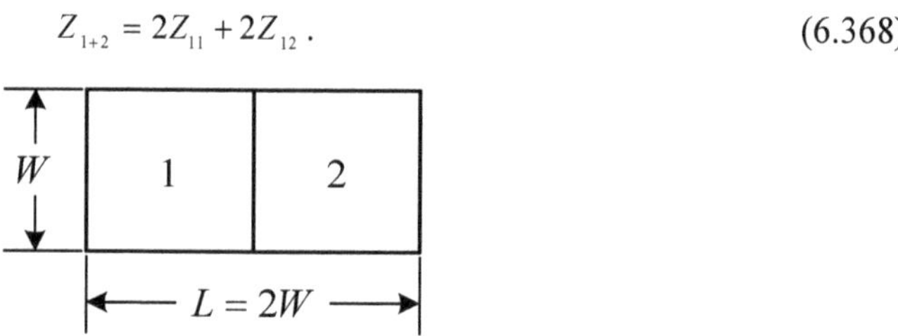

Figure 6.68: Adjacent square pistons forming rectangular piston with aspect ratio $L/W = 2$.

Values of normalized nondimensional coefficients for $Z_{11} = Z_1$ and Z_{1+2} (as the self-radiation impedances at aspect ratios $kW = 1$ and 2) are known and presented in Table 6.7.

Table 6.7: Values of the nondimensional coefficients of the mutual radiation impedances for adjacent radiating surfaces.

kd		1	2	3	4	5	6
d	α_{12}/α_{11}	0.84	0.47	0.11	-0.09	-0.07	-0.01
	β_{12}/α_{11}	0.54	-0.13	-0.26	-0.14	-0.03	0.02
d	α_{12}/α_{11}	0.80	0.46	0.12	-0.02	-0.015	<0.005
	β_{12}/α_{11}	0.66	-0.08	-0.22	-0.15	-0.028	<0.05

Remembering that

$$Z_1 = S(\alpha_{11} + j\beta_{11}) \text{ and } Z_{1+2} = 2S(\alpha_{1+2} + j\beta_{1+2}) \tag{6.369}$$

from Eq. (6.368) will be obtained

$$\frac{\alpha_{12}}{\alpha_{11}} = \frac{\alpha_{1+2}}{\alpha_{11}} - 1, \tag{6.370}$$

$$\frac{\beta_{12}}{\alpha_{11}} = \frac{\beta_{1+2}}{\alpha_{11}} - 1. \tag{6.371}$$

Values of these functions vs. $kW = kd$ are presented in Table 6.7. Values of analogous coefficients for the adjacent circular disks are also presented in the Table for comparison.

6.6.4.4 Interaction between the Cylinders Embedded in the Rigid Cylindrical Baffle

The mutual radiation impedances between cylinders embedded in the rigid cylindrical baffle were considered in Ref. 10. It was shown that the active and reactive components of the mutual impedance between identical cylinders vibrating in phase can be calculated from expressions

$$r_{12} = \frac{16\omega\rho}{\pi k^3}\int_0^{\pi/2} A\cos(kd_{12}\sin\theta)d\theta \tag{6.372}$$

and

$$x_{12} = \frac{8a\omega\rho}{k^2}\left[\int_0^{\pi/2} B\cos(kd_{12}\sin\theta)d\theta + \int_0^{\infty} C\cos(kd_{12}\cosh\psi)d\psi\right], \tag{6.373}$$

where functions A, B, and C are determined by formulas (6.192), (6.194) and (6.195), respectively.

In application to the cylindrical transducers two ranges of cylinders wave sizes are of special interest in relation to different transducer types: transducers made of PZT ceramic compositions that employ extensional vibrations, and transducer of the flexural type including slotted cylinders. Typical values of ka for the cylindrical transducers of extensional type operating around their resonance frequencies may be in the range $ka \approx 2.2 \pm 0.2$. Rational height of a single ring comprising transducer must be chosen from consideration of avoiding a harmful effect of the coupled vibrations. From this point of view, it is desirable to have the aspect ratio of the ring $(h/2a) < 0.5$, i.e., $(h/\lambda) < (0.35 \div 0.40)$. For the transducers of flexural type typically $ka < (0.2 \div 0.3)$, and $(h/\lambda) < (0.1 \div 0.15)$.

Results of calculations performed by formulas (6.372) and (6.373) for cylinders at $ka = 2.0$ and $(h/\lambda) = 0.36$ are plotted in Figure 6.69. They show fairly good agreement with those for the simple sources, though the model for cylinders suggests them vibrating in the rigid cylindrical baffle, whereas for simple sources – in free space. Obviously, for the cylinders that have smaller wave size the agreement should be even better. In order to check how close the models used are to the real situation, experimental verification of results of calculations was made using one of the methods of determining the mutual impedances that is described in Section 6.4.6. In Figure 6.69 and Figure 6.70 results of comparison between calculated and measured values of the normalized mutual resistances and reactances for two couples of the cylinders having different wave sizes are presented.

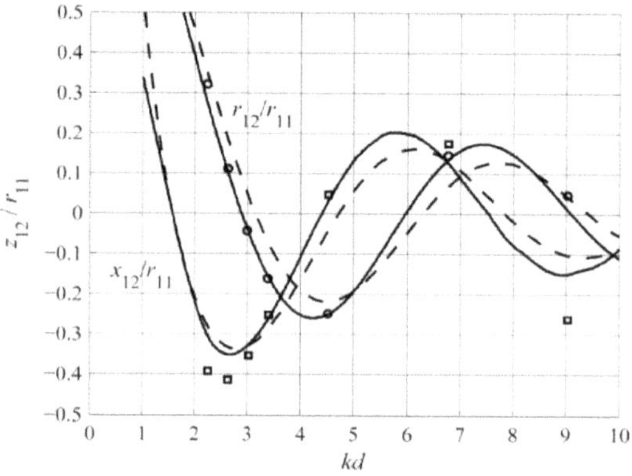

Figure 6.69: Comparison of normalized mutual radiation impedances of two cylindrical transducers calculated according to the used models and measured in free space: the model in the rigid baffle (solid line), simple sources (dashed line), experimental data (circles and squares): $ka = 2.2$, $h/\lambda = 0.36$.

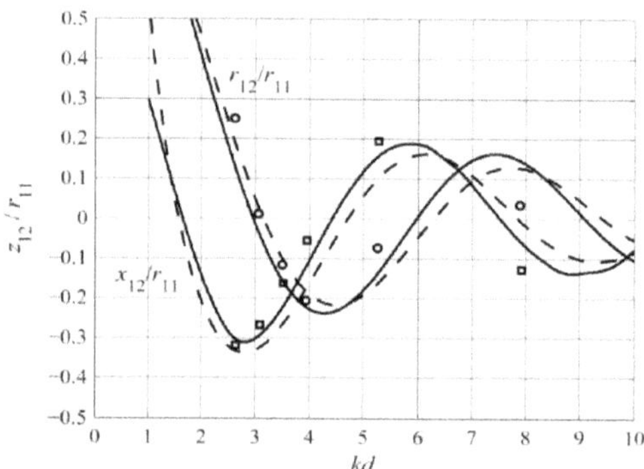

Figure 6.70: Comparison of normalized mutual radiation impedances of two cylindrical transducers calculated according to the used modes and measured in free space: the model in the rigid cylindrical baffle (solid line), simple sources (dashed line), experimental data (circles and squares): $ka = 2.6$, $h/\lambda = 0.42$.

Though all the data are close, a little better agreement exists between experimental data and those calculated for the cylinders in the rigid cylindrical baffle. The conclusion can be made that in majority of practical needs the model of interaction of simple sources can be used for transducers designing.

6.6.5 Methods for Measuring the Mutual Impedances

6.6.5.1 Measuring the Mutual Impedance between Two Transducers in the Free Field by the Z method

Consider two simultaneously operating transducers. Their equivalent circuits in general can be represented as shown in Figure 6.62. Assume that the transducers are electromechanically identical. All the electromechanical parameters of the transducers including internal mechanical impedance Z_m^E may be considered as known. Otherwise, they can be determined by common measurements performed on unloaded transducer (in air). The equivalent electromechanical circuits of the transducers may be reduced to the form shown in Figure 6.71 (a), where the mechanical branches are transformed into electrical side. In the Figure the following notations are introduced for the transducers numbered 1 and 2: $C = C_m^E n^2$ for both transducers.

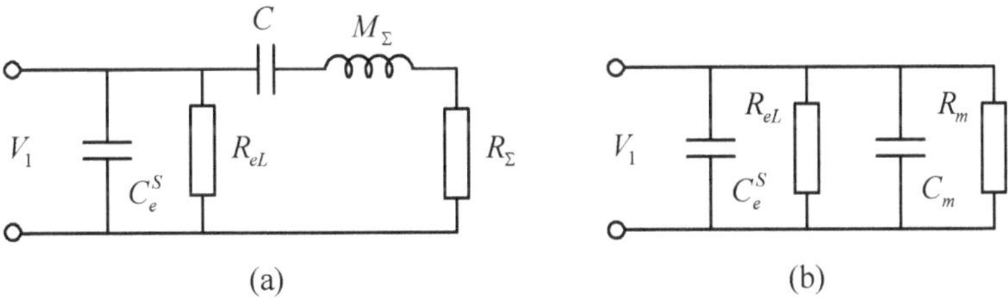

Figure 6.71: The equivalent circuit of interacting transducers with the mechanical branch transformed into electrical side: (a) with series motional impedance, and (b) after converting the motional impedance into admittance.

$$R_{\Sigma 1} = R_{11} + R_{12} + R_{mL}, \quad M_{\Sigma 1} = M + X_1 / \omega ; \tag{6.374}$$

$$R_{\Sigma 2} = R_{11} + R_{21} + R_{mL}, \quad M_{\Sigma 2} = M + X_2 / \omega \tag{6.375}$$

for transducer number 1 and 2, respectively. In the expressions for masses

$$X_1 = X_{11} + X_{12}, \quad X_2 = X_{11} + X_{21} . \tag{6.376}$$

Further we denote the electrical analogs of the acoustic impedances in the same way as their acoustic counterparts are denoted but without subscript "ac". Thus, for example, $Z_{il} = R_{il} + jX_{il}$ is the electrical analog of the acoustic impedance Z_{acil}. The relation between acoustic impedances and their electrical analogs is $Z_{il} = Z_{acil} / n^2$.

Note that the normalized introduced impedances, being determined on the electrical side of a transducer, are the same as the normalized introduced impedances on acoustic side, namely, $Z_{12} / R_{11} = Z_{ac12} / R_{ac11}$ and for the purpose of estimating the mutual impedances it is not necessary to know the absolute values of the impedances in acoustic units.

The self-impedance, which is denoted as Z_{11} for both transducers, may not be considered as known. It differs from the radiation impedance of a single transducer measured in the free space due to presence of an interacting transducer. Moreover, by definition the self-radiation impedance has to be determined under the condition that the interacting transducer is blocked, i.e., in the equivalent circuits of Figure 6.62 (b) $U_2 = 0$ and $Z_{12} = z_{12}U_2 / U_1 = 0$.

The electrical analogs of acoustic quantities can be obtained by measuring the input impedances of the transducers. After converting of the series motional impedance into admittance the electrical circuit of transducer input may be represented in the form shown in Figure 6.71 (b), which is convenient for interpreting the results of measuring the transducer parameters by an impedance analyzer. In Figure 6.71 (b) C_m is the motional "capacitance" that may become inductance at some frequencies. Parameters of circuits in Figure 6.71 can be expressed through results of measuring the components of admittance as follows,

$$R_{\Sigma i} = \frac{G_m}{(\omega C_m)^2 + G_m^2}, \quad X_{\Sigma i} = -\frac{\omega C_m}{(\omega C_m)^2 + G_m^2}, \tag{6.377}$$

where

$$X_{\Sigma i} = \omega M_{\Sigma i}[1 - (f_r / f)^2]. \tag{6.378}$$

When measuring in air, $M_{\Sigma i} = M = M_{eqv} / n^2$ and $f_r = f_a$. When measuring in water, $M_{\Sigma i} = M + X_i / \omega$ and $f_r = f_w$. Directly measured quantities are $G = G_{eL} + G_m$ and $C_p = C_e^S + C_m$. Typical plots from which the motional capacitance, C_m, and conductivity, G_m, vs. frequency may be determined are depicted in Figure 6.72. At the resonance frequency $C_m = 0$, and $C_p = C_e^S$. Thus, the clamped capacitance C_e^S can be considered as known. If a measurement transducer may be approximated as having one mechanical degree of freedom, this quantity also can be obtained as $C_e^S = C_{Lf}(1-k_{eff}^2)$, where C_{Lf} and k_{eff} are the capacitance measured at low frequency and effective coupling coefficient of the transducer. The motional conductivity, $G_m = G_p - G_{eL}$, in general can be found as shown in the figure, although very often G_{eL} is much smaller than G_m and can be neglected.

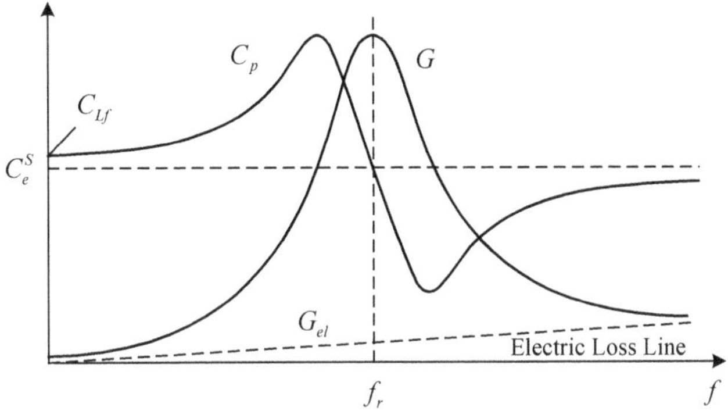

Figure 6.72: Qualitative illustration of a typical plot for the input parameters of a transducer in the parallel circuit representation

Given that $C_m(f)$ and $G_m(f)$ are measured in a frequency range around the resonance frequency, $R_{\Sigma i}$ and $X_{\Sigma i}$ can be calculated by formulas (6.377), and using the relations (6.378) and (6.374) through (6.376) the wanted self- and mutual radiation impedances may be found. Further we illustrate the technique of determining these quantities at resonance frequency, in which case they can be obtained without complicated calculations.

The reactive components of radiation impedance may be determined in two alternative ways. Formula (6.378) can be represented as

$$X_{\Sigma 1} = \omega M(1 - f_a^2 / f^2) + X_{11} + X_{12} = \omega(M + X_1 / \omega)(1 - f_w^2 / f^2). \quad (6.379)$$

From Eq. (6.379) follows that at frequency $f = f_w$

$$X_{11} + X_{12} = 2\pi f_w M(f_a^2 / f_w^2 - 1) = 2\pi M\, F(f_w), \quad (6.380)$$

where it is denoted for brevity

$$F(f_w) = f_w(f_a^2 / f_w^2 - 1). \quad (6.381)$$

Taking into account equations (6.379) and (6.377) we obtain that at the frequency $f = f_a$,

$$X_{11} + X_{12} = -\omega_a C_m(f_a) / \{[(\omega_a C_m(f_a)]^2 + G_m^2(f_a)\}. \quad (6.382)$$

Remembering that $C_m(f_w) = 0$, the active component of radiation impedance will be obtained from Eq. (6.377) in the form

$$R_{\Sigma 1} = R_{11} + R_{12} + R_{mL} = 1/G_m(f_w) \tag{6.383}$$

To determine all the components of impedances Z_{11} and Z_{12} two more equations in addition to the equations (6.380) or (6.382) and (6.383) must be obtained. For this purpose the following experiments can be done:

Experiment I

Measuring the input impedance of one of the transducers, while equal voltages are applied to both in phase, $V_2 = V_1$. In this case $U_1 = U_2$ and $Z_{ac12} = z_{ac12}$ due to symmetry. At resonance frequency f_{wI} of the transducer measured in water we obtain according to formulas (6.380) and (6.383) that

$$R_{11} + r_{12} + R_{mL} = 1/G_{mI} \tag{6.384}$$

and

$$X_{ac11} + x_{ac12} = 2\pi M F_I . \tag{6.385}$$

It is denoted in equations (6.384) and (6.385)

$$G_{mI} = G_m(f_{wI}), \quad F_I = F(f_{wI}) . \tag{6.386}$$

(Subscripts made by roman numbers here and further correspond to the number of an experiment.)

Experiment II

Measuring the input impedance of one of the transducers, while applied voltages are equal by magnitude and opposite in phase, $V_2 = -V_1$. In this experiment $U_2 = -U_1$ and $Z_{ac12} = -z_{ac12}$. At the resonance frequency f_{wII} we have

$$R_{11} - r_{12} + R_{mL} = 1/G_{mII} \tag{6.387}$$

and

$$X_{ac11} - x_{ac12} = 2\pi M F_{II} . \tag{6.388}$$

Combining the results of measurements expressed by formulas (6.384) and (6.387) we arrive at

$$R_{11} + R_{mL} = (G_{mI} + G_{mII}) / 2G_{mI}G_{mII} \tag{6.389}$$

and

$$r_{12} = (G_{mI} - G_{mII}) / 2G_{mI}G_{mII}\,. \tag{6.390}$$

In the carefully designed transducers the resistance of mechanical loss, R_{mL}, is much smaller than the radiation resistance and it can be neglected (at least such transducers must be chosen for investigating acoustic interactions). Otherwise, the resistance of loss can be determined separately by measuring the conductivity $G_m = 1/R_{mL}$ in air. After this note is made, the resistance of mechanical loss further will be neglected for the sake of brevity. Thus, it follows from equations (6.389) and (6.390) that

$$r_{12} / R_{12} = (G_{mI} - G_{mII}) / (G_{mI} + G_{mII})\,. \tag{6.391}$$

From formulas (6.385) and (6.388) it follows that

$$X_{11} = \pi M (F_I + F_{II}), \tag{6.392}$$

$$x_{12} = \pi M (F_I - F_{II}), \tag{6.393}$$

and

$$x_{12} / X_{11} = (F_I - F_{II}) / (F_I + F_{II})\,. \tag{6.394}$$

Calculating the absolute values of the reactances by formulas (6.392) and (6.393) requires knowing the equivalent mass of the transducer. Therefore it can be advantageous to use formula (6.382) for this purpose, although externally it looks more complicated. In this case all the quantities needed for calculation are available through experimentation. Practically both experiments can be accomplished by measuring the input impedance of transducers, when connected in phase and anti-phase, accordingly.

The Z–method of evaluation of radiation impedances is based on their comparison with the internal impedance of a measurement transducer. The results obtained are less accurate if the ratio of components of radiation impedance that must be measured to the corresponding parameters of comparison is small. The situation becomes especially critical for the acoustic reactance, which have to be compared with large quantity ωM_{eqv} at relatively small deviation from the resonance frequency. Some relief can be achieved by a proper selection of the measurement transducers intended for investigating the acoustic interaction. Thus, measures should be taken to minimize the equivalent mass of the transducers (for example, in the case that a ring transducer is concerned it is better to use thinner rings). More radically, the above-described experimental technique can be modified in order to increase the accuracy of measurement of the mutual impedances. In this case the input impedance of one of the transducers must be measured while another transducer is operating under a larger applied voltage. Suppose that $V_2 \gg V_1$. Qualitatively it is clear that in this case it should be $U_2 \gg U_1$, and therefore $Z_{ac12} \gg z_{ac12}$. In this way the increased value of Z_{12} can be measured with greater accuracy. But in order to calculate z_{12} from thus obtained results the exact value of ratio of velocities $K_U = U_2 / U_1$ must be known. A peculiarity of in this case is that the introduced impedances in the mechanical branches of the equivalent circuits in Figure 6.62 become different, namely, $Z_{ac21} = Z_{ac12} / K_U^2$. Therefore $U_2 / U_1 \neq V_2 / V_1$, and the value of K_U must be determined separately. This can be done through the following experiment.

Experiment III

Measuring the mutual impedance between the two transducers under the condition that voltages applied to them are in phase but have different magnitudes and $V_2 \gg V_1$. The self-radiation impedance of the transducer can be considered as known being obtained from experiments I and II by formulas (6.389) and (6.392). Components of the introduced impedance $Z_{12} = z_{12} K_U$ may be determined as a result of performing the same procedures as in the Experiment I. Namely, at the resonance frequency f_{wIII}

$$R_{11} + R_{12} = 1 / G_{mIII}, \tag{6.395}$$

and

$$X_{11} + X_{12} = 2\pi M F_{III}. \tag{6.396}$$

Considering expressions (6.389) and (6.392) for R_{11} and X_{11} (remember that resistance R_{mL} is neglected), we arrive at

$$R_{12} = 1 / G_{mIII} - (G_{mI} + G_{mII}) / 2G_{mI} G_{mII} \tag{6.397}$$

and

$$X_{12} = \pi M (2F_{III} - F_I - F_{II}). \tag{6.398}$$

Now the mutual resistance and reactance can be found from relation

$$r_{12} + jx_{12} = (R_{12} + jX_{12}) / K_U , \tag{6.399}$$

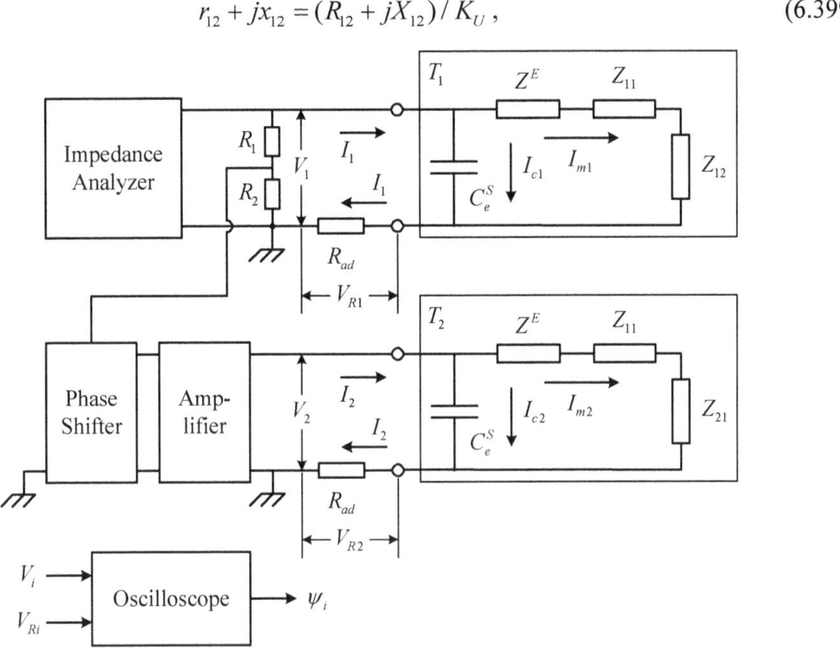

Figure 6.73: Electrical circuit of an experimental setup for measuring the mutual impedances by Z method in a general case that different voltages are applied to the interacting transducers, $Z^E = Z_m^E / n^2$.

but still the ratio of velocities K_U remains to be determined. One of the possible setups for fulfillment experiment III and for determining K_U is represented in Figure 6.73, where the equivalent circuits of interacting transducers with their mechanical branches transformed into the electrical sides are included. Condition $(R_1 + R_2) \gg |Z_{input}|$ must be fulfilled to exclude influence of the voltage divider on results of measuring the transducer impedance.

Currents flowing through the transducers are denoted in Figure 6.73 as follows: I_i is the total current through the transducer i ($i = 1, 2$), I_{Ci} is the current through the blocked transducer and I_{mi} is the motional current through the electrical analog of the mechanical branch. The motional current is proportional to the vibration velocity of mechanical system $I_{mi} = n\,U_i$, therefore

$$K_U = U_2 / U_1 = I_{m2} / I_{m1} \tag{6.400}$$

and the ratio of velocities may be determined experimentally as the ratio of the motional currents I_{m2} and I_{m1}. The total current through a transducer,

$$I_i = I_{Ci} + I_{mi} \tag{6.401}$$

can be measured as shown in Figure 6.73. Namely,

$$I_i = V_{Ri} / R_{ad}, \tag{6.402}$$

where V_{Ri} is the voltage across the known additional resistance R_{ad} connected in series with the transducer (it must be $R_{ad} \ll |Z_{input}|$ in order not to change voltage applied to the transducer). The phase angle between the total current (or voltage V_{Ri}) and applied voltage V_i must be measured simultaneously. We denote this angle as

$$\arg (V_{Ri} / V_i) = \psi_i . \tag{6.403}$$

As the transducers are assumed to be electromechanically identical and with their parameters predetermined, the blocked capacitances can be considered as equal, $C_{e1}^S = C_{e2}^S$, and known. (If the capacitances were not exactly equal, they could be equalized by adding a capacitance in parallel to the transducer with smaller C_e^S.) Thus, the current through the capacitance can be calculated as

$$I_{Ci} = j\omega C_e^S V_i . \tag{6.404}$$

After the currents I_i and I_{Ci} are determined, the motional currents I_{mi} may be calculated following the procedure, that is illustrated for clearness by the vector diagram shown in Figure 6.74. (Note that in this experiment voltages applied to the transducers are in phase.)

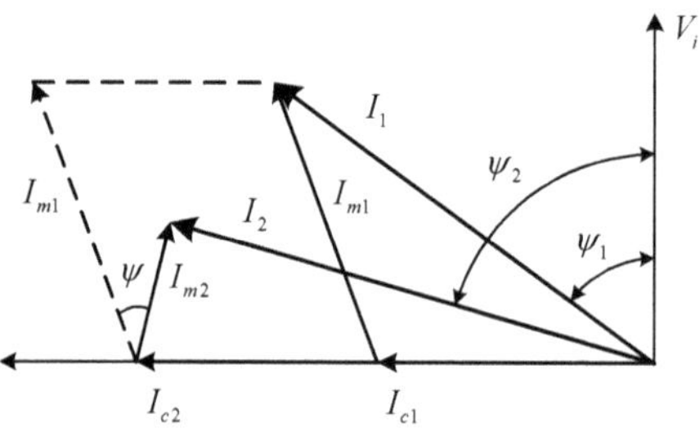

Figure 6.74: The vector diagram for evaluation of the motional currents

Knowing I_{m2} and I_{m1}, we arrive at the required ratio of the velocities of the transducers

$$K_U = I_{m2} / I_{m1} = |I_{m2} / I_{m1}| e^{j\psi}, \tag{6.405}$$

where ψ is the phase angle between vectors I_{m2} and I_{m1}. For increasing accuracy of calculation of the motional currents from results of measuring the total currents, the blocked capacitances can be tuned at the measurement frequency by inserting inductors in parallel. In this case $I_{mi} = I_i$. If the transducers are not identical but their parameters are known, this does not change the matter in principle, it just complicates the calculations.

After K_U is determined by formula (6.405), the components r_{12} and x_{12} of the mutual impedance may be calculated from Eq. (6.399) as follows

$$r_{12} = (R_{12}\cos\psi + X_{12}\sin\psi)\left|\frac{I_{m1}}{I_{m2}}\right|, \tag{6.406}$$

$$x_{12} = (X_{12}\cos\psi - R_{12}\sin\psi)\left|\frac{I_{m1}}{I_{m2}}\right|. \tag{6.407}$$

In this case voltages V_2 and V_1 are in phase, and it is likely that the phase shift ψ is small, i.e. $\cos\psi \cong 1,\ \sin\psi \cong 0$ and

$$r_{12} \cong R_{12} / K_U, \quad x_{ac12} \cong X_{ac12} / K_U, \tag{6.408}$$

where R_{12} and X_{ac12} are given by formulas (6.397) and (6.398).

As noted previously, the results of measuring the introduced reactances are less accurate than that of the introduced resistances. If to produce the phase shift $\psi = -\pi/2$ between U_2 and U_1, then $x_{12} = R_{12}/|K_U|$,(i.e., the mutual reactance becomes the introduced resistance, as follows from relation (6.407) and the accuracy of determining this quantity could be greatly increased. This can be achieved by setting

$$V_2 \cong jV_1. \tag{6.409}$$

Some additional phase shift resulting in a less accurate approximation may occur between the velocities U_2 and U_1 due to asymmetric acoustic loading of the transducers by the introduced impedances. The phase shifter is included in the measurement circuit in Figure 6.74 to provide the needed phase shift between the voltages applied to the transducers. After the phase shift of $\pi/2$ is insured between the motional currents I_{m2} and I_{m1}, the procedure of measuring the introduced resistance R_{12} and thus of determining of the mutual reactance $x_{12} = R_{12}/|K_U|$ is the same, as it is demonstrated in Experiment III.

The experimental data presented in Figure 6.69 and Figure 6.70 were obtained with system of two coaxially oriented cylindrical transducers in free space by employing the Z method.

6.6.5.2 Measuring the Mutual Impedance by the V Method

Returning to the equivalent circuits Figure 6.62 (b), the following relations can be obtained between the output voltage of transducer 2 and velocity U_1 of vibration of transducer 1 depending on position "*oc*" or "*sc*" of the switch:

$$V_{2oc} = z_{ac21} U_1 n / j\omega C_e^s (Z_{moc}^E + Z_{ac11}), \tag{6.410}$$

$$V_{2sc} = I_{m2} R = z_{ac21} U_1 n / (Z_m^E + Z_{ac11}). \tag{6.411}$$

In these relations $Z_{moc}^E = Z_m^E + n^2 / j2\pi f C_e^S$ is the mechanical impedance of the open circuited transducer, resistance R is assumed to be much smaller than $1/\omega C_e^S$ and I_{m2} is the motional current through transducer 2. Note that $U_2 n = I_{m2}$ and therefore Eq. (6.411) can be rewritten as

$$z_{ac21} / (Z_m^E + Z_{ac11}) = I_{m2} R / I_{m1} = V_{sc} / I_{m1}. \tag{6.412}$$

The procedure of determining the motional current I_{m1} is described in the experiment III and therefore it can be considered as available for calculating the mutual impedance z_{ac21} from Eq. (6.412). The mutual impedance can be determined from Eq. (6.412) around the resonance frequency band. This holds so far as all the parameters of the transducer are known including the self-radiation impedance Z_{ac11}, which can be obtained from the above-described Z method. At the resonance frequency of transducer in air, f_a, $Z_m^E \cong 0$ and Eq. (6.412) becomes

$$(z_{ac21} / Z_{ac11})_{f_a} = (V_{sc} / I_{m1})_{f_a}. \tag{6.413}$$

For frequencies below the resonance frequency, at which the mechanical system of a transducer may be considered as stiffness controlled, $Z_{ac11} \ll Z_m^E$, $z_{ac21} \ll Z_m^E$, $Z_m^E \cong 1/ j2\pi f C_m^E$, and from Eq. (6.410) we obtain

$$z_{ac21} = (V_{2oc} / V_1)[(1 - k_{eff}^2) / j2\pi f C_m^E k_{eff}^2]. \tag{6.414}$$

It is considered here that $n^2 C_m^E / C_e^s = k_{eff}^2 / (1 - k_{eff}^2)$. After the ratio

$$V_{2oc} / V_1 = |V_{2oc} / V_1| e^{j\psi_{1oc,2}}, \tag{6.415}$$

where $\psi_{1oc,2}$ is the phase angle between the voltages, is measured both the active and reactive components of the mutual impedance can be determined from equation (6.414).

The block-diagram of an experimental setup for implementing V method is shown in Figure 6.75.

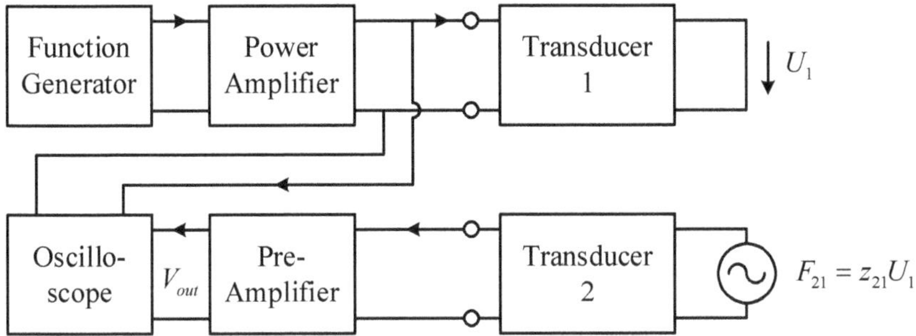

Figure 6.75: The block-diagram of an experimental setup for implementing the V method.

The V method is especially advantageous for estimating the relative change of mutual impedance versus separation between transducers. Suppose that voltage applied to radiating transducer is kept constant ($V_1 = \text{const}$), and output voltages of transducer 2 are measured at separation y and at separation d, which corresponds to the position of transducer 2 adjacent to the radiating transducer. Then by using expressions (6.413) and (6.414) we obtain

$$\left|z_{ac21} / Z_{ac11}\right|_y / \left|z_{ac21} / Z_{ac11}\right|_d = \left|V_{2sc} / I_{m1}\right|_y / \left|V_{2sc} / I_{m2}\right|_d \qquad (6.416)$$

at the resonance frequency and

$$\left|z_{ac21}(y)\right| / \left|z_{ac21}(d)\right| = \left|V_{2oc}(y)\right| / \left|V_{2oc}(d)\right| \qquad (6.417)$$

at frequencies below the resonance.

An important feature of V method is that it makes it possible using the measurement transducers operating far below their resonance frequency for investigating the mutual impedances. The only requirement is that configuration of the radiating surfaces of the measurement transducers must be the same as configuration of the actual transducers, for which the results of the investigation are intended.

The Z and V methods may be considered as complementary to each other. The Z method is advantageous for determining the absolute values of self- and mutual radiation impedances at the frequencies close to resonance of the transducers. The V method can be used for measuring in a frequency range below resonance frequency of the transducers used for investigating the mutual impedances. But it has some shortcomings, when it is used for measuring absolute values of the impedances around the resonance frequencies. Moreover, for implementing V method at these frequencies the self-radiation impedance of a transducer must be known in advance.

6.6.5.3 Determining the Mutual Impedances between Transducers in an Array

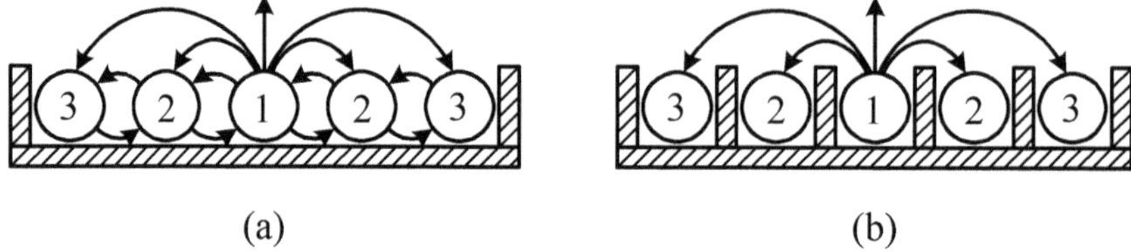

Figure 6.76: Examples of array configurations, for which the mutual radiation impedances between transducers have to be determined: (a) array of the parallel cylindrical transducers of flexural (slotted ring included) or extensional type; (b) the same array with baffles installed between the transducers.

The typical examples of transducer and array configurations, in which case the only reasonable way for determining the mutual impedances between transducers is through experimenting are shown in Figure 6.76. The experimental methods for investigating acoustic interactions are considered in Ref. 27. Here the main results of this work are presented.

The technique described above for measuring the mutual radiation impedances between two transducers in the free space cannot be applied in a straightforward way to analogous measurements in an array. When measuring in array one has to deal with a number of simultaneously vibrating transducers, although the interaction between only two of them must be measured.

Due to the general property of the mutual impedances to decrease with increasing the separation between transducers, contribution of sufficiently remote transducers to results of measuring the two transducers under investigation can be neglected, and often only a part of an array needs to be tested. For example, in array of transducers shown schematically in Figure 6.76 the objective is to determine the mutual impedances between transducers z_{ac21}, z_{ac31} and so on until the mutual impedance reaches a level that may be regarded negligible, i.e., the ratio $|z_{aci1}|/|z_{aci1}|$ becomes small enough. In this case the number of transducers that can be considered as representative for conducting the measurements may be restricted by a group of i members of the array from each side of the central transducer # 1.The most advantageous for achieving this goal is to fulfill an estimation of the mutual impedances by V method in a frequency range below the resonance frequency of the measurement transducers. Certain limitations of this method arise from the fact that in practice we are interested in knowing the values of mutual impedances in the frequency range around the resonance frequency of the actual transducers in the array. Therefore, the measurement transducers having radiating surface of the same shape as the actual transducers must have a higher resonance frequency.

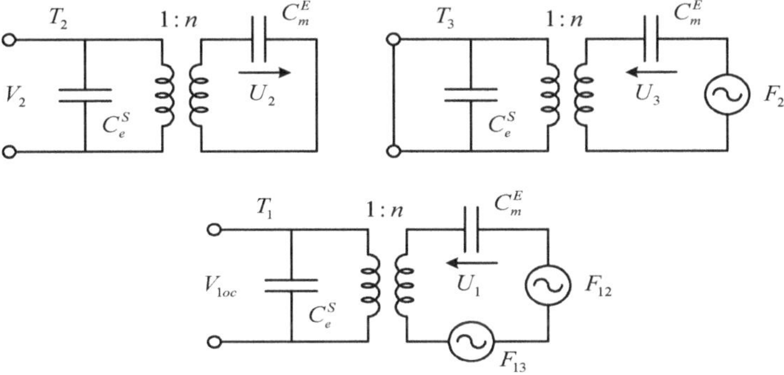

Figure 6.77: To application of the V method for determining mutual impedances in an array

So far as arrays populated by transducers with flat radiating surfaces are concerned (such, for example, as Tonpilz or flexural plate transducers), the size of the surface can be independent of the resonance frequency of the transducer. In the case that the ring transducers populate an array, the diameter of the radiating surface is inversely proportional to the resonance frequency. Therefore, the results obtained below the resonance frequency of a measurement transducer can be applied to an actual transducer in the frequency range around its resonance only in the case that the resonance frequency of the measurement transducer is higher. Thus, the results obtained below the resonance of measurement transducers made of PZT ceramic may be applicable for the ring transducers made of a material with smaller sound speed. Since the frequency range of measurements is below the resonance frequency of the measurement transducers, their mechanical systems may be considered as being stiffness controlled, i.e., $Z_m^E \approx 1/j\omega C_m^E$, and all the radiation impedances can be neglected. The effects of interaction between transducer 1 and transducers 2 and 3 can be represented by the forces $F_{ac12} = 2z_{ac12}U_2$ and $F_{ac13} = 2z_{ac13}U_3$

inserted in their equivalent circuits. After this we will assume that the central transducer #1 will operate in the receive mode and voltages are applied interchangeably to transducers 2 and 3. An outline of application of V method is as follows. Assume that voltage V_2 is applied to transducers 2. Inputs of transducers 3 is short circuited and the output voltage V_{1oc} of transducer 1 is measured. From the equivalent circuits in Figure 6.77 we obtain

$$U_2 = V_2 n\, j\omega C_m^E, \quad U_3 = z_{ac12} U_2 j\omega C_m^E . \tag{6.418}$$

Upon substituting these values of velocities U_2 and U_3 into expressions for forces F_{ac12} and F_{ac13} we obtain for the total acoustic force acting on transducer 1

$$F_{ac1} = F_{ac12} + F_{ac13} = 2z_{ac12}\left[1 + z_{ac13} j\omega C_m^E\right] j\omega C_m^E n\, V_2 . \tag{6.419}$$

The second term in the brackets can be neglected, because the mutual radiation impedance is much smaller then $1/\,j\omega C_m^E$. From this point the analysis becomes analogous to those for the two transducers in the free space, and finally we arrive at the following expressions for acoustomotive force F_{ac1}, for the output voltage of transducer 1 under acting of this force and for the mutual impedance:

$$F_{ac1} = 2z_{ac12} j\omega C_m^E\, nV_2 , \tag{6.420}$$

$$(V_{1oc})_2 = 2z_{ac12} j\omega C_m^E V_2\, k_{eff}^2 / (1 - k_{eff}^2) , \tag{6.421}$$

$$z_{ac12} = [(V_{1oc})_2 / V_2](1 - k_{eff}^2) / j2k_{eff}^2 \omega C_m^E . \tag{6.422}$$

In order to determine the mutual impedance z_{13}, voltage V_3 must be applied to the transducers 3, outputs of transducers 2 have to be short circuited and the output voltage of transducer 1 has to be measured. After using exactly the same procedure as in the previous case, for the mutual impedance z_{ac13} we will obtain expression analogous to expression (6.422), namely,

$$z_{ac13} = [(V_{1oc})_3 / V_3](1 - k_{eff}^2) / j2k_{eff}^2 \omega C_m^E . \tag{6.423}$$

Based on formulas (6.422) and (6.423) a general conclusion can be drawn that if the equal voltages are applied to the consecutive transducers, while measuring the output voltage of transducer 1, $(V_{1oc})_i$, then

$$|z_{ac1i}| / |z_{ac12}| = |(V_{1oc})_i| / |(V_{1oc})_2| . \tag{6.424}$$

This relation is convenient for estimating a comparative contribution of the mutual impedances between transducers in array to the total radiation impedance of a single transducer, while formulas of the type of formulas (6.420) and (6.421) can be used for evaluating the components of the mutual impedances.

In order to determine the self- and mutual radiation impedances of transducers in an array in the frequency band around the resonance frequency, the Z method can be used. We will assume that all the transducers under consideration are located inside the array and therefore they have equal self-radiation impedances. The transducers located at the edges of the array generally may have different self-radiation impedances. This may cause an additional error in determining the mutual impedances. To avoid an "edge effect" and to keep the self- impedances of the measurement transducers approximately equal, at least one more transducer must be added from each side to a group of transducers chosen for the measurements, although these transducers are not intended to actively participate in the measuring procedure. Thus, only a restricted group of transducers may be used to model a real situation in array in terms of interaction between neighboring transducers. The assumption that the transducers are identical by their electromechanical properties remains in place.

The quantities Z_{ac11}, z_{ac12}, and z_{ac13} may be determined by measuring the input impedance of transducer 1. The equivalent circuit of this transducer, in which effects of interaction with transducers 2 and 3 are accounted for by the introduced impedances $Z_{ac12} = 2z_{ac12}U_2 / U_1$ and $Z_{ac13} = 2z_{ac13}U_3 / U_1$, is shown in Figure 6.78. Consider the following experiments for determining the wanted radiation impedances.

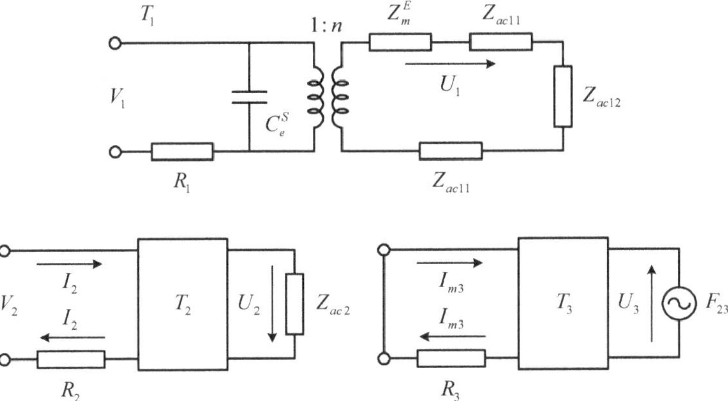

Figure 6.78: To application of the Z method for measuring the mutual impedances in an array

Experiment Ia

Voltages V_1 and V_2 such as $|V|_2 \gg |V_1|$ are applied in phase to transducers 1 and 2. Outputs of transducers 3 are short circuited through a small resistance R_3. Under action of acoustomotive force $F_{32} = z_{ac12}U_2$ the motional current I_{m3} through the electrical side of the transducer 3 will be $I_{m3} = U_3 n$. As it follows from the equivalent circuit in Figure 6.78 at $R_3 \ll 1/\omega C_e^S$

$$I_{m3} = V_{R_3} / R_3, \tag{6.425}$$

where V_{R_3} is the voltage across resistance R_3. The radiation impedance of transducer 1 may be represented as

$$Z_{acIa} = Z_{ac11} + 2z_{ac12}(U_2 / U_1)_{Ia} + 2z_{ac13}(U_3 / U_1). \tag{6.426}$$

The ratios of velocities can be replaced by ratios of the motional currents, I_{mi}, on the electrical side of transducers. Thus, $U_2 / U_1 = I_{m2} / I_{m1}$ and $U_3 / U_1 = I_{m3} / I_{m1}$. We represent the ratio of motional currents as

$$I_{mi} / I_{m1} = |I_{mi} / I_{m1}| e^{j\psi_{1i}}, \tag{6.427}$$

where ψ_{1i} is a phase shift between the currents. In the same way as in Experiment I we obtain the following equations analogous to Eq. (6.384) and (6.385) at the resonance frequency of transducer 1 in water, which in this experiment we denote as f_{wIa}:

$$R_{\Sigma Ia} = R_{11} + 2\,\mathrm{Re}\{z_{12} |I_{m2} / I_{m1}|_{Ia} e^{j(\psi_{12})_{Ia}} + z_{13} |I_{m3} / I_{m1}|_{Ia} e^{j(\psi_{13})_{Ia}}\} = 1/G_{mIa}, \tag{6.428}$$

$$X_{\Sigma Ia} = X_{11} + 2\,\mathrm{Im}\{z_{12} |I_{m2} / I_{m1}|_{Ia} e^{j(\psi_{12})_{Ia}} + z_{13} |I_{m3} / I_{m1}|_{Ia} e^{j(\psi_{21})_{Ia}}\} = 2\pi M F_{Ia}. \tag{6.429}$$

The total input resistance $R_{\Sigma Ia}$, resonance frequency f_{wIa} and the motional currents I_{m1}, I_{m2} can be measured by using the experimental setup illustrated in Figure 6.74 and procedures described in experiment III. Current I_{m3} can be calculated from formula (6.425) after voltage V_{R_3} is measured. In order to get equations necessary for calculating of all of the radiation impedances, several more experiments have to be fulfilled.

Experiment II a

Voltage V_2 of the same magnitude as in experiment I *a* is applied to transducers 2 in anti-phase, while transducers 3 remain short circuited. As $V_1 / V_2 \ll 1$ and hence $U_1 / U_2 \ll 1$, the introduced impedances

$$Z_{ac21} = z_{ac12} U_1 / U_2, \quad Z_{ac23} = z_{ac12} U_3 / U_2 \tag{6.430}$$

in the equivalent circuit of transducer 2 can be neglected in comparison with its self-radiation impedance. Thus, it can be concluded that $(U_2)_{IIa} \cong -(U_2)_{Ia}$. Since $U_3 \sim z_{ac12} U_2$, the relation follows $(U_3)_{IIa} \cong -(U_3)_{Ia}$. As the result we arrive at

$$(I_{m2})_{IIa} \cong -(I_{m2})_{Ia}, \quad (I_{m3})_{IIa} \cong -(I_{m3})_{Ia} \, . \tag{6.431}$$

The currents $(I_{m1})_{Ia}$ and $(I_{m1})_{IIa}$ have different magnitudes but are in phase with voltage V_1 at the resonance frequency. Therefore, the conclusion can be made that

$$(\psi_{12})_{Ia} \cong (\psi_{12})_{IIa}, \quad (\psi_{13})_{Ia} \cong (\psi_{13})_{IIa} \, . \tag{6.432}$$

Taking into account relations (6.431) and (6.432), the input resistance and reactance of transducer 1 at resonance frequency f_{wIIa} may be represented as

$$R_{\Sigma IIa} = R_{11} - 2\,\mathrm{Re}\{z_{12} \left|I_{m2}\right|_{Ia} e^{j(\psi_{12})_{Ia}} / \left|I_{m1}\right|_{IIa} + z_{13} \left|I_{m3}\right|_{Ia} e^{j(\psi_{13})_{Ia}} / \left|I_{m1}\right|_{IIa}\} = 1/G_{mIIa}, \tag{6.433}$$

$$X_{\Sigma IIa} = X_{11} - 2Jm\{z_{12} \left|I_{m2}\right|_{Ia} e^{j(\psi_{12})_{Ia}} / \left|I_{m1}\right|_{IIa} + z_{13} \left|I_{m3}\right|_{Ia} e^{j(\psi_{13})_{Ia}} / \left|I_{m1}\right|_{IIa}\} = 2\pi M F_{IIa} \, . \tag{6.434}$$

After multiplying Eqs. (6.428) and (6.433) by $\left|I_{m1}\right|_{Ia}$ and $\left|I_{m1}\right|_{IIa}$, respectively, and summing and subtracting them we obtain

$$R_{11} = (R_{\Sigma Ia} \left|I_{m1}\right|_{Ia} + R_{\Sigma IIa} \left|I_{m1}\right|_{IIa}) / (\left|I_{m1}\right|_{Ia} + \left|I_{m1}\right|_{IIa}), \tag{6.435}$$

$$\begin{aligned} &4\,\mathrm{Re}\{z_{12} \left|I_{m2}\right|_{Ia} e^{j(\psi_{12})_{Ia}} + z_{13} \left|I_{m3}\right|_{Ia} e^{j(\psi_{13})_{Ia}}\} = \\ &= R_{\Sigma Ia} \left|I_{m1}\right|_{Ia} - R_{\Sigma IIa} \left|I_{m1}\right|_{IIa} - R_{11} (\left|I_{m1}\right|_{Ia} - \left|I_{m1}\right|_{IIa}). \end{aligned} \tag{6.436}$$

Similarly, from Eqs. (6.429) and (6.434) will be obtained

$$X_{11} = 2\pi M (F_{Ia} \left|I_{m1}\right|_{Ia} + F_{IIa} \left|I_{m1}\right|_{IIa}) / (\left|I_{m1}\right|_{Ia} + \left|I_{m1}\right|_{IIa}), \tag{6.437}$$

$$\begin{aligned} &4\,\mathrm{Im}\{z_{12} \left|I_{m2}\right|_{Ia} e^{j(\psi_{12})_{Ia}} + z_{13} \left|I_{m3}\right|_{Ia} e^{j(\psi_{13})_{Ia}}\} = \\ &= 2\pi M [F_{Ia} \left|I_{m1}\right|_{Ia} - F_{IIa} \left|I_{m1}\right|_{IIa}] - X_{11} (\left|I_{m1}\right|_{Ia} - \left|I_{m1}\right|_{IIa}). \end{aligned} \tag{6.438}$$

Thus, the self-radiation impedance of the transducers can be calculated from Eqs. (6.435) and (6.437). Two more equations are needed in addition to Eqs. (6.436) and (6.438) for determining the four components of the mutual impedances To obtain these equations the following experiments may be made: Experiment III *a*, in which case voltage $V_3 \gg V_1$ is applied to transducers 3, and transducers 2 are short circuited; Experiment IV *a,* in which case the voltage of the same magnitude is applied to transducers 3 in anti-phase to voltage V_1. The two missing equations analogous to Eqs. (6.436) and (6.438) will be obtained by repeating the procedure used in Experiments *I a* and *II a*. It may be well expected that the angles ψ_{12} and ψ_{13} are small at the resonance frequency of transducer 1, at which measurements take place. In this case $\sin\psi_{1i} \approx 0$, $\cos\psi_{1i} \cong 1$ and equations of the type (6.436) and (6.438) will be greatly simplified.

It must be noted that the described experiments and related calculations of the mutual radiation impedances are not very exact. But great accuracy in their determining often is not required, because the mutual radiation impedances are only a part of the self-impedance of a transducer.

6.7 References

1. L. E. Kinsler, A. R. Frey, A. B. Coppens, J. V. Sanders, *Fundamentals of Acoustics*, 4th edn. (John Wiley & Sons, New York, 2000)
2. A. D. Pierce, *Acoustics: An Introduction to Its Physical Principles and Applications* (Published by the Acoustical Society of America through the American Institute of Physics, Woodbury, 1994)
3. P. M. Morse, H. Feshbach, *Methods of Theoretical Physics*, Part I (McGraw-Hill, New York, 1953)
4. P. M. Morse, K. U. Ingard, *Theoretical Acoustics* (McGraw-Hill, New York, 1968)
5. M. Abramowitz, I. A. Stegun (ed.), *Handbook of Mathematical Functions* (National Bureau of Standards Applied Mathematical Series, Washington, DC, 1964)
6. G. N. Watson, *A Treatise on the Theory of Bessel Functions* (Cambridge Univ. Press, Cambridge, 1995)
7. M. I. Karnovskii, V.G. Lozovik, "Acoustic field of an infinitive circular cylindrical radiator with complex boundary conditions on the surface". Sov. Phys. Acoust. **10**(3), 265-268 (1965)
8. J. L. Butler, "Solution of acoustical-radiation problems by boundary collocation". J. Acoust. Soc. Am. **48**, 325-335 (1970)
9. E. L. Shenderov, *Wave Problems in Hydroacoustics* (Sudostroyeniye, Leningrad, 1972). English translation: Joint Publications Research Service, Report JPRS-58146 (National Technical Information Service, Springfield, VA, 1973)
10. D. H. Robey, "On the radiation impedance of an array of finite cylinders". J. Acoust. Soc. Am. **27**, 706-710 (1955)
11. D. P. Laird, H. Cohen, "Directionality patterns for acoustic radiation from a source on a rigid cylinder". J. Acoust. Soc. Am. **24**, 46-49 (1952)
12. J. L. Butler, A. L. Butler, "Fourier series solution for the radiation impedance of a finite cylinder". J. Acoust. Soc. Am. **104**(5), 2773-2778 (1998)
13. V. A. Kozyrev, E. L. Shenderov, "Radiation impedance of a cylinder of finite height". Sov. Phys. Acoust. **26**(3), 230-236 (1980)
14. V. A. Kozyrev, E. L. Shenderov, "Radiation impedance of a cylinder with acoustically compliant ends". Sov. Phys. Acoust. **27**(3) (1981)
15. E. L. Shenderov, *Radiation and Diffraction of Sound* (Sudostroenie, Leningrad, 1989) (in Russian)
16. C. L. Bachand, *Electroacoustic Transducer and Transmit System Modeling for Performance Prediction*, Ph. D. Dissertation, Univ. of Massachusetts, Dartmouth, 2012
17. J. W. S. Rayleigh, *The Theory of Sound* (Dover, New York, 1945)

18. A. Sauter, Jr., W. W. Soroka, "Sound transmission through rectangular slots of finite depth between reverberant rooms". J. Acoust. Soc. Am. **47**(1A), 5-11 (1970)
19. D. S. Bernett, W. W. Soroka, "Tables of rectangular piston radiation impedance functions, with application to sound transmission loss through deep apertures". J. Acoust. Soc. Am. **51**(5B), 1618-1623 (1972)
20. L. Ya. Gutin, "A sound field of the piston-like projectors", Zhurnal Tekhnicheskoi Fiziki, **7**(10), (1937). Selected works in Shipbuilding (Sudostroenie, Leningrad, 1977), p. 95 (in Russian)
21. T. Mellow, L Karkkainen, "On the sound field of an oscillating disk in a finite open and closed circular baffle". J. Acoust. Soc. Am. **118**(3) 1311-1325 (2005)
22. C. H. Sherman, J. L. Butler, *Transducers and Arrays for Underwater Sound* (Springer, New York, 2006)
23. R. V. DeVore, D. B. Hodge, R. G. Kouyoumjian, "Radiation by finite circular pistons imbedded in a rigid circular baffle". I. Eigenfunction solution. J. Acoust. Soc. Am. **48**(5B), 1128-1134 (1970)
24. R. L. Pritchard, "Mutual acoustic impedance between radiators in an infinite rigid plane". J. Acoust. Soc. Am. **32**(6), 730-737 (1960)
25. D. T. Porter, "Self- and mutual-radiation impedance and beam patterns for flexural disks in a rigid plane". J. Acoust. Soc. Am. **36**(6), 1154-1161 (1964)
26. E. M. Arise, "Mutual radiation impedance of square and rectangular pistons in a rigid infinite baffle". J. Acoust. Soc. Am. **36**(8), 1521-1525 (1964)
27. B. S. Aronov, "Experimental methods for investigating the acoustical interaction between transducers". J. Acoust. Soc. Am. **119**(6), 3822-3830 (2006)

LIST OF SYMBOLS

Symbol	Description
A	radius
B	bulk modulus
c, c_c, c_w	sound speed, speed of sound in ceramic composition and in water
c^E_{mi}	elastic stiffness of a piezoceramics at constant electric field
C, C^S_e	capacitance, capacitance of blocked transducer
C, C^E_{eqv}	compliance, equivalent compliance of a mechanical system at constant electric field
d, d_{mi}	separation, distance; piezoelectric constant
D	diameter, flexural rigidity $D = Yh^3/12(1-\sigma^2)$
D_i, D^E_i	charge density, charge density at constant electric field
e^E_{mi}	piezoelectric constant, $e_{mi} = d_{mj}c^E_{ji}$, j =1…6
E, E_{op}, E_p	electric field, operating field, permissible field
Ef	effectiveness
f, f_r, f_{ar}, Δf	frequency, resonance frequency, antiresonance frequency, deviation of frequency
f_{ip}	partial resonance frequencies of a coupled system
F, F_{eqv}	force, equivalent force
G	torsional rigidity
h	height
$H(\theta,\varphi)$	directional factor
I	current
I_L, I_C, I_m	current through inductance, current through capacitance, motional current
J, J_p	moment of inertia, polar moment of inertia
k; k_c, k_{eff}; k_{dif}	wave number $k = \omega / c$; electromechanical coupling coefficient, effective coupling coefficient; diffraction coefficient
k_E, k_T	reserves of the electrical and mechanical strength coefficients
K, K^E_{eqv}, K_{il}	rigidity, equivalent rigidity of a mechanical system, mutual rigidity of coupled systems
ΔK	additional rigidity term that characterizes electrical interaction

Symbol	Description
	between elements in nonuniformly deformed piezoelectric body
l, t, w	length, thickness, width
L ; L_p, L_s	Lagrangian, inductance; parallel and series inductances
ms_w	Mismatch coefficient, $ms_w = r_{ac} / r_{opt}$
ms_i	mode shape coefficient
M ; M_{eqv}, M_{il}	Moment, total mass; equivalent mass, mutual mass of coupled systems
n	turns ratio, electromechanical transformation coefficient,
N, N_i	Number of segments in segmented mechanical system, electromechanical transformation coefficients, $i = 1,3$.
o	subscript that denotes a reference point
P , P_o ; P_h	sound pressure, sound pressure of simple source; hydrostatic pressure
Q, Q_e, Q_m	quality factor, $Q = W_{kin} / W_{Loss}$; electrical and mechanical quality factors
r, $\boldsymbol{r}$	distance, radius vector
r, r_{mL}; r_{ac}, r_{opt}	resistance, resistance of mechanical loss; radiation resistance, optimal value of the radiation resistance
R, R_{eL}	resistance, resistance of electrical loss
s_{mi}^E	elastic compliance of piezoceramics at constant electric field
S , S_{ik}, S_i	deformation, tensor of deformation $(i, k = 1,2,3)$, tensor of deformation $(i = 1,..,6)$
S_Σ, S_{av}, S_{eff}	surface area, average surface area, effective surface area
T , T_{ik}, T_i	stress, stress tensor $(i, k = 1,2,3)$, stress tensor $(i = 1,..,6)$
T_{op}, T_p	operating stress, permissible stress
u, U ; U_o, U_i	Velocity; velocity of reference point, velocity of reference point in i^{th} mode of vibration
$U_{\tilde{V}}$	volume velocity
v, V	voltage
$\tilde{V}$	volume
w; w_{int}, w_e, w_{mch}, w_{em}	width, energy density; densities of the internal, electrical, mechanical, and electromechanical energies
W , $\dot{W}$, $\bar{\dot{W}}$	energy, energy flux (power), complex power
W_{el}, W_e^S	total electrical energy, electrical energy stored in a blocked

Symbol	Description
	piezoelement
W_{int}, W_m, W_{em}, W_{ac}	internal, mechanical, electromechanical, and acoustic energies
W_{kin}, W^E_{pot}	kinetic energy, potential energy at constant electrical field
W_{eL}, W_{mL}	energies of electrical and mechanical loss
$\dot{W}_{mE}$, $\dot{W}_{mT}$	maximum power electric field limited and mechanical stress limited
ΔW	additional energy term that characterizes electrical interaction between elements in nonuniformly deformed piezoelectric body
x; x_{ac}	coordinate; reactance of acoustic radiation
y; $y = \delta / t$	coordinate; ratio of thickness of active layer to total thickness of mechanical system
Y, $Y^E_i = 1/s^E_{ii}$	Young's modulus, Young's modulus of piezoceramics (i =1, 3)
Y^E_a, Y_p	Young's moduli of active and passive materials
Y_σ	$Y_\sigma = Y/(1-\sigma^2)$
z; z_{il}	Coordinate; mutual impedance between modes of vibration
Z, $Z_{il} = z_{il} U_i / U_l$	impedance, introduced impedance
Z_m, Z^E_m, Z_{in}	mechanical impedance, impedance at constant electric field, input impedance
Z_{ac}	radiation impedance
α_{ac}	nondimensional coefficients of the radiation resistance
$\alpha_c = n^2 C^E_m / C^S_e$	coefficient related to effective coupling coefficient, $k^2_{eff} = \alpha_c / (1+\alpha_c)$
β_{ac}	nondimensional coefficient of the radiation reactance
$\beta = f_{1p} / f_{2p}$	detuning factor between partial frequencies of a coupled system
γ, γ_m, γ_k,	coefficient of coupling between partial systems, coefficients of inertial and elastic coupling
γ_Y	$\gamma_Y = Y_p / Y^E_a$
γ_ρ	$\gamma_\rho = \rho_p / \rho_a$
η; η_{em}, η_{ma}, η_{ea}	efficiency; electromechanical, mechanoacoustic, electroacoustic efficiencies
δ	separation between electrodes,
δ_e, δ_m	angles of dielectric and mechanical losses, $\tan\delta_e = 1/Q_e$, $\tan\delta_m = 1/Q_m$
ε; ε^T_{ik}, ε^S_{ik}	dielectric constant; tensors of dielectric constants of piezoceramics

Symbol	Description
	at free and clamped conditions
θ; $\theta(\mathbf{r})$	angle, mode shape
λ	wavelength, Lame constant
μ	Lame constant (share modulus)
ξ, ξ_o	displacement, displacement of reference point
ρ, ρ_a, ρ_p	density, density of the active and passive materials
σ, σ_i^E	Poisson's ratio; Poisson's ratio of piezoceramics, $\sigma_1^E = -s_{12}^E / s_{11}^E$, $\sigma_3^E = -s_{13}^E / s_{33}^E$
Σ	surface in general
φ	angle
χ	diffraction function
ω, ω_r, ω_{ar}	angular frequency, resonance and antiresonance frequencies
$\Omega = f^2 / f_{1p}^2$	nondimensional frequency factor
$\Omega = 2\Delta f / f_r$	normalized bandwidth

1. Vectors are displayed in bold letters.
2. Low case letters denoting the time dependent quantities indicate instantaneous values; the capital letters are values in rms.
3. An overbar on a capital letter denotes a complex quantity.

APPENDIX A. Properties of Passive Materials

Table A.1: Elastic properties of the passive materials *).

Material	Y (gpa)	σ	$\rho \cdot 10^{-3}$ (kg/m^3)	c (m/s)	$\rho c \cdot 10^{-6}$ (kg/m^{2}s)
Aluminum	71	0.33	2.7	5130	13.5
Alumina	300	0.21	3.7	9000	33.3
Beryllia, (BeO)	345	0.26	3.0	10,700	32.1
Beryllium Bronze, (BeCu)	125	0.30	8.2	3900	32,0
Brass	97	0.31	8.5	3400	29.0
Corprene	0.23	0.43	1.1	460	0.51
Glass	62	0.24	2.3	5200	12
G-10	24	0.14	1.8	3600	6,6
Invar	148	0.3	8.0	4300	34
Lead	16.5	0.44	11,3	1200	13.6
Macor	67	0.29	2.5	5180	13
Pyrex	64	0.24	2.3	5300	12
Stainless steel	193	0.28	7.9	4940	39
Tin	50	0.36	7.3	2600	19
Titanium	104	0.36	4.5	4810	21.6
*) Bulk modulus $B = Y/3(1-2\sigma)$. Shear modulus $\mu = Y/2(1+\sigma)$					

Table A.2: Properties of the fluids at room conditions

Liquid	Air	Water	Seawater	Castor oil	Motor oil SAE-30	Hydraulic fluid ISO 32	Silicon oil
B, GPa	$142 \cdot 10^{-6}$	2.15	2.34	2.1	1.5	1.8	2.1
$\rho \cdot 10^3$ kg/m3	$1.2 \cdot 10^{-3}$	1.0	1.02	0.96	0.88	0.86	0.97
c, m/s	340	1500	1500	1470	1300	1450	1500

Table A.3: Properties of the polyurethanes

Property		ρ, kg/m^3	c, m/s	B, GPa	G, MPa
PR1547	4°C	1.05	1650	2.9	6
	34°C		1500	2.3	4
GS960PU, 20°C		1.08	1700	3.3	1.2

APPENDIX B. Properties of Piezoelectric Ceramics

Table B.1: Piezoelectric constants.

Property	PZT-4 Type I	PZT-5A Type II	PZT-8 Type III	PZT-5H Type VI
s_{11}^{E}, 10^{-12} m²/N	12.3	16.4	11.5	16.5
s_{33}^{E}	15.5	18.8	13.5	20.7
s_{13}^{E}	−5.31	−7.22	-4.8	-8.45
s_{12}^{E}	−4.05	−5.74	-3.7	-5.7
s_{44}^{E}	39.0	47.5	31.9	-4.78
s_{11}^{D}	10.9	14.4	10.1	15.5
s_{33}^{D}	7.9	9.46	8.5	9.0
s_{13}^{D}	−2.1	−2.98	-2.5	-3.0
s_{12}^{D}	−5.42	−7.71	-4.5	-7.3
s_{44}^{D}	19.3	25.2	22.6	23.7
s_{66}	32.7	44.3	30.4	48.5
c_{11}^{E}, 10^{10} N/m²	13.9	12.1	14.9	12.6
c_{33}^{E}	11.5	11.1	13.2	11.7
c_{13}^{E}	7.43	7.52	8.11	8.41
c_{12}^{E}	7.78	7.54	8.11	7.95
c_{44}^{E}	2.56	2.11	3.13	2.3
c_{11}^{D}	14.5	12.6	15.2	11.7
c_{33}^{D}	15.9	14.7	16.9	15.7
c_{13}^{D}	6.09	6.52	7.03	7.22
c_{12}^{D}	8.39	8.09	8.41	8.18
c_{44}^{D}	5.18	3.97	4.46	4.22
c_{66}	3.06	2.26	3.40	2.26
$d_{31}, 10^{-12}\ C/N$	-123	-171	-97	-274
d_{33}	289	374	225	593
d_{15}	496	584	330	741
e_{31}, C/m²	−5.2	−5.4	-4.1	-6.5

Property	PZT-4 Type I	PZT-5A Type II	PZT-8 Type III	PZT-5H Type VI
e_{33}	15.1	15.8	14.0	–23.3
e_{15}	12.7	12.3	10.3	17
$\varepsilon_{11}^{S} / \varepsilon_0$ [1)]	730	916	900	–
$\varepsilon_{33}^{S} / \varepsilon_0$	635	830	600	1470
$\varepsilon_{11}^{T} / \varepsilon_0$	1475	1730	1290	3130
$\varepsilon_{33}^{T} / \varepsilon_0$	1300	1700	1000	3400
k_{31}	0.334	0.344	0.30	0.39
k_{33}	0.7	0.705	0.64	0.75
k_{15}	0.71	0.685	0.55	0.52
k_p	0.58	0.60	0.51	0.65
k_t	0.513	0.486	0.48	0.50
ρ, 10^3 kg/m^3	7.5	7.75	7.6	7.5
$\tan \delta_e$	0.004	0.02	<0.002	0.02
Q_m	500	75	1000	65

[1)] $\varepsilon_0 = \varepsilon_0 = 8.85 \cdot 10^{-12}\ F/m$

The values of all the parameters are presented at small signals and at room temperature. For dependencies of their values from the operating and environmental conditions see Chapter 11.

APPENDIX C. Special Functions

In the Appendix some data regarding the properties of special functions that are required for treating the radiation and vibration problems related to the cylindrical and spherical transducers are summarized. More details regarding properties of the functions and their numerical values can be found [1, 2], which are the primary sources of the information and where these functions are tabulated. Some of the integral relations that include the special functions are presented from a source [3] where much more particular useful relations can be found.

C.1 Cylindrical Bessel Functions

Definition

Cylindrical functions $Z_n(x)$ are the solutions to Bessel equation

$$\frac{d^2 Z_n}{dx^2} + \frac{1}{x}\frac{dZ_n}{dx} + \left(1 - \frac{n^2}{x^2}\right) = 0 . \qquad \text{(C.1)}$$

Partial solutions to this equation are the Bessel functions (cylindrical functions of the first kind) $J_n(x)$, Neumann functions (cylindrical functions of the second kind) $N_n(x)$, and Hankel functions (cylindrical functions of the third kind) $H_n^{(1)}(x)$ and $H_n^{(2)}(x)$, where $H_n^{(1)}(x) = J_n(x) + jN_n(x)$ and $H_n^{(2)}(x) = J_n(x) - jN_n(x)$. The functions $H_n^{(1)}(x)$ or $H_n^{(2)}(x)$ are used alternatively according to the time dependence $e^{-j\omega t}$ or $e^{j\omega t}$ (the latter is accepted in our treatment). In course of this treatment it will be assumed that n is the natural integer number and for the cylindrical coordinates $x = kr$. Thus, it will be used form of

$$H_n^{(2)}(kr) = J_n(kr) - jN_n(kr) . \qquad \text{(C.2)}$$

Properties

$$H_{-n}^{(2)}(x) = J_{-n}(x) - jN_{-n}(x) , \qquad \text{(C.3)}$$

where

$$J_{-n}(x) = (-1)^n J_n(x), \quad N_{-n}(x) = (-1)^n N_n(x) . \qquad \text{(C.4)}$$

Series representation

$$J_n(x) = \frac{1}{0!n!}\left(\frac{x}{2}\right)^2 - \frac{1}{1!(n+1)!}\left(\frac{x}{2}\right)^{n+2} + \frac{1}{2!(n+2)!}\left(\frac{x}{2}\right)^{n+4} - \ldots \quad \text{(C.5)}$$

$$J_0(x) = 1 - \frac{x^2}{2^2} + \frac{x^4}{2^2 \cdot 4^2} - + \frac{x^6}{2^2 \cdot 4^2 \cdot 6^2} - \ldots \quad \text{(C.6)}$$

$$J_1(x) = \frac{x}{2} - \frac{2x^3}{2 \cdot 4^2} + \frac{3x^5}{2 \cdot 4^2 \cdot 6^2} - \ldots \quad \text{(C.7)}$$

Approximations at small argument $x < 1$ (low frequency approximations at $x = kr \ll 1$)

$$J_0(x) \approx 1 - \frac{x^2}{4}, \quad J_1(x) \approx \frac{x}{2} - \frac{x^3}{16}, \quad \text{(C.8)}$$

$$N_0 \approx \frac{2}{\pi}(\ln x - 0.11), \quad N_1 \approx -\frac{2}{\pi} \cdot \frac{1}{x}, \quad \text{(C.9)}$$

$$H_1^{(2)}(x) \approx \frac{x}{2} + j\frac{2}{\pi x}, \quad H_1^{(2)\prime}(x) \approx -j\frac{2}{\pi (x)^2}. \quad \text{(C.10)}$$

At large arguments $x \gg 1$ (high frequency approximation, large distances from a cylinder at $x = kr \to \infty$)

$$H_n^{(2)}(x) \to \sqrt{\frac{2}{\pi x}} e^{-j\left(x - \frac{\pi n}{2} - \frac{\pi}{4}\right)}, \quad \text{(C.11)}$$

$$J_n(x) \to \sqrt{\frac{2}{\pi x}} \cos\left(x - \frac{n\pi}{2} - \frac{\pi}{4}\right), \quad \text{(C.12)}$$

$$N_n(x) \to \sqrt{\frac{2}{\pi x}} \sin\left(x - \frac{n\pi}{2} - \frac{\pi}{4}\right). \quad \text{(C.13)}$$

Functional equations

$$Z_{n-1}(x) + Z_{n+1}(x) = \frac{2n}{x} Z_n(x) \quad \text{(C.14)}$$

$$N_{n-1} J_n - N_n J_{n-1} = \frac{2}{\pi x} \quad \text{(C.15)}$$

Differential formulas

$$\frac{dZ_n}{dx} = -\frac{n}{x} Z_n + Z_{n-1} = \frac{n}{x} Z_n - Z_{n+1} = \frac{1}{2}(Z_{n-1} - Z_{n+1}) \quad \text{(C.16)}$$

$$Z_0' = -Z_1, \quad Z_1' = Z_0 - \frac{1}{x}Z \tag{C.17}$$

Integral formulas

$$\int x^{-n+1} Z_n(x)dx = -x^{-n+1} Z_{n-1}(x), \quad \int x^{n+1} Z_n(x)dx = x^{n+1} Z_{n+1}(x) \tag{C.18}$$

$$\int Z_1(x)dx = -Z_0(x), \quad \int xZ_0(x)dx = xZ_1(x) \tag{C.19}$$

$$\int J_n^2(x)xdx = \frac{x^2}{2}[J_n^2(x) - J_{n-1}(x)J_{n+1}(x)] \tag{C.20}$$

Integral representation

$$J_n(x) = \frac{1}{2\pi j^n} \int_0^{2\pi} e^{jx\cos\varphi} \cdot e^{jn\varphi} d\varphi \tag{C.21}$$

$$J_0(x) = \frac{1}{\pi} \int_0^{\pi} e^{jx\cos\varphi} d\varphi = \frac{1}{\pi} \int_0^{\pi} \cos(x\sin\varphi)d\varphi = \frac{2}{\pi} \int_0^{\pi/2} \cos(x\sin\varphi)d\varphi \tag{C.22}$$

Also tabulated are functions Struve that are solutions to one of variations of the Bessel equation [1, 2]:

$$S_0(x) = \frac{2}{\pi} \int_0^{\pi/2} \sin(x\cos\varphi)d\varphi\,, \tag{C.23}$$

$$S_1(x) = \frac{4}{\pi} \int_0^{\pi/2} \sin(x\cos\varphi)\sin^2\varphi d\varphi\,. \tag{C.24}$$

There series representations are

$$S_0(x) = \frac{2}{\pi}\left[x - \frac{x^3}{1^2 \cdot 3^2} + \frac{x^5}{1^2 \cdot 3^2 \cdot 5^2} - \ldots\right], \tag{C.25}$$

$$S_1(x) = \frac{2}{\pi}\left[\frac{x^2}{1^2 \cdot 3} - \frac{x^4}{1^2 \cdot 3^2 \cdot 5} + \frac{x^6}{1^2 \cdot 3^2 \cdot 5^2 \cdot 7} - \ldots\right], \tag{C.26}$$

$$\int xS_0(x)dx = xS_1(x)\,. \tag{C.27}$$

Modified Bessel functions (Bessel functions of imaginary values of argument), $I_n(x)$ and $K_n(x)$

The modified functions are the partial solutions to the equation (Compare with Eq. (C.1))

$$\frac{d^2 Z_n}{dx^2} + \frac{1}{x}\frac{dZ_n}{dx} - \left(1 + \frac{n^2}{x^2}\right) = 0 . \tag{C.28}$$

The modified functions are defined by equations:

$$I_n(x) = j^n J_n(-jx) \tag{C.29}$$

for the first kind and

$$K_n(x) = -\frac{j\pi}{2} e^{-j\frac{\pi n}{2}} H_n^{(2)}(-jx) \tag{C.30}$$

for the second kind, with

$$K_0(x) = -\frac{j\pi}{2} H_0^{(2)}(-jx), \quad K_1(x) = -\frac{\pi}{2} H_1^{(2)}(-jx) . \tag{C.31}$$

The properties of these functions can be obtained from formulations of the corresponding properties of functions $J_n(x)$ and $H_n^2(x)$ by replacing $x \to -jx$. In particular

$$I_{-n}(x) = I_n(x), \quad K_{-n}(x) = K_n(x) , \tag{C.32}$$

$$K_0'(x) = -K_1(x). \tag{C.33}$$

C.2 Spherical Bessel Functions

The partial solutions to equation

$$\frac{d^2 R}{dz^2} + \frac{2}{z}\frac{dR}{dz} + \left[1 - \frac{m(m+1)}{z^2}\right] R = 0 , \tag{C.34}$$

where $z = kr$, are the spherical Bessel functions (or *Bessel* functions for the spherical coordinates). Spherical Bessel functions of order *m* of the first kind are defined as

$$j_m(z) = \sqrt{\pi / 2z} J_{m+1/2}(z) ; \tag{C.35}$$

of the second kind (spherical Neumann functions) as

$$y_m(z) = \sqrt{\pi / 2z} N_{m+1/2}(z) ; \tag{C.36}$$

and of the third kind (spherical Hankel functions) as $h_m(z)$. For outgoing wave

$$h_m^{(2)}(z) = j_m(z) - j y_m(z) = \sqrt{\pi / 2z} H_{m+1/2}^{(2)}(z) \tag{C.37}$$

In particular,

$$\left.\begin{aligned} j_0(z) &= \frac{\sin z}{z}, \quad y_0(z) = -\frac{\cos z}{z}; \\ j_1(z) &= \frac{\sin z}{z^2} - \frac{\cos z}{z}, \quad y_1(z) = -\frac{\sin z}{z} - \frac{\cos z}{z^2}; \\ j_2(z) &= \left(\frac{3}{z^3} - \frac{1}{z}\right)\sin z - \frac{3}{z^2}\cos z, \quad y_2(z) = -\frac{3}{z^2}\sin z - \left(\frac{3}{z^3} - \frac{1}{z}\right)\cos z. \end{aligned}\right\} \tag{C.38}$$

Functions j_m and y_m are tabulated [1] at $z < 0.3$ as

$$\left.\begin{aligned} j_m(z) &\approx \frac{(z)^m}{1\cdot 3\cdot 5\cdots(2m+1)}, \quad y_m(z) \approx -\frac{1\cdot 3\cdot 5\cdots(2m-1)}{(z)^{m+1}}, \\ h_1^{(2)} &\approx \frac{1}{3}\left[z + j\frac{3}{(z)^2}\right], \quad h_1^{(2)\prime}(z) \approx \frac{1}{3}\left[1 - j\frac{6}{(z)^3}\right]; \end{aligned}\right\} \tag{C.39}$$

and at $z \to \infty$

$$\left.\begin{aligned} j_m(z) &\to \frac{1}{z}\cos\left(z - \frac{m+1}{2}\pi\right), \quad y_m(z) \to \frac{1}{z}\sin\left(z - \frac{m+1}{2}\pi\right), \\ h_m^{(2)} &\to \frac{1}{z}e^{-j\left(z - \frac{m+1}{2}\pi\right)}, \end{aligned}\right\} \tag{C.40}$$

$$y_{m-1}(z)j_m(z) - y_m(z)j_{m-1}(z) = z^{-2}. \tag{C.41}$$

The following properties are the same for the functions j_m, y_m and h_m that will be collectively denoted as f_m.

Recurrent relations

$$f_{m-1}(z) + f_{m+1}(z) = (2m+1)z^{-1}f_m(z) \tag{C.42}$$

$$mf_{m-1}(z) - (m+1)f_{m+1}(z) = (2m+1)\frac{d}{dz}f_m(z) \tag{C.43}$$

$$h_1^{(2)\prime}(z) = \frac{1}{3}[h_0^{(2)}(z) - 2h_2^{(2)}(z)] \tag{C.44}$$

$$\frac{d}{dz}[z^{m+1}f_m(z)] = z^{m+1}f_{m-1}(z), \quad \frac{d}{dz}[z^{-m}f_m(z)] = -z^{-m}f_{m-1}(z) \tag{C.45}$$

Integral formulas

$$\int f_1(z)dz = -f_0(z), \quad \int f_0(z)z^2dz = z^2 f_1(z) \tag{C.46}$$

$$\int f_m^2(z)z^2dz = \frac{z^2}{2}[f_m^2(z) - f_{m-1}(z)f_{m+1}(z)] \tag{C.47}$$

C.3 Legendre Polynomials

The partial solutions to Legendre equation

$$\frac{d}{dx}\left[(1-x^2)\frac{dP}{dx}\right]+m(m+1)P=0 \text{ or } (x^2-1)\frac{d^2P}{dx^2}+2x\frac{dP}{dx}-m(m+1)P=0\,. \qquad \text{(C.48)}$$

at m integer and $x=\cos\theta$ are the Legendre polynomials of the order m

$$P_m(x)=\frac{1}{2^m\,m!}\frac{d^m}{dx^m}(x^2-1)^m\,. \qquad \text{(C.49)}$$

In particular,

$$\left.\begin{aligned} &P_0(x)=1,\\ &P_1(x)=x=\cos\theta,\\ &P_2(x)=\frac{1}{2}(3x^2-1)=\frac{1}{4}(3\cos 2\theta+1),\\ &P_3(x)=\frac{1}{2}(5x^2-3z)=\frac{1}{8}(5\cos 3\theta+3\cos\theta).\end{aligned}\right\} \qquad \text{(C.50)}$$

$$P_m(-x)=(-1)^m P_m(x),\quad P_m(x)=P_{-(m+1)}(x) \qquad \text{(C.51)}$$

Recurrent relation

$$P_{m-1}(x)=\frac{2m+1}{m}xP_m(x)-\frac{m+1}{m}P_{m+1}(x) \qquad \text{(C.52)}$$

Differential formulas

$$mP_m(x)=xP_m'(x)-P_{m-1}'(x) \qquad \text{(C.53)}$$

$$(2m+1)P_m(x)=\frac{d}{dx}[P_{m+1}(x)-P_{m-1}(x)] \qquad \text{(C.54)}$$

Orthogonality

$$\int_{-1}^{1}P_n(x)P_m(x)dx=\begin{cases}0 & n\neq m\\ 2/(2m+1) & n=m\end{cases} \qquad \text{(C.55)}$$

Useful integrals with Legendre polynomials

$$\int_0^1 P_m(x)P_n(x)dx = \begin{cases} 1/(2m+1) & m=n \\ \dfrac{(-1)^{0.5(m+n+1)}}{2^{m+n+1}} \times \ldots & m \neq n, \\ \dfrac{m!n!}{(m+n)(m+n+1)\{[(m/2)![(n-1)/2]!\}^2} & m>n, \\ & (m-n) \text{ even} \\ 0 & m \text{ even}, \\ & n \text{ odd} \end{cases} \tag{C.56}$$

$$\int_0^1 P_{2m}(x)dx = 0, \quad \int_0^1 P_{2m+1}(x)dx = \frac{(-1)(-3)\cdots(-2m+1)}{(2m+2)\cdot 2m\cdot(2m-1)} \tag{C.57}$$

$$\int_{-1}^1 x^b P_m(x)dx = 0 \text{ at } b<m \tag{C.58}$$

$$\int_{-1}^1 [P_m'(x)]^2 dx = m(m+1) \tag{C.59}$$

$$\int_{-1}^1 (1-x^2)[P_m'(x)]^2 dx = \frac{2m(m+1)}{2m+1} \tag{C.60}$$

$$\int_{-1}^1 (1-x)^{-1/2} P_m(x)dx = \frac{2^{3/2}}{2m+1} \tag{C.61}$$

C.4 References

1. M. Abramowitz, I. A. Stegun (ed.), *Handbook of Mathematical Functions* (National Bureau of Standards Applied Mathematical Series, Washington, DC, 1964)
2. G. N. Watson, *A Treatise on the Theory of Bessel Functions* (Cambridge Univ. Press, Cambridge, 1995)
3. I. S. Gradshtein, I. M. Ryzhik, *Tables of integrals, Series, and Products* (Elsevier/Academic Press, Amsterdam, 2007)

www.ingramcontent.com/pod-product-compliance
Ingram Content Group UK Ltd.
Pitfield, Milton Keynes, MK11 3LW, UK
UKHW050446210526
471278UK00007B/226